Simple Models of Magnetism

Simple Models of Magnetism

Ralph Skomski

Department of Physics and Astronomy
and
Nebraska Center for Materials and Nanoscience
University of Nebraska

OXFORD
UNIVERSITY PRESS

Great Clarendon Street, Oxford OX2 6DP

Oxford University Press is a department of the University of Oxford.
It furthers the University's objective of excellence in research, scholarship,
and education by publishing worldwide in

Oxford New York

Auckland Cape Town Dar es Salaam Hong Kong Karachi
Kuala Lumpur Madrid Melbourne Mexico City Nairobi
New Delhi Shanghai Taipei Toronto

With offices in

Argentina Austria Brazil Chile Czech Republic France Greece
Guatemala Hungary Italy Japan Poland Portugal Singapore
South Korea Switzerland Thailand Turkey Ukraine Vietnam

Oxford is a registered trade mark of Oxford University Press
in the UK and in certain other countries

Published in the United States
by Oxford University Press Inc., New York

First published 2008

British Library Cataloguing in Publication Data
Data available

Library of Congress Cataloging in Publication Data
Data available

Typeset by Newgen Imaging Systems (P) Ltd., Chennai, India
Printed in Great Britain
on acid-free paper by
Biddles Ltd., King's Lynn, Norfolk

ISBN 978–0–19–857075–2 (Hbk)

1 3 5 7 9 10 8 6 4 2

In den Wissenschaften ist viel Gewisses, sobald man sich von den Ausnahmen nicht irremachen läßt und die Probleme zu ehren weiß.

There is much certainty in science, unless one gets confused by exceptions and is unable to honor the problems.

Johann Wolfgang von Goethe

follow no path ... all paths lead where
truth is here

e. e. cummings

Contents

List of abbreviations

AFM	antiferromagnetic
AMR	anisotropic magnetoresistance
ATP	adenosine triphosphate
bcc	body-centered cubic
BCS	Barden-Cooper-Schrieffer
CI	configuration interaction
CMR	colossal magnetoresistance
CPA	coherent-potential approximation
DM	Dzyaloshinski-Moriya
DMS	dilute magnetic semiconductor
DOS	density of states
EA	Edwards-Anderson
ESD	elongated single-domain
ESR	electron spin resonance
fcc	face-centered cubic
FI	ferrimagnetic
FM	ferromagnetic
FMR	ferromagnetic resonance
FOMP	first-order magnetization process
FORC	first-order reversal curve
GMR	giant magnetoresistance
hcp	hexagonal close packing
HOMO	highest occupied molecular orbital
LCAO	linear combination of atomic orbitals
LSDA	local spin density
LUMO	lowest unoccupied molecular orbital

MFM	magnetic force microscopy
MO	molecular orbitals
MRAM	magnetoresistive random-access memory
MTJ	magnetic tunnel junction
PM	paramagnetism
RE-TM	rare-earth transition-metal
RG	renormalization-group
RKKY	Ruderman-Kittel-Kasuya-Yoshida
RW	random-walk
SAW	self-avoiding walk
SCF	self-consistent field
SK	Sherrington-Kirkpatrick
SPD	single-point detection
SRT	spin-reorientation transition
SWR	spin-wave resonance
TMR	tunnel magnetoresistance

Panels and tables

Panels

Tables

Preface

Magnetic models help us to understand magnetism and have also influenced other branches of science, such as quantum mechanics, statistical mechanics, metallurgy, and biology. One reason is the transparent phase space, which yields a clear relationship between model assumptions and physical results. For example, the Ising model considers two magnetization states per atom, $\uparrow$ (spin-up) and $\downarrow$ (spin-down). Interatomic interactions as well as local and global magnetic fields yield a rich zero- and finite-temperature physics, and extensions of the Ising model are being used in many areas of science and society. Human brain-cell activity (firing or quiescent), nonmagnetic alloys (site occupancy by copper or zinc in brass), and the social role of movie stars (good or bad) can all be cast in form of an Ising model. For instance, the relatively trivial cinematographic analog of the magnetic field is an external force, introduced by the movie director and trying to change the nature of the characters from bad to good. In a slightly less poetic context, magnetic models play an important role in the improvement of magnetic systems and materials, from computer hard disks to permanent magnets in motors.

This book is an introduction to atomic, mesoscopic, and macroscopic models of magnetism. The style and presentation is kept as transparent as possible, with many examples and explicit discussions of illustrative limits. This and the absence of lengthy calculations are designed to make it accessible to graduate and advanced undergraduate students, to experimentalists with little specific interest in theoretical details, and to nonspecialists interested in the interdisciplinary aspects of magnetic modeling. An important point is that the magnetism community consists of many subcommunities, such as soft magnetism, permanent magnetism, magnetic recording, oxides, spin electronics, magnetic alloys, micromagnetic simulations, first-principle calculations, and theory of phase transitions. There is considerable overlap among some of these subfields, but there are also big gaps, and one aim of this book is to bridge these gaps.

As the title suggests, emphasis is on *simple* models of magnetism. Toy models are included if they contain substantial physics, but little attention is paid to simplistic models and to phenomenological models whose main aim is to fit experimental data. Space limitations preclude the discussion of complicated models used in numerical and complex analytical calculations. However, there are no sharp boundaries, and some phenomenological and numerical models are mentioned or briefly discussed. Some complicated models, such as the Hubbard model, are based on simple assumptions and therefore included, but with emphasis on transparent limit and usually marked by an asterisk. It is often intriguing to apply simple models to complicated magnetic, thereby investigating both the model and the system. If an unreasonable model assumption causes a simple model to fail, then it also leads to the failure of complicated models and numerical calculations. Note that the focus of the book is on specific models, not on the methodology of scientific models. In some cases, there are no sharp boundaries between models and approximations. An example is the mean-field model, which is often defined as an approximation to a microscopic Hamiltonian.

The chapters and sections have a self-explanatory building-block structure, but cross-references are used to elucidate the hierarchy of magnetic models and to elaborate interdisciplinary connections. Chapter 2 deals with the quantum-mechanical origin of magnetism and focuses on the magnetic moment, whereas Chapter 3 is concerned with the zero-temperature spin structure of magnets. Chapters 4, 5, and 6 are devoted to models of anisotropy, magnetic hysteresis, and finite-temperature magnetism, respectively, whereas Chapter 7 deals with disordered magnetic structures. Chapter 8 is concerned with time-dependent effects and, finally, Chapter 9 discusses a few special topics and interdisciplinary models. All sections contain easy-to-follow case studies and discussion of the models' applicability. The latter is of considerable importance, because a given set of magnetic data is often compatible with two or more models. For example, hysteresis loops are sometimes reproduced by physically contradictory magnetic models, and independent experiments, such as crystallography, micrography, and magnetization dynamics, are necessary to understand the system. To invoke Kant's "Zur Kritik der reinen Vernunft", *Vernunft*—thinking, reasoning, modeling—is meaningful only if linked to practical experience.

This book has benefited from countless interactions with fellow scientists, ranging from scientific discussions at conferences and collaborations with colleagues to conversations about specific aspects of the book's presentation. This includes but is not limited to Ch. Binek, J. M. D. Coey, P. A. Dowben, S. Ducharme, A. Enders, P. Fulde, D. Givord, G. C. Hadjipanayis, K. Hono, S. S. Jaswal, A. Kashyap, R. D. Kirby, J. Kirschner, H. Kronmüller, D. Leslie-Pelecky, J. P. Liu, S.-H. Liou, M. E. McHenry, S. Michalski, H. Mireles, O. N. Mryasov, K.-H. Müller, M. O'Shea, R. F. Sabiryanov, D. Sander, T. Schrefl, W. Soffa, A. Solanki, K. D. Sorge, A. F. Starace, G. M. Stocks, E. Y. Tsymbal, X.-H. Wei, J. Zhang, and J. Zhou. Support by NSF MRSEC, the W. M. Keck foundation, DOE, INSIC, and NCESR is gratefully acknowledged, as is the pleasant and helpful cooperation with S. Adlung of OUP.

Particular thanks are due to D. J. Sellmyer, director of NCMN, formerly CMRA, for infrastructural support, numerous scientific remarks, and suggestions on the presentation of this work. It is fair to say that this book could not have been realized without his support. Above all, I would like to thank my wife, Verona Skomski, for her help with the preparation of the final manuscript, and both her and my son Daniel for their patience during the preparation and writing of this work.

Lincoln, November 2006

1

Introduction: The simplest models of magnetism

During the last 100 years, magnetism has made a giant step forward. By the second half of the nineteenth century, Maxwell's equations had established the relation between different electromagnetic fields, and scientists and engineers were aware of the dipolar character of magnetostatic forces and interactions. Figure 1.1 illustrates the state of the art during that time. However, many questions remained unanswered or were not even asked at that time. What is the atomic origin of the magnetization, and how does it involve quantum mechanics and relativistic physics? What determines

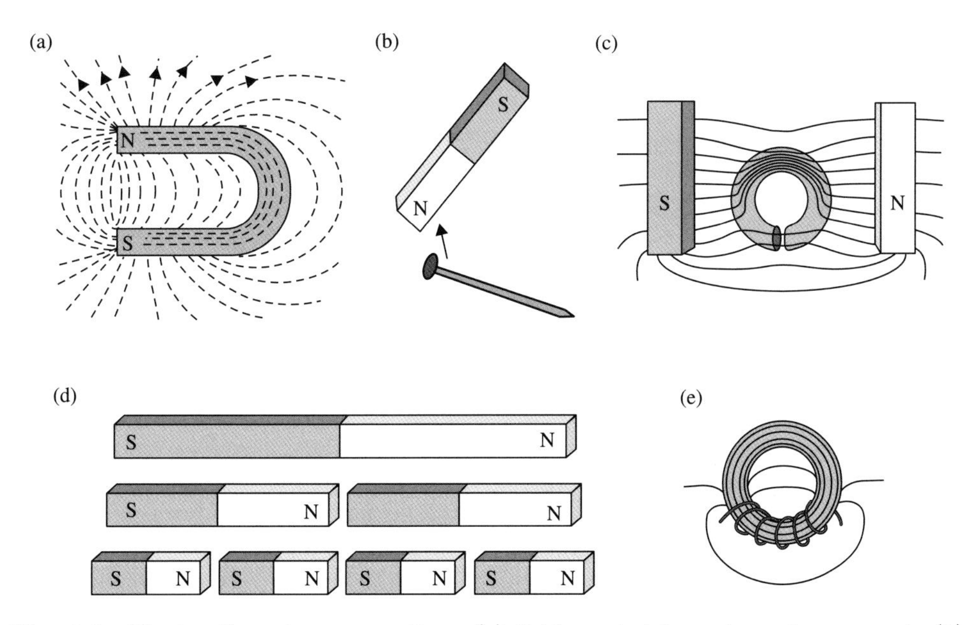

Fig. 1.1 Nineteenth-century magnetism: (a) field created by a horseshoe magnet, (b) mechanical force on a ferromagnetic body, (c) flux lines in a magnetic medium, (d) dipole character of magnetism, corresponding to the absence of magnetic monopoles, and (e) currents as one source of magnetic fields. Note that the Earth has magnetic poles that change sign every few 100,000 years. At present, the South Pole is near the geographical North Pole, so that the vectors of the magnetic field H point towards the Arctic.

the hard or soft character of a steel magnet? How to explain the Curie temperature, and why can't we ascribe it to magnetostatic interaction between atomic dipole moments? How can magnetic properties by tuned by systematically varying crystal structure, chemical composition, and nanostructure? Which ways are there to exploit magnetism in computer science and in other areas of advanced technology? Myriads of questions like these have arisen every decade and turned magnetism into a field of intense research. The modeling of magnetic phenomena and materials is a crucial aspect of this research.

To provide an introduction to magnetism and to magnetic modeling, we start with some well-known and extremely simple models. In fact, some aspects of the models are simplistic rather than simple, and this introductory chapter may also be called, *What is wrong with the simplest models of magnetism?*

However, in spite of their very limited applicability, these models are not useless. First, they hint at typical problems encountered in magnetism and provide a basis and motivation for the models in the main chapters of the book. Second, even the simplest models describe a piece of reality if used in an adequate context.

1.1 Field and magnetization

Let us start with some basic concepts. Magnetized bodies are characterized by their dipole moments $\mathbf{m} = \mathbf{M}V$, where $\mathbf{M}$ is the (average) magnetization and V is the volume of the magnet. The dipole moment is probed most easily by putting the magnet into an external magnetic field $\mathbf{H}$. The interaction of the moment with the external magnetic field $\mathbf{H}$ is described by the energy

$$E = -\mu_o \mathbf{m} \cdot \mathbf{H} \tag{1.1}$$

where the magnetic field constant μ_o ensures that the energy has the correct dimension. (In SI units, $\mu_o = 4\pi \times 10^{-7}\,\mathrm{J/A^2\,m}$.) Depending on the context, the energy (1.1) is also known as the Zeeman energy.

Figure 1.2 shows the simplest case of a compass needle or small particle in a homogeneous magnetic field. The magnetic energy $E = -\mu_o\, m\, H\, \cos\theta$, where θ is the angle of the compass needle relative to the magnetic field. The mechanical torque, $\Gamma = -\,\mathrm{d}E/\mathrm{d}\theta$, is equal to $-\mu_o\, m\, H\, \sin\theta$, so that the lowest energy is obtained for $\theta = 0$, or $\mathbf{m}||\mathbf{H}$.

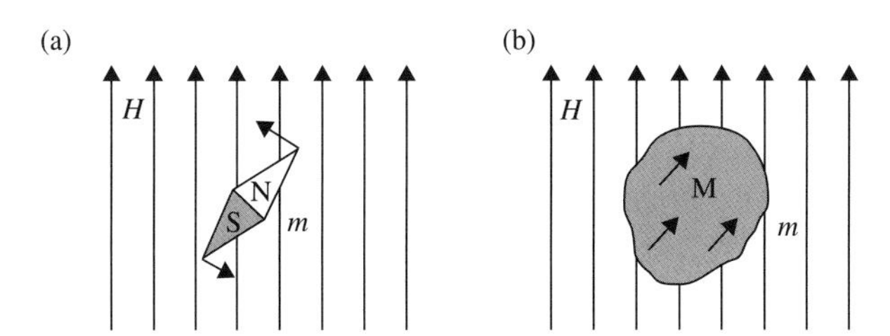

Fig. 1.2 Magnetized bodies in a magnetic field: (a) compass needle and (b) homogeneously magnetized particle. Typical field values are 0.05 mT (geomagnetic field), 0.1 T (low-grade fridge magnet), and 2 T (strong electromagnet).

Equation (1.1) describes two special cases of great practical interest: (i) small magnets in a homogeneous or inhomogeneous magnetic field and (ii) magnets of arbitrary size in a homogeneous magnetic field. Large magnets in inhomogeneous fields are described by

$$E = -\mu_o \int \mathbf{M}(\mathbf{r}) \cdot \mathbf{H}(\mathbf{r})\, \mathrm{d}V \tag{1.2}$$

or, in terms of sums over atomic moments $\mathbf{m}_i = \mathbf{m}(\mathbf{r}_i)$,

$$E = -\mu_o \sum_i \mathbf{m}_i \cdot \mathbf{H}(\mathbf{r}_i) \tag{1.3}$$

Throughout the book, we will change between the notations (1.1–3), making a suitable choice for each individual system.

A big challenge in magnetism is to find the magnetization M and the magnetic field H. Maxwell's equations yield H and the related flux density $\mathbf{B} = \mu_o(\mathbf{H} + \mathbf{M})$ from the magnetization, but they do not explain the origin of the magnetization. A popular and formally correct equation is $M = \chi H$, where χ is the magnetic susceptibility. For small M, the equation of state $M = M(H)$ can be linearized and yields a classification into paramagnets ($\chi > 0$) and diamagnets ($\chi < 0$).

However, considering χ as a constant is inappropriate for most materials. First, the relationship between M and H is generally nonlinear, approaching a finite saturation magnetization M_s. A more precise definition of the susceptibility is therefore $\chi = \mathrm{d}M/\mathrm{d}H$, measured at $H = 0$. Second, M is not necessarily a unique function of H. This phenomenon, illustrated in Fig. 1.3, is known as *magnetic hysteresis*. Key parameters of the hysteresis loop are the coercivity H_c, at which the magnetization is zero, and the remanent magnetization or *remanence* M_r. Coercivity and remanence are complemented by parameters describing the loop shape and the area under the loop. Hysteresis is caused by magnetic anisotropy and means that the so-called micromagnetic susceptibility $\partial M/\partial H$ depends on the magnetization history; Maxwell's equations do not explain hysteresis.

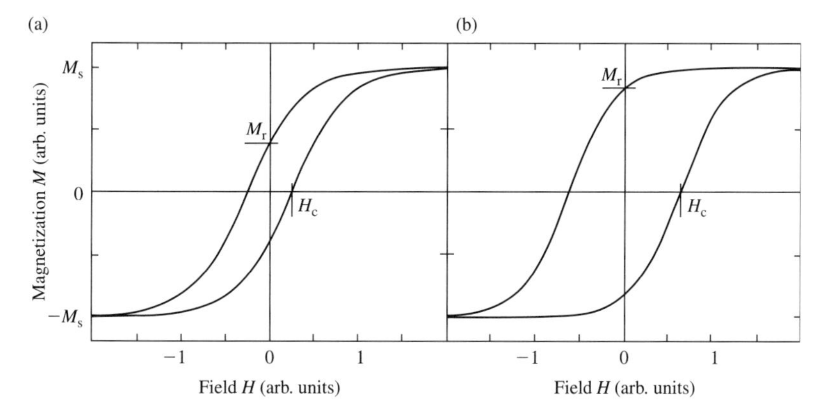

Fig. 1.3 Typical hysteresis loops: (a) soft magnets and (b) hard or permanent magnets. The motion on the loops is counter-clockwise.

Atomically, the magnetism of solids nearly exclusively originates from electrons. Nuclear moments contribute very little to the magnetization but are important, for example, in resonance imaging. Saturation means that all available atomic moments are aligned parallel to the magnetic field. As a rule of thumb, one electron per atom corresponds to an atomic moment of one Bohr magneton ($1\mu_B = 9.274 \times 10^{-24}$ J/T) and to a magnetization of 1 tesla ($\mu_o M = 1$ T). For example, elemental iron has a room-temperature magnetization of 2.15 T, corresponding to about 2 electrons per atom. Comparing these values with the total number 26 of electrons per iron atom, we see that only a small fraction of the electrons contributes to the magnetization. Most materials are actually nonmagnetic, indicating that the moment contributions of electrons in solids tend to cancel each other. The origin, size, and orientation of the magnetic moment is a key problem in magnetism.

1.2 The circular-current model

A very simple model ascribes the magnetic moment to a circular motion of electrons around atomic nuclei. The mechanism is very similar to the creation of a magnetic moment in a coil, as shown in Fig. 1.4(a). The starting point is the equation $\oint \mathbf{H} \cdot \mathrm{d}\mathbf{r} = NI$, where I is the current and N is the number of windings (Section A.4). The field is large inside the coil and near the poles but small elsewhere in free space. Performing the integral on the path C in Fig. 1.3(a) yields $H = I/L$, where L is the length of the coil. The moment HV of an empty coil of volume $\pi R^2 L$ is therefore $m = \pi R^2 I$. Since the moment is independent of the length L of the coil, this equation can also be used to describe a single current loop, as in Fig. 1.4(b). Next, we make the assumption that the current loop contains a single electron of velocity v, so that $I = ev/2\pi R$ and

$$m = \tfrac{1}{2}\, e\, v\, R \tag{1.4}$$

Here the charge of the electron $e = 1.602 \times 10^{19}$ As.

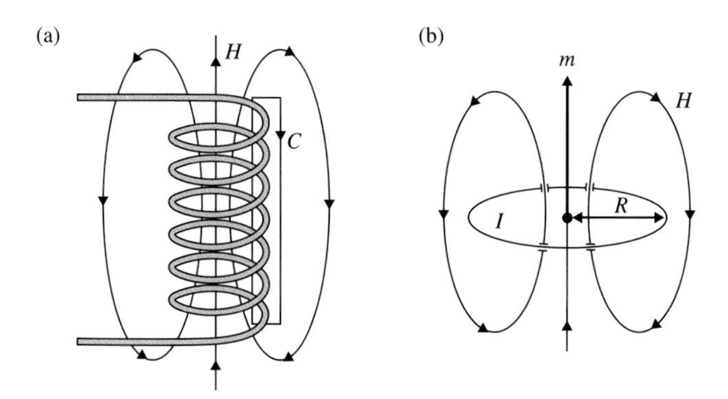

Fig. 1.4 The orbital moment **m** of an atom (left) is created by the circular motion of electrons, similar to the magnetic field created by a solenoidal coil (right).

According to (1.3), an external magnetic field could lower the energy of a solid by creating a magnetic moment m. Does this mechanism explain the moment of magnetic solids? The answer is no. Orbital moments require a nonzero electron velocity v, which costs kinetic energy. Adding the kinetic energy $\frac{1}{2}m_e v^2$ to the magnetic energy $-\mu_o mH$ and using (1.4) leads to $E = 2m_e m^2/e^2R^2 - \mu_o mH$. Minimizing this energy with respect to m yields the estimate $\mu_o He^2R^2/4m_e$ for the orbital moment created by the external magnetic field. For $\mu_o H = 1\,T$ and typical atomic radii of order $1\,\mathring{A} = 10^{-10}$ m, the corresponding magnetization is about 10^{-5} T. This is far too small to explain ferromagnetism.

The next step is to ask whether and how electrons in solids acquire a velocity v that would yield a magnetic moment. The answer comes from *quantum mechanics* (Section A.3). Electrons can be described as waves and the corresponding boundary conditions fix the velocity in an atom or solid. To determine the product $v\,R$ in (1.4), we recall that the quantity $L = m_e\,R\,v$ is the *angular momentum.* Quantum mechanics shows that the angular momentum is quantized in units of $\hbar = 1.054 \times 10^{-34}$ Js. Physically, a rotating electron of mass m_e is described by an angular wave function $\psi(\phi)$ subject to the boundary condition $\psi(\phi + 2\pi) = \psi(\phi)$. As a consequence, the moment is quantized in units of the *Bohr magneton*

$$\mu_B = \frac{e\,\hbar}{2m_e} \tag{1.5}$$

Its numerical value, $\mu_B = 0.927 \times 10^{-23}$ J/T, yields indeed the correct order of magnitude, indicating that the Bohr magneton is a key quantity in magnetism.

If each electron carried a moment of order μ_B, we would observe a large saturation magnetization in virtually every material. In reality, even in magnetic materials, only a small fraction of the electrons contributes to the moment of magnetic solids. For example, the atomic moment of iron, about 2.2 μ_B, is much smaller than the number 26 of electrons per iron atom. We will see that this is due to interactions with the atomic cores and between electrons.

A specific shortcoming of the circular-current model is that the current loops of Fig. 1.3 may be easily destroyed by electrostatic interactions with neighboring atoms. For example, only about 5% of the iron moment of 2.2 μ_B is due to circular currents. The orbital moment is said to be *quenched* by the crystal field. In fact, about 95% of the magnetization of iron originates from the *spin* of the electrons. Figure 1.4 illustrates the difference between orbital moment (a) and spin moment (b). The spin is unrelated to the orbital motion of the electrons and fully survives in a crystal field. It is usually denoted by ↑ (spin-up) and ↓ (spin-down), and each spin corresponds to a moment $m = \mu_B$. This explains why (1.5) predicts the correct order of magnitude for the magnetization, in spite of the limited applicability of the circular-current model.

The spin is unrelated to the orbital motion of the electron (Fig. 1.5). If a classical electron really spinning about an axis, then the velocity needed to reproduce its moment would exceed the velocity of light. In fact, the spin is of relativistic origin and can be considered as a kind of a magnetic analog of the rest energy $m_e c^2$ in the energy expression $E = m_e c^2 + \frac{1}{2}m_e v^2 + \mathrm{O}(1/c^4)$). As the rest energy, the spin remains nonzero for $v = 0$, while both the kinetic energy $\frac{1}{2}m_e v^2$ and the orbital moment (1.5)

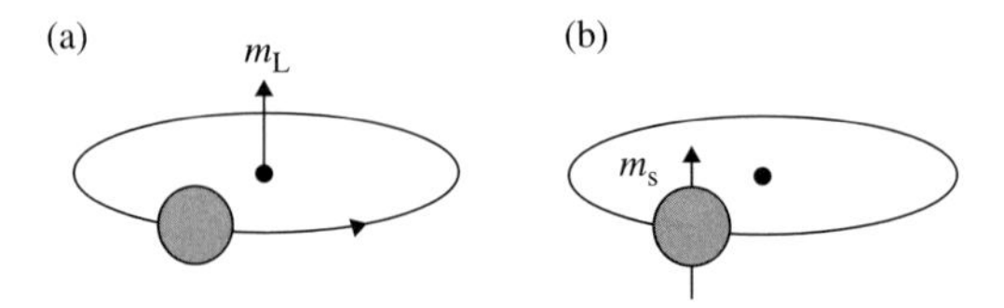

Fig. 1.5 Origin of the magnetic moment: (a) orbital moment and (b) spin moment. In many ferromagnets, including Fe, the magnetization largely originates from the spin.

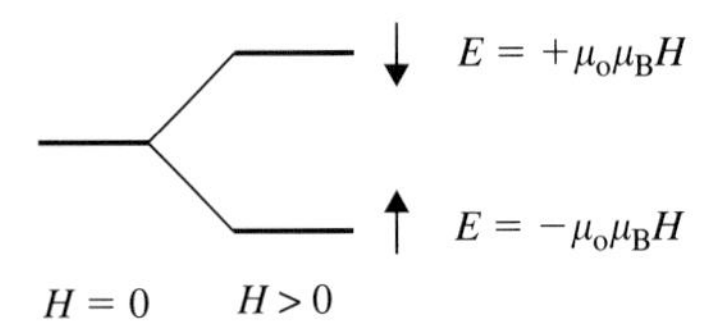

Fig. 1.6 Zeeman interaction of an electron with an external magnetic field H.

vanish. The main difference is that spin interactions, like other magnetic interactions, are very small, proportional to v^4/c^4 in the orbital-moment analogy.

1.3 Paramagnetic spins

Solid-state magnetism reflects interactions between magnetic atoms, but it is instructive to start with noninteracting moments. The model is also known as the "paramagnetic gas", because it can be used describe noninteracting magnetic impurities, ions in solutions, and transition-metal atoms in the gas phase. Let us assume that each atom carries one spin and that the orbital moment is zero. In agreement with quantum mechanics (Section A.3.2), the spin has the character of an operator or matrix.

For a spin σ and a field in the z-direction, $\mathbf{H} = H\,\mathbf{e}_z$, the quantum-mechanical analog to (1.5) is the Hamiltonian

$$\mathsf{H} = -\mu_o\mu_B H\sigma_z \tag{1.6}$$

where

$$\sigma_z = \begin{pmatrix} 1 & 0 \\ 0 & -1 \end{pmatrix} \tag{1.7}$$

is known as the diagonal *Pauli matrix*. The matrix has the two eigenvalues 1 and -1, corresponding to the spin directions $\uparrow$ and $\downarrow$, and the energy eigenvalues $E_\pm = \pm\,\mu_o\mu_B H$. In other words, the positive external field makes the $\uparrow$ configuration energetically more favorable. This *Zeeman splitting* is illustrated in Fig. 1.6.

The Zeeman energy favors spin alignment parallel to the external magnetic field. In thermal equilibrium (Section 5.1), the respective probabilities $p_\pm$ of realizing the $\uparrow$ and $\downarrow$ states are $\exp(-E_\pm/k_B T)$, that is

$$p_\pm = \frac{1}{Z}\exp\left(\pm\frac{\mu_o\mu_B H}{k_B T}\right) \tag{1.8}$$

where the partition function $Z = \exp(\mu_o\mu_B H/k_B T) + \exp(-\mu_o\mu_B H/k_B T)$ ensures the normalization of the probability, $p_+ + p^- = 1$. The thermally averaged moment is equal to $\langle m\rangle = (p_+ - p_-)\mu_B$. Using (1.8) and taking into account that $(e^x - e^{-x})/(e^x + e^{-x}) = \tanh(x)$ we obtain

$$\langle m\rangle = \mu_B \tanh\left(\frac{\mu_o\mu_B H}{k_B T}\right) \tag{1.9}$$

This equation shows that an external magnetic field creates a temperature-dependent spin polarization. At zero temperature, $\langle m\rangle = \pm\,\mu_B$, depending on the field direction $H_z = \pm\,H$. This corresponds to full spin polarization or magnetization. At high temperature, we can exploit the $\tanh(x) \approx x$ for small arguments and obtain $\langle m\rangle/V = \chi H$, where

$$\chi = \frac{\mu_o\mu_B^2}{k_B T V} \tag{1.10}$$

is the *Curie susceptibility* of the system and V is the volume per spin. (The volume doesn't affect $\langle m\rangle$ but makes χ dimensionless, which is convenient for many purposes.)

Equation (1.9) shows that a magnetic field creates a magnetic moment by aligning existing spins. We also see that the moment decreases with increasing temperature, in agreement with experiment. Can this mechanism explain ferromagnetism? The answer requires a more detailed look at the temperature dependence of the moment. Figure 1.6 shows the average moment in a typical laboratory-scale field of 1 T. The moment is continuous but strongly reduced above temperatures of about 1 K, as contrasted to the Curie temperature T_c of elemental iron, 1043 K. This indicates that (1.6) is unable to predict ferromagnetism. Furthermore, the smooth temperature dependence in Fig. 1.7 is at odds with the existence of a sharp Curie temperature.

The reason for the low-temperature character of (1.9) is the above-mentioned smallness of the magnetic interactions. The temperature equivalent of the Bohr magneton, $\mu_B/k_B = 0.672$ k/T means that spin alignment due to external fields is very effectively overcome by thermal excitations. The same is true for magnetostatic interaction fields

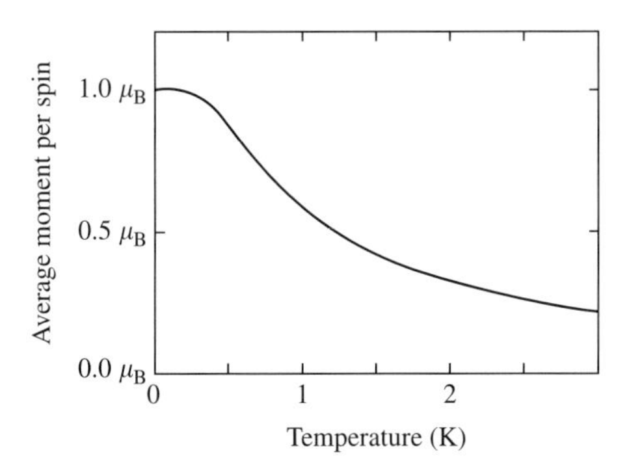

Fig. 1.7 Magnetization of an $S = \frac{1}{2}$ ion (or electron) in a field of 1 T.

between atomic moments, which establish the *Weber-Ewing model* of ferromagnetism (Bozorth 1951). Note that two compass needles may interact on a macroscopic scale, where $m = M_s V$ is much larger than μ_B, but atomic-scale dipole interactions are small.

A similar result is obtained by analyzing the magnetic susceptibility. The Curie susceptibility (1.10) is unable to explain observed room-temperature moments, as are most other susceptibilities. For example, the argumentation below Fig. 1.4 is essentially a susceptibility calculation, leading to $\chi \sim 10^{-5}$, and susceptibilities $\chi \sim \pm 10^{-5}$ are frequently encountered in paramagnetic materials such as Al and Mg ($\chi > 0$) and diamagnets such as Cu and B ($\chi < 0$ due to inductive currents opposing the applied magnetic field). The widespread occurrence of such small susceptibilities reflects the relativistic character of magnetic interactions. Kinetic and electrostatic energies of electrons scale as $\frac{1}{2} m_e v^2$, whereas magnetic interactions are proportional to mv^4/c^4. In solids, $v \sim \alpha c$, where $\alpha = 4\pi\varepsilon_o e^2/\hbar c \approx 1/137$ is Sommerfeld's fine-structure constant. Susceptibilities imply a competition between magnetic and mostly electrostatic forces, so that $\chi \sim \alpha^2$ (exercise on magnetic susceptibilities)

1.4 Ising model and exchange

We have seen that magnetic fields yield some spin polarization but are unable to explain the observed magnetic moments. In fact, even in the absence of magnetic field, ferromagnets possess a nonzero *spontaneous magnetization*. It is created and stabilized by strong interactions known as *exchange* (Heisenberg 1928, Bloch 1929). These interactions distinguish the toy magnets on our fridges from typical "nonmagnetic" materials, such as glass, copper, and wood. The quantum-mechanical origin of exchange will be discussed in Chapter 2, but it is instructive to outline how the exchange yields a finite-temperature spontaneous magnetization in zero magnetic field. The starting point is the *Ising model* (1925) defined by

$$H = -\sum_{i>j} J_{ij} s_i s_j - \mu_o \mu_B \sum_i H_i s_i \tag{1.11}$$

where $s_i = \pm 1$ is the i-th spin and the J_{ij} is the exchange interaction between the i-th and j-th spins. Figure 1.8 illustrates that positive and negative J favor parallel and antiparallel spin alignment, respectively.

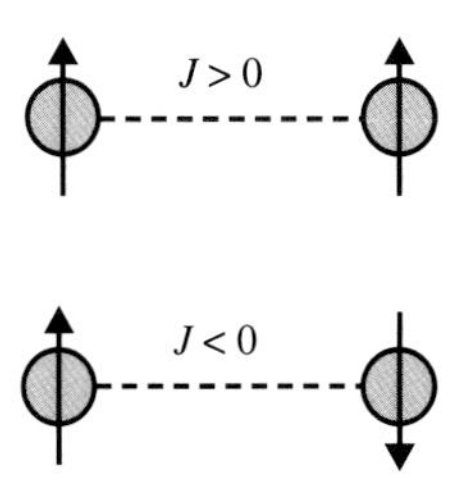

Fig. 1.8 Ferromagnetic exchange ($J > 0$) and antiferromagnetic exchange ($J < 0$).

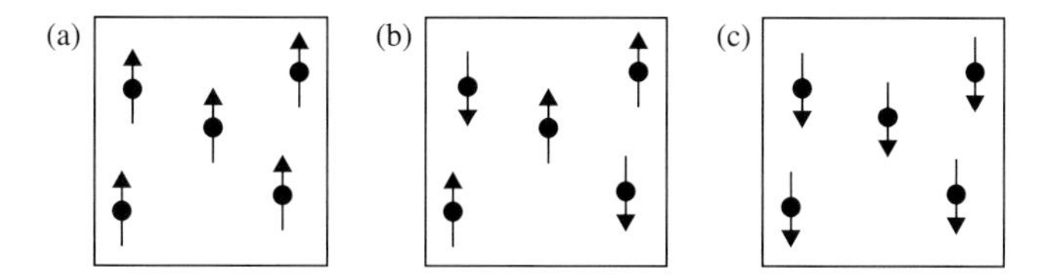

Fig. 1.9 Ensemble of Ising spins with different average normalized moments and magnetizations $m = \langle s \rangle$: (a) $m = +1$, (b) $m = 0$, and (c) $m = -1$.

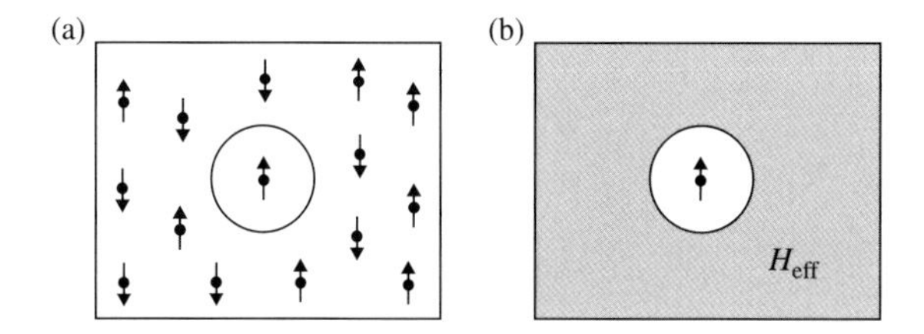

Fig. 1.10 Mean-field approximation: (a) exact environment and (b) mean-field description. The effective field H_{eff} contains the external field and the exchange field.

Unlike the paramagnetic spin of (1.6), Ising spins are *classical* entities. In (1.6), the spin may have x and y components, even if they do not enter the Hamiltonian. Physically, the spin may precess around the z-axis, and this precession requires a quantum-mechanical treatment. By definition, the Ising model excludes spin configurations others than $s_i = \pm 1$ or $\uparrow$ and $\downarrow$. Figure 1.9 illustrates this point by showing some spin structures with various average magnetizations.

For each spin configuration $s_\mu = (s_1, s_2, \ldots s_N)$, the energy is given by (1.11), and the thermal average of the magnetization of the i-th spin is obtained by summation over all spin configurations

$$\langle s \rangle = \frac{1}{Z} \sum_\mu \exp\left(\frac{-E_\mu}{k_B T}\right) \tag{1.12}$$

In spite of the simple character of (1.11–12), the solution of the Ising model is a complicated problem. The main reason is the large number of involved spin configurations, $\mu = 1, 2, 3, \ldots 2^N$, where N is the number of spins. In Chapter 5 we will see that this has far-reaching consequences for finite-temperature magnetism.

Here we consider a simplified version of the Ising model, namely the *mean-field* model defined by

$$\mathsf{H} = -(\mu_o \mu_B H + J_{MF} \langle s \rangle) \sum_i s_i \tag{1.13}$$

The interactions J_{ij} are now incorporated into an effective or mean field, $H_{\text{eff}} = H + J_{MF} \langle s \rangle / \mu_o \mu_B$, and (1.13) describes a noninteracting system, or paramagnetic gas. Figure 1.10 illustrates the meaning of the mean-field model.

As the physically different paramagnetic ion of (1.6), the mean-field Ising model has two eigenstates per spin, and using the procedure leading to (1.9) we obtain the important relation

$$\langle s \rangle = \tanh\left(\frac{\mu_o \mu_B H + J_{MF}\langle s \rangle}{k_B T}\right) \tag{1.14}$$

We will rederive and discuss this equation in Chapters 2 and 5, but would like to mention two striking features. First, (1.14) predicts ferromagnetism, $\langle s \rangle \neq 0$, below a sharp critical or Curie temperature T_c. This must be compared to the smooth temperature dependence of Fig. 1.7. Second, in contrast to (1.9), a nonzero moment $\langle m \rangle = \mu_B \langle s \rangle$ may be obtained for $H = 0$. Both features are in agreement with the observation of a spontaneous magnetization below T_c.

Let us consider $H = 0$ and start with the low-temperature limit, where $1/T = \infty$ and the argument in the hyperbolic tangent is large. The function $\tanh(x)$ is then equal to $\mathrm{sgn}(x) = \pm 1$, and both $\langle s \rangle = 1$ (all spins up) and $\langle s \rangle = -1$ (all spins down) are solutions of (1.14). This is known as spontaneous symmetry breaking, and $\langle s \rangle$ is, essentially, the spontaneous magnetization responsible for the sticking of our toy magnets on the fridge. High temperatures are described by small arguments $x = J_{MF}\langle s \rangle / k_B T$ and correspond to $\langle s \rangle \approx 0$ near the Curie temperature. In this limit, $\tanh(x) = x$ and $\langle s \rangle = J_{MF}\langle s \rangle / k_B T$, so that the division by the small though unknown quantity $\langle s \rangle$ yields $1 = J_{MF}/k_B T$ and $T_c = J_{MF}/k_B$.

The exchange J, as introduced in this subsection and discussed in Chapter 2, is the key to the understanding of many magnetic phenomena, including the spontaneous magnetization below the Curie temperature. Unfortunately, the present models are simplistic in other regards. For example, Fig. 1.10 fixes the magnetization in the $\pm\mathbf{e}_z$ direction, so that all ferromagnets should be permanent magnets, characterized by a strong anisotropy in the z-direction. In fact, most ferromagnets are rather soft, and neither the Ising model nor its isotropic equivalent, the Heisenberg model, are able to predict magnetic anisotropy. A more fundamental problem is that the mean-field model predicts ferromagnetism for virtually any magnetic system with positive (ferromagnetic) exchange $J > 0$. A famous counterexample is the exact solution of the one-dimensional Ising model, which predicts paramagnetism at any nonzero temperature.

1.5 The viscoelastic model of magnetization dynamics

An important aspect of magnetism is *hysteresis* (Fig. 1.3). What is its physical origin? A very simple approach is to consider magnetic bodies as linear media, in analogy to viscoelastic models widely used in rheology, polymer physics, and metallurgy. The idea is that an external force f or field H creates a response such as an elongation x or a magnetization M. In the simplest mechanical case, the linear response is described by the spring-and-dashpot model shown in Fig. 1.11. The model consists of a spring of spring constant K and a dashpot described by the viscosity constant η. The parallel coupling of the elements means that the total force f is the sum of elastic and viscous forces.

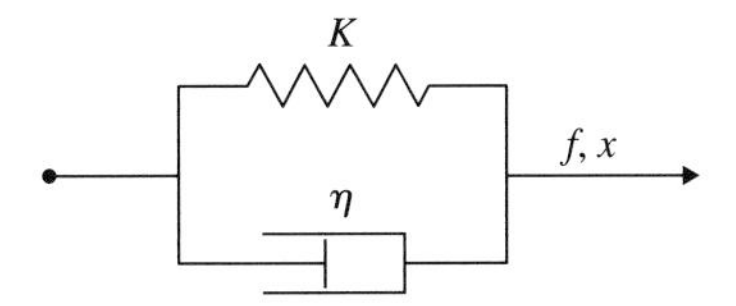

Fig. 1.11 Spring and dashpot model of viscoelasticity. A force f creates an elongation x that depends on K and M.

There are two contributions, an elastic contribution due to the spring and a viscous contribution due to the dashpot. The elastic contribution $f = Kx$ is a straight line without hysteresis, but the total force is the sum of the elastic force Kx and the viscous force $\eta \mathrm{d}x/\mathrm{d}t$

$$f = Kx + \frac{\eta \mathrm{d}x}{\mathrm{d}t} \tag{1.15}$$

There are several ways of solving this equation. For example, a constant force f_o, switched on at $t = 0$, yields

$$x(t) = \frac{f_\mathrm{o}}{K}\left(1 - \exp\left(\frac{-t}{\tau}\right)\right) \tag{1.16}$$

where the *relaxation time* $\tau = \eta/K$ describes the approach to the equilibrium value $x = f_\mathrm{o}/K$. A straightforward but somewhat cumbersome general method of solving (1.16) is to approximate $f(t)$ as a sum of steps, $f(t + \Delta t) = f(t) + \Delta f$. The total response is then obtained as a sum of the contributions of the type (1.16).

A more elegant approach is to consider the force $f(t)$ as a superposition of sinoidally oscillating functions (Fourier transform), as contrasted to a superposition of step functions (Laplace transform). A force $f = f_\mathrm{o}\sin(\omega t)$ yields a response $x = x_\mathrm{o}\sin(\omega t - \phi_\mathrm{o})$, where ϕ_o is the phase shift of the system. Due to the phase shift, the dependence of x on f is hysteretic (Lissajous figure). Putting $f(t)$ and $x(t)$ into (1.16) leads to

$$x_\mathrm{o} = f_\mathrm{o}\frac{\cos(\phi_\mathrm{o})}{K} \tag{1.17}$$

and

$$\tan\phi_\mathrm{o} = \frac{\eta\omega}{K} \tag{1.18}$$

An alternative way to derive these relations is to consider the spring constant as a complex quantity $K* = K' + \mathrm{i}K''$, similar to the complex description of alternating-current circuits. Putting $x = x_\mathrm{o}\exp(i\omega t)$ then yields $K* = K + \mathrm{i}\omega\eta$.

In the *magnetic analogy*, f and x correspond to H and M, respectively, and the spring constant has the character of an inverse susceptibility, $K = 1/\chi$. Figure 1.12 shows two loops for $\phi_\mathrm{o} = 15\%$ (left) and $\phi_\mathrm{o} = 50\,\%$ (right). The phase-shift angle obeys

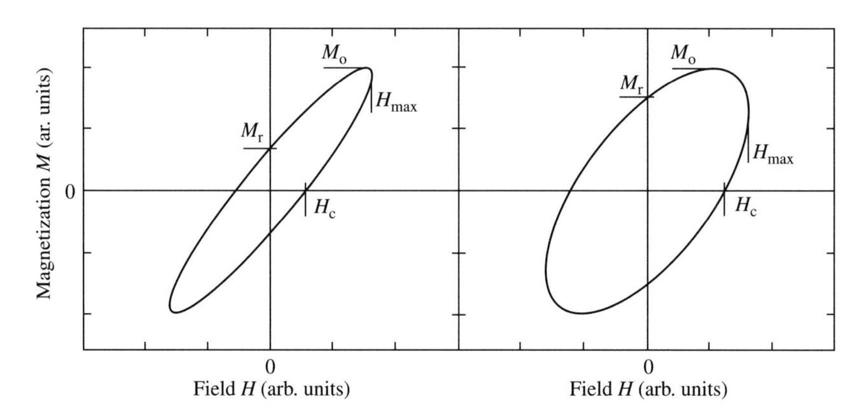

Fig. 1.12 Hysteresis loops predicted from the viscoelastic model of magnetic hysteresis. The parameters used are $\phi_o = 15°$ (left) and $\phi_o = 50°$ (right).

$\tan\phi_o = \eta\omega\chi$, and the hysteresis-loop parameters are

$$M_o = \cos\phi_o \, \chi_o \, H_o \tag{1.19}$$

$$M_r = \sin\phi_o \, M_o \tag{1.20}$$

$$H_c = \sin\phi_o \, H_o \tag{1.21}$$

Since the angle ϕ_o increases with ω, remanence (M_r) and coercivity (H_c) are largest for fast magnetization processes. In the opposite limit of very slow magnetization processes, $\phi_o \sim \omega$ and both H_c and M_r approach zero, as expected for equilibrium. The energy loss per cycle (or hysteresis-loop area)

$$\frac{\Delta E}{V} = \pi \sin\phi_o \mu_o M_o \, H_o \tag{1.22}$$

Here μ_o and the volume V of the magnet ensure that ΔE has the dimension and physical meaning of an energy.

Comparison of Figs. 1.3 and 1.12 shows that the viscoelastic model reproduces some basic features of magnetic hysteresis, but a detailed analysis shows that the applicability of the model is very limited. First, (1.19) predicts that the maximum magnetization M_o is a linear function of the magnetic field, in agreement with the linear character of (1.15). However, experimental hysteresis loops are highly nonlinear. Figure 1.3 indicates that the magnetization approaches a finite high-field value, namely the saturation magnetization M_o. Saturation means that all atomic moments are aligned parallel, so that any additional external field does not translate into a magnetization change. Second, the predicted ellipsoidal loop shape is different from experimental loops.

Third, the parameterization of the loop in terms of η has no sound physical basis. This is somewhat different from the mechanical analogy, where the relaxation time $\tau \sim \eta$ tends to have a well-defined physical meaning. For example, a mechanism realized in steel is the Snoek damping, where the relaxation time τ describes the interstitial diffusion of carbon atoms. In most magnets, there is no such simple mechanism, and the model description in terms of η or τ is simplistic. Fourth, eqs. (1.17–18) predict an approximately linear dependence of H_c and M_r on the frequency ω. In fact, experiments probing the time dependence of hysteretic properties tend to yield logarithmic laws.

Some quantitative improvement is achieved by using more complicated spring-and-dashpot models. The parallel connection shown in Fig. 1.11 is also known as the Voigt or Kelvin model, as compared to the Maxwell model (series connection). There are also models involving three or more elements, but the models all suffer from the main shortcomings of the approach, namely from the restriction to linear response and from the phenomenological character of the involved parameters.

Exercises

1. ***Magnetization of solids.*** Typical ferromagnetic metals have 10^{29} atoms per m^3 and atomic moments of order $1\,\mu_B$ per atom. Calculate the zero-temperature magnetiztion and compare the result with the magnetizations of Fe, Co, and Ni. What are the reasons for the relatively low magnetization of many oxides?
2. ***Magnetic moment of the Earth.*** Estimate the Earth's magnetic moment from the from the geomagnetic field of order 0.05 mT. If the Earth's magnetic field were created by a homogeneously magnetized sphere of iron, with a magnetization of 2.2 T, what would be the diameter of the iron sphere?
   Answer: 8×10^{22} Am^2, corresponding to an average magnetization of 0.09 mT.
3. ***Direction of circulation on a hysteresis loop.*** Is the circulation on a typical ferromagnetic hysteresis loop clockwise or anticlockwise?
4. ***Paramagnetic moment of single spin.*** Calculate the moment of a single spin in a magnetic field of 0.5 T at 4.2 K and at 300 K.
5. ***Susceptibility of iron.*** The susceptibility of a soft-magnetic iron piece is equal to $\chi = 500$. Use the relation $M = \chi H$ to calculate the magnetization in a field of 300 mT and discuss the result.
6. ***Magnetostatic field components.*** Determine the field (H_x, H_y, H_z) at $\mathbf{r} = \mathbf{R}$ created by a magnetic dipole (M_x, M_y, M_z) at $\mathbf{r} = 0$.
7. ***Field in iron-cored solenoid.*** Estimate the magnetic field created by a long iron-cored solenoid of diameter 1 cm: (a) directly at the pole and (b) at a distance of 2 cm.
8. ***Mechanical force of a permanent magnet.*** Estimate the maximum force excerted by a Sm-Co permanent magnet of cross-section area 1 cm^2. Can this result also be used for soft magnets?
   Answer: The force scales as $f = \mu_o M^2 A/L^2 2$, where L^2 is the cross-section area of the magnet. Taking $\mu_o M = 1$ T we obtain $f/L^2 = 400\,\mathrm{kN/m}^2$ and $f/g \approx 4$ kg. In soft magnets, domain formation strongly reduces the net magnetization.
9. ***Magnetization contribution due to nucleons.*** Compared to electrons, the magnetization contribution due to protons and neutrons is very small. Why?
10. ***Viscoelastic loops.*** Show that the viscoelastic model of magnetization dynamics yields ellipsoidal hysteresis loops.

11. ***Magnetic field created by a long wire.*** Calculate the field created by a current I in a long wire, determine the total magnetostatic field energy, and discuss the result.
    Answer: The field is $H = I/2\pi r$ (see Appendix 4), so that the energy $\frac{1}{2}\mu_o \int H^2 \mathrm{d}V$ diverges. This means that one must put a lot of electrical energy into the wire to create the field. In practice, all wires are closed, and one must consider a current loop rather than a linear wire. (Note that the direction of the field is given by the right-hand rule: When the current is in the direction of the thumb, then the field is in the direction of the curved fingers. For a solenoid: a current in the direction of the right hand's fingers creates a magnetic field in the direction of the right thumb.)
12. ***Susceptibilities.*** Show that typical magnetic susceptibilities are of order α^2, where $\alpha = 1/137$, and discuss a few exceptions.
    Hint: Analyze the ground-state energy of the hydrogen atom and take into account that $e^2/4\pi\varepsilon_o a_o \approx m_e\alpha^2c^2$.
    Answer: As in Section 1.2, the calculation involves the following steps: $I = e/t$, $v = 2\pi R/t$, $A = \pi R^2$, $m = eRv/2$, $R = 4\pi\varepsilon_o\hbar^2/m_e e^2$, $v = e^2/4\pi\varepsilon_o\hbar$. The calculation ascribes the susceptibility to the competition between Zeeman energy and atomic energies. It does not apply when the main competition is between Zeeman and thermal energies, as in ferromagnets near T_c and in paramagnetic gases.
13. ***Magnetic poles and magnetic field.*** Express the magnetic field in terms of $\nabla \cdot \mathbf{M}$.
    Answer: $\mathbf{H}(\mathbf{r}) = -(1/4\pi) \int (\mathbf{r} - \mathbf{r}')\nabla \cdot \mathbf{M}(\mathbf{r}')/|\mathbf{r} - \mathbf{r}'|^3 \mathrm{d}V' + \mathbf{H}_o$, where $\mathbf{H}_o$ is the external field.
14. ***Coercivity in the viscoelastic analogy.*** Consider and determine the coercivity as a function of the maximum field H_o and maximum sweep rate $\mathrm{d}H/\mathrm{d}t$ in the viscoelastic model of hysteresis.
    Hint: Take into that $\max(\mathrm{d}H/\mathrm{d}t) = \omega H_o$.

2
Models of exchange

Ferromagnetism requires a strong force to create the atomic moments and to stabilize the parallel orientation of neighboring moments. As mentioned in the introduction, the applied magnetic field is unable to compete against interatomic energies, such as the kinetic energy of the electrons, and temperatures of the order of 1 K are sufficient to destroy ferromagnetic order established by magnetostatic dipole interactions ($\mu_B/k_B = 0.672\,\mathrm{K/T}$). The strong force responsible for the moment and magnetization is the *exchange* interaction. Figure 2.1 shows typical spin structures caused by exchange interactions. Elemental Fe, Co, and Ni, as well as many alloys and some oxides, are ferromagnetic (a), whereas antiferromagnetism (b) is observed, for example, in Mn and NiO. This chapter deals with the interatomic exchange responsible for long-range magnetic order, but also with the intra-atomic exchange which creates atomic moments. Emphasis is on atomic models of exchange, continuum models will be treated in Section 4.2.3.

The competition between interatomic exchange and thermal disorder leads to the vanishing of the spontaneous magnetization at a well-defined sharp Curie temperature T_c. The total interatomic exchange per atom does not exceed about 0.1 eV, corresponding to $T_c \approx 1000\,\mathrm{K}$. This is much smaller than the intra-atomic exchange, which is of the order of 1 eV, so that atomic moments at T_c remain close to their zero-temperature values and typical magnetization processes in solids are caused by magnetization rotations.

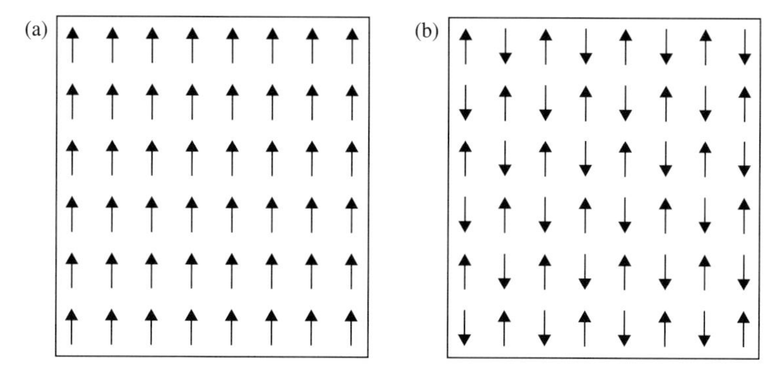

Fig. 2.1 Basic types of magnetic order: (a) ferromagnetism and (b) antiferromagnetism. Each arrow represents one atomic spin.

In the simplest case, it is sufficient to consider two electrons. The starting point is the *Pauli principle*, which forbids the double occupancy of a quantum state by fermions. By definition, fermions are particles with half-integer spin, such as electrons and protons. Since the spin is a quantum number, double occupancy of an orbital is possible for antiparallel spins (↑↓) but forbidden for parallel spins (↑↑). This explains, for example, the periodic table of the elements: the 1s orbital of helium is occupied by a ↑↓ electron pair, and an additional electron is not allowed to gain energy by jumping into the 1s shell. To realize a parallel spin orientation, as of interest in magnetism, one electron must occupy an excited one-electron orbital. The necessary energy comes from the Coulomb interaction

$$V_C = \frac{e^2}{4\pi\epsilon_o|\mathbf{r} - \mathbf{r}'|} \tag{2.1}$$

between the two electrons at $\mathbf{r}$ and $\mathbf{r}'$. The Coulomb interaction is spin-independent but larger for electrons in a common orbital (↑↓) than for electrons in different orbitals (↑↑). In other words, the Coulomb interaction favors parallel spin alignment but competes against an increase in one-electron energy. This true for all types of exchange: the intra-atomic exchange between electrons in one atom, as in Fig. 2.2(a), the interatomic exchange between localized spins on different atoms, as in Fig. 2.2(b), and the itinerant exchange in metals such as iron, which combines both intra- and interatomic features. For the moment, our focus is on the *interatomic* exchange; intra-atomic and itinerant exchange interactions will be discussed in Sections 2.2 and 2.4, respectively.

From (2.1), the exchange is obtained by comparing the total energies for ferromagnetic (FM) and antiferromagnetic (AFM) two-electron wave functions $\Psi_{FM}(\mathbf{r}, \mathbf{r}')$ and $\Psi_{AFM}(\mathbf{r}, \mathbf{r}')$. The original *Heisenberg model*, which describes the exchange between two neighboring atoms, is closely related to the Heitler-London approximation in chemistry. It assumes that

$$\Psi_{FM} \sim \phi_l(\mathrm{r})\phi_r(\mathbf{r}') - \phi_r(\mathrm{r})\phi_l(\mathbf{r}') \tag{2.2}$$

and

$$\Psi_{AFM} \sim \phi_l(\mathrm{r})\phi_r(\mathbf{r}') + \phi_r(\mathrm{r})\phi_l(\mathbf{r}') \tag{2.3}$$

where $\phi_l(\mathrm{r})$ and $\phi_r(\mathrm{r})$ are the respective atomic wave functions of the "left" and "right" atoms. Evaluating the energy $E = \int \Psi^*(\mathbf{r}, \mathbf{r}')\mathsf{H}(\mathbf{r}, \mathbf{r}')\Psi(\mathbf{r}, \mathbf{r}')\mathrm{d}V\,\mathrm{d}V'$ for (2.2–3) yields the *exchange constant* $J = (E_{AFM} - E_{FM})/2$. A positive J means parallel or ferromagnetic spin coupling (↑↑), whereas an negative J means that the spins are coupled antiparallel or antiferromagnetic (↑↓).

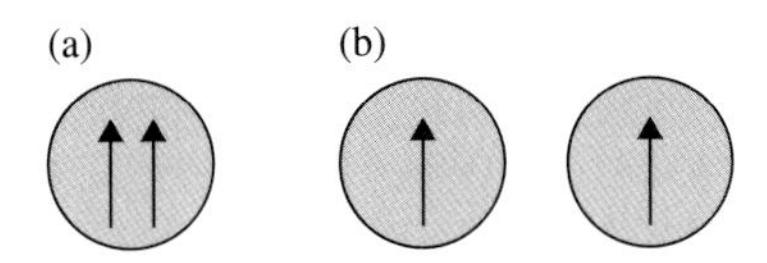

Fig. 2.2 Ferromagnetic exchange: (a) intra-atomic and (b) interatomic. Both types of exchange have the same origin, but intra-atomic exchange tends to be stronger, of order 1 eV or 10000 K, as compared to typical interatomic exchange of about 0.1 eV or 1000 K.

There are two arguments behind the choice of (2.2) and (2.3). First, each of the two atoms is assumed to be occupied by one electron. This explains why the terms in (2.2–3) contain both ϕ_l and ϕ_r. Second, the Pauli principle forbids the double occupancy of any orbital by electrons of parallel spin. For example, the wave function $\phi_\mathrm{o}(\mathrm{r})\phi_\mathrm{o}(\mathbf{r}')$ has two electrons in the same orbital ϕ_o and implies a $\uparrow\downarrow$ occupancy. More generally, symmetric and antisymmetric real-space wave functions $\Psi(\mathbf{r}', \mathbf{r}) = \Psi(\mathbf{r}, \mathbf{r}')$ and $\Psi(\mathbf{r}', \mathbf{r}) = -\Psi(\mathbf{r}, \mathbf{r}')$ correspond to $\uparrow\downarrow$ and $\uparrow\uparrow$ configurations, respectively. This explains the assignment of ferromagnetism and antiferromagnetism in (2.2–3).

The original Heisenberg model yields J as a function of integrals over various energy terms and provides a qualitative understanding of exchange. In particular, (2.1) is a relatively strong interaction of electrostatic origin, as compared to the weak magnetic forces considered in the introduction. On the other hand, the choice of (2.2–3) is intuitive, based on the idea that each atom is occupied by one electron. Experience shows that electrons are likely to hop to neighboring atoms, but the corresponding antiferromagnetic configurations $\phi_\mathrm{l}(\mathrm{r})\phi_\mathrm{l}(\mathbf{r}')$ and $\phi_\mathrm{r}(\mathrm{r})\phi_\mathrm{r}(\mathbf{r}')$ are ignored in (2.2–3). To make matters worse, atomic wave functions $\phi_\mathrm{l}(\mathrm{r})$ and $\phi_\mathrm{r}(\mathrm{r})$ between neighboring atoms are generally nonorthogonal. The corresponding overlap integral $S_\mathrm{o} = \int \phi_\mathrm{l}(\mathrm{r})\phi_\mathrm{r}(\mathrm{r})\,\mathrm{d}V$ amounts to off-diagonal matrix elements between the atomic wave functions and yields some hopping. This is inconsistent and means that neither one-electron properties nor two-electron properties are described correctly. Below, we will solve this problem by starting from exact one-electron wave functions that are orthogonal.

2.1 Atomic origin of exchange

Summary The competition between the kinetic and Coulomb energies decides whether the spin state is ferromagnetic ($\uparrow\uparrow$) or antiferromagnetic ($\uparrow\downarrow$). The motion of the electrons is realized by interatomic hopping and accompanied by an energy reduction due to hybridization. The Pauli principle forbids the occupancy of low-lying or "bonding" states by electrons with parallel spin, so that the corresponding ground-state spin structure is $\uparrow\downarrow$. Ferromagnetism is a many-electron effect and means that some electrons occupy excited one-electron states. The necessary energy is supplied by the Coulomb repulsion between electrons, which punishes $\uparrow\downarrow$ occupancies of one-electron orbitals. The corresponding exchange constant J, defined as half the energy difference between the $\uparrow\downarrow$ and $\uparrow\uparrow$ states, depends on parameters such as the interatomic distance and the number of electrons per atom. An exactly solvable two-electron model considers two atoms and one atomic orbital per atom. Aside from direct exchange, which is always positive, J reflects the relative strength of the Coulomb integral compared to the hopping integral. As a rule, pronounced interatomic hopping destroys the parallel spin alignment. Finally, a two-electron model is used to introduce and discuss various models and approximations, including the Heisenberg, Hubbard, and Kondo models, and to separate independent-electron or Hartree-Fock contributions from correlation corrections.

To see how exchange arises from the Coulomb interaction (2.1), we must start from a suitable set of one-electron wave functions $\psi(\mathrm{r})$. Atomic wave functions are a poor choice, because they are not orthogonal and do not form a set of one-electron eigenfunctions. Physically, the overlap between atomic wave functions causes the electrons to hop onto neighboring atoms, and the atomic energy levels are no longer representative of the system. It is therefore convenient to use one-electron eigenfunctions constructed from the atomic wave functions. For simplicity, the focus of this section is on hydrogen-like 1s electrons, but the same principles apply to other electrons, such as 3d electrons in iron. The only differences are the involvement of several orbitals per atom (Section 2.2) and an angular dependence of the exchange (Section 2.3).

2.1.1 One-electron wave functions

Well-separated atoms are described by atomic wave functions, but in solids the wave functions overlap and remix. The atomic wave functions are solutions of the Schrödinger equation

$$E\Psi = -\frac{\hbar^2}{2\mathrm{m}}\nabla^2\Psi + V_\mathrm{o}(\mathrm{r})\Psi \tag{2.4}$$

where $V_\mathrm{o}(\mathrm{r}) \sim Z/r$ is the negative (attractive) potential from the nucleus. For hydrogen, $Z = 1$, but magnetic atoms have effective nuclear charges $Z > 1$. Ignoring atomic excitations, (2.4) yields wave functions $\phi_\mathrm{o} \sim \exp(-r/R_\mathrm{o})$ and some atomic energy E_at. The range R_o of the wave function depends on Z and, more generally, on whether s, p, d, or f electrons are considered (Section 2.3). Figure 2.3 shows two s-type wave functions centered on neighboring atoms.

In a molecule or solid, the potential $V(\mathrm{r}) = V_\mathrm{o}(\mathrm{r})$ must be replaced by a sum over all atoms, $V(\mathbf{r}) = \sum_\mathrm{i} V_\mathrm{o}(|\mathbf{r} - \mathbf{R}_\mathrm{i}|)$, where R_i is the position of the i-th atom. For two atoms located at $\mathbf{R}_\mathrm{i} = 0$ and $\mathbf{R}_\mathrm{i} = \mathbf{R}$,

$$E\Psi = -\frac{\hbar^2}{2m}\nabla^2\Psi + V_\mathrm{o}(\mathbf{r})\Psi + V_\mathrm{o}(|\mathbf{r} - \mathbf{R}|)\Psi \tag{2.5}$$

Both the energy levels and the eigenfunctions of this equation differ from those of (2.4). In lowest-order approximation, the perturbed wave functions are linear combinations

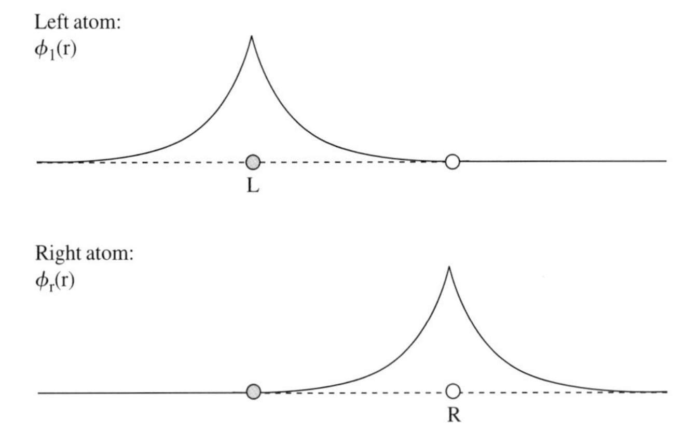

Fig. 2.3 Atomic wave functions of the diatomic pair model.

of $\phi_\mathrm{l}(\mathrm{r}) = \phi_\mathrm{o}(\mathrm{r})$ and $\phi_\mathrm{r}(\mathrm{r}) = \phi_\mathrm{o}(|\mathbf{r} - \mathbf{R}|)$. This leads to a 2×2 matrix Hamiltonian (Section A.2.3 and Section A3.2)

$$\mathsf{H} = \begin{pmatrix} E_\mathrm{o} & t \\ t & E_\mathrm{o} \end{pmatrix} \tag{2.6}$$

Here $E_\mathrm{o} = E_\mathrm{at} + \Delta E$ is the atomic energy in the molecule and t is the *hopping integral*

$$t = \int \phi_\mathrm{o}^*(\mathrm{r}) V_\mathrm{o}(|\mathbf{r} - \mathbf{R}|) \phi_\mathrm{o}(|\mathbf{r} - \mathbf{R}|) \, \mathrm{dV} \tag{2.7}$$

The hopping character of t becomes clear by rewriting this equation as

$$t = -\int \phi_\mathrm{o}^*(\mathrm{r}) \mathsf{T} \phi_\mathrm{o}(|\mathbf{r} - \mathbf{R}|) \, \mathrm{dV} \tag{2.8}$$

where $\mathsf{T} = -\hbar^2 \nabla^2 / 2m$ is the kinetic-energy operator (exercise on hopping). Equation (2.8) means that an electron initially located at $\mathbf{r} = \mathbf{R}$ is transferred by the operator T to a new position (*) around $\mathbf{r} = 0$. The hopping integral (2.8) decreases with increasing interatomic distance R.

Diagonalization of (2.6) yields a symmetric eigenfunction $\psi_\mathrm{s}(\mathrm{r}) \sim \phi_\mathrm{l}(\mathrm{r}) + \phi_\mathrm{r}(\mathrm{r})$ and an antisymmetric eigenfunction $\psi_\mathrm{a}(\mathrm{r}) \sim \phi_\mathrm{l}(\mathrm{r}) - \phi_\mathrm{r}(\mathrm{r})$. Figure 2.4 compares the superposed or hybridized functions $\psi_\mathrm{s}(\mathrm{r})$ and $\psi_\mathrm{a}(\mathrm{r})$. In chemistry, the superposed functions are also known as molecular orbitals (MO), and their construction from atomic orbitals is known as the LCAO (linear combination of atomic orbitals) or LCAO-MO approach. The energy of the symmetric or bonding state is lower than that of the antisymmetric or antibonding state. Noting that t is negative for s-states, we find the respective energies $E_\mathrm{o} \pm t$ for bonding and antibonding states. Since t decreases with increasing interatomic distance, the level splitting is smallest for well-separated atoms.

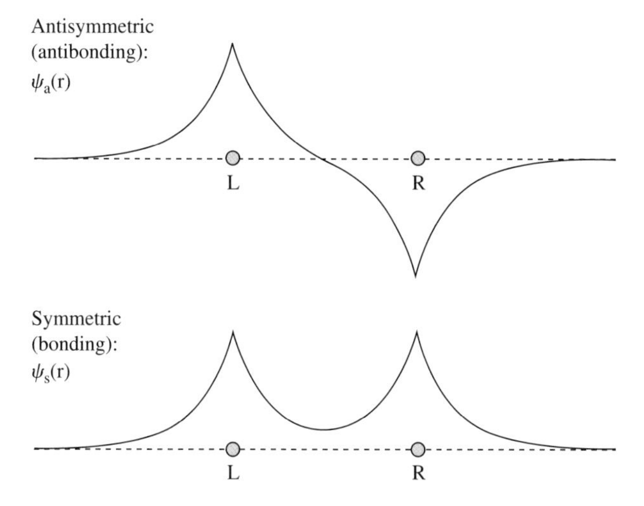

Fig. 2.4 Symmetric and antisymmetric wave functions. These functions are the starting point for the calculation of the exchange.

Energy splittings due to interatomic hybridization are widespread and of great importance in chemistry and solid-state science. They affect not only the magnetism, as of interest in this book, but also the chemical bonding. For example, Fig. 2.4 provides a direct explanation of the chemical bonding in the H_2^+ ion. The bonding character of the symmetric wave function is seen from the electron density $n(\mathbf{r}) = \psi^*(\mathbf{r})\psi(\mathbf{r})$, which is relatively large between the atoms, as compared to zero in the antibonding state. Alternatively, rapidly changing wave functions have a higher curvature $\nabla^2\psi(\mathrm{r})$ and therefore a higher kinetic energy. In solids, the hybridization involves more than two atoms, but the basic physics remains unchanged. For example, hybridization is responsible for the formation of energy bands in metals, and the bandwidth W is analogous to the level splitting $2|t|$ (Section 2.4).

The wave functions $\psi_\mathrm{s}(\mathrm{r})$ and $\psi_\mathrm{a}(\mathrm{r})$ provide an adequate description of the system but are somewhat counterintuitive. It would be more convenient to have orthogonal wave functions that are centered around a given atom, similar to the atomic wave function in Fig. 2.3. Such wave functions are known as *Wannier functions*.

$$\phi_\mathrm{L} = \frac{1}{\sqrt{2}}(\phi_\mathrm{s} + \phi_\mathrm{a}) \quad \text{and} \quad \phi_\mathrm{R} = \frac{1}{\sqrt{2}}(\phi_\mathrm{s} - \phi_\mathrm{a}) \tag{2.9}$$

Inverting these expressions yields

$$\psi_\mathrm{s} = \frac{1}{\sqrt{2}}(\phi_\mathrm{L} + \phi_\mathrm{R}) \quad \text{and} \quad \psi_\mathrm{a} = \frac{1}{\sqrt{2}}(\phi_\mathrm{L} - \phi_\mathrm{R}) \tag{2.10}$$

As an example, Fig. 2.5 compares the Wannier functions ϕ_L with the corresponding atomic wave function ϕ_l. Wannier functions are somewhat less localized than atomic wave functions, which may require the inclusion of an extended neighborhood in calculations. However, in this section we consider well-separated atoms, where atomic and Wannier wave function are very similar.

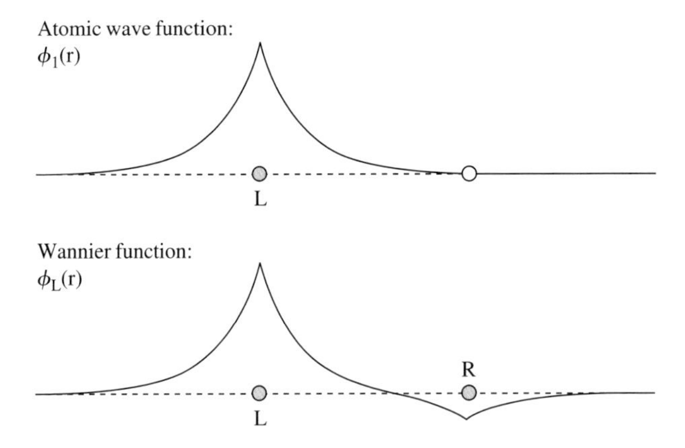

Fig. 2.5 Example of a diatomic Wannier function.

Panel 1 Normalization of atomic wave functions

Unlike Wannier functions, atomic wave functions are nonorthogonal. This complicates the calculation of energy levels and wave functions in solids. The reason is the involvement of the overlap integral $S_o = \int \phi_r^*(\mathbf{r})\phi_l(\mathbf{r})\, d\mathbf{r}$. Compare, for example, the normalizations

$$\psi_s = \frac{1}{\sqrt{2(1+S_o)}}(\phi_l + \phi_r) \quad \text{and} \quad \psi_a = \frac{1}{\sqrt{2(1-S_o)}}(\phi_l - \phi_r)$$

with the corresponding Wannier expression $1/\sqrt{2}$. Equation (2.9) yields an explicit relation between the atomic wave functions and the Wannier functions:

$$\phi_L = \frac{1}{2}\left(\frac{1}{\sqrt{1-S_o}} + \frac{1}{\sqrt{1+S_o}}\right)\phi_l - \frac{1}{2}\left(\frac{1}{\sqrt{1-S_o}} - \frac{1}{\sqrt{1+S_o}}\right)\phi_r$$

$$\phi_R = \frac{1}{2}\left(\frac{1}{\sqrt{1-S_o}} + \frac{1}{\sqrt{1+S_o}}\right)\phi_r - \frac{1}{2}\left(\frac{1}{\sqrt{1-S_o}} - \frac{1}{\sqrt{1+S_o}}\right)\phi_l$$

Since the original Heisenberg model is based on atomic wave functions, the exchange depends on the overlap integral. However, this is an arbitrary choise, and for Wannier functions, the exchange is actually independent of S_o. In the next section, we will see that exchange involves the hopping integral, which is related to the overlap integral but has a well-defined physical meaning.

Exercises

1. Show that $\int \phi_R^*(\mathbf{r})\phi_L(\mathbf{r})d\mathbf{r} = 0$ and $\int \phi_s^*(\mathbf{r})\phi_a(\mathrm{r})d\mathbf{r} = 0$, in spite of a nonzero overlap integral $S_o = \int \phi_r^*(\mathbf{r})\phi_l(\mathbf{r})d\mathbf{r}$.
2. Determine ϕ_L and ϕ_R from ϕ_l and ϕ_r for well-separated atoms, where S_o is small.

2.1.2 Two-electron wave functions

The wave functions $\psi_s(\mathbf{r})$ and $\psi_a(\mathbf{r})$ provide an adequate description of the one-electron problem but do not explain exchange. In fact, if (2.5) was the only consideration, then two electrons would occupy the bonding orbital in Fig. 2.4 and form a $\uparrow\downarrow$ pair. Due to the Pauli principle, $\uparrow\uparrow$ configurations mean that one electron occupies an energetically unfavorable antisymmetric state. Parallel spin alignment is caused by the Coulomb energy (2.1), which describes interactions between electrons. For two electrons, it is necessary to solve the Schrödinger equation $E\Psi = \mathbf{H}\Psi$ or, explicitly,

$$E\Psi = \mathsf{H}_o(\mathrm{r})\Psi + \mathsf{H}_o(\mathbf{r}')\Psi + V_C(\mathbf{r}, \mathbf{r}')\Psi \tag{2.11}$$

Here $\mathsf{H}_o(\mathrm{r}) = -\hbar^2\nabla^2/2m + V_o(\mathrm{r}) + V_o(|\mathbf{r} - \mathbf{R}|)$ is the one-electron Hamiltonian in (2.5) and $\Psi(\mathbf{r},\, \mathbf{r}')$ is the *two-electron* wave function.

To solve (2.11), we exploit the fact that any many-electron wave function can be expanded into products of orthogonal one-electron states. For example, the two-electron wave function $\phi_L(\mathbf{r})\phi_R(\mathbf{r}')$ means that the first electron $(\mathbf{r})$ is on the left atom, whereas the second atom $(\mathbf{r}')$ is on the right atom. This wave function can also be written as $|LR\rangle = |L\rangle|R\rangle$. The total two-electron wave function is L

$$\Psi(\mathbf{r}, \mathbf{r}') = c_{\mathrm{I}}\phi_{\mathrm{L}}(\mathbf{r})\phi_{\mathrm{L}}(\mathbf{r}') + c_{\mathrm{II}}\phi_{\mathrm{L}}(\mathbf{r})\phi_{\mathrm{R}}(\mathbf{r}') + c_{\mathrm{III}}\phi_{\mathrm{R}}(\mathbf{r})\phi_{\mathrm{L}}(\mathbf{r}') + c_{\mathrm{IV}}\phi_{\mathrm{R}}(\mathbf{r})\phi_{\mathrm{R}}(\mathbf{r}') \quad (2.12)$$

Alternative notations are $|\Psi\rangle = c_{\mathrm{I}}|LL\rangle + c_{\mathrm{II}}|LR\rangle + c_{\mathrm{III}}|RL\rangle + c_{\mathrm{IV}}|RR\rangle$ and, in vector form, $\Psi = (c_{\mathrm{I}}, c_{\mathrm{II}}, c_{\mathrm{III}}, c_{\mathrm{IV}})$. For a comparison of these notations, see Section A.2.1.

In the expansion, the use of Wannier functions $|L\rangle$ and $|R\rangle$ is convenient but not necessary, and other orthogonal sets of one-electron wave functions can equally well be used. For example, $\phi_{\mathrm{a}}(\mathbf{r})\phi_{\mathrm{s}}(\mathbf{r}')$ means that the first and second electrons are in the antibonding and bonding states, respectively. For N electrons and N_{o} one-electron states, there are altogether $N_{\mathrm{o}}^{\mathrm{N}}$ many-electron wave functions. The many-electron wave functions are superpositions of the $N_{\mathrm{o}}^{\mathrm{N}}$ product wave functions, which introduces a considerable complexity even in relatively small molecules (Fulde 1991, Senatore and March 1994).

2.1.3 Hamiltonian and spin structure

Since we restrict ourselves to the four wave functions $|LL\rangle$, $|LR\rangle$, $|RL\rangle$, and $|RR\rangle$, the total Hamiltonian has the form of a 4×4 energy matrix E_{ij}. An explicit example is $E_{12} = \int\int \phi_{\mathrm{L}}^*(\mathbf{r})\phi_{\mathrm{L}}^*(\mathbf{r}')\mathsf{H}(\mathbf{r}, \mathbf{r}')\phi_{\mathrm{L}}(\mathbf{r})\phi_{\mathrm{R}}(\mathbf{r}')\mathrm{d}V\mathrm{d}V'$. From (2.11) we see that each matrix element is a sum of $3 + 3 + 1 = 7$ terms, so that the matrix contains $16 \times 7 = 112$ integrals. However, the left and right atoms are equivalent, and (2.11) is symmetric with respect to $\mathbf{r}$ and $\mathbf{r}'$. As exemplified by $\langle LL|\mathsf{H}|LR\rangle = \langle LL|\mathsf{H}|RL\rangle$, this leads to a considerable reduction of the number of integrals.

Compared to the one-electron case, the addition of V_{C} has two effects on the remaining integrals. First, it modifies the one-electron expressions E_{o} and t, which should therefore be treated as effective parameters. Second, it yields new terms without one-electron equivalents. Straightforward evaluation of the energy matrix elements yields

$$\mathsf{H} = 2E_{\mathrm{o}} + \begin{pmatrix} U & t & t & J_{\mathrm{D}} \\ t & 0 & J_{\mathrm{D}} & t \\ t & J_{\mathrm{D}} & 0 & t \\ J_{\mathrm{D}} & t & t & U \end{pmatrix} \quad (2.13)$$

Here

$$U = \int\int \phi_{\mathrm{L}}^*(\mathbf{r})\phi_{\mathrm{L}}^*(\mathbf{r}')V_{\mathrm{C}}(\mathbf{r}, \mathbf{r}')\phi_{\mathrm{L}}(\mathbf{r})\phi_{\mathrm{L}}(\mathbf{r}')\,\mathrm{d}V\mathrm{d}V' \quad (2.14)$$

is the Coulomb integral and

$$J_{\mathrm{D}} = \int\int \phi_{\mathrm{L}}^*(\mathbf{r})\phi_{\mathrm{R}}^*(\mathbf{r}')V_{\mathrm{C}}(\mathbf{r}, \mathbf{r}')\phi_{\mathrm{R}}(\mathbf{r})\phi_{\mathrm{L}}(\mathbf{r}')\,\mathrm{d}V\mathrm{d}V' \quad (2.15)$$

is the direct exchange.

Using the electron density $n_{\mathrm{L}} = \psi_{\mathrm{L}}^*\psi_{\mathrm{L}}$ and introducing the "mixed" density $n_{\mathrm{mix}} = \psi_{\mathrm{L}}^*\psi_{\mathrm{R}}$ we obtain

$$U = \int\int n_{\mathrm{L}}(\mathbf{r})V_{\mathrm{C}}(\mathbf{r}, \mathbf{r}')n_{\mathrm{L}}(\mathbf{r}')\,\mathrm{d}V\mathrm{d}V' \quad (2.14\mathrm{a})$$

$$\text{and } J_{\mathrm{D}} = \int\int n_{\mathrm{mix}}(\mathbf{r})V_{\mathrm{C}}(\mathbf{r}, \mathbf{r}')n_{\mathrm{mix}}(\mathbf{r}')\,\mathrm{d}V\mathrm{d}V' \quad (2.15\mathrm{a})$$

Both integrals are *positive*, because they describe the self-interaction of real or fictitious charge densities. The Coulomb integral describes the strong repulsion between two electrons in an atom and is equal to energy necessary to add an electron to an already occupied localized orbital. By comparison, the direct exchange has no classical equivalent. Since n_{mix} is relatively small, $J_{\mathrm{D}} \ll U$. Typical orders of magnitude are a few 0.01 eV for J_{D} and a few eV for U.

The eigenfunctions and eigenenergies of the two-electron problem are obtained by diagonalizing the interaction matrix (2.13). The exchange is then obtained by identifying the spin structure of the eigenstates and comparing the energies of the lowest-lying ferromagnetic and antiferromagnetic states. By direct calculation, we convince ourselves that (0, 1, −1, 0), or $|LR\rangle - |RL\rangle$, is an eigenstate of energy $E_{\mathrm{o}} - J_{\mathrm{D}}$. A second eigenstate, (1, 0, 0, −1) or $|LL\rangle - |RR\rangle$, has the energy $E_{\mathrm{o}} + U - J_{\mathrm{D}}$. The third and fourth eigenstates are mixtures of (1, 0, 0, 1) and (0, 1, 1, 0). Explicitly, the eigenfunctions are

$$|1\rangle = \frac{1}{\sqrt{2}}|LR\rangle - \frac{1}{\sqrt{2}}|RL\rangle \tag{2.16a}$$

$$|2\rangle = \frac{1}{\sqrt{2}}|LL\rangle - \frac{1}{\sqrt{2}}|RR\rangle \tag{2.16b}$$

$$|3\rangle = \frac{\sin\chi}{\sqrt{2}}(|LL\rangle + |RR\rangle) + \frac{\cos\chi}{\sqrt{2}}(|LR\rangle + |RL\rangle) \tag{2.16c}$$

$$|4\rangle = \frac{\cos\chi}{\sqrt{2}}(|LL\rangle + |RR\rangle) - \frac{\sin\chi}{\sqrt{2}}(|LR\rangle + |RL\rangle) \tag{2.16d}$$

where $\tan(2\chi) = -4t/U$, and the corresponding energies

$$E_1 = 2E_{\mathrm{o}} - J_{\mathrm{D}} \tag{2.17a}$$

$$E_2 = 2E_{\mathrm{o}} + U - J_{\mathrm{D}} \tag{2.17b}$$

$$E_3 = 2E_{\mathrm{o}} + \frac{U}{2} + J_{\mathrm{D}} - \sqrt{4t^2 + \frac{U^2}{4}} \tag{2.17c}$$

$$E_4 = 2E_{\mathrm{o}} + \frac{U}{2} + J_{\mathrm{D}} + \sqrt{4t^2 + \frac{U^2}{4}} \tag{2.17d}$$

The Pauli principle implies that symmetric and antisymmetric real-space wave functions $\Psi(\mathbf{r}, \mathbf{r}') = \pm\Psi(\mathbf{r}', \mathbf{r})$ describe antiferromagnetic and ferromagnetic spin configurations, respectively. In (2.16), only $|1\rangle$ is antisymmetric and therefore ferromagnetic, whereas the other three eigenfunctions are symmetric (antiferromagnetic). Note that the present model does not distinguish between antiferromagnetism and paramagnetism.

The direct exchange J_{D} is much smaller than U, but t and U are generally of comparable magnitude. In magnets, t depends not only on the radius of the orbital but also on the interatomic distance, and it is convenient to consider the exchange as a function of the hopping integral t. For small t, the wave functions are localized, and the eigenfunctions are reminiscent of atomic orbitals. For large t, the behavior of the

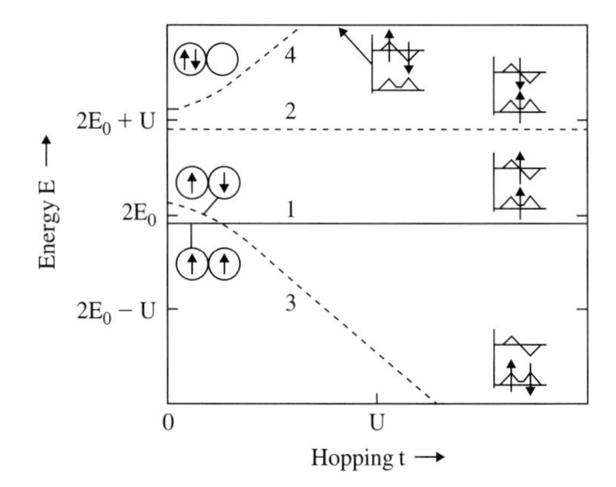

Fig. 2.6 Energy levels (2.17) and schematic two-electron wave functions (2.16) for the diatomic pair model. The figures show the occupancy of atomic orbitals in the limit of weak hopping (left) and that of bonding and antibonding orbitals for strong hopping (right). All functions are symmetric with respect to interchanging left and right atoms, but for clarity, only a part of the configurations is shown. Using independent-electron levels (strong hopping) to describe strongly interacting systems (weak hopping) yields quasiparticles with finite lifetimes (Fermi liquid).

system is determined by the one-electron level splitting $\pm t$, and the electrons occupy bonding or antibonding states. Figure 2.6 shows the energies (2.17) and illustrates the wave functions (2.16). In practice, ferromagnetism is caused by transition-metal atoms (3d, 4d, 5d, 4f, 5f), due to the near-degeneracy of the inner shells and the relatively large correlations. However, this is a rule rather than a rigid statement, and there are exceptions for non-transition-metal clusters with nearly degenerate states.

Since $E_2 > E_1$ and $E_4 > E_3$, the exchange is determined by the competition between the lowest-lying ferromagnetic state (solid line 1 in Fig. 2.6) and the lowest-lying antiferromagnetic state (dashed line 3). The exchange constant $J = \frac{1}{2}(E_1 - E_3)$ is given by

$$J = J_{\mathrm{D}} + \frac{U}{4} - \sqrt{t^2 + \frac{U^2}{16}} \tag{2.18}$$

From this equation, we can draw two conclusions. First, the direct exchange J_{D} is not the only consideration, and the occasionally encountered equating of J with the "exchange integral" J_{D} is a poor approximation. Second, the trend towards ferromagnetism decreases with increasing hopping, that is, with decreasing interatomic distance. This is indeed observed in some magnets, where mechanical pressure changes the magnetization state from ferromagnetic to paramagnetic or antiferromagnetic.

It is instructive to compare (2.18) with the Curie temperature, which is proportional to J. As a crude rule, the exchange obeys the semiphenomenological Bethe-Slater-Néel curve, which predicts antiferromagnetism in the case of small interatomic distances, ferromagnetism at intermediate distances, and the absence of magnetic order in the limit of very large interatomic distances. This trend is indeed reproduced by the present model (Panel 2).

Panel 2 The Bethe-Slater-Néel curve

Equation (2.18) predicts the exchange J as a function of a few parameters that depend on the interatomic distance and, therefore, on the atomic number of the element. As recognized long ago by Sommerfeld and Bethe (1933), this leads to a characteristic dependence of the Curie temperature T_c on the atomic number. The relation can be expressed in various ways, for example by plotting T_c as a function of the distance between neighboring atoms or between neighboring 3d orbitals. The figure plots T_c as a function of the atomic number, parameterized by the number n of 3d electrons of double-positive ions. The fit is obtained from (2.18), by assuming that both the hopping and direct exchange integrals obey an exponential dependence on $R \sim n$.

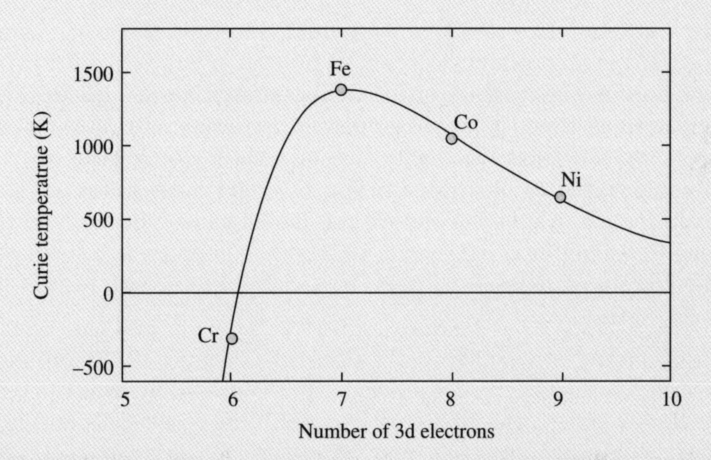

The fit reproduces the experimental behavior very well. However, this does not mean that (2.18) provides a quantitative description of J and T_c. First, the parameters are not obtained from numerical simulations or model calculations. Second, the dependence on the number of electrons n contains only aspect of the problem, namely that the interatomic distance varies across the 3d series. The present model contains one electron per orbital, corresponding to half-filled shells ($n = 5$). Half-filled shells tend to yield a strong trend towards antiferromagnetism (Section 2.4), which is compensated in the fit by an overestimation of J_D and U by a factor of order 2. Third, the curve ignores the structural dependence of the Curie temperature. For example, fcc iron exhibits a pronounced trend towards antiferromagnetism (or paramagnetism), and the somewhat smaller interatomic distance of fcc iron, compared to bcc iron, is not sufficient to explain the difference.

2.1.4 Heisenberg model

The basic idea behind the Heisenberg model (1928) is to ignore ionic configurations $|LL\rangle$ and $|RR\rangle$, because they correspond to a strong and energetically unfavorable intra-atomic Coulomb repulsion. This approach is related to the Heitler-London approximation in molecular science. In terms of (2.13), it is realized by putting $U = \infty$. The energy levels are $E_o \pm J_D$, corresponding to $J = J_D$, and the eigenfunctions are proportional to $|LR\rangle \pm |RL\rangle$. This is different from the "naïve" Heisenberg model,

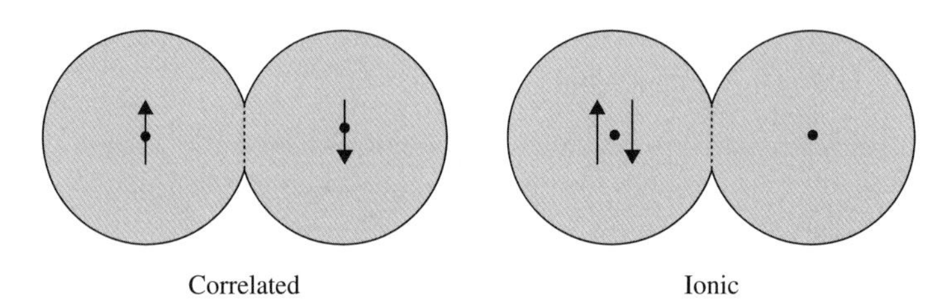

Fig. 2.7 Formation of temporary ionic states due to interatomic hopping. The Heisenberg model assumes correlated wave functions (left) and ionic states are, in lowest order, ignored.

where atomic wave functions $|l\rangle$ and $|r\rangle$ are used, and from the exact solution (2.16), where $|LR\rangle - |RL\rangle$ is an eigenfunction but $|LR\rangle + |RL\rangle$ is mixed with $|LL\rangle + |RR\rangle$.

The Heisenberg model works best for well-separated atoms, where the number of electrons per atom is fixed. The corresponding materials are often insulators with well-localized wave functions. In reality, interatomic hopping leads to a temporary formation of ionic states, as illustrated in Fig. 2.7, and these states are ignored. This is important in metals, where the Heisenberg model cannot be used without careful consideration of hopping. It is instructive to treat the hopping as a small perturbation and to ideally correlated Heisenberg limit. Expanding (2.18) with respect to the small parameter t/U yields

$$J = J_\mathrm{D} - \frac{2t^2}{U} \tag{2.19}$$

This equation and its extensions are widely used to explain the sign of the exchange. The direct exchange is ferromagnetic, but hopping yields an antiferromagnetic contribution.

A different approach to the Heisenberg model considers individual electrons as spins $\mathrm{S} = \frac{1}{2}$ whereas parallel and antiparallel spin configurations are characterized by the respective total spins $\mathrm{S} = 0$ and $\mathrm{S} = 1$. Since (2.11) is spin-independent, we can write the corresponding two-electron wave functions as $\Psi(\mathbf{r}, \mathbf{r}')\chi(\sigma, \sigma')$. The function $\Psi(\mathbf{r}, \mathbf{r}')$ has been treated above, so that we can now focus on $\chi(\sigma, \sigma')$. For a symmetric real-space wave function $\Psi(\mathbf{r}, \mathbf{r}') = \Psi(\mathbf{r}', \mathbf{r})$, the antisymmetry of the total wave function requires the spin function to be antisymmetric, so that $\chi(\sigma_1, \sigma_2) = -\chi(\sigma_1, \sigma_2)$, and vice versa. An example is the $1s^2$ configuration of helium, which is symmetric in real space, $\psi_\mathrm{s}(\mathbf{r}')\psi_\mathrm{s}(\mathrm{r})$, but antisymmetric in spin space.

However, the antiparallel spin functions $\uparrow\downarrow$ and $\downarrow\uparrow$ cannot be used in their original form, because they are neither symmetric nor antisymmetric. Analyzing the symmetry of the combinations $\uparrow\downarrow \pm \downarrow\uparrow$ we find that $\mathrm{S} = 0$ corresponds to a antisymmetric spin function $\uparrow\downarrow - \downarrow\uparrow$, whereas $S = 1$ yields a triplet with the antisymmetric wave functions $\uparrow\uparrow$ ($S_\mathrm{z} = 1$), $\uparrow\downarrow + \downarrow\uparrow$ ($S_\mathrm{z} = 0$), and $\downarrow\downarrow$ ($S_\mathrm{z} = 1$). More generally, a spin S has $2S + 1$ spin orientations $S_\mathrm{z} = -S, \ldots S-1, S$, and the magnitude of the spin vector is given by $\mathbf{S}^2 = S(S+1)$. Our shorthand notation $\uparrow\uparrow$ for $S = 1$ and $\uparrow\downarrow$ for $S = 0$ actually implies a small field in z-direction, which ensures that $\uparrow\uparrow$ is the only occupied $S = 1$ state and no confusion arises between $S = 0$ and $S = 1$.

Let as now assume an interaction $\mathsf{H} = -2J\mathbf{s} \cdot \mathbf{s}'$ between two spins $\mathbf{s}$ and $\mathbf{s}'$ and find the eigenvalues of the system. The total spin $\mathbf{S} = \mathbf{s} + \mathbf{s}'$ obeys $\mathbf{S}^2 = \mathbf{s}^2 + 2\mathbf{s} \cdot \mathbf{s}' + \mathbf{s}'^2$. Since $\mathbf{S}^2 = S(S+1)$ for any spin, both $\mathbf{s}^2$ and $\mathbf{s}'^2$ are equal to $\frac{1}{2}(1 + \frac{1}{2}) = \frac{3}{4}$, and $2\,\mathbf{s} \cdot \mathbf{s}' = \mathbf{S}^2 + 3/2$. For $S = 0$ and $S = 1$, this means $2\mathbf{s} \cdot \mathbf{s}' = 3/2$ and $2\mathbf{s} \cdot \mathbf{s}' = 7/2$, respectively, and the energy level splitting between the AFM and FM states is $2J$, as in the derivation of (2.18) from (2.17).

More generally, it is possible to define the Heisenberg model in terms of the *Heisenberg Hamiltonian*

$$\mathbf{H} = -2\sum_{i>j} J_{ij}\, \mathbf{s}_i \cdot \mathbf{s}_j - g\,\mu_o\mu_B \sum_i \mathbf{H}_i \cdot \mathbf{s}_i \tag{2.20}$$

where the summation includes all atomic spins and $\mathbf{H}_i$ is the local magnetic field acting on the i-th spin. This equation is an example of a spin Hamiltonian, where electronic quantities, such as hopping and Coulomb interaction, are mapped onto spin variables. Note that some authors use different normalizations for J, which yield factors such as 2 and $\frac{1}{2}$. This depends, for example, on whether the summation $\sum_{ij}$ is limited to pairs of spins, $\sum_{i>j}$. From a quantum-mechanical point, (2.20) is an approximation, but the model works surprisingly well for a broad range of materials and phenomena. In practice, the J_{ij} are often treated as phenomenological parameters.

Throughout the following sections and chapters, we will explore and exploit the predictions of the Heisenberg model (2.20). Typical problems are the determination of the J_{ij} for different classes of magnetic solids (Section 2.3) and to define the Heisenberg model for metals such as Fe and Co (Section 2.4).We will also discuss extensions and modifications of (2.20), such as XY and other n-vector models, including the Ising model (Section 5.2).

2.1.5 Independent-electron approximation

The weak-correlation limit of negligibly small Coulomb interactions, $U = 0$, yields the eigenfunctions $|LR\rangle - |RL\rangle$ (ferromagnetic) and $|LL\rangle + |LR\rangle + |RL\rangle + |RR\rangle$ (antiferromagnetic). Using (2.9), these functions can also be written as $|s\,a\rangle - |a\,s\rangle$ and $|s\,s\rangle$, respectively. This corresponds to the limit of metallic ferromagnetism, because $|s\rangle$ and $|a\rangle$ are delocalized states, Fig. 2.4. Figure 2.8 shows the occupancy of the corresponding one-electron levels of energy $E_o \pm t$. As expected for noninteracting electrons, the energies are additive and equal to $2E_o - 2|t|$ for the antiparallel ground state and $2E_o$ for the excited parallel configuration.

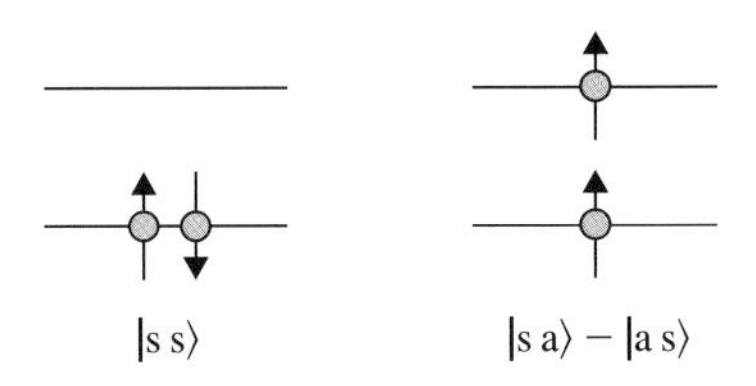

Fig. 2.8 Energy levels of noninteracting electrons. Ignoring the Coulomb and direct-exchange integrals, the energies are $E_o \pm t$.

The next step is to treat U and J_D as small perturbations. In lowest order, the energies are obtained by using the unperturbed eigenfunctions $|s\,s\rangle$ and $|s\,a\rangle - |a\,s\rangle$ to evaluate the full Hamiltonian (2.13). In other words, the wave function of the interacting electrons is approximated by one-electron wave functions. To obey the Pauli principle, the one-electron wave functions form antisymmetric combinations, that is, the total wave function (real-space and spin) has the character of a Slater determinant. For two electrons 1 and 2, the Slater determinant is $|\psi_1\sigma_1\rangle|\psi_2\sigma_2\rangle - |\psi_2\sigma_2\rangle|\psi_1\sigma_1\rangle$. This expression is zero for $|\psi_2\sigma_2\rangle = |\psi_1\sigma_1\rangle$, thereby realizing the Pauli principle.

The use of one-electron wave functions in the form of a single Slater determinant is the essence of the *independent-electron* or Hartree-Fock approximation. The method is related to the LCAO approximation in chemistry and to the Stoner model (Section 2.4) in magnetism. Implicit examples of Slater determinants are the wave functions in Fig. 2.8. Since our Hamiltonian is spin-independent, we can exploit $|\psi\sigma\rangle = |\psi\rangle|\sigma\rangle$ and rewrite the determinant as $|\psi_1\psi_2\rangle|\sigma_1\sigma_2\rangle - |\psi_2\psi_1\rangle|\sigma_2\sigma_1\rangle$. For parallel spins, $|\sigma_1\rangle = |\sigma_2\rangle$, the real-space part of the wave function is $|\psi_1\psi_2\rangle - |\psi_2\psi_1\rangle$, as compared to the symmetric spin part $|\sigma_1\sigma_1\rangle$. For antiparallel spins, $|\psi_1\rangle = |\psi_2\rangle$ yields the real-space part $|\psi_1\psi_1\rangle$ and the spin part $|\sigma_1\sigma_2\rangle - |\sigma_2\sigma_1\rangle$. Slater determinants are zero when two electrons are in the same state (orbital *and* spin), so that Hartree-Fock many-electron states be described by occupation numbers $n_i = 0$ or 1 where i labels the one-electron states. Considering the occupation numbers as eigenvalues of a fermionic occupation-number operator makes it possible to abstract from details of the one-electron wave functions and leads to the picture of "second quantization" (Anderson 1965). It is convenient to write $\mathsf{n} = \mathsf{a}^+\mathsf{a}$, where a^+ and a are creation and destruction operators, respectively. Using these operators, one-electron averages can be written as $\mathsf{a}^+\mathsf{A}\mathsf{a}$, which is very similar to the "ordinary" quantum-mechanical average $\int \psi + \mathsf{A}\psi \mathrm{d}V$. For the algebra of the operators involved and their use to describe electron-electron interactions and spin waves, see Section 2.1.7 and Section 6.1.3, respectively.

The spin structures of Fig. 2.8 are obtained by specifying $|\psi_1\rangle = |s\rangle$ and $|\psi_2\rangle = |a\rangle$. Applied to (2.13), the corresponding independent-electron wave functions $|s\,s\rangle$ and $|s\,a\rangle - |a\,s\rangle$ yield the exchange

$$J = \frac{U}{4} + J_D - |t| \tag{2.21}$$

Since U is much larger than J_D, this amounts to a competition between Coulomb repulsion and interatomic hopping. Consistent with our previous findings, interatomic hopping destroys ferromagnetic spin alignment, but (2.21) overestimates the trend towards ferromagnetism, especially for large U. Since J_D and $|t|$ decrease with increasing interatomic distance but U remains constant, (2.21) predicts ferromagnetism above some interatomic distance. This trend is indeed observed in some materials, but a famous counter-example is the hydrogen molecule, which is nonferromagnetic at all distances. The reason for this failure is the assumption of one-electron wave functions $|s\rangle$ and $|a\rangle$, which are formed from $|\mathrm{L}\rangle$ and $|\mathrm{R}\rangle$ by interatomic hopping. For large U, hopping is energetically unfavorable, because the corresponding double occupancy of atomic orbitals costs much energy.

The independent-electron approximation works best for itinerant transition-metal elements and alloys, such as Fe and Co, and also for some oxides. These materials exhibit pronounced interatomic hopping and are often conductors with delocalized wave functions. Typical methods are the LCAO approach (linear combination of atomic orbitals) and electronic band-structure calculations. The approach is very powerful when the one-electron potential $V_o(\mathrm{r})$ contains the self-consistently determined interactions with the other electrons in the solid. This self-consistent-field (SCF) approach treats all Coulomb interactions on a mean-field level. It is the basis for the treatment of the itinerant magnetism of iron-series metallic magnets (Section 2.4).

2.1.6 Correlations

The one-electron approximation maps electron-electron interactions onto a self-consistent interaction field. This is an example of a quantum-mechanical *mean-field approach*. The main shortcoming of mean-field models (Panel 3) is the neglect of correlations. By definition, *correlation* effects go beyond the Hartree-Fock approximation, or the use of a single Stater determinant. For example, in the independent-electron approximation, the lowest-lying antiferromagnetic (AFM) state, namely $|s\,s\rangle \sim |LL\rangle + |LR\rangle + |RL\rangle + |RR\rangle$, contains both single-occupied and double-occupied or "ionic" atomic orbitals. The electrons do not distinguish whether a given atomic orbital is already occupied by another electron and occupy the left and right atoms with equal probability. In reality, ionic states cost Coulomb energy and are partially suppressed.

The opposite limit of Heisenberg interactions (Section 2.1.4) corresponds to the overcorrelated antiferromagnetic wave function $|LR\rangle + |RL\rangle$, where hopping and ionic states are suppressed completely. This is equally unrealistic, because interatomic hopping lowers the energy and gives rise to some ionicity. In fact, the exact AFM wave function (2.16c) is intermediate between the undercorrelated independent-electron limit $|LL\rangle + |LR\rangle + |RL\rangle + |RR\rangle$ and the overcorrelated Heisenberg limit $|LR\rangle + |RL\rangle$. By contrast, the lowest-lying ferromagnetic state (2.16a), $|LR\rangle - |RL\rangle = |s\,a\rangle - |a\,s\rangle$, is independent of U, indicating that ferromagnetic configurations are essentially unaffected by correlations. This is because the Pauli principle forbids ferromagnetic $|LL\rangle$ and $|RR\rangle$ configurations for any ratio t/U. Figure 2.9 provides a pictorial and very general interpretation of correlation effects by comparing the Hartree-Fock exchange hole (a) with the correlation hole (b). The former is a consequence of the Pauli principle and excludes electrons of parallel spin, whereas the latter affects electrons of antiparallel spin, which are allowed by the Pauli principle but electrostatically unfavorable.

The Hartree-Fock or independent-electron approximation obeys the Pauli principle and contains many-body interactions in an approximate form. The magnetic properties of many systems are well reproduced by numerical electronic-structure calculations based on the one-electron approximation. For example, density-functional calculations (Kohn and Sham 1965), which are conceptually related to the Thomas-Fermi approach (see below), have developed into a useful tool to calculate the magnetic moments of itinerant magnets (Coehoorn 1989). In its most general form, density-functional theory reproduces the correct ground-state energy, but excitations are much less well-reproduced (Fulde 1991). Furthermore, the magnitude of correlation corrections depends on the considered material. One example is the LSDA+U approximation (local spin density plus U), where electron interactions in localized orbitals (U) are treated on a mean-field or independent-electron level. This approach works fairly well for many systems but fails to describe specific correlation effects.

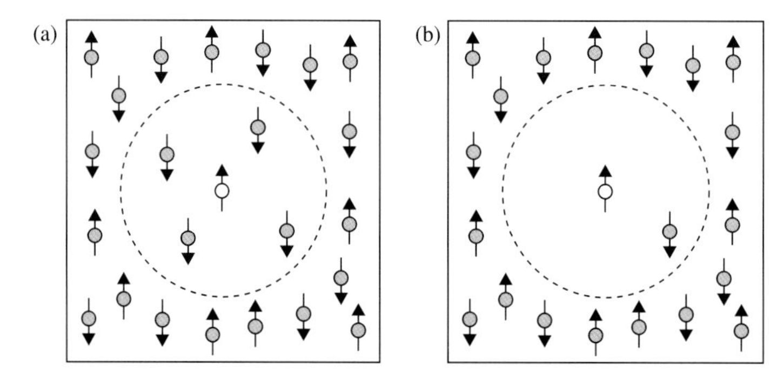

Fig. 2.9 Correlations in an electron gas: (a) exchange hole and (b) correlation hole. A given electron (white) repels electrons in its immediate neighborhood. In the Hartree-Fock or independent-electron picture (a), the Pauli principle excludes parallel spins but leaves electrons with antiparallel spins unaffected. The real situation is similar to (b), where antiparallel spins are allowed by the Pauli principle but discouraged by the Coulomb repulsion. On a mean-field level, the correlation hole (b) is accounted for by adding a Hubbard-type repulsive U.

For strongly correlated systems, such as 4f ions, the independent-electron picture breaks down nearly completely (Fulde 1991) and Hund's rules are a better starting point for the understanding of the involved physics. Correlation corrections are also important in many 3d oxides, where band-structure calculations predict insulating behavior only when the Fermi level lies in a band gap. This greatly overestimates the conductivity of oxides such as CoO, where the band structure suggests metallicity but correlations cause the electrons to localize and yield an insulating state. Pictorially, the electrons are "owned" by individual atoms, and electron hopping onto sites that are already occupied costs Coulomb energy. In the one-electron picture, this problem does not arise, because the interaction is mapped onto a mean field and the electrons lose their individuality. Third, correlations are sometimes important in itinerant 3d magnets. For example, local spin density (LSDA) band-structure calculations can often be used to determine the anisotropy of 3d metals but fail in the case of Ni (Trygg *et al.* 1995).

For the dense homogeneous electron gas, or *jellium*, the correlation energy has been calculated to some accuracy (Ashcroft and Mermin 1976). The energy per electron is $2.21/r_s^2 - 0.916/r_s + 0.0622\ln(r_s) - 0.096$, where the dimensionless interelectronic distance r_s is defined by $1/n = (4\pi/3)r_s^3/a_o^3$ and the energy is measured in rydberg (A.1.2). The $1/r_s^2$ and $1/r_s$ terms are Hartree-Fock contributions of kinetic and electrostatic origin, respectively, whereas the logarithmic and constant terms are the lowest-order correlation corrections. In the low-density limit (r_s large), the electrons become localized and probably form a lattice known as Wigner crystal (Senatore and March 1994).

An important electron-electron interaction effect is *screening*. A charged particle in a metal attracts or repels conduction electrons, thereby reducing or "screening" the particle's electrostatic field. The Thomas-Fermi model (Fermi 1928) considers a homogeneous electron gas in a slowly varying potential $V(\mathbf{r})$. The electron density n obeys $\nabla^2 V = -en/\varepsilon_o$ (Coulomb interaction, Section A.4.1) and $E_F - V = \hbar^2(3\pi^2 n)^{2/3}/2m_e$

(free electron gas, Section 2.4.1) The linearization of the problem yields an exponential screening with the decay length $1/\kappa$ given by $\kappa^2 = 4k_F/a_o$, where k_F is the Fermi wave vector (Section 2.4.1). Since k_F increases with n, metallic screening is short-range ($1/\kappa \approx 0.55\,\text{Å}$ in Cu) —in contrast to semiconductors, where n is small. Note that the Thomas-Fermi model deals with Coulomb interactions but does not address correlation effects. It is, in fact, an independent-electron model and closely related to the density-functional approach (Section 2.4.2). Another shortcoming is the restriction to smooth changes in $V(\mathbf{r})$. For point-like potentials, the wave character of the electrons comes into play, and one must use quantum-mechanical perturbation theory. Depending on the context, this is known as Lindhard, Friedel, or RKKY screening (Section 2.3.2).

Panel 3 Mean-field approaches

Mean-field models, also known as self-consistent-field or molecular-field models, are an important tool to approximate many-body interactions. As mentioned in Section 1.4, they replace the many-body interactions by the interaction with an effective field. The idea is to rewrite products of operators (or classical functions) AB as

$$\mathsf{AB} = \mathsf{A}\langle\mathsf{B}\rangle + \langle\mathsf{A}\rangle\mathsf{B} - \langle\mathsf{A}\rangle\langle\mathsf{B}\rangle + \mathsf{C}_{\mathrm{AB}}$$

where

$$\mathsf{C}_{\mathrm{AB}} = (\mathsf{A} - \langle\mathsf{A}\rangle)(\mathsf{B} - \langle\mathsf{B}\rangle)$$

Ignoring C_{AB}, we obtain

$$\mathsf{AB} = \mathsf{A}\langle\mathsf{B}\rangle + \langle\mathsf{A}\rangle\mathsf{B}$$

Note that the term $\langle\mathsf{A}\rangle\ \langle\mathsf{B}\rangle$ can safely be ignored, because it amounts to a physical unimportant shift of the zero-point energy.

Mean-field models are widely used to treat Coulomb interactions, finite-temperature effects, and structural disorder. Examples are the Ising model introduced in Section 1.4, and the Heisenberg interaction $-\sum_{\mathrm{ij}} J_{\mathrm{ij}}\mathbf{s}_{\mathrm{i}} \cdot \mathbf{s}_{\mathrm{j}}$, where $J_{\mathrm{ji}} = -J_{\mathrm{ij}}$ yields the mean-field expression $-2\sum_{\mathrm{ij}} J_{\mathrm{ij}}\mathbf{s}_{\mathrm{i}} \cdot \langle \boldsymbol{s}_{\mathrm{j}}\rangle$. In this section, we consider the product $\mathsf{n}_\uparrow\mathsf{n}_\downarrow$, where the n_σ are electron densities and the mean-field solution corresponds to the independent-electron approximation. The applicability of mean-field models depends on the relative role of the fluctuations (correlations) C_{AB}. In the Heisenberg analogy, the state of neighboring spins must be close to the average spin. As will see in Chapter 5, this condition is violated near the Curie temperature.

A striking feature of mean-field models is that they *overestimate* the trend towards magnetic order. For example, mean-field Curie temperatures are larger than the exact predictions, and the independent-electron approximation may incorrectly predict ferromagnetism. This is due to the assumption of a mean field. In reality, there may be correlated spin blocks with some kind of local order but zero net magnetization.

Exercise

Derive the mean field $\mathbf{H}_{\mathrm{MF}}$ for the Heisenberg model with nearest-neighbor interactions $J_{\mathrm{ij}} = J$ and specify the result for simple cubic, bcc, and fcc crystals. Discuss the result in terms of the number z of nearest neighbors.

Another consequence of correlations is deviations from Fermi-liquid behavior. In a one-electron picture, the Pauli principle means that electrons fill the available one-electron states as liquid is poured into a jar, with a sharp Fermi surface of energy E_F (Section 2.4.1). In lowest order, electron-electron interactions lead to quasiparticles, which have long lifetimes due to $E_F \ll k_B T$ and behave similarly to noninteracting particles (Fulde 1991, Schofield 1999). However, the effective mass m^* of a quasiparticle differs from the bare electron mass not only by the (one-electron) lattice contribution but also by an additional interaction correction. With increasing interaction strength, the lifetime decreases, and the zero-temperature discontinuity at the Fermi level narrows.

The correlation hole leads to an intriguing question: Can we approximate correlation effects by choosing one-electron wave functions that avoid the hole by being confined to some region in space? Depending on the ratio t/U, the one-electron wave function would resemble an atomic function, $|\mathrm{L}\rangle$ or $|\mathrm{R}\rangle$, or be similar to the "delocalized" function $|\mathrm{L}\rangle + |\mathrm{R}\rangle$. Indeed, the ground state of the hydrogen molecule can be written as $|\mathrm{AB}\rangle + |\mathrm{BA}\rangle$ (Coulson and Fischer 1949, Falicov and Harris 1969). In our notation, the lowest-lying antiferromagnetic state obeys $|A\rangle = |L\rangle + \lambda|R\rangle$ and $|B\rangle = |R\rangle + \lambda|L\rangle$, where λ is a mixing parameter varying from $\lambda = 0$ (Heisenberg limit) to $\lambda = 1$ (independent electrons). Note that the corresponding two-electron wave function

$$|\Psi\rangle = (1 + \lambda^2)(|LR\rangle + |RL\rangle) + 2\lambda(|LL\rangle + |RR\rangle) \tag{2.22}$$

and the corresponding exchange are exact. This is a rather specific result, which cannot be generalized to more complicated electron-electron interaction problems.

Approaches such as the use of Coulson-Fischer wave functions are known as *unrestricted* Hartree-Fock approximations. The basic idea is to use wave functions whose symmetry is lower than that of the Hamiltonian. For example, (2.22) assumes that the electrons are predominantly assigned to different atoms. This reduces the electrostatically unfavorable admixture of ionic configurations. Correlations are therefore partly taken into account, in spite of the one-electron character of the approach. In general, the treatment of many-electron correlations has remained a demanding task (Fulde 1991). It requires approaches such as the configuration interaction (CI) method, where several Slater determinants are used to find the eigenstates and energy levels.

2.1.7 *Hubbard model

Since $U \gg J$, it is tempting to ignore J while keeping the hopping integral. This assumption is the basis for the *Hubbard model.* By putting $J_D = 0$ in Section 2.1.3 we obtain

$$J = \frac{\mathrm{U}}{4} - \sqrt{\mathrm{t}^2 + \frac{\mathrm{U}^2}{16}} \tag{2.23}$$

that is, $J = -2t^2/U$ for large U and $J = U/4 - |t|$ for small U. The Hubbard model is widely used to investigate correlation effects (Jones and March 1973, Fulde 1991).

For example, it describes the transition from the metallic conductivity to insulating behavior when U exceeds some threshold. The corresponding localization of the electron wave functions is known as *Mott localization.* This Mott-Hubbard transition must be distinguished from Anderson localization of one-electron wave functions, which is caused by disorder (Chapter 7.1). Strong Coulomb interactions yield an energy gap separating low-lying states with one electron per atom from states with two electrons per atom, and this level splitting is modified by interatomic exchange.

An alternative and widely used approach to the Hubbard model is to write the Hamiltonian as a sum of local interaction terms $U\mathsf{n}_\uparrow\mathsf{n}_\downarrow$, where n_σ is the particle number operator for an electron of spin σ (second quantization, Section 2.1.5). In matrix form,

$$\mathbf{n}_\sigma = \begin{pmatrix} 0 & 0 \\ 0 & 1 \end{pmatrix} \tag{2.24}$$

with the eigenvalues 1 (particles present) and 0 (vacuum state). It is convenient to introduce the creation operator

$$\mathsf{a}_\sigma^+ = \begin{pmatrix} 0 & 0 \\ 1 & 0 \end{pmatrix} \tag{2.25a}$$

and the annihilation operator

$$\mathbf{a}_\sigma = \begin{pmatrix} 0 & 1 \\ 0 & 0 \end{pmatrix} \tag{2.25b}$$

so that $\mathsf{n}_\sigma = \mathsf{a}_\sigma^+ \mathsf{a}_\sigma$. Equation (2.25) realizes the Pauli principle, because the application of the creation operator a_σ^+ to an occupied state, (0 1), yields zero. By defining a_σ^+ and a_σ for different sites i and j, it is easy to include interatomic hopping, so that

$$\mathsf{H} = \Sigma_{\mathrm{ij}}\, t_{\mathrm{ij}}\, \mathsf{a}_{\mathrm{i}\uparrow}^+ \mathsf{a}_{\mathrm{j}\uparrow} + \Sigma_{\mathrm{ij}}\, t_{\mathrm{ij}}\, \mathsf{a}_{\mathrm{i}\downarrow}^+ \mathsf{a}_{\mathrm{j}\downarrow} + U \Sigma_{\mathrm{i}}\, \mathsf{n}_{\mathrm{i}\uparrow}\, \mathsf{n}_{\mathrm{i}\downarrow} \tag{2.26}$$

Here t_{ij} is the hopping integral between sites i and j. For example, $t_{\mathrm{ij}}\mathsf{a}_{\mathrm{i}\uparrow}^+\mathsf{a}_{\mathrm{j}\uparrow}$ describes the hopping of a ↑ electron from the site j to the site i.

The Coulomb energy $U\mathsf{n}_\uparrow\mathsf{n}_\downarrow$ provides a simple interpretation of ferromagnetism. Writing the magnetization as $\mathsf{m} = \mathsf{n}_\uparrow - \mathsf{n}_\downarrow$ and the total number of electrons as $\mathsf{n} = \mathsf{n}_\uparrow + \mathsf{n}_\downarrow$ yields $U\mathsf{n}_\uparrow\mathsf{n}_\downarrow = U(\mathsf{n}^2 - \mathsf{m}^2)/4$. Since the number n of electrons per atom is fixed, the Coulomb interaction favors the formation of a magnetic moment. Note that this argument applies not only to half-filled systems such as the two-electron model discussed above (one electron per level) but also to arbitrary electron concentrations. However, the average hopping energy depends on the electron concentration, too, so that U is not the only consideration.

Treating the Hubbard model in the mean-field or independent-electron approximation (Panel 3) replaces $U\mathsf{n}_\uparrow\mathsf{n}_\downarrow$ by $U\mathsf{n}_\uparrow\langle n_\downarrow\rangle + U\langle n_\uparrow\rangle\mathsf{n}_\downarrow$. In other words, ↑ electrons interact with a "sea" of ↓ electrons (and vice versa), and the Coulomb interaction amounts to a ferromagnetic mean-field contribution to the single-electron energy $E(\mathbf{k})$. However, as in all mean-field approaches, the trend towards ferromagnetism is overestimated. This can be seen by comparing the independent-electron result $U/4 - |t|$

with the exact two-electron exchange (2.23): the former is positive for sufficiently large U, but the exact solution is always negative.

Panel 4 Merits and limitations of the two-electron model

The model (2.13) provides a qualitatively correct explanation of ferromagnetism. It is exactly solvable, yields the exchange constant J as a function of interatomic hopping t and Coulomb energy U, and properly accounts for correlation effects. It is also a convenient tool for discussing more complex models, such as the Hubbard and Kondo models.
However, in solids the *number of electrons* is infinite, and the restriction to two electrons is a far-reaching assumption. A major difference is the formation of energy bands (Section 2.4), as compared to the discrete level splitting in Fig. 2.8. Furthermore, the two-electron model does not distinguish between paramagnetism and antiferromagnetism. In a Pauli paramagnet, the atomic moments are zero, whereas antiferromagnets exhibit nonzero atomic moments but antiparallel orientations of neighboring atomic moments. In other words, the present model treats intra- and interatomic exchange on equal footing, which is a rather crude approximation.
In addition, the consideration is limited to *one orbital per atom*. The following examples illustrate some implications. The first point is that ferromagnetic materials tend to exhibit pronounced *intra*-atomic exchange between electrons in partially filled inner transition-metal shells (Section 2.2). Second, several types of electrons may be involved in a given atom, such as localized inner-shell electrons and conduction electrons. Third, half-filled levels are a very special case, characterized by a pronounced trend towards antiferromagnetism (Section 2.4). Examples are the hydrogen molecule and some metals, such as manganese. The filling also affects the conductivity, as rationalized by the t-J model (Section 7.2.8).

2.1.8 *Kondo model

So far, we have assumed that our two orbitals are equivalent. Applying the model to systems where one orbital is localized and one orbital is delocalized yields new physics. An important example is the *Kondo model*. It was originally developed to explain the striking resistance minimum due to rare-earth impurities in nonmagnetic metals. Normally, the metallic resistivity exhibits a monotonic increase with temperatures, because thermally excited lattice distortions (phonons) scatter the electrons. In Kondo systems, there is a temperature (Kondo temperature) below which the resistivity starts to rise again. This indicates a very effective scattering mechanism at low temperatures.

In a broad sense, the Kondo effect amounts to the interaction between localized and extended electron states. In spin-dilute magnetic transition-metal alloys (spin glasses), such as $Cu_{1-x}Fe_x$, the effect involves delocalized 4s and localized 3d electrons (Fischer and Hertz 1991) and a wave-vector summation with a strong long-wavelength contribution. Heavy-fermion compounds, such as UPt_3 and $CeAl_2$, can be considered as Kondo lattices where conduction electrons interact with localized 4f or 5f electrons (Fulde 1991). Another example is manganites, where the interaction involves two types of Mn 3d electrons: localized t_{2g} electrons and extended e_g electrons (Section 3.3.3).

Let us assume, for simplicity, that a single conduction electron, described by a delocalized orbital $|c\rangle$, interacts with a localized rare-earth 4f state $|f\rangle$. The small radius of the 4f shells (Section 2.2) yields large Coulomb integrals, so we ignore the relatively small Coulomb interaction in the delocalized orbital and the direct exchange. However, the atomic energy of the 4f electron is lower than that of the delocalized electron by some energy difference ΔE. In terms of the wave functions $|ff\rangle$, $|fc\rangle$, $|cf\rangle$, and $|cc\rangle$, this yields the Hamiltonian:

$$\mathbf{H} = \begin{pmatrix} U - \Delta E & t & t & 0 \\ t & 0 & 0 & t \\ t & 0 & 0 & t \\ 0 & t & t & \Delta E \end{pmatrix} \tag{2.27}$$

In the absence of hybridization ($t = 0$), the ground state is degenerate, $|f\,c\rangle \pm |c\,f\rangle$, and both states have the energy $E = 0$. The first excited antiferromagnetic state, $|dd\rangle$, has the energy $E = \Delta E$, because one electron is moved from the $|f\rangle$ orbital to the $|c\rangle$ orbital.

Hopping leaves the ferromagnetic state $|f\,c\rangle - |d\,f\rangle$ unchanged but reduces the energy of the antiferromagnetic state, $|f\,c\rangle + |c\,f\rangle$, by some admixture of $|\mathrm{cc}\rangle$ character. The corresponding ground-state energy is $-2t^2/\Delta E$. At low temperatures, $T < 2t^2/k_\mathrm{B}\Delta E$, the delocalized electron is in the ground state and coupled to the 4f electron, thereby realizing the Kondo effect. Of course, to obtain quantitative predictions for metals, it is necessary to treat the interaction of localized electrons with an infinite number of conduction electrons. This is done by solving more elaborate versions of the Kondo model or by considering related models, such as the Anderson model, which focuses on the formation of localized moments in itinerant magnets.

The many-body character of the Kondo model is realized by $U = \infty$ in (2.27). This ensures that the low-lying levels do not interact with the $|ff\rangle$ state. The remaining three levels have various properties that are commonly encountered in strongly correlated systems. First, the relatively small splitting between the antiferromagnetic singlet and ferromagnetic triplet states must be contrasted to the one-electron energy ΔE. The low-lying excitation involves spin degrees of freedom ($\uparrow\uparrow$ vs. $\uparrow\downarrow$), whereas ΔE refers to a charge degree of freedom, namely the hopping of an electron from the $|f\rangle$ orbital to the $|c\rangle$ orbital. This phenomenon is known as *spin-charge separation* and is widely encountered in correlated electron systems.

Second, the interaction of the localized impurity spin yields a quasiparticle. At low temperatures, the impurity surrounds itself with conduction electrons of opposite spin. In turn, the many-electron interactions modify independent-electron properties and mean that the "bare" or "naked" independent electrons carry a "dress" or "cloud" due to many-body interactions. A simple example of such a quasiparticle is an electron surrounded by the correlation hole (Fig. 2.9.) Third, the Kondo effect is related to the behavior of heavy-fermion systems such as $CeAl_3$ (Fulde 1991) These systems are conducting but subject to relatively strong Coulomb repulsion, and at low temperatures they behave like metals with very small hopping integrals—or "heavy" electrons—and with intriguing behavior of the Fermi surface (Paschen *et al.* 2004).

2.2 Magnetic Ions

Summary The magnetism of solids, including transition-metal magnets, retains many features of atomic or ionic magnetsim. Ionic magnetic moments are created by the intra-atomic exchange between inner-shell electrons. The electrons are described by the quantum numbers n, l, and m, as obtained from the Schrödinger equation, and obey the corresponding angular-momentum commutation rules. For practical reasons, magnetic ions are divided into iron-series (3d), palladium-series (4d), platinum-series (5d), and actinide (5f) ions, but other electrons, such as 2p electrons, may also be involved. Atomic moments contain both spin and orbital contributionsm but the survival of ionic moments in a crystalline environment depends on the considered element. As a rule, Hund's-rules 4f moments are conserved, whereas 3d orbital moments are largely quenched by the crystal field. The orbital moment contributes to magnetization and is an important requirement for magntocrystalline anisotropy. Electrons in molecules, clusters, and solids occupy states reminiscent of atomic orbitals, even in metals. In most magnetic solids, the moment reflects the partially filled inner shells of transition-metal elements, such as the iron-series 3d shells and rare-earth 4f shells. In addition, many ions and some molecules exhibit paramagnetism due to other electrons. For example, each O_2 molecule has a 2p moment of $2\mu_B$. In the previous section, we have seen how atomic orbitals hybridize and yield interatomic exchange. This section focuses on the symmetry and size of the atomic orbitals and on the corresponding atomic moments.

2.2.1 Atomic orbitals

A simple but powerful approach to atomic orbitals is the hydrogen-like or hydrogenic model. It is defined by the Schrödinger equation

$$-\frac{\hbar^2}{2m}\nabla^2\psi - \frac{Ze^2}{4\pi\varepsilon_o r}\psi = E\psi \tag{2.28}$$

where Z is an effective nuclear charge. It is well known that the solutions of this equation form atomic shells characterized by the principal quantum number n and the energy

$$E_\mathrm{n} = -\frac{1}{n^2}\frac{m_\mathrm{e}}{2}Z^2\,\alpha^2\,c^2 \tag{2.29}$$

The n-th level contains $2n^2$ degenerate states described by the orbital quantum number l and the magnetic quantum number $m = m_\mathrm{z}$.

Each shell contains $n-1$ subshells labeled as $s\,(l=0)$, $p\,(l=1)$, $d\,(l=2)$, and $f\,(l=3)$, and each subshell contains $2l + 1$ magnetic quantum states $\mathrm{m} = -l, \ldots, 0, \ldots, l$. Figure 2.10 illustrates the angular dependence of the wave function. In addition, each orbital $\{n,\, l,\, m\}$ can accommodate a pair of $\uparrow$ and $\downarrow$ electrons. Equation (2.29) predicts subshells with the same principal quantum number to have the same energy. This degeneracy with respect to l is a consequence of the $1/r$ potential.

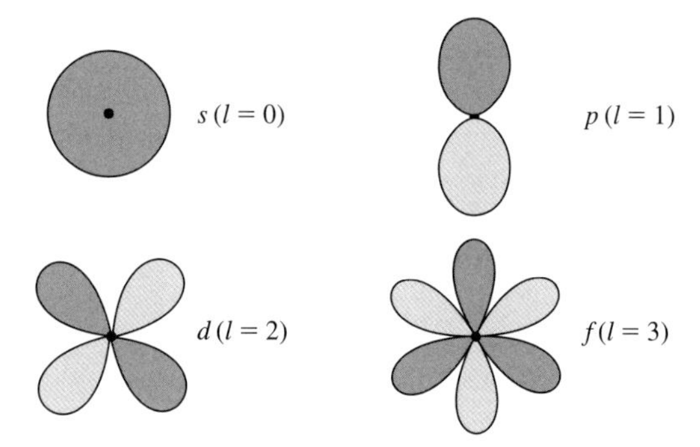

Fig. 2.10 Orbital quantum number l and angular dependence of the wave function ψ. The schematic figure is a top view on the maximally $2l$ lobes with positive or negative sign of ψ (dark and bright regions, respectively). Note that the spherical symmetry of s-state wave functions excludes magnetic anisotropy.

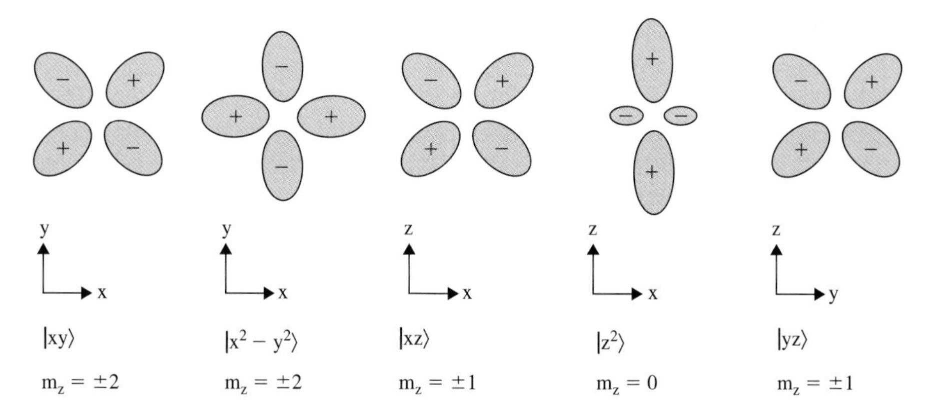

Fig. 2.11 3d wave functions (schematic). The bottom row shows the magnetic quantum numbers with respect to the quantization axis (z-axis).

In reality, interactions between electrons make Z dependent on r, so that the potential loses its simple $1/r$ character and the different subshells have different energies. For example, the filling of the 4s shell starts before the 3d shell is completely filled. This is the reason for the existence of the partly filled electron shells responsible for ferromagnetism.

The periodic table of the elements is obtained by successive filling of the shells and subshells with electrons. Completely filled inner shells and subshells are magnetically inert, aside from a small diamagnetic contribution. This is because both the orbital moments $m = -l, \ldots, \ldots, l$ and the spins ($\uparrow$ and $\downarrow$) add up to zero. Aside from exceptions such as O_2, valence and conduction electrons also form electron pairs with zero net moment. In most cases, the magnetic moment originates from the partially filled inner shells of transition metals, classified as iron-series (3d), palladium-series (4d), platinum-series (5d elements), rare-earths (4f), and actinide (5f) elements.

The magnetism of iron-series transition-metal magnets originates from the 3d electrons. The 3d subshell is defined by the quantum numbers $n=3$ and $l=2$, contains five orbitals $m=0, \pm 1, \pm 2$, and may be occupied by up to 10 electrons. Figure 2.11 shows one orthogonal set of 3d orbitals. The explicit equations for the shown orbitals are

$$|xy\rangle = N_o R_{3d}\,(\mathrm{r}) \sin^2\theta \sin 2\theta \tag{2.30a}$$

$$|x^2 - y\rangle = N_o R_{3d}\,(\mathrm{r}) \sin^2\theta \cos 2\theta \tag{2.30b}$$

$$|xz\rangle = 2N_o R_{3d}\,(\mathrm{r}) \sin\theta \cos\theta \cos\phi \tag{2.30c}$$

$$|z^2\rangle = N_o R_{3d}\,(\mathrm{r})(3\sin^2\theta - 1) \tag{2.30d}$$

$$|yz\rangle = 2N_o R_{3d}\,(\mathrm{r}) \sin\theta \cos\theta \sin\phi \tag{2.30e}$$

where $N_o = \sqrt{15/16\pi}$. Aside from this real set of wave functions, there exist complex wave functions of the type $\exp(\pm im\phi)$. The sets of wave functions are linear combinations of each other, and both are solutions of the Schrödinger equation. However, they are nonequivalent with respect to orbital moment and magnetic anisotropy. We will return to this important point in Section 3.3. The radial part of the wave function

$$R_{3d} = \frac{4Z^{5/2}r^2}{81a_o^2\sqrt{30a_o^3}} \exp\left(-\frac{Zr}{a_o}\right) \tag{2.31}$$

For both sets of wave functions. Bohr's hydrogen radius, $a_o = 0.529$ Å, is the fundamental length.

Palladium-series 4d and platinum-series 5d electrons, which carry a net moment in alloys such as PdFe and PtCo, have the same angular dependence as iron-series 3d electrons and the same total number of orbitals (10). However, the radial part of the wave function, $R(r)$, is different for each series (Z). The 4f and 5f shells of the lanthanides (rare earths) and actinides contain up to 14 electrons per atom. Figure 2.9 shows that the angular dependence differs from that of d electrons.

Rare-earth atoms tend to form tripositve ions, such as Sm^{3+} and Gd^{3+}, in both metals and insulators. By contrast, iron-series elements have two or more possible oxidation states in oxides and fractional occupancies in metals (Section 2.4). Oxides of early 3d elements (T) frequently contain T^{3+} or T^{4+} ions, whereas the late 3d elements prefer to form T^{2+} or T^{3+} ions. A good example is iron, where Fe^{2+} (ferrous iron, $3d^6$) and Fe^{3+} (ferric iron, $3d^5$) coexist in many oxides. Here the notation $3d^n$ mean that the 3d shell contains n electrons. In a *very crude* approximation, one may think of late 3d atoms in metals as T^+ ions, but this is a poor representation of the actual physical situation.

An important feature of the partially filled inner shells is their comparatively small radius. This has far-reaching implications for moment formation, exchange, and magnetic anisotropy. In particular, it reduces the interaction with the atomic environment and makes it possible to approximate atoms as magnetic ions. Figure 2.12 compares the sizes of some atoms and shells. The inner-shell character is most pronounced for the rare-earth 4f shells, where the respective atomic (R) and ionic (R^{3+}) radii of about 1.8 Å and 1.0 Å are significantly larger than the 4f-shell radius of approximately 0.5 Å.

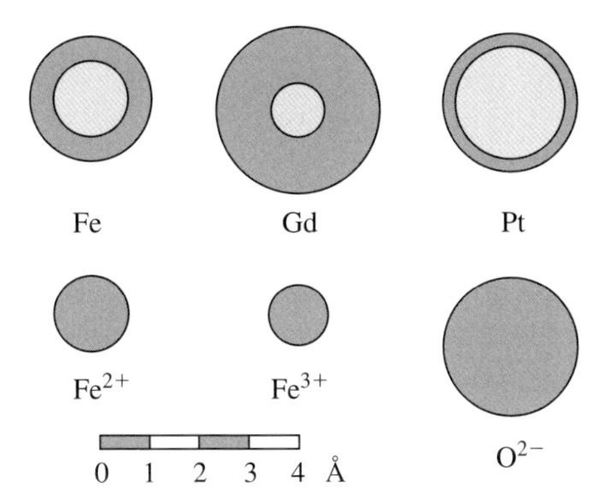

Fig. 2.12 Approximate sizes of some neutral atoms and ions. The dark areas in the neutral atoms show the size of the partially filled inner shells (Fe 3d, Gd 4f, and Pt 5f). Note that $10\,\text{Å} = 1\,\text{nm} = 10^{-9}\,\text{m}$.

2.2.2 Angular-momentum algebra

Spins have the character of quantum-mechanical quantities or operators (Section A.3). Single electrons, characterized by spins $S = \frac{1}{2}$, are described by *spin operators* $\mathbf{s} = \frac{1}{2}\boldsymbol{\sigma}$, where

$$\sigma = \begin{pmatrix} 0 & 1 \\ 1 & 0 \end{pmatrix} \mathbf{e}_\mathrm{x} + \begin{pmatrix} 0 & -\mathrm{i} \\ \mathrm{i} & 0 \end{pmatrix} \mathbf{e}_\mathrm{y} + \begin{pmatrix} 1 & 0 \\ 0 & -1 \end{pmatrix} \mathbf{e}_\mathrm{z} \tag{2.32}$$

is the vector formed by the Pauli matrices σ_x, σ_y, and σ_z. In terms of $\boldsymbol{\sigma}$, the Zeeman energy (1.1) of an electron has the simple form

$$\mathsf{H} = -\mu_\mathrm{o}\mu_\mathrm{B}\boldsymbol{\sigma} \cdot \mathbf{H} \tag{2.33}$$

A less common dimensionless version of this equation is $\mathsf{H} = -\boldsymbol{\sigma} \cdot \mathbf{H}$. It is often convenient to choose $\mathbf{H} = H\mathbf{e}_\mathrm{z}$, so that the z axis is a quantization axis and the eigenvectors (1, 0) and (0, 1) of σ_z have the character of $\uparrow$ and $\downarrow$ states, respectively.

The Pauli matrices σ_x, σ_y, and σ_z have some remarkable properties. First, each of the matrices has the eigenvalues ± 1, corresponding to spin eigenvalues $\pm\frac{1}{2}$ in units of $\hbar$. Second, the square of the spin, $\langle \mathrm{s}^2 \rangle = 3/4$, is *larger* than the square of the projection $\pm\frac{1}{2}$ in any given direction. This means that there is a contribution to $\langle s^2 \rangle$ due to spin fluctuations related to Heisenberg's uncertainty relation. Third, together with the unit matrix, Pauli's spin matrices define a four-dimensional space, indicating that the spin is a relativistic phenomenon (Section A.3.5).

Spins are angular-momentum quantities and obey the corresponding algebra. Starting from the classical angular-momentum definition $\mathbf{L} = \mathbf{r} \times \mathbf{p}$ and exploiting the fact that $\mathbf{p} = -\mathrm{i}\hbar\nabla$ we obtain

$$\mathbf{L} = \mathrm{i}\hbar \left(z\frac{\partial}{\partial y} - y\frac{\partial}{\partial z} \right) \mathbf{e}_\mathrm{x} + \mathrm{i}\hbar \left(x\frac{\partial}{\partial z} - z\frac{\partial}{\partial x} \right) \mathbf{e}_\mathrm{y} + \mathrm{i}\hbar \left(y\frac{\partial}{\partial x} - x\frac{\partial}{\partial y} \right) \mathbf{e}_\mathrm{z} \tag{2.34}$$

The z-component of this expression can also be written as $\mathsf{L}_\mathrm{z} = -\mathrm{i}\hbar\partial/\partial\phi$, indicating that angles and angular momenta are quantum-mechanically conjugate, similar to z

and $\mathsf{p}_z = -\,\mathrm{i}\hbar\partial/\partial z$. The eigenvalues of $\mathbf{L}^2$ and L_z are $L(L+1)$ and $L_z\hbar = m\hbar$, where $-L \leq m \leq L$ is known as the *magnetic quantum number*. Physically, it corresponds to the orbital-moment projection onto the z axis.

The angular-momentum components L_x, L_y, and L_z exhibit a number of interesting properties. First, the easy-to-verify example $[\mathsf{L}_x,\ \mathsf{L}_y] = \mathsf{L}_x\mathsf{L}_y - \mathsf{L}_y\mathsf{L}_x = \hbar\mathsf{L}_z$ shows that the components do not commute. A general and compact expression for the commutation behavior of angular-momentum components is $\mathsf{L}\times\mathsf{L} = \mathrm{i}\hbar\mathsf{L}$ or, in units of $\hbar$, $\mathsf{L}\times\mathsf{L} = \mathrm{i}\mathsf{L}$. This translates into a quantum dynamics reminiscent of classical precession (Section 5.1 and Section A.3.1). It is straightforward to show that $\mathbf{s} = \frac{1}{2}\boldsymbol{\sigma}$ obeys $\boldsymbol{\sigma}\times\boldsymbol{\sigma} = \mathrm{i}\boldsymbol{\sigma}$, meaning that the orbital motion of the electron and the spin obey the same angular-momentum rules. In particular, L and S have $2L+1$ and $2S+1$ respective projections $L_z = -L, \ldots L-1, L$ and $S_z = -S, \ldots, S-1, S$ onto a given quantization axis $\mathbf{e}_z$.

The squares of the angular momenta are $\langle\mathbf{S}^2\rangle = S(S+1)$ and $\langle\mathbf{L}^2\rangle = L(L+1)$, as exemplified by $\langle\mathbf{S}^2\rangle = 3/4$ for $S = \frac{1}{2}$. Similar expressions exist for anisotropic interactions. For example, in Section 3.4 we will use the term $\frac{1}{2}(3\mathsf{S}_z^2 - S(S+1))$ to describe uniaxial anisotropy. When $\mathbf{S}$ and $\mathbf{L}$ are coupled, then the total angular momentum $\mathbf{J}$ obeys $J_z = -J, \ldots, J-1, J$ and $\langle\mathsf{J}^2\rangle = J(J+1)$. Figure 2.13 illustrates the components of $\mathbf{S}$, $\mathbf{L}$, and $\mathbf{J}$. A physical difference between orbital and spin operators is that $\mathbf{L}$ acts of the real-space wave function, whereas $\mathbf{S}$ acts on the spin function.

From (2.32) we see that the eigenfunctions of σ_z are the vector columns (1, 0) and (0, 1), corresponding to the eigenvalues ± 1 (or $\uparrow$ and $\downarrow$). Rotations in spin space are described by the unitary SU(2) matrix, which is closely related to the three-dimensional rotation matrix O(3). For a clockwise rotation by an angle α around a unit vector $\mathbf{n}$, the unitary matrix is $\mathsf{U} = \exp(-\mathrm{i}\alpha\mathbf{n}\cdot\boldsymbol{\sigma}/2) = \cos(\alpha/2) - \mathrm{i}\mathbf{n}\cdot\boldsymbol{\sigma}\sin(\alpha/2)$. Explicitly,

$$\mathsf{U}(\phi,\theta) = \begin{pmatrix} \cos\left(\frac{\theta}{2}\right)\exp\left(\frac{-\mathrm{i}\phi}{2}\right) & -\sin\left(\frac{\theta}{2}\right)\exp\left(\frac{-\mathrm{i}\phi}{2}\right) \\ \sin\left(\frac{\theta}{2}\right)\exp\left(\frac{-\mathrm{i}\phi}{2}\right) & \cos\left(\frac{\theta}{2}\right)\exp\left(\frac{+\mathrm{i}\phi}{2}\right) \end{pmatrix} \tag{2.35}$$

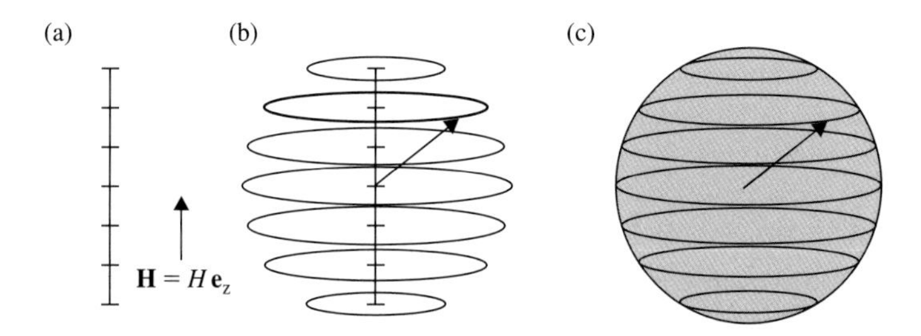

Fig. 2.13 Angular-momentum components of $\mathbf{S}$ (or $\mathbf{L}$ or $\mathbf{J}$) parallel and perpendicular to the quantization axis $\mathbf{e}_z$: (a) Zeeman energy levels in a magnetic field $\mathbf{H} = H\mathbf{e}_z$, (b) circles indicating the x and y components, and (c) meaning of $\langle\mathbf{S}^2\rangle = \mathrm{S(S+1)}$. The Zeeman-energy levels are proportional to S_z and correspond to (a), whereas the classical limit of a continuous vector is reproduced by assuming a zero level spacing in (c).

Applying this transformation to $|\uparrow\rangle = (1,0)^{\mathrm{T}}$, where T indicates that the spin function is a vector column, yields the striking spin function $(\cos(\theta/2)\exp(-\mathrm{i}\phi/2),\ \sin(\theta/2)$ $\exp(+\mathrm{i}\phi/2))^{\mathrm{T}}$. The reason is the transformation behaviour of the Pauli spin matrices: spin-space rotations by an angle π, that is between parallel and antiparallel spins, correspond to real-space rotations by an angle $\pi/2$.

2.2.3 Vector model and Hund's rules

The magnetism of most atoms involves two or more d or f electrons, and each electron possesses a spin and, generally, an orbital momentum or moment. The addition of these contributions is the scope of the *vector model* of atomic magnetism. In the *L*-*S* or Russell-Saunders coupling scheme, the total orbital momentum $\mathbf{L} = \Sigma_{\mathrm{i}}\mathbf{L}_{\mathrm{i}}$ and the total spin momentum $\mathbf{S} = \Sigma_{\mathrm{i}}\mathbf{S}_{\mathrm{i}}$ combine to yield the total momentum $\mathbf{J} = \mathbf{L} + \mathbf{S}$. The moments obey the matrix-operator rules outlined above, such as and $\mathbf{S}^2 = S(S+1)$, $\mathbf{L}^2 = L(L+1)$, and $\mathbf{J}^2 = J(J+1)$. Russell-Saunders levels characterized by the quantum numbers L and S form a *term* denoted by ${}^{2\mathrm{S}+1}L$. For example, the term symbol ${}^2F$ means $L = 3$ and $S = \frac{1}{2}$. Spin-orbit coupling causes ionic terms to split into *multiplets* denoted by ${}^{2\mathrm{S}+1}L\mathrm{J}$, where $|L-S| \leq J \leq |L+S|$. For instance, the ${}^2P$ term splits into a ${}^2P\frac{1}{2}$ doublet and a ${}^2P_{3/2}$ quartet. Note that the *L*-*S* coupling considered here differs from the j-j coupling in very heavy elements ($\mathrm{Z} > 75$), where the strong spin-orbit interaction dismantles the total ionic spin and orbital momenta and $\mathbf{J}_{\mathrm{i}}^2 = J_{\mathrm{i}}(J_{\mathrm{i}}+1)$ for each electron.

The ground state of rare-earth ions obeys *Hund's rules* (1925). The rules are emipirical but have a sound physical basis. First, subject to the Pauli principle, the total spin S is maximized. Second, subject to the Pauli principle and to Hund's first rule, the orbital moment L is maximized. Third, L and S couple parallel, $J = |L+S|$, in more-than-half filled 4f shells and antiparallel, $J = |L-S|$, in less-than-half filled 4f shells. The last rule reflects spin-orbit coupling. Hund's rules are well satisfied for most rare-earths, because the 4f-shell radii of about 0.5 Å are much smaller than the atomic radii of about 1.8 Å. This enhances the spin-orbit-coupling, reduces the crystal-field interaction, and leaves the orbitals nearly unquenched. According to Hund's third rule, the ground-state multiplets of rare-earth ions obey $J = |L \pm S|$. Excited multiplets have relatively high energies, with the notable exceptions of Eu^{3+} and Sm^{3+}, where the splitting between the lowest-lying multiplets is about 0.1 eV (Taylor and Darby 1972). The remaining $(2J+1)$-fold intramultiplet degeneracy is removed by interactions such as Zeeman coupling and interatomic exchange. This is of importance for the temperature dependence of the magnetic anisotropy (Chapter 5.5).

2.2.4 Spin and orbital moment

According to the vector model, the addition of spin and orbital angular momenta gives rise to the total angular momentum $J = L + S$. However, this is only one aspect of the addition of L and S. In a magnetic field, the moment belonging to a spin is $2\mu_{\mathrm{B}}S$ rather than $\mu_{\mathrm{B}}S$, corresponding to a *Landé* or *g*-factor 2 and reproducing the Zeeman energy (2.33). Rigid coupling between L and S yields $2J+1$ Zeeman levels, consistent with $\mathbf{m} = g\mu_{\mathrm{B}}\,\mathbf{J}$ and $\mathsf{H} = -g\mu_{\mathrm{o}}\mu_{\mathrm{B}}\mathbf{J}\cdot\mathbf{H}$. For pure orbital and spin-only moments $g = 1$ and $g = 2$, respectively.

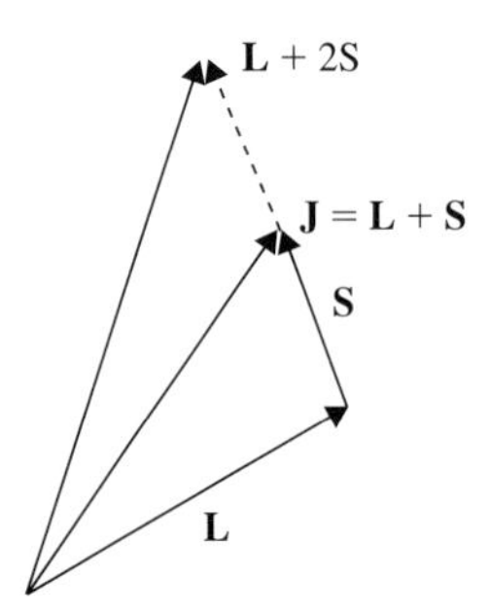

Fig. 2.14 Total angular momentum $\mathbf{J}=\mathbf{L}+\mathbf{S}$ and the term $\mathbf{L}+2\mathbf{S}$ entering the Zeeman energy.

For arbitrary L and S, the interaction with an external field involves $\mathbf{L}+2\mathbf{S}$ rather than $\mathbf{J}=\mathbf{L}+\mathbf{S}$. It is therefore necessary to project $\mathbf{L}+2\mathbf{S}$ onto $\mathbf{J}$, so that $(\mathbf{L}+2\mathbf{S})\cdot\mathbf{J}=g\mathbf{J}^2$. Figure 2.14 illustrates the meaning of this projection. Since

$$J(J+1) = L(L+1) + 2\mathbf{L}\cdot\mathbf{S} + S(S+1) \tag{2.36}$$

the g-factor

$$g = \frac{3}{2} + \frac{1}{2}\frac{S(S+1) - L(L+1)}{J(J+1)} \tag{2.37}$$

The first and second halves of the lanthanide series exhibit $g=1-S/(J+1)$ and $g=1+S/J$, respectively.

Exchange between magnetic ions is a spin-only interaction, so that the projection $\mathbf{S}\cdot\mathbf{J}=(g-1)\mathbf{J}^2$ must be used. This ensures zero exchange for $g = 1$ (orbital magnetism). The corresponding *de Gennes* factor $G=(g-1)^2J(J+1)$ is important for the finite-temperature behavior of rare-earth magnets, affecting, for example, the Curie temperature.

Table 2.1 lists some ground-state properties of tripositive rare-earth ions. Spectroscopic and magnetic measurements indicate that Hund's rules are well satisfied by rare-earth ions. Table 2.2 shows the Hund's-rule ground states of free 3d ions. Compared to rare-earth 4f electrons, Hund's rules—notably the second and third rules—are poorly obeyed by 3d shells. For example, $g\approx 2$ for iron-series atoms in both metallic and nonmetallic crystalline environments. This is due to the quenching of the orbital moment of 3d electrons in (Section 3.3.4). In 3d magnets, only the first rule applies. In metals, it is known as the "Stoner" rule, but orbital-polarization or second-rule and spin-orbit or third-rule corrections can be incorporated into numerical electronic-structure calculations (Eriksson *et al.* 1990, Eschrig *et al.* 2005).

Aside from contributing to the total magnetization, the orbital moment is closely linked to the magnetocrystalline anisotropy. In fact, $g=2$ rather than $g\approx 2$ means spin-only magnetism and zero magnetic anisotropy (Section 3.2–3). This a concern for 3d ions, whereas 4f ions have stable orbital moments. By comparison, 4d, 5d, and 5f ions are intermediate between 3d and 4f ions, with often substantial orbital moments.

As we have seen in Section 2.1, the trend towards ferromagnetism reflects the competition between one-electron energies and Coulomb interactions (exchange). Hund's

Table 2.1 Hund's-rules ground states of 4f ions. The numbers in the header row are orbital quantum numbers. Note that the "nonmagnetic" rare earth Y, La, and Lu have $S = L = J = 0$

	L:	+3	+2	+1	0	−1	−2	−3	S	L	J	g	gJ
$4f^1$	Ce^{3+}	↑							$\frac{1}{2}$	3	5/2	6/7	15/7
$4f^2$	Pr^{3+}	↑	↑						1	5	4	4/5	16/5
$4f^3$	Nd^{3+}	↑	↑	↑					3/2	6	9/2	8/11	36/11
$4f^4$	Pm^{3+}	↑	↑	↑	↑				2	6	4	3/5	12/5
$4f^5$	Sm^{3+}	↑	↑	↑	↑	↑			5/2	5	5/2	2/7	5/7
$4f^6$	Eu^{3+}	↑	↑	↑	↑	↑	↑		3	3	0	–	0
$4f^7$	Gd^{3+}	↑	↑	↑	↑	↑	↑	↑	7/2	0	7/2	2	7
$4f^8$	Tb^{3+}	↑↓	↑	↑	↑	↑	↑	↑	3	3	6	3/2	9
$4f^9$	Dy^{3+}	↑↓	↑↓	↑	↑	↑	↑	↑	5/2	5	15/2	4/3	10
$4f^{10}$	Ho^{3+}	↑↓	↑↓	↑↓	↑	↑	↑	↑	2	6	8	5/4	10
$4f^{11}$	Er^{3+}	↑↓	↑↓	↑↓	↑↓	↑	↑	↑	3/2	6	15/2	6/5	9
$4f^{12}$	Tm^{3+}	↑↓	↑↓	↑↓	↑↓	↑↓	↑	↑	1	5	6	7/6	7
$4f^{13}$	Yb^{3+}	↑↓	↑↓	↑↓	↑↓	↑↓	↑↓	↑	$\frac{1}{2}$	3	7/2	8/7	4

Table 2.2 Hund's-rules ground states of 3d ions. In reality, $L \approx 0$, $J \approx S$ and $g \approx 2$ (quenching). The numbers in the header row are orbital quantum numbers.

	L:	+2	+1	0	−1	−2	S	L	J	g	gJ
$3d^1$	Ti^{3+}, V^{4+}	↑					$\frac{1}{2}$	2	3/2	4/5	6/5
$3d^2$	V^{3+}, Cr^{4+}	↑	↑				1	3	2	2/3	4/3
$3d^3$	Cr^{3+}, Mn^{4+}	↑	↑	↑			3/2	3	3/2	2/5	3/5
$3d^4$	Cr^{3+}, Mn^{3+}	↑	↑	↑	↑		2	2	0	–	0
$3d^5$	Mn^{2+}, Fe^{3+}	↑	↑	↑	↑	↑	5/2	0	5/2	2	5
$3d^6$	Fe^{2+}, Co^{3+}	↑↓	↑	↑	↑	↑	2	2	4	3/2	6
$3d^7$	Co^{2+}, Ni^{3+}	↑↓	↑↓	↑	↑	↑	3/2	3	9/2	4/3	6
$3d^8$	Ni^{2+}, Co^{+}	↑↓	↑↓	↑↓	↑	↑	1	3	4	5/4	5
$3d^9$	Cu^{2+}, Ni^{+}	↑↓	↑↓	↑↓	↑↓	↑	$\frac{1}{2}$	2	5/2	6/5	3

rules assume that the exchange is sufficiently strong to ensure maximum spin parallelity. This is indeed the case for most bulk 3d oxides, where the crystal-field interaction (Section 3.3.3) is weaker than the intra-atomic exchange. However, strongly anisotropic crystalline environments may mean that some of the 3d ↑ electrons acquire very high one-electron energies and reverse their spin to jump into low-lying ↓ orbitals. This is known as a *high-spin low-spin transition.* For example, in Co^{2+} (Table 2.2) the spin changes from 3/2 to 1/2 if the ↑ electron with the highest energy jumps into the lowest unoccupied ↓ orbitals. We will return to this effect in Section 3.3.3, in the context of the $e_g - t_{2g}$ crystal-field splitting. In fact, the high-spin low-low spin transition is an ionic equivalent to the collapse of the itinerant magnetization as described by the Stoner criterion (Section 2.1.5 and Section 2.4.3), except that the collapse reflects the crystal field rather than the bandwidth. Note that the presence of 3d, 4d, 5d, 4f or 5f atoms is a practical rather than fundamental condition for ferromagnetism. If s and p orbitals form degenerate or nearly degenerate states, then the electrons' relatively weak intra-atomic exchange may be sufficient to create atomic moments with ferromagnetic coupling.

2.3 Exchange between local moments

Summary Interatomic exchange between magnetic moments localized on individual atoms is well described by Heisenberg interactions J_{ij} between atomic spins of constant magnitude. Depending on the respective positive or negative sign of the J_{ij}, the exchange favors parallel or antiparallel alignment of neighboring spins, which often translates into ferromagnetic (FM), ferrimagnetic (FI), or antiferromagnetic (AFM) order. Ferri- and antiferromagnetic spin structures involve the formation of magnetic sublattices, which may be spontaneous (AFM) or indicative of nonequivalent crystallographic sites (FI). In addition, there exist noncollinear or incommensurate spin structures due to competing exchange interactions, for example in many rare-earth elements. Noncollinear spin configurations may also be caused by external magnetic fields, as exemplified by the spin-flop tranistion in antiferromagnets. Oxides are often ferri- or antiferromagnetic, and typical exchange mechanism are superexchange and double exchange. Interactions between local moments in metals are fairly well approximated by the RKKY model, which yields an oscillatory exchange mediated by conduction electrons. This section deals with exchange interactions between local magnetic moments in metals and nonmetals. Emphasis is on the sign of the exchange and the resulting zero-temperature spin structure. Finite-temperature effects will be considered in Section 5.2–4, whereas changes in the spin structure due to magnetic fields are discussed in Chapter 4, especially in Section 4.1.3.

2.3.1 Exchange in oxides

The magnetic moment of iron-series transition-metal or 3d oxides, such as Fe_3O_4 and CrO_2 is given by the *spin*, so that the moment, measured in μ_B, is equal to the number of unpaired spins. This includes both the atomic moments (intra-atomic exchange) and the net moment (interatomic exchange). As in other 3d-based magnets, such as Fe and Co, the orbital moment is often quenched by the crystal field, so that $g \approx 2$ and $L \approx 0$. Since the quenching affects not only the moment but also the magnetic anisotropy, it will be discussed in Section 3.34. In good approximation, the atomic spins then follow from Hund's first two rules, Table 2.2. For example, ferric iron (Fe^{3+}) has a half-filled shell ($3d^5$), and the five ↑ electrons yield $m = 5\mu_B$. Ferrous iron (Fe^{2+}) has six 3d electrons per ion, five ↑ and one ↓, so that $m = 4\mu_B$.

Depending on the sign of the interatomic exchange constant J_{ij}, neighboring atomic spins tend to align antiferromagnetic (AFM) or ferromagnetic (FM). In transition-metal oxides, the exchange is often but not always antiferromagnetic. Many oxides, such as NiO and CoO, are therefore antiferromagnets with zero net moment. However, the numbers of ↑ and ↓ spins are not necessarily equal, because the moments are associated with specific crystallographic sites. This incomplete spin compensation is known as *ferrimagnetism*. For example, each formula unit of magnetite Fe_3O_4, or $FeO \cdot Fe_2O_3$, contains two Fe^{3+} ions with antiparallel moments and one Fe^{2+} ion. This yields 4 μ_B per formula unit, close to the experimental value of 4.1 μ_B. The idea that of atoms on different sites may carry different magnetic moments is the basis of the *sublattice model* of magnetism. For example, magnetite has two antiparallel sublattices A and B. One Fe^{3+} ion per formula unit belongs to the sublattice A, while the second Fe^{3+} ion and the Fe^{2+} ion form the sublattice B.

Sublattice formation is one reason for the structural and magnetic diversity of magnetic oxides. An example of a permanent magnet oxide is barium ferrite, $BaFe_{12}O_{19}$ or $BaO \cdot (Fe_2O_3)_6$, where four $\uparrow$ and two $\downarrow Fe_2O_3$ units yield a moment of 20 μ_B per formula unit, as compared the experimental value of 19.6 μ_B. Garnets, such as $Y_3Fe_5O_{12}$ or $\frac{1}{2}(Y_2O_3)_3(Fe_2O_3)_5$, are of interest as soft magnets for microwave applications. The three $\uparrow$ and two $\downarrow Fe_2O_3$ units correspond to a moment of 5 μ_B per formula unit, very close to the experimental value. Aside from determining the net moment, the sublattice structure has a far-reaching effect on magnetic properties at both zero and at finite temperatures (Section 2.3.3 and Section 5.3.6).

In some oxides, the exchange is ferromagnetic. An example is CrO_2 where the Cr^{4+} ion yields a moment of 2 μ_B per formula unit. What determines the sign of the exchange in oxides? There are various and generally competing exchange mechanisms in oxides. An important indirect-exchange mechanism is the *superexchange* mediated by the 2p electrons of the O^{2-} anions separating the iron cations (White 1970). It describes, for example, the $Mn^{2+} - Mn^{2+}$ exchange in MnO. To discuss the sign of the exchange, we take into account that interatomic hopping favors antiferromagnetic exchange (2.18). However, hopping requires a nonzero overlap between the involved atomic orbitals. For p and d orbitals, the hopping depends on the bond angles. This is epitomized by the Goodenough-Kanamori rules (Goodenough 1963, Anderson 1963), which states the exchange is usually antiferromagnetic, but ferromagnetic due to direct exchange if the overlap is zero by symmetry. Figure 2.15 illustrates the corresponding angles. In magnetite, the A-O-B superexchange is realized by a 125° bond and is antiferromagnetic. Note that the superexchange mechanism is not restricted to oxides but also applies to halides, such as MnF_2. An important mechanism in oxides with mixed valence is *double exchange*. An example is Fe_3O_4, where the interacting Fe^{2+}

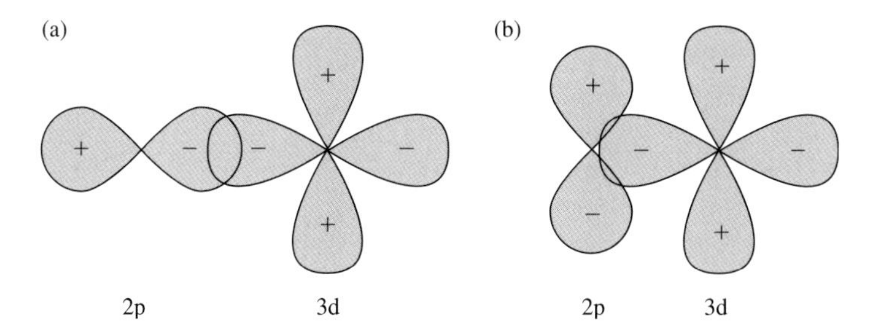

Fig. 2.15 Overlap and exchange: (a) nonzero overlap (180°) and zero overlap (90°). In (a), the hopping integral is nonzero, corresponding to antiferromagnetic indirect exchange, but in (b) the hopping integral is zero by symmetry.

and Fe^{3+} ions can be considered as a pair of Fe^{3+} ions plus an extra electron. The extra electrons hop onto neighboring Fe^{3+} ions, where they are subject to ferromagnetic Hund's-rules exchange. This mechanism is called hopping with spin memory and leads to metallic conductivity. It is similar to the model of Section 2.1, especially to the Kondo exchange between localized and delocalized moments (Section 2.1.8). In a slightly different context, it is also known as the Zener model (1951).

Double exchange is exploited in magnetoresistive perovskites (CMR manganites). The parent compound, $LaMnO_3$ or $\frac{1}{2}La_2O_3 \cdot Mn_2O_3$ contains Mn^{3+} ions only and is an antiferromagnetic insulator. Doping with strontium amounts to replacing the tripositive lanthanum ions by Sr^{2+}, which is compensated by the formation of Mn^{4+} ions. The resulting mixed valence, Mn^{3+} and Mn^{4+}, leads to ferromagnetic double exchange and to metallic conductivity. In both the tri- and quadipositive ions, the low-lying t_{2g} triplets are occupied by three well-localized 3d electrons, but in Mn^{3+}, there is an additional e_g electron. The Hund's-rule interaction yields a ferromagnetic alignment between the e_g and t_{2g} electrons, and due to hybridization with the oxygen 2p electrons, the e_g electron experiences a pronounced interatomic hopping. The corresponding spin transfer yields both ferromagnetism and conductivity, and the manganites are magnetoresistive (Section 7.2.7). In the double-exchange model, the effective hopping integral $t = t_o \cos(\theta/2)$, that is maximum hopping for ferromagnetic spin alignment ($\theta = 0$) and zero hopping for antiferromagnetic spin alignment ($\theta = \pi$). For an approximate independent-electron interpretation of this effect, see Section 2.4.4. In the Kondo picture (Section 2.1.8), where correlations are taken into account, the t_{2g} and e_g levels correspond to the localized and extended orbitals although the e_g and t_{2g} orbitals in one Mn atom are orthogonal to each other, whereas the model of Section 2.1.8 focuses on interatomic hopping.

2.3.2 Ruderman-Kittel exchange

Local moments in metals are coupled by an indirect exchange involving conduction electrons. Examples are magnetic 3d impurities in nonmagnetic metallic hosts and 4f ions in rare-earth elements and alloys. The role of the conduction electrons is similar to that of oxygen 2p electrons in oxides, but the details of the interaction are different. The basic idea is to embed two local moments in a gas of conduction electrons and to calculate the energy of the perturbed electron gas for parallel and antiparallel spin configurations. In the simplest case, one considers the conduction electrons as a free-electron gas (Section 2.4.1) and treats the magnetic perturbation by second-order perturbation theory (Section A3.3). This mechanism is known as Ruderman-Kittel-Kasuya-Yoshida or *RKKY* interaction.

The free-electron RKKY interaction is obtained by considering magnetic perturbations of the type $\pm V_o \delta(\mathbf{r})$, where the sign denotes the spin direction of the local moment relative to the considered conduction electron and V_o describes the strength of the corresponding interaction. The local magnetic moment creates a wave-like local perturbation, similar to a stone thrown into water, Fig. 2.16(a), and yields an oscillating interaction. As we will analyze in Section 2.4.1, the electrons have a maximum wave vector k_F, so that distances smaller than about $2\pi/k_F$ cannot be resolved. The net interaction $J_{ij} = J(|\mathbf{R}_i - \mathbf{R}_j|)$ is obtained by summation over all conduction electrons.

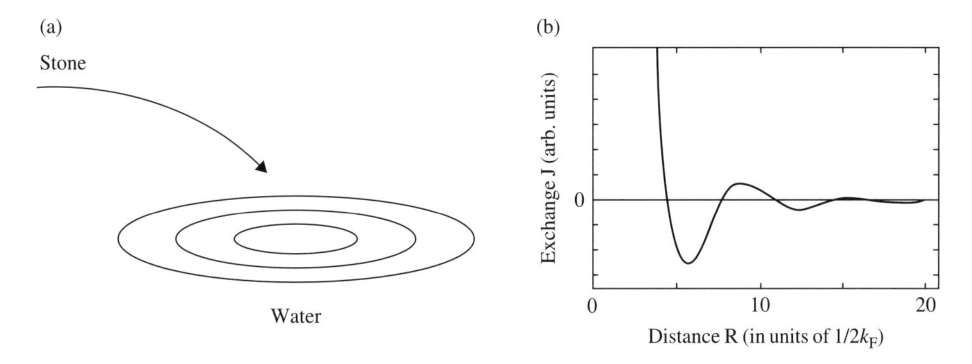

Fig. 2.16 RKKY interactions: (a) mechanical analogy and (b) distance dependence. In nonmagnetic metals, the oscillations are known as Fridel oscillations.

Figure 2.16(b) shows the result,

$$J(R) = J_o \frac{2k_F R \cos(2k_F R) - \sin(2k_F R)}{(2k_F R)^4} \tag{2.38}$$

In metals, k_F is large (Section 2.4.1) and the oscillation period does not exceed a few Å.

Equation (2.38) works fairly well for rare-earth elements, where localized 4f moments are embedded in a sea of 5d and 6s conduction electrons. It also works for localized 3d moments in hosts such as Cu and 4f moments in rare-earth transition-metal alloys, where the RKKY interaction is mediated by 4s electrons. The magnetism of transition-metal-rich rare-earth intermetallics, such as $Nd_2Fe_{14}B$, is largely determined by the transition-metal sublattice. The interaction between the rare-earth atoms is practically negligible, but the interaction between the rare-earth and transition-metal sublattices is larger and contributes to magnetization, Curie temperature and, especially, anisotropy. In the 3d ferromagnets Fe, Co, and Ni, and in exchange-enhanced Pauli-paramagnets such as Pd and Pt, there are asymptotic oscillations reminiscent of (2.38), although the moments in these systems are itinerant rather than localized (Section 2.4.3 and Section 5.2.5).

Aside from atomic-scale RKKY interactions in alloys, spin glasses, and other magnetic materials, one encounters nanoscale RKKY interactions between thin-film layers and embedded particles (Section 7.4.2). These interactions are obtained from (2.38) by summation over pairs of atoms. Note that RKKY oscillations require a sharp Fermi surface (Fig. 2.23). At finite temperatures, the thermal smearing of the Fermi surface yields an exponential decay of the oscillations, with a decay length of $\hbar k_F / 2\pi m_e k_B T$.

Equation (2.38) describes exchange interactions mediated by *free* electrons, but the underlying perturbation theory can also be used to treat arbitrary electron systems, such as tight-binding electrons in metals (Mattis 1965) and electron clusters in dilute magnetic semiconductors (Skomski *et al.* 2006b). Figure 2.17 illustrates the exchange between localized magnetic moments in a dilute magnetic semiconductor (DMS) with shallow nonmagnetic donors (or acceptors). In this case, $J(R_{ij})$ changes sign on a length scale comparable to the average distance between the donors.

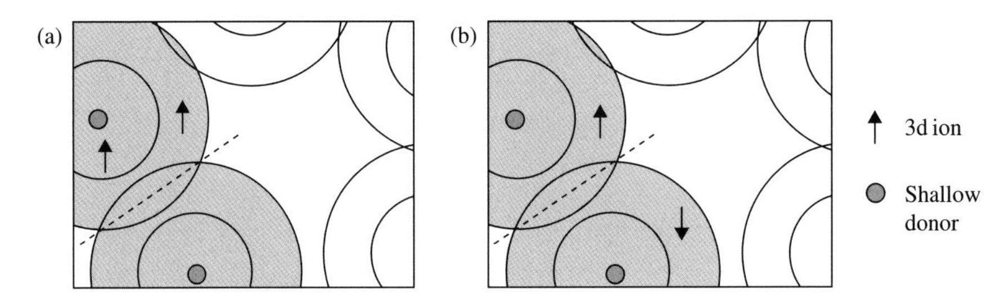

Fig. 2.17 Indirect exchange in dilute magnetic semiconductors with shallow donors or acceptors: (a) FM and (b) AFM coupling.

An alternative, though somewhat less accurate, explanation is that k_F decreases with the electron density (Yu and Cardona 1999), so that the low carrier densities in semiconductors translate into RKKY oscillations that are long-range by atomic standards, larger than $10\,\text{Å} = 1\,\text{nm}$. A related effect is quantum confinement in semiconductor nanodots, where the role of the localized spins is played by the boundaries of the dots. We will return to these phenomena in Section 2.4.1 and Section 7.2.8.

2.3.3 Zero-temperature spin structure

Interatomic exchange leads to several types of magnetic order, such as ferromagnetism, ferrimagnetism, antiferromagnetism, and noncollinear spin structures. Figure 2.18 shows some examples. In a broader context, spin structures have several aspects. First, this subsection focuses on zero-temperature magnetic order in perfect crystals. Here we focus on magnetic order of atomic origin, which not be confused with micromagnetic features such as domains, which are realized on larger length scales and obeys different laws (Chapter 4). For the moment, we also ignore finite-temperature magnetic order, as epitomized by the Curie transition (Chapter 5), and noncollinearities associated with spin dynamics, such as spin waves (Chapter 6). Finally, there are various types of order in imperfect crystals, such as random-anisotropy magnets and spin glasses (Chapter 7).

The spin structures shown in Fig. 2.18 all derive from exchange interactions $J_{ij} = J(\mathbf{R}_j - \mathbf{R}_i)$ between atoms located at $\mathbf{R}_i$ and $\mathbf{R}_j$. A simple but powerful model approach is to describe spin structures in terms of atomic spin variables $\mathbf{s}_i = \mathbf{M}(\mathbf{r}_i)/M_s$, where $M_s = |\mathbf{M}(\mathbf{r})|$. In simple ferromagnets, all atoms are equivalent, so that $N_s = 1$ and $\mathbf{s}_i = \mathbf{s}$. In general, it is possible to divide the crystal into *sublattices* $\mathbf{s}_i (i = 1 \ldots N_s)$, where $1 \leq N_s \leq \infty$ is the number of sublattices. We have already encountered this approach in the discussion of the magnetic moment of oxides (Section 2.3.1).

Sublattice formation may be spontaneous, as in typical antiferromagnets or linked to the atomic composition, as in ferrimagnets. Figures 2.18(b–e) contain two sublattices each. In some sense, sublattice models go beyond classical physics, because the local magnetization directions can be considered as quantum-mechanical averages of the type $\langle\psi|\mathbf{S}|\psi\rangle$. However, they parameterize rather than predict the exchange and are unable to explain quantum phenomena that explicitly involve the wave functions $|\psi\rangle$.

In ferromagnets, such as Fe, Co, and $Nd_2Fe_{14}B$, the exchange is positive, the spins are all parallel, and the atomic moments add. Ferrimagnets, such as $DyCo_5$,

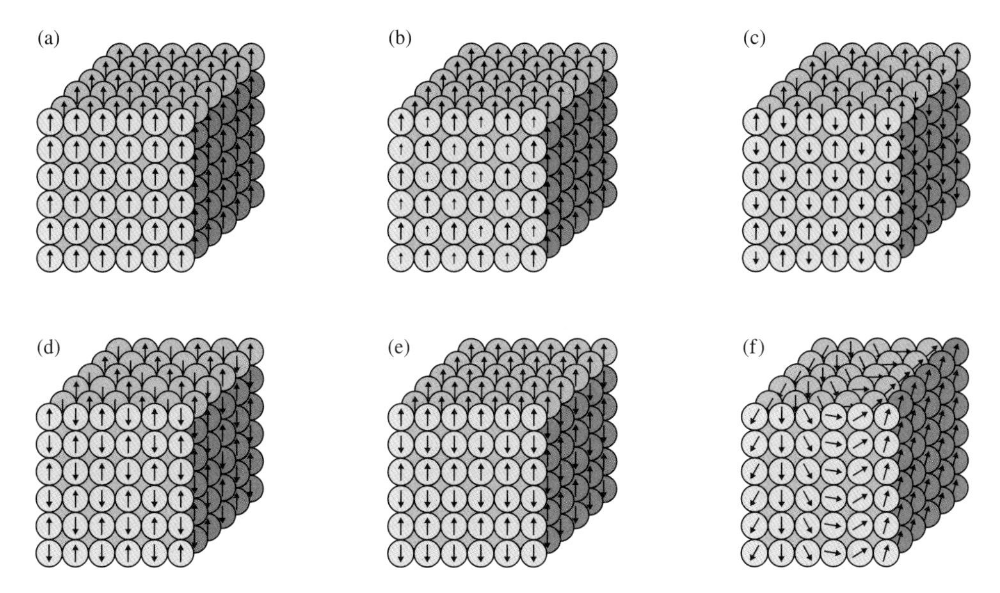

Fig. 2.18 Zero-temperature magnetic order: (a) one-sublattice ferromagnet, (b) two-sublattice ferromagnet, (c) ferrimagnet, (d–e) antiferromagnet, and (f) noncollinear magnet.

Fe_3O_4 and $BaFe_{12}O^{19}$ and antiferromagnets, such as CoO and MnF_2, involve negative exchange and exhibit two or more sublattices with opposite moments. This amounts to a ferrimagnetic reduction or antiferromagnetic absence of a net moment. Typical antiferromagnets have crystallographically equivalent but magnetically nonequivalent sites, and there are often two or more competing configurations, such as (d) and (e). A striking feature is that periodic exchange constants J_{ij} may give rise to aperiodic or incommensurate spin structures, as in Fig. 2.18(f). Examples are the helimagnetism of many rare-earth elements below room temperature (Moorjani and Coey 1984) and the spin-density-wave antiferromagnetism in Cr.

In fair approximation, the spin structures of Fig. 2.18 are described by a Heisenberg-type model Hamiltonian derived from (2.20)

$$H = -\sum_{\mathrm{m<n}} J_{\mathrm{mn}} \mathbf{s}_{\mathrm{n}} \cdot \mathbf{s}_{\mathrm{m}} - \sum_{\mathrm{m}} \mathbf{h}_{\mathrm{m}} \cdot \mathbf{s}_{\mathrm{m}} \qquad (2.39)$$

Here $\mathbf{s}_{\mathrm{m}}$ is the magnetization of the m-th sublattice, J_{mn} is the exchange between the m-th and n-th sublattices, and $\mathbf{h}_{\mathrm{m}} = 2\mu_{\mathrm{o}}\mu_{\mathrm{B}}\mathbf{H}_{\mathrm{m}}$ describes the local field. The J_{mn} are exchange energies per atom. They may be taken from experiment, by measuring the temperature dependence of the magnetic properties, or as a sum over atomic exchange interactions, $J_{\mathrm{mn}} = \sum_{\mathrm{i}\in\mathrm{n}} J(\mathbf{R}_{\mathrm{i}} - \mathbf{R}_{\mathrm{m}})$.

The determination of the spin structure from (2.39) involves two steps. First, the number and geometry of the sublattices must be known. This is quite demanding for antiferromagnets and for more complicated structures. Second, the energy must be minimized, by comparing different configurations and by varying $\mathbf{s}_{\mathrm{m}}$.

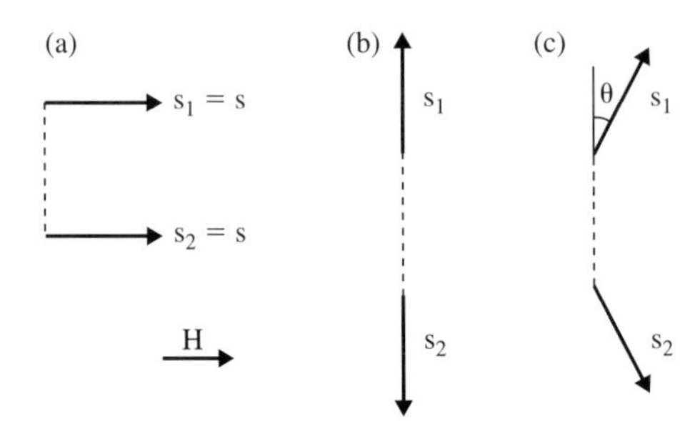

Fig. 2.19 Spin structure of isotropic ferromagnets and antiferromagnets: (a) ferromagnet, (b) AFM in zero magnetic field and (c) AFM in nonzero magnetic field.

A simple example is a *two-sublattice magnet* in a homogeneous magnetic field,

$$E = J^{*}\mathbf{s}_1 \cdot \mathbf{s}_2 - \mathbf{h} \cdot (\mathbf{s}_1 + \mathbf{s}_2) \tag{2.40}$$

where $J^* = - J_{12}$. Figure 2.19 shows some cases. In terms of the angle ϕ, the energy

$$E = J^* \cos(2\theta) - 2h \cdot \sin(\theta) \tag{2.41}$$

Figure 2.19(a) shows the ferromagnetic (FM) case, where $J^* < 0$. Antiferromagnetism is realized for $J^* > 0$, and minimizing (2.41) with respect to θ yields the magnetization component $\sin\theta = h/2J^*$ parallel to the applied field. Figure 2.19(b) and (c) show the respective spin states for zero and nonzero fields. The spin state (c) is also known as the spin-flop state (Section 4.1.4).

The angle ϕ in (c) increases with the magnetic field, which corresponds to a nonzero *susceptibility* $\partial s_z/\partial h = 1/J^*$. The ferromagnetic susceptibility is zero, because the present model considers magnetization rotations only and the spins in Fig. 2.19(a) are already parallel to the field. In reality, the magnitude of the magnetization is weakly field-dependent, which is observed as a small high-field susceptibility. Typical orders of magnitude are $\chi \sim 10^{-4}$ for paramagnetic (or diamagnetic) high-field susceptibilities and about $\chi \sim 10^{-2}$ for processes such as that shown in Fig. 2.19(c). This classification excludes susceptibility singularities near phase transitions, such as the divergence of the ferromagnetic susceptibility at the Curie point and the sharp but finite susceptibility maximum of antiferromagnets at the Néel temperature T_N. The large slope $\chi \sim 1$ encountered in hysteresis loops is no true atomic susceptibility but reflects micromagnetic magnetization processes.

The model of (2.40) is limited to nearest-neighbor interactions and parameterized by a single exchange constant J^*. However, the oscillatory character of the RKKY interactions (2.38) indicates that more distant neighbors may be important. The competition between nearest-neighbor and next-nearest-neighbor interactions is the origin of *noncollinear* spin structures, such as that in Fig. 2.18(f). The rare-earth elements, especially the heavy rare earths, tend to form noncollinear spin structures of various types. The wave vectors of the spin states are generally unrelated to the lattice spacing, or *incommensurate*.

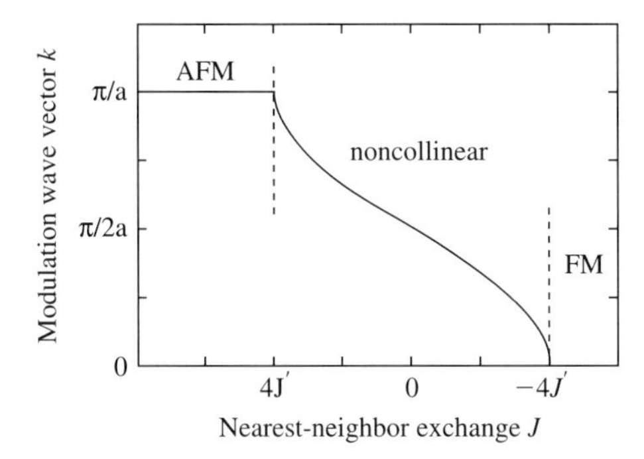

Fig. 2.20 Helical spin state, as realized for negative nearest-neighbor exchange between atomic layers. The geometrical meaning of k is illustrated in Fig. 2.18(f).

In Fig. 2.18(f), the sublattices have the character of layers $\mathbf{s}_\mathrm{n}$, and the number N_s of relevant layers is generally infinite. Denoting the magnetization angle of the n-th layer by θ_n, we obtain the total energy

$$E = -J \sum_\mathrm{n} \cos(\theta_\mathrm{n+1} - \theta_\mathrm{n}) - J' \sum_\mathrm{n} \cos(\theta_\mathrm{n+2} - \theta_\mathrm{n}) \tag{2.42}$$

Here J and J' are exchange constants between nearest and next-nearest layers. The *ansatz* $\theta_\mathrm{n+1} = \theta_\mathrm{n} + \delta$ yields the energy per atom

$$E/N = J^* \cos(\delta) - J' \cos(2\delta) \tag{2.43}$$

and δ is obtained by energy minimization, $\partial E/\partial\delta = 0$. The corresponding equation of state

$$(J^* + 4J' \cos\delta) \sin\delta = 0 \tag{2.44}$$

has ferromagnetic ($\delta = 0$), antiferromagnetic ($\delta = \pi$), and noncollinear ($0 < \delta < \pi$) solutions. For $J' > 0$, there is a trivial FM-AFM transition at $J = 0$, but for $J' < 0$, the FM and AFM configurations are separated by a noncollinear or *helimagnetic* phase. The noncollinear state is characterized by the angle $\delta = \arccos(-J/4J')$ and the modulation wave vector $k = \delta/a$, where a is the layer spacing. Figure 2.20 shows the dependence of k on J for negative values of J'. Helimagnetic order of this type is encountered in Tb, Dy, and Ho, in a temperature window between between ground-state ordering (very low temperatures) and paramagnetism. For example, Dy is ferromagnetic below 85 K and helimagnetic between 85 and 178 K.

2.4 Itinerant magnetism

Summary The magnetism of Fe, Co, and Ni, as well as that of typical transition-metal alloys, is delocalized or *itinerant*. In a simple one-electron picture, the electrons fill the available delocalized states until the Fermi level is reached. This explains the widespread occurrence of noninteger magnetic moments in metallic ferromagnets. Nonmagnetic metals have two equally populated ↑ and ↓ subbands; and an applied magnetic field transfers a few electrons from the ↓ band to the ↑ band. This is known as Pauli paramagnetism, but the corresponding spin polarization is small, typically less than 0.1%. In itinerant ferromagnets, the atomic orbitals hybridize and form bands. The corresponding one-electron energies, as epitomized by the bandwidth, compete against Hund's-rules intra-atomic exchange, and ferromagnetism is realized in *narrow* bands. The simplest model of itinerant ferromagnetism is the Bloch model, where the intra-atomic exchange is evaluated for free electrons. A more sophisticated model is the Stoner model, which relates the onset of ferromagnetism to the density of states (DOS) at the Fermi level. The density of states exhibits a strong dependence on the crystal structure, which makes it difficult to predict the ferromagnetic moment from the atomic composition. Itinerant magnets with approximately half-filled bands exhibit a strong trend towards antiferromagnetism, because the hybridization energy of half-filled ↑ and ↓ bands is lower than that of completely-filled ↑ bands.

The magnetism of the iron-series transition-metal elements is caused by extended, delocalized, or *itinerant* electrons. The itinerant character is epitomized by the non-integer spin moments per atom, such as the 2.2 μ_B for Fe, 1.7 μ_B for Co, and 0.6 μ_B for Ni. Since each spin carries a moment of 1 μ_B, these noninteger values cannot be of ionic origin but reflect the interatomic hopping of the moment-carrying electrons. Each delocalized electron is owned by all atoms, so that the moment per atom is not necessarily integer.

Itinerant magnetism is not restricted to Fe, Co, and Ni but also occurs in many alloys, such as $Fe_{1-x}Ni_x$, PtCo, and the low-T_c intermetallic $ZrZn_2$. Notable exceptions include metallic rare-earth magnetism, where the rare-earth 4f electrons remain localized, and heavy-fermion compounds such as $CeAl_3$, where the electrons are barely delocalized. Rare-earth transition-metal intermetallics such as $SmCo_5$ and $Nd_2Fe_{14}B$ exhibit both itinerant (3d) and localized (4f) features. The magnetic properties of itinerant alloys may change drastically on chemical substitution. For example, the cubic Laves-phase compound YFe_2 is ferromagnetic, whereas YNi_2 is a Pauli paramagnet. On the other hand, $ZrZn_2$, MnBi, $CrBr_3$ are made from nonferromagnetic elements but are ferromagnetic. In fact, the record room-temperature polarization 2.43 T is found in $Fe_{65}Co_{35}$, significantly higher than the magnetizations of Fe and Co. Predicting a magnet's behavior from the atomic composition and structure is therefore a demanding task.

In iron-series transition metals, there are two types of delocalized electrons: 4s electrons and 3d electrons. Both 4s and 3d electrons contribute to transport properties, such as electrical and thermal conductivities, but the magnetic moment largely originates from the 3d electrons. However, 4s help to realize an RKKY-type exchange between the atomic moments. This is different from the rare-earth elements, where the metallic conductivity is due to delocalized 5d and 6s electrons but the magnetic moment originates from localized 4f electrons. As in other transition-metal magnets, the orbital moment of the itinerant 3d electrons is largely *quenched.* Typical orbital moments of itinerant 3d electrons are of order $0.1\,\mu_B$, so that the Landé g-factor is close to 2 and the moment is equal to the number of unpaired spins.

Panel 5 Exchange in metals and alloys

There is a rich variety of itinerant magnets, including systems that involve both localized and delocalized features.

Iron-series transition metals and their alloys. Fe, Co, and Ni are typical itinerant ferromagnets. Mn and Cr have complicated nonferromagnetic ground states, whereas the early transition-metal elements are Pauli paramagnets. Since alloying has a strong effect on the density of states, there is no simple general rule predicting the magnetic structure. $Fe_{65}Co_{35}$ is a ferromagnet with a record magnetization of 2.43 T, but the presence or absence of ferromagnetic elements is no reliable criterion. MnBi and $ZrZn_2$ are ferromagnets with Curie temperatures of 633 and 17 K, respectively, but YNi_2 is paramagnetic at all temperatures.

Rare-earth magnets. Simplifying somewhat, the magnetism of the rare-earth elements, such as Sm, Gd, and Dy, reflects RKKY-type interactions between the rare-earths' localized 4f shells. Various types of order are encountered, such as ferromagnetism, antiferromagnetism, ferrimagnetism, and noncollinear order. In rare-earth transition-metal intermetallics, such as $TbFe_2$, $Nd_2Fe_{14}B$, and $SmCo_5$, the rare-earth 4f shells remain well localized but couple to the itinerant 3d electrons of the transition-metal sublattice. Intermetallics containing light and heavy rare earths are usually ferromagnetic and ferrimagnetic, respectively. As a rule, the anisotropy of rare-earth transition-metal intermetallics originates from the rare-earth sublattice, whereas the magnetization is largely provided by the transition-metal sublattice.

Alloys containing heavy transition metals. The 4d (palladium series), 5d (platinum series), and the early 5f (actinide) elements are Pauli paramagnets at room temperature, but exhibit various types of itinerant magnetism at low temperature and in compounds. Examples are $L1_0$ magnets, such as CoPt, and US (uranium sulfide, $T_c = 177\,K$). The strong spin-orbit coupling of 4d, 5d, and 5f elements amounts to significant orbital-moment contributions and to high magnetic anisotropies. For example, the *US* zero-temperature moment of $1.55\,\mu_B$ involves moments of about $+3.5\,\mu_B$ (spin) and $-2\,\mu_B$ (orbital). By comparison, 3d orbital moments are of the order of $+0.1\,\mu_B$.

Metallic oxides and related materials. Oxides are often insulating or semiconducting, but some are metallic or half-metallic. In half-metallics, such as CrO_2 and NiMnSb, there are two different spin channels, a conducting channel and a (nearly) insulating channel (Section 7.2.8). Magnetic perovskites (manganites) such as $La_{1-x}Sr_xMnO_3$ exhibit FM-AFM transitions that are accompanied by significant conductivity changes.

Gases in Metals. Interstitial gas atoms especially H and N, drastically alter exchange-related and other magnetic properties of some elements and alloys (Section 7.2.6). The changes reflect both the increase in interatomic distances (lattice expansion) and modifications of the electronic structure. Examples are PtH_x, NiH_x, and $Sm_2Fe_17N_3$.

In Section 2.1 we have seen that ferromagnetism reflects the competition between one-electron energy-level splittings and Coulomb interactions. Let us first focus on the *energy levels.* Each atomic orbital contributes one level, and in metallic solids, the energy levels broaden into bands described by the density of states $D(E)$. Figure 2.21

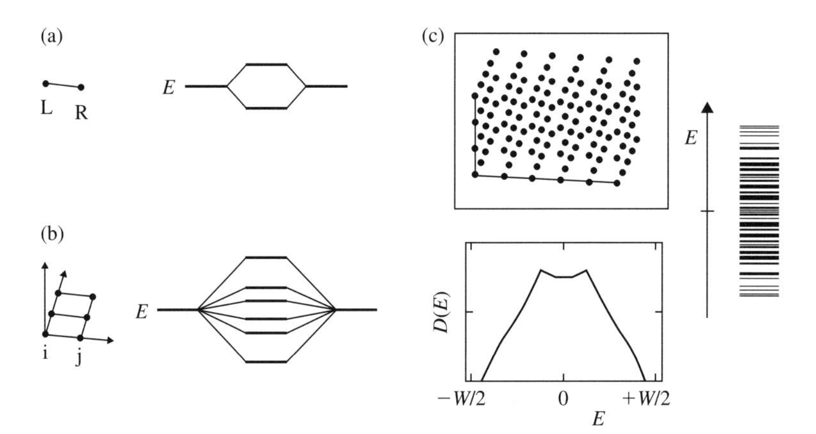

Fig. 2.21 Energy levels and density of states (DOS): (a) diatomic model, (b) small cluster and (c) transition to solid. The DOS maxima correspond to regions where the energy levels are particularly dense.

illustrates this how the levels develop on going from a diatomic pair to a solid. A measure of level splitting is the *bandwidth* W. Since the total number of states $\int D(E)\mathrm{d}E \sim D(E_\mathrm{F})W$ is fixed by the number of atoms, $W \sim 1/D(E)$. As in the diatomic model, the level splitting is proportional to the hopping integral, so that $W \sim |t|$. In practice, large interatomic distances mean small hopping and narrow bands.

The delocalized character of the 3d electrons makes it possible to approximate them as independent electrons. Equation (2.21) then predicts ferromagnetism for sufficiently weak hopping, that is, for narrow bands. A simple model is the Bloch model, which is based on free electrons (Section 2.4.1). The applicability of the model is limited, but it provides a qualitatively correct explanation of itinerant ferromagnetism and a criterion for the transition from paramagnetism to ferromagnetism. Section Section 2.4.2 deals with the electronic structure of metals, whereas Section 2.4.3 is devoted to the Stoner model of itinerant ferromagnetism. Finally, in Section 2.4.4 we discuss the origin of antiferromagnetism in itinerant magnets.

2.4.1 Free electrons, Pauli susceptibility, and the Bloch model

The simplest approach is to ignore the crystal potential $V(\mathbf{r})$ and to treat the solid as a "jellium". This is known as the free-electron model or, in the context of itinerant exchange, as the *Bloch model.* Consider free electrons in a volume $L \times L \times L$, for which the one-electron Schrödinger equation yields the well-known particle-in-a-box states

$$\psi(x, y, z) = \psi_\mathrm{o} \sin\left(\frac{\pi n_\mathrm{x} x}{L}\right) \sin\left(\frac{\pi n_\mathrm{y} y}{L}\right) \sin\left(\frac{\pi n_\mathrm{z} z}{L}\right) \tag{2.45a}$$

and the energy levels

$$E(n_\mathrm{x}, n_\mathrm{y}, n_\mathrm{k}) = \frac{\pi^2 \hbar^2}{2m_\mathrm{e} L^2}(n_\mathrm{x}^2 + n_\mathrm{y}^2 + n_\mathrm{z}^2) \tag{2.45b}$$

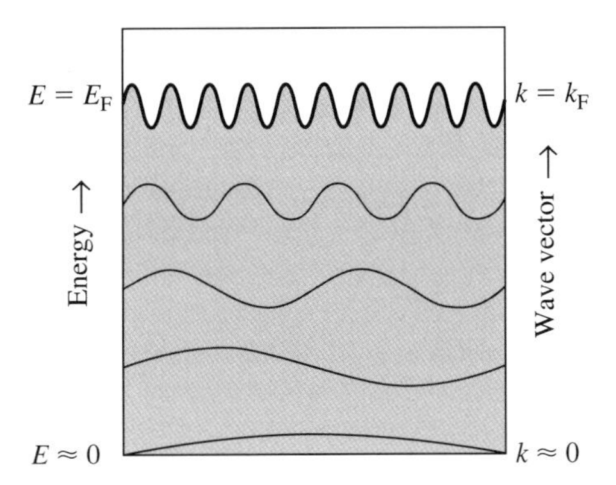

Fig. 2.22 Pictorial real-space interpretation of a free-electron Fermi liquid.

For finite sizes L, the energy levels are discrete. Such levels are encountered in semiconductor quantum dots and in metallic magnetic molecules. In the latter case, the uneven level spacing leads to magic numbers of structural stability. One example is Cr encapsulated in Si (Khanna, Rao and Jena 2002). Cubic particles, as assumed in (2.45), are characterized by the lowest magic numbers are 2 (all $n_i = 1$) and 8 (one $n_i = 2$). This is in close analogy to the filling of electron shells in atoms, although higher-order magic numbers generally differ from the atomic predictions. However, itinerant ferromagnetism is a macroscopic phenomenon, and for $L = \infty$, the numbers n_x, n_y, and n_z form a continuum and $E = \hbar k^2/2m_e$.

Since the electrons are noninteracting, they fill the available lowest lying orbitals with ↑ and ↓ electrons like water is poured into a jar. This "Fermi-liquid" behavior is illustrated in Fig. 2.22. The energy of the highest occupied levels is known as the *Fermi energy* $E_F = \hbar^2 k_F^2/2m_e$, where k_F is the corresponding Fermi wave vector. In this section, we ignore correlation effects, so that the Fermi-liquid picture carries over to arbitrary wave functions. Correlations, as included in the weak-hopping limit of Fig. (2.5), render the concept of individually occupied one-electron levels inadequate and mean that some electrons have wave vectors larger than k_F.

It is convenient to look at free electrons in wave-vector space. Each set of quantum numbers n_x, n_y, and n_z corresponds to a wave vector $\mathbf{k} = k_x\ \mathbf{e}_x + k_y \mathbf{e}_y + k_z \mathbf{e}_z$ whose components k_i may be positive or negative and are multiple integers of the cell size $\Delta k = 2\pi/L$. In the limit of infinite solids, $L = \infty$ yields $\Delta k = 0$, and the k-space is filled continuously until k_F is reached. Figure 2.23(a) illustrates that this filling mechanism results in a Fermi surface of radius k_F and energy E_F. The Fermi energy E_F increases with the electron density $n = N/V$, where $V = L^3$. Taking into account that each k-space cell is occupied by a pair of ↑↓ electrons, we obtain $N = 2\, 4\pi(k_F/\Delta k)3/3$, $k_F = (3\pi^2 n)^{1/3}$ and

$$E_F = \frac{\hbar^2}{2m}(3\pi^2 n)^{2/3} \tag{2.46}$$

For simple metals, which are reasonably well described by the free-electron theory, E_F is of the order of a few eV. The same result is obtained by integrating the density of states $D(E)$ over all energies $E \leq E_F$. Figure 2.23(b) shows $D(E) \sim \mathrm{d}n/\mathrm{d}E$, which is

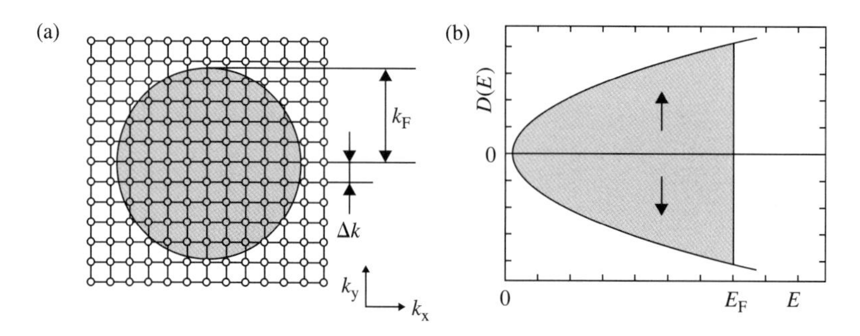

Fig. 2.23 Free electrons in k-space: (a) Fermi sphere, centered around $k_x = k_y = 0$, and (b) density of states. The density of k points, $1/\Delta k^3$, increases with the size of the metallic body, but the Fermi level depends on the electron density only.

obtained by inverting the function (2.46) and yields $n(E) \sim E^{3/2}$ and $D(E) \sim E^{1/2}$. Explicit expressions for the density of states of free electrons are $D(E) = 3N(E)/2E_F$ and $D(E) = 3nE^{1/2}/2E_F^{3/2}$.

Figure 2.23(b) explains why simple metals are nonferromagnetic (paramagnetic): the available low-lying orbitals are occupied by $\uparrow\downarrow$ electron pairs. However, what happens if we apply an external magnetic field? Are magnetic fields of about 1T sufficient to create ferromagnetism by transferring electrons from occupied $\downarrow$ to unoccupied $\uparrow$ states? Let us consider the case of full spin polarization, where all $\downarrow$ electrons move into $\uparrow$ orbitals. The transfer corresponds to a replacement of n in (2.46) by $2n$ and changes the average energy $\langle E \rangle$ from $3E_F/5$ to $0.952\,E_F$. Comparing this energy difference of about 1 eV with typical Zeeman energies of the order of 0.1 meV confirms our previous finding that magnetostatic fields are unable to create atomic-scale ferromagnetic order. In fact, itinerant ferromagnetism reflects *exchange fields* rather than magnetostatic fields.

In the Bloch model, the itinerant exchange is obtained by evaluating the Coulomb interaction $e^2/4\pi\varepsilon_o|\mathbf{r} - \mathbf{r}'|$ for a Slater determinant constructed from free-electron states $\exp(i\mathbf{k}\cdot\mathbf{r})$. The calculation involves the Fourier transformation $|\mathbf{r} - \mathbf{r}'| \to 1/k^2$ (Ashcroft and Mermin 1976), but here we will restrict ourselves to an approximate solution. Free electrons are described by the electron density $n = 6/\pi d_e^3 = \frac{3}{4}\pi r_s^3$, and the effective interelectronic distance $d_e = 2r_s$ is the only length involved. The kinetic energy (2.46) scales as $1/d_e^2$, as compared to the Coulomb energy proportional to $1/|\mathbf{r} - \mathbf{r}'| \sim 1/d_e$. In dense systems, d_e is small and the kinetic energy dominates. With increasing d_e, the Coulomb term becomes more important, and at some interelectronic distance d_o the paramagnetic state becomes unstable with respect to ferromagnetism. The Bloch model yields the correct trend, although the actual value $d_o = 5.8$ Å has little to do with the situation in transition metals. In fact, itinerant magnetism is caused by 3d electrons, as contrasted to the quasi-free 4s conduction electrons, and interelectronic distances of 5.8 Å go far beyond the scope of free-electron theory.

As described in the last part of Section 2.1.7, the independent-electron approximation corresponds to the introduction of an exchange field $H = H_J$. In lowest order,

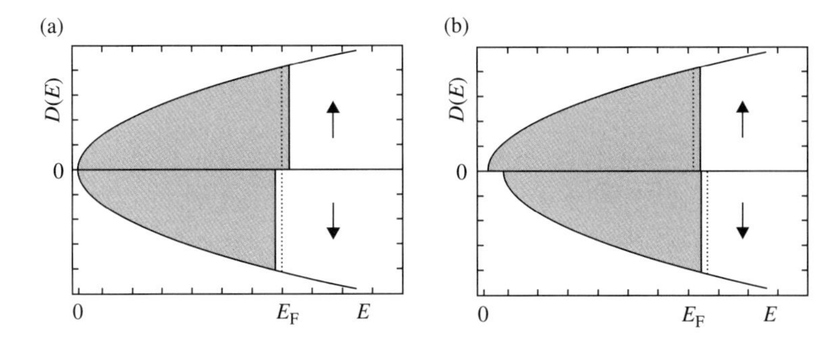

Fig. 2.24 Pauli susceptibility: (a) electron transfer from the (↓) to the (↑) band and (b) adjustment of Fermi level.

the corresponding magnetization $M = \mu_B(n_\uparrow - n_\downarrow)$ obeys $M = \chi_p H$, where χ_p is the *Pauli susceptibility*. This linear relation corresponds to the quadratic energy density

$$\frac{E}{V} = \frac{\mu_o \mu_B^2}{2\chi_p}(n\uparrow - n\downarrow)^2 - \mu_o \mu_B (n\uparrow - n\downarrow)H \tag{2.47a}$$

and is reproduced by $\partial E/\partial(n_\uparrow - n_\downarrow) = 0$. The first term in (2.47a) is the increase in kinetic energy due to the transfer of ↓ electrons into ↑ states. In terms of the densities of states (DOS) per spin, $D_\uparrow + D_\downarrow = D$,

$$\frac{E}{N} = \int_{-\infty}^{E\uparrow} ED\uparrow(E)\,\mathrm{d}E + \int_{-\infty}^{E\downarrow} ED\downarrow(E)\,\mathrm{d}E - \mu_o \mu_B (n\uparrow - n\downarrow)H \tag{2.47b}$$

Figure 2.24(a) shows the meaning of this energy. In the linear regime, $n_\uparrow - n_\downarrow$ is small and the $D_{\uparrow,\downarrow}(\mathrm{E}) = D_\mathrm{s}(\mathrm{E})$ can be replaced by $D_\mathrm{s}(E_\mathrm{F})$. Writing $E_{\uparrow,\downarrow} = E_\mathrm{F} \pm \delta E$ and exploiting that $n_\uparrow - n_\downarrow = 2D_\mathrm{s}(E_\mathrm{F})\delta\mathrm{E}$ we evaluate the two integrals and obtain, by comparison with (2.47a),

$$\chi_\mathrm{p} = 2\mu_o \mu_B^2 D_\mathrm{s}(E_\mathrm{F}) \tag{2.48}$$

The larger the density of states, the higher the susceptibility and the magnetization created by a given field. Since this equation derives from (2.47a), it applies not only to free electrons but also to metals with arbitrary densities of states $D_\mathrm{s}(E)$.

At this point, it is in order to compare the Pauli susceptibility, which describes the transfer of electrons from ↓ to ↑ states, with other susceptibilities. By definition, the magnetic susceptibility $\chi = \mathrm{d}M/\mathrm{d}H$ describes the magnetization change due to an applied or external magnetic field. Typically, one restricts the consideration to magnetization changes of atomic origin, thereby excluding micromagnetic susceptibilities associated with domains and hysteresis (Chapter 4). There are several types of paramagnetic ($\chi > 0$) and diamagnetic ($\chi < 0$) susceptibilities. The strongly temperature-dependent Curie-Langevin paramagnetism of paramagnetic gases, $\chi \sim 1/T$, reflects the competition between the Zeeman and thermal energies. In ferromagnets, this expression must be replaced by $\chi \sim 1/|T - T_\mathrm{c}|$ (Fig. 5.6).

Diamagnetism is a consequence of Lenz's law (A.4.2). The magnetic field causes the electrons to move and the induced magnetic field opposes the flux change by shielding the applied field. The diamagnetic susceptibility due to localized electrons $\chi = -ne^2\mu_o\langle r^2\rangle/6m_e$, where n is the electron density, $\mu_o = 1/\varepsilon_o c^2$, and $\langle r^2\rangle$ is the electrons' mean square distance from the nucleus. For free electrons, the diamagnetic and Pauli-paramagnetic contributions are $-n\mu_o\mu_B^2/2E_F$ and $+3n\mu_o\mu_B^2/2E_F$, respectively, where $\mu_B = e\hbar/2m_e$.

It is instructive to rewrite the susceptibilities in terms of dimensionless units. The basic atomic length is Bohr's hydrogen radius $a_o = 4\pi\varepsilon_o\hbar^2/m_e e^2$, or $a_o = 0.529$ Å, whereas the strength of electrostatic interactions is described by the dimensionless electromagnetic coupling constant $\alpha = e^2/4\pi\varepsilon_o \hbar c = 1/1/37$ (Sommerfeld's fine-structure constant). Typical velocities and energies of electrons in solids are $v = \alpha c$ and $E = \frac{1}{2} m\alpha^2 c^2$ (13.6 eV), and the magnitude of diamagnetic and Pauli-paramagnetic susceptibilities is of order α^2. This indicates that magnetism is a relativistic effect, scaling as a power of v/c. An elementary derivation of this relationship is actually provided by the circular-current model of Section 1.2, where the relativistic character of the interaction is hidden in the Zeeman term (Section 3.3.1). Aside from the divergence of χ at $T = 0$ (paramagnetic gases) and $T = T_c$ (ferromagnets), a similar hierarchy is encountered for Curie-Langevin paramagnets.

2.4.2 Band structure

Itinerant magnetism reflects the competition between kinetic energy and Coulomb repulsion. The kinetic energy is epitomized by the Pauli susceptibility, which contains the kinetic energy in form of the density of states. The larger $D_s(E)$ and χ_P, the easier it is for the Coulomb interaction spin-polarize the electrons. To quantify this effect, we must determine $D_s(E)$ for realistic one-electron wave functions. The major problem is the involvement of the periodic crystal potential $V(\mathbf{r})$. The lattice periodicity makes it possible to simplify the problem by Fourier transformation, so that the states and energies have the character of Bloch states and can be labeled by a wave vector $\mathbf{k}$. In a sense, the symmetric and antisymmetric diatomic wave functions of Fig. 2.3 can be considered as rudimentary Bloch wave functions having $k = 0$ and $k = \pi/a$, respectively.

In solids, there are two limits, namely quasi-free electrons in weak crystal potentials and tight-binding electrons in strong potentials. For nearly free electrons, the dispersion relation E_k is similar to the free-electron expression $\hbar k^2/2m_e$, but different bands are separated by gaps. In simple hypercubic lattices (chain, square lattice, simple cubic), the first gap occurs at $k = \pi/a$, where a is the lattice parameter. Figure 2.25 explains the origin of the gap for a one-dimensional model potential. At the gap, the two wave functions $\psi_-(x)$ and $\psi_+(x)$ have the same wave vector but different energies, because the maxima of the electron densities $n(x) = \psi_\pm^*(x)\psi_\pm(x)$ are located differently with respect to the lattice charges. In Fig. 2.25, the function $\psi_{(-x)}$ has the lower energy, because the maxima of $n(x)$ are closer to the positive atomic cores. The gap increases with the strength of the potential.

The free-electron approach works surprisingly well for many metals if the jellium includes not only the nuclei but also the core electrons. For example, Na has the electron configuration [Ne]3s, where [Ne] or $1s^2 2s^2 2p^6$ is a completely filled shell.

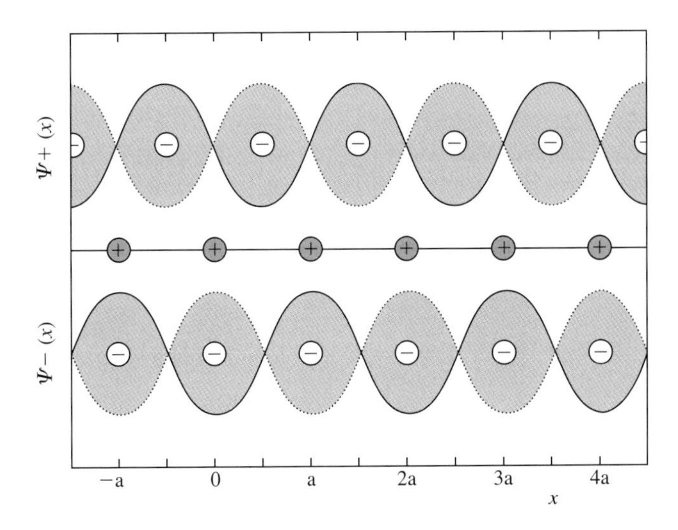

Fig. 2.25 Origin of the band gap for a chain of positive nuclei (+). The top and bottom parts of the figure shows electron wave functions that oscillate between the solid and dashed lines. Both wave functions have the same wave vector $k = \pi/a$ but are shifted by $a/2$, corresponding to $\sin(kx + \pi/2) = \cos(kx)$. Since the electron charge (−) is concentrated near the maxima of the wave functions, the attractive interaction with the nuclei and therefore the energy levels depend on the shift. In this figure, the state ψ_- has the lower energy.

This leaves us with Na^+ ions embedded in a gas of 3s electrons. The attractive core potential isn't necessarily small, and on approaching the ions, the 3s electrons gain kinetic energy. However, this enhanced kinetic energy, which amounts to rapid local oscillations of the wave function, is canceled by the reduced potential energy near the core. Projecting the rapid atomic oscillations onto core states $|\phi\rangle$ yields a *pseudopotential* which is often much weaker than the true potential (Sutton 1993) and explains why the free-electron approach works reasonably well.

The inner-shell 3d electrons of iron-series transition metals experience strong potentials and cannot be described as nearly free electrons. A better starting point is the tight-binding approximation, whose idea is outlined in Panel 6. To describe the 3d electrons, one considers atomic orbitals $\phi_\mu(\mathbf{r} - \mathbf{R}_\mathrm{i})$, where $\mathbf{R_i}$ is the position of the i-th atom and the band index μ labels the involved atomic orbitals, especially the five 3d orbitals. In this *tight-binding* model, the wave functions are superpositions of the atomic orbitals,

$$\psi_{\mathbf{k}\mu}(\mathbf{r}) = \sum_\mathrm{i} \exp(\mathrm{i}\mathbf{k} \cdot \mathbf{R}_\mathrm{i})\phi_\mu(\mathbf{r} - \mathbf{R}_\mathrm{i}) \tag{2.49}$$

Figure 2.26 compares the band structures for nearly free electrons and the tight-binding model, which is also known as the LCAO approximation (linear combination of atomic orbitals).

Inserted into the Schrödinger equation, wave functions of different μ are mixed by the crystal potential. The corresponding hopping integrals, which depend on the coordination angles of the atomic neighbors, have been analyzed by Slater and Koster

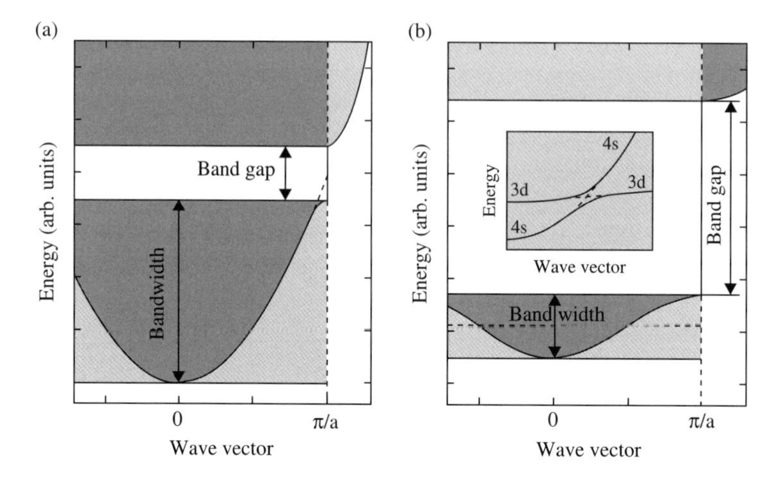

Fig. 2.26 Simple band-structure approximations: (a) nearly free electrons with broad bands and small gaps, and (b) tightly bound electrons with narrow bands and broad gaps (see Fig. 2.25 for the origin of the band gap at $k = \pi/a$). The schematic inset illustrates how free and tightly bound electrons hybridize. This hybridization contributes to the rather complicated densities of states of transition metals, Fig. 2.32.

(1954). Ignoring s and p electrons, there are three fundamental hopping integrals, $V_{\mathrm{dd}\sigma}$, $V_{\mathrm{dd}\pi}$, and $V_{\mathrm{dd}\delta}$, where σ, π, and δ refer to the respective continuous, twofold, and fourfold symmetries of the bond. As a consequence, the solution of the Schrödinger equation amounts to the diagonalization of an interaction matrix $V_{\mu\mu'}$ for each k-point.

The size of the matrix $V_{\mu\mu'}$ depends on the total number of nonequivalent atomic orbitals. In other words, the wave vector $\mathbf{k}$ takes care of the infinite size of the solid, but the local environment must be treated separately. This is of importance in magnetic alloys, especially in permanent magnets (Skomski and Coey 1999). For example, $Y_2Fe_{14}B$ contains 68 atoms per unit cells and altogether 280 3d orbitals. The quantum-mechanical treatment of these orbitals amounts to diagonalizing a 280×280 matrix for each k-point.

We see that band-structure calculations are quite involved, even if we restrict ourselves to tight-binding wave functions (2.49). However, for some purposes it is sufficient to replace the correct densities of states (DOS) by model functions. Well-known examples are Gaussian, rectangular (Friedel), and semicircular (Hubbard) DOS. Hybridization does not affect the band's center of gravity, so that the essential parameter of these functions is the bandwidth. Let us consider the n-th moments of the DOS, defined as

$$\mu_{(\mathrm{n})} = \int_{-\infty}^{\infty} E^{\mathrm{n}} D(E)\, \mathrm{d}E \tag{2.50}$$

The zeroth moment, $\int D(E)\mathrm{d}E$, is equal to the total number of states and therefore fixed. The first moment is equal to the average energy, $\int E\, D(E)\mathrm{d}E = \langle E \rangle$. Without loss of generality we can adjust the energy zero so that $\langle E \rangle = 0$. The second moment, $\mu_{(2)} = \int E^2 D(E)\mathrm{d}E$, determines the bandwidth W. For example, rectangular bands are characterized by $W^2 = 12\mu_{(2)}$. Figure 2.27(a) shows some model DOS.

Panel 6 Tight-binding and LCAO models

The tight-binding model assumes that the one-electron states are superpositions of atomic wave functions. In chemistry, the approach is commonly known as linear combination of atomic orbitals (LCAO). The wave functions are constructed directly from atomic orbitals (extreme tight binding) or from better adapted functions, such as Wannier functions. The model reproduces essential features of the electronic structures of molecules, clusters, surfaces, nanostructures, and solids, but fails to provide reliable results when the overlap between the atomic orbitals is large.

An example is the single-band tight-binding model, where the wave functions are superpositions of atomic wave functions $\phi|(\mathbf{r} - \mathbf{R}_{\rm i})| = |i\rangle$. The model is usually associated with s electrons, but it can also be used for p and d electrons in cases where subband hybridization is negligible. Written as a matrix, the model Hamiltonian $\langle i|\mathbf{H}|j\rangle = t_{\rm ij}$, where $t_{\rm ij}$ is the hopping integral between orbitals centered at $\mathbf{R}_{\rm i}$ and $\mathbf{R}_{\rm j}$. The time-independent Schrödinger equation assumes the form

$$E|i\rangle = \sum_{\rm j} t_{\rm ij}|j\rangle$$

The energies E_μ and eigenfunctions $|\mu\rangle = \sum_{\rm i} c_{\rm i}(\mu)|i\rangle$ are obtained by diagonalizing the matrix $t_{\rm ij}$. For *periodic* structures, the solution simplifies to $|\mathbf{k}\rangle = \sum_{\rm i} \exp({\rm i}\mathbf{k}\cdot\mathbf{R}_{\rm i})|i\rangle$. An example is the infinite atomic chain, where $\mathrm{E}(k) = 2t\cos(ka)$.

In general, structures with $N_{\rm o}$ nonequivalent sites require the diagonalization of an $N_{\rm o} \times N_{\rm o}$ matrix. For example, at a surface, one must distinguish between surface atoms, subsurface atoms, and different sites deeper in the bulk. A similar involvement of nonequivalent sites occurs in other problems involving matrix diagonalization, such as the determination of the mean-field Curie temperature (Section 5.3) and the spin-wave-spectrum (Section 6.1) from the exchange matrix $J_{\rm ij}$.

To determine the second moment, we consider two types of wave functions: the unknown eigenfunctions $|q\rangle = |k_\mu\rangle$ and the known local orbitals $|p\rangle = |i_\mu\rangle$. Each state $|p\rangle$ contributes a delta-peak or "needle" to the DOS, as indicated in Fig. 2.21, so that $D(E) = \sum_{\rm q}\delta(E - E_{\rm q})$ and $\mu_{\rm (n)} = \sum_{\rm q} E_{\rm q}^{\rm n}$. Using $\mathsf{H}|q\rangle = E_{\rm q}|q\rangle$ we convince ourselves that $\mathsf{H}^{\rm n}|q\rangle = E_{\rm q}^{\rm n}|q\rangle$ and obtain $\mu_{\rm (n)} = \sum_{\rm q}\langle q|\mathsf{H}^{\rm n}|q\rangle$. Inserting $\sum_{\rm p}|p\rangle\,\langle p| = 1$ and $\sum_{\rm p'}|p'\rangle\,\langle p'| = 1$, makes it possible to perform the summation over q and yields $\mu_{\rm (n)} = \sum_{\rm p}$

$\langle p|H^n|p\rangle$. Alternatively, $\sum_q \langle q|H^n|q\rangle$ has the character of a trace and does not change on using a different orthogonal set of wave functions. Next, we write the second moment as $\sum_{pp'} \langle p|H|p'\rangle \langle p'|H|p\rangle = t_{pp'} t_{p'p}$, where the $t_{pp'} = \langle i\mu|H|j\mu'\rangle$ are the hopping integrals. The summation $\sum_{pp'} = \sum_{ij\mu\mu'}$ includes interatomic hopping ($i \neq j$) between any orbitals ($\mu = \mu'$ and $\mu \neq \mu'$). In fair approximation, we can restrict ourselves to nearest-neighbor hopping between the i-th atom and its z neighbors at $\mathbf{R}_j$, and the summation over μ and μ' reduces to a trace over the square of the matrix $t(i,j)_{\mu\mu'}$. The procedure yields, for rectangular bands, the *exact* tight-binding result

$$\mu_{(2)} = z(V_{dd\sigma}^2 + V_{ds\sigma}^2 + 2V_{dd\pi}^2 + 2V_{dd\delta}^2) \tag{2.51}$$

Relative values of the fundamental hopping integrals are $V_{dd\sigma}:V_{dd\pi}:V_{dd\delta} = -6:+4:-1$, so that there is only one independent d-band parameter. The parameter $V_{ds\sigma}$ describes the s-d hybridization in the tight-binding approximation. Figure 2.27(b) illustrates that this hybridization leads to a broadening of the 3d band by about 10%. Note that (2.51) is a realization of the moments theorem (Cyrot-Lackmann 1968, Heine 1980), which relates the $\mu_{(n)}$ to real-space features such as the number z of nearest neighbors.

The bandwidths of the late iron-series transition metals are about 5 eV. Naturally, both the exact bandwidth and the DOS depend on details of the band structure, in spite of the exact result (2.51). A particularly simple model is the Friedel model, where the DOS is approximated by a rectangle of width W. In this case, the density of states per spin and atom $D_s = 5/W$, corresponding to five 3d orbitals per atom.

From a practical point of view, there is no point in refining the tight-binding approach for transition metals, because advanced first-principle calculations (density-functional calculations) are now able to overcome the inherent simplifications of the

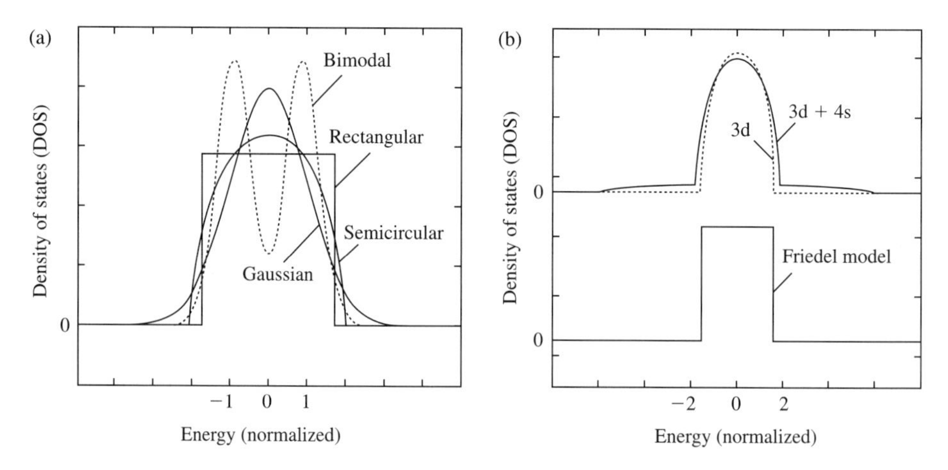

Fig. 2.27 Second-moment description of 3d bands: (a) densities of states and (d) effect of s-d hybridization. The DOS in (a) are all characterized by the same moments $\mu_{(1)}$, $\mu_{(2)}$, and $\mu_{(3)}$. The DOS of bcc iron is bimodal, similar to the dashed line in (a).

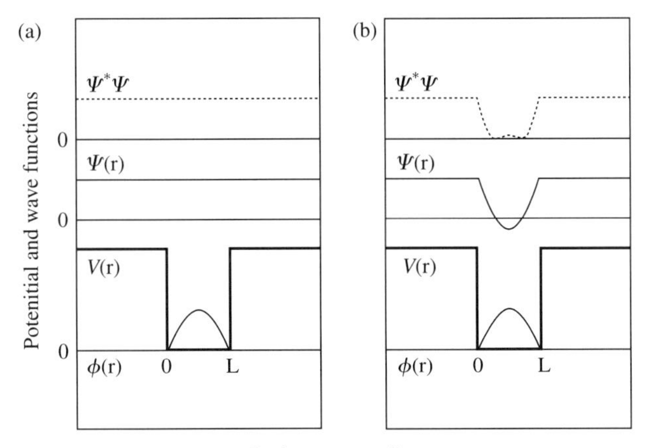

Fig. 2.28 Pseudopotentials: (a) original wave functions and (b) effect of pseudopotential for a one-dimensional model atom of size L. In (a), the delocalized wave function ψ covers both in the interstitial area (or "free space") and the space where the atom is located. This is unphysical, because the delocalized wave function must be orthogonal to the atomic wave functions. This leads to an orthogonality hole $\psi^*\psi$, as shown in (b).

model. This includes not only the use of improved one-electron wave functions but also the above introduced pseudopotential (Fig. 2.28) and the self-consistent calculation of the one-electron potential. However, tight-binding calculations have remained useful for very large numbers of nonequivalent crystallographic sites (Section 7.1).

2.4.3 Stoner model and beyond

The knowledge of the density of states, $D_s(E) = \frac{1}{2}D(E)$, enables us to predict the onset of ferromagnetism. The starting point is the analysis leading to the Pauli susceptibility. In the one-electron approximation, the Coulomb energy translates into an effective exchange field (Section 2.1.7) and the corresponding energy $U(n_\uparrow - n_\downarrow)^2/4$ must be added to (2.47a):

$$\frac{E}{V} = \frac{\mu_o \mu_B^2}{2\chi_P}(n\uparrow - n\downarrow)^2 - \frac{U}{4}(n\uparrow - n\downarrow)^2 - \mu_o \mu_B (n\uparrow - n\downarrow)H \tag{2.52}$$

No fancy quantum mechanics is necessary at this stage, because mean-field products such as $\mathbf{n}_\uparrow\langle n_\downarrow\rangle$ yield trivial averages $\langle n_\uparrow\rangle\langle n_\downarrow\rangle = n_\uparrow n_\downarrow$. Using (2.48), we obtain

$$\frac{E}{V} = \frac{\mu_o \mu_B^2}{2\chi}(n\uparrow - n\downarrow)^2 - \mu_o \mu_B (n\uparrow - n\downarrow)H \tag{2.53}$$

where

$$\chi = \frac{\chi_P}{1 - U D_s(E_F)} \tag{2.54}$$

is the *exchange-enhanced* Pauli susceptibility.

Equation (2.54) is of key importance for the understanding of itinerant ferromagnetism. With increasing $D_s(E_F)$, the susceptibility increases, eventually diverges, and for

$$UD_s(E_F) > 1 \tag{2.55}$$

the paramagnetic state is unstable. This is the famous *Stoner criterion* for itinerant ferromagnetism (Stoner 1938). The intra-atomic Coulomb parameter U is often treated as a phenomenological constant, known as the Stoner parameter, and then frequently denoted by I. This is meaningful, because the conduction electrons screen the $1/|\mathbf{r}-\mathbf{r}'|$ interaction and reduce the effective Coulomb interaction to $I \approx 1\,\text{eV}$.

Figure 2.29 compares the Stoner criterion with the two-electron model of Section 2.1. In both cases, the hopping energy ($\Delta E \sim |t| \sim W \sim 1/D_s$) competes against U and (2.21) predicts ferromagnetism for $U/4 + J_D > |t|$, in close analogy to the Stoner criterion. For iron-series transition-metals, where $W \approx 5\,\text{eV}$, the rectangular-band model predicts $D_s = 5/W$ and $UD_s(E_F) \approx 1$. This indicates that iron-series transition-metal elements may or may not satisfy the Stoner criterion. The late iron-series transition-metal elements Fe, Co, and Ni, are ferromagnetic, because their relatively large effective nuclear charges Z_{eff} reduce the interatomic hopping. Aside from the three iron-series elements, itinerant ferromagnetism is observed in many alloys. Examples are YCo_5, MnBi FeCo, and $La_2Fe_{14}B$. Very weak itinerant ferromagnets, such as $ZrZn_2$ and Sc_3In, barely satisfy the Stoner criterion and are ferromagnetic at low temperatures.

Some elements, such as Sc, Y, Pd, and Pt, do not satisfy the Stoner criterion but are close to ferromagnetism. According to (2.54), they are exchange-enhanced Pauli paramagnets, and the relatively large susceptibility means that they are easily spin polarized by neighboring magnetic atoms. This leads to "giant magnetic moments" per polarizing atom, for example in the $Pd_{1-x}Fe_x$ system (Fischer and Hertz 1991). A related phenomenon is giant magnetic anisotropy (Section 3.4.4), which has its origin in the anisotropy contribution of the spin-polarized atoms. In some materials, the trend

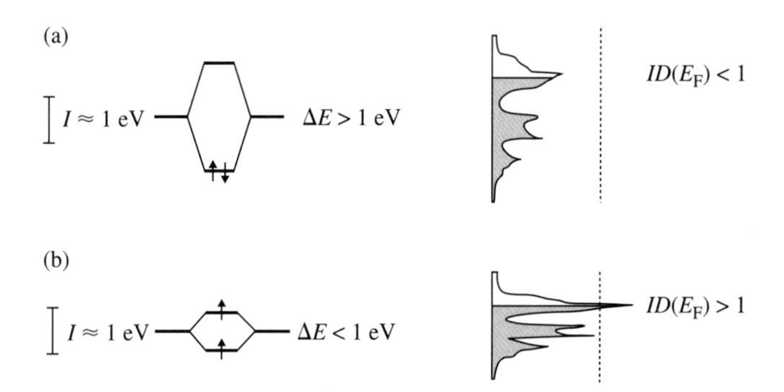

Fig. 2.29 Stoner criterion: (a) paramagnetism and (b) ferromagnetism. Ferromagnetism occurs if the Coulomb interaction, as parameterized by I (or U), is larger than the one-electron level splitting $\Delta E \sim 1/D$. In the Stoner model, this means $D(E_F) > 1/I$. The dotted line describes the onset of ferromagnetism, $D(E) = 1/I$.

(a)

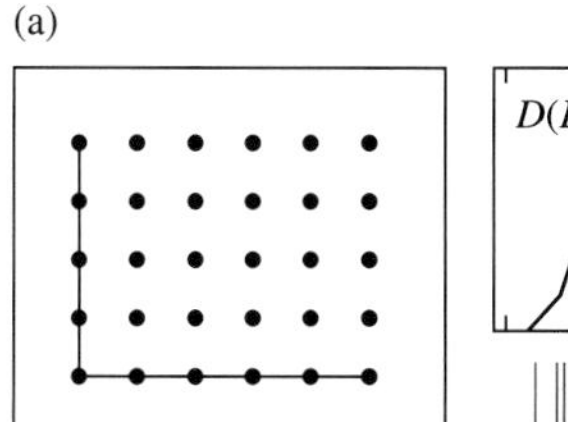

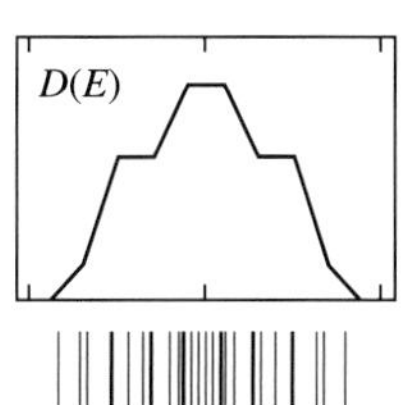

(b)

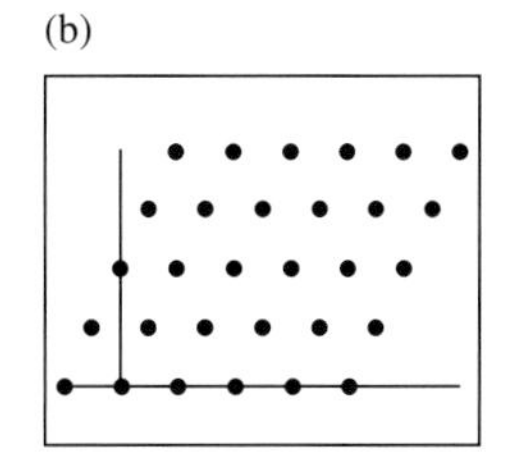

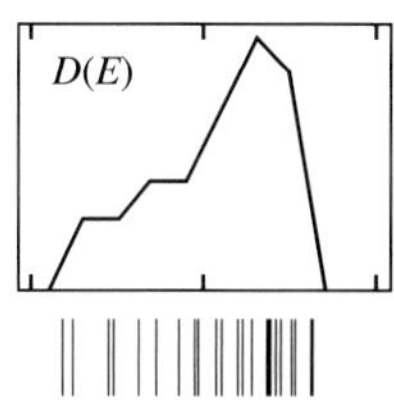

Fig. 2.30 Density of states in (a) less dense packed and (b) dense packed structures. The two curves are s-band tight-binding predictions, but a similar skewing occurs in more complicated systems. In fcc and hcp structures, the main antibonding peak is close to the upper end of the d band, whereas the bcc peak is more central.

ferromagnetism is strengthened or created by band narrowing due to mechanical strain or substitutional nonmagnetic atoms.

In general, the bandwidth is not the only consideration. The Stoner criterion involves the density of states at the Fermi level rather than the average DOS, and it is advantageous to have the Fermi level close to a peak. The position of the peaks depends on the crystal structure. Figure 2.30 compares the density of states in square and triangular model lattices. For the triangular lattice, the DOS is skewed towards the upper band edge, and a similar effect is encountered in three-dimensional magnets. The reason is the presence of hopping loops having the form of equilateral triangles, which can be shown to skew the densities of states by yielding a nonzero third moment $\mu_{(3)}$. Triangular hopping loops are characteristic of dense-packed systems, such as face-centered cubic (fcc) and hexagonal metals, and also of many rare-earth transition-metal intermetallics (see Appendix A.5). Since the band filling increases from Fe to Co and Ni, dense-packed iron magnets suffer most from this shift. Examples are metastable fcc Fe, which is nonferromagnetic unless mechanically strained, and the otherwise excellent permanent-magnet alloy $Nd_2Fe_{14}B$, which has a Curie temperature of only 312° C.

When the Stoner criterion is satisfied, the paramagnetic state becomes unstable, and the system jumps into a ferromagnetic state with nonzero moment. The magnitude of the moment depends on the details of the DOS. A simple model is the *rigid band* model. Strong ferromagnets, such as Co and Ni, are close to complete band filling, and since all ↑ levels are occupied the moment to equal the number of holes in the 3d band. Figure 2.31 illustrates the meaning of this prediction. In practice, the d-band filling is tuned by alloying or by adding interstitial hydrogen, which is protonic in most metals (H^+) and adds about one electron per hydrogen atom to the 3d band. In weak ferromagnets, such as iron, the situation is more complicated, because some of the ↑ orbitals remain unoccupied.

Figure 2.32 shows the density of states for bcc iron. In the paramagnetic state (a), the Fermi level is close to the main peak of the DOS, and the Stoner criterion is satisfied. The character of the ferromagnetic moment is illustrated by the "spin-transfer" picture (b), which is similar to Fig. 2.24(a). It is customary to include the exchange field in the Hamiltonian, rather than treating it as an external field, so that ↑ and ↓ have a common Fermi level. In the rigid-band model, the Fermi level adjustment

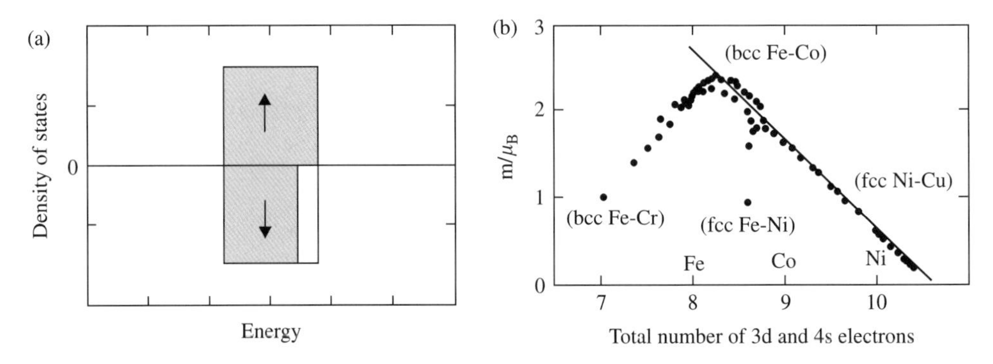

Fig. 2.31 Itinerant moment: (a) nearly filled 3d band and (b) Slater-Pauling curve. The solid line corresponds to strong ferromagnetism.

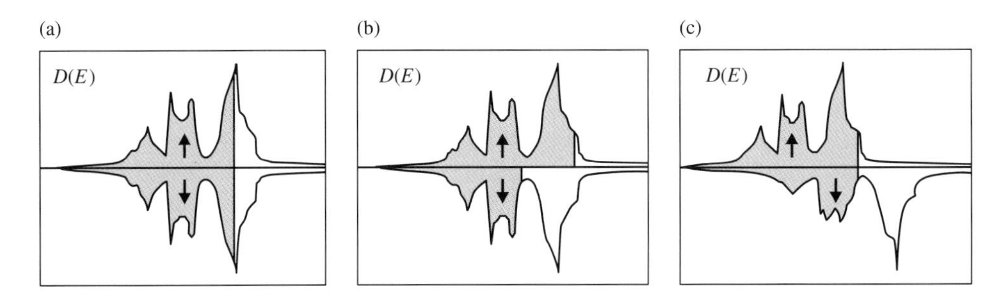

Fig. 2.32 Density of states in bcc iron: (a) paramagnetic density of states, (b) spin transfer, and (c) adjustment of Fermi level.

is realized by a relative shift of the $\uparrow$ and $\downarrow$ DOS, as in Fig. 2.24(b). The corresponding moment m is closely related to the exchange splitting of band: $E_F(\uparrow) - E_F(\downarrow) = m\,I$. In reality, the $\uparrow$ and $\downarrow$ bands undergo some distortion (c), and the shift of the 4s band is smaller than that of the 3d band. It also is necessary to keep track of the 4s electrons that go into the 3d band. Table 2.3 summarizes the electronic structure of Fe, Ni, and Co.

2.4.4 *Itinerant antiferromagnets

Some itinerant magnets are *antiferromagnetic*, especially in the case of nearly half-filled 3d bands. Examples are Mn and Cr below 95 and 312 K, respectively. Itinerant antiferromagnetism is easily explained by writing the Stoner exchange in form of a spin-dependent exchange $-V(\mathbf{r})$ for $\uparrow$ and $V(\mathbf{r})$ for $\downarrow$ electrons. This term corresponds to $Un_\uparrow\langle n_\downarrow\rangle + U\langle n_\uparrow\rangle n_\downarrow$ in Section 2.1.7 and means that intra-atomic exchange favors the formation of a magnetic moment. In ferromagnets, $V = V_o$ is positive, but in antiferromagnets $V = \pm V_o$, depending on the sublattice. Figure 2.33 compares the potentials V for ferromagnets and antiferromagnets. In antiferromagnets, each second spin has the "wrong" sign, which affects the energy gain on interatomic hopping. To determine whether ferromagnetic or antiferromagnetic order is more favorable, one

Table 2.3 Some intrinsic properties of elemental 3d ferromagnets (bcc Fe, fcc Ni, and hcp Co).

	Unit	Fe	Co	Ni
m	μ_B	2.217	1.753	0.616
M_s(RT)	T	2.16	1.76	0.61
T_c	K	1044	1360	627
$D_s(E_F)$	1/eV	1.54	1.72	2.02
I	eV	0.93	0.99	1.01
$n_{3d} + n_{4s}$	–	8	9	10
$n_{3d}(\uparrow)$	–	4.8	5.0	5.0
$n_{3d}(\downarrow)$	–	2.6	3.3	4.4
$n_{4s}(\uparrow)$	–	0.3	0.35	0.3
$n_{4s}(\downarrow)$	–	0.3	0.35	0.3
3d↑ holes	–	0.2	0.0	0.0
3d↓ holes	–	2.4	1.7	0.6

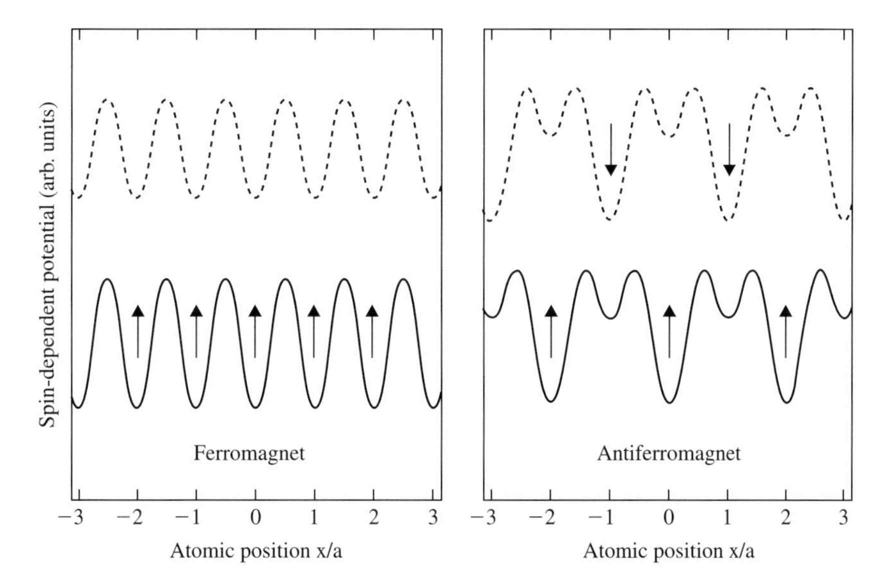

Fig. 2.33 Spin-dependent potentials. The Coulomb interactions favor parallel intra-atomic spin alignment, so the hopping onto a site of opposite spin costs energy.

puts the electrons in the spin-dependent potential and calculates the total energy for both cases.

Let us assume that U is sufficiently strong to ensure a magnetic moment m_o, so that we can ignore Pauli paramagnetism and focus on whether the interatomic coupling is ferro- or antiferromagnetic. For well-separated atoms, where $|t| = 0$, the FM and AFM configurations have the same energy, $-Um_o$, corresponding to zero interatomic exchange. However, for finite $|t|$, the energy difference depends on the band filling. In ferromagnets, the hopping contribution is large, and adding electrons to an empty band, yields a large energy gain. With increasing $\uparrow$ band filling, the total energy reaches a minimum and then reaches its original value. This is because the center of gravity of a band does not change on hybridization. For example, in the two-level system in Fig. 2.29(b), one electron occupies a low-lying state, but the addition of a second electron, corresponding to a half-filled band (or filled $\uparrow$ band) reduces the energy gain to zero. As a consequence, nearly empty bands exhibit a strong trend towards ferromagnetism, and the same is true for nearly filled bands. This explains the frequent occurrence of ferromagnetism in the late transitions metals Fe, Co, and Ni (exercise on ferromagnetism across the 3d series).

Both ferro- and antiferromagnets gain energy on hybridization, but the AFM gain is small, because it involves the energetically unfavorable hopping onto sublattices with the "wrong" sign of the potential $\pm V_o$ (right-hand side of Fig. 2.33). This is often accompanied by a small conductivity in the AFM state, as exemplified by CMR manganites (Section 7.2.7). The large level splitting $\pm V_o$ in antiferromagnets is therefore only weakly affected by interatomic hopping (Section A2.3). This mechanism favors ferromagnetism, except for half-filled bands, which exhibit a trend toward antiferromagnetism. Half-filled FM bands do not benefit from hybridization, because the $\uparrow$ band is complete filled and its energy is equal to the center-of-gravity energy, as in Fig. 2.29(b), whereas the AFM energy gain remains nonzero. Figure 2.34 shows a phase diagram for bands with rectangular DOS, Note that the actual calculation of the moment m involves the self-consistent determination of the potential $V_o \sim Im$.

From Fig. 2.34 we see that antiferromagnetism (AFM) competes not only against ferromagnetism (FM) but also against paramagnetism (PM). Transitions of the type PM-AFM-FM, as well as more complicated itinerant spin structures, such as non-collinear configurations and incommensurate spin-density waves, are indeed observed

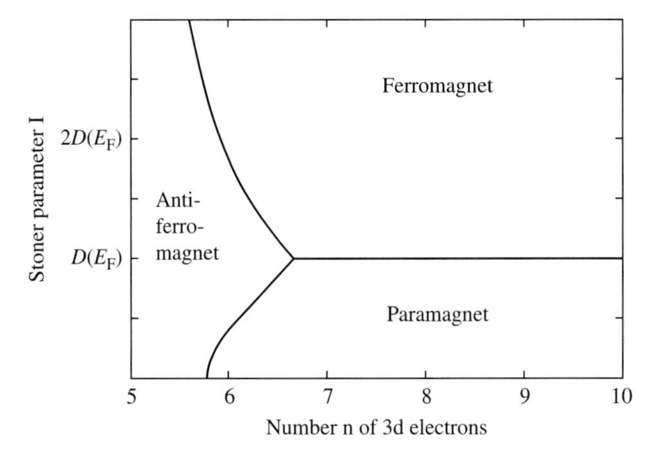

Fig. 2.34 Itinerant antiferromagnetism in an independent-electron approximation. Correlations distort the phase diagram and make ferromagnetism less favorable.

in many of materials. Figure 2.34 is easily generalized to arbitrary spin structures, but the calculation is numerically demanding (Liechtenstein *et al.* 1987, Gubanov, Liechtenstein, and Postnikov 1992) and yields a complicated physical picture, especially in alloys (McHenry *et al.* 2001, Kashyap *et al.* 2004). In addition, correlation effects (Section 2.1.6) mean that the calculations outlined above overestimate the trend towards ferromagnetism.

Exercises

1. ***Symmetry of heteronuclear wave functions.*** Are heteronuclear real-space wave functions symmetric or antisymmetric?
2. ***Angular momentum operators.*** Use p_x and x to show that $\mathsf{L}_z = -i\hbar\partial/\partial\phi$.
3. ***Exchange and kinetic energy.*** Refute or confirm the argument that magnetic modeling in terms of interatomic exchange constants J is approximate, because J refers to fixed lattice sites and ignores the motion of the electrons.
4. ***Energy levels for three simple atoms.*** Calculate the one-electron energy levels for three hydrogen atoms forming a triangle with equal sides (H_3^{2+}). Compare the result with the diatomic model.
5. ***Moments of oxides.*** Estimate the magnetic moment per transition-metal atom for Fe_2O_3, CoO, NiO, Fe_3O_4, CrO_2, and $Y_3Fe_5O_{12}$.
   Hint: The 3d orbital moment is quenched, and the Y^{3+} ion nonmagnetic. In many Fe oxides, there exist both 2+ and 3+ sites with different moments.
6. ***High-spin low-spin transitions.*** In contrast to Fe^{2+} and Co^{2+}, there is no high-spin low-spin transition for Ni^{2+} ions in an octahedral crystal field. Why?
   Answer: All t_{2g} levels are occupied.
7. ***Angular momentum and Pauli matrices.*** Angular momentum operators $\mathbf{J}$ (both orbital momentum $\mathbf{L}$ and spin momentum $\mathbf{S}$) obey the commutation rules $\mathsf{J}_x\mathsf{J}_y - \mathsf{J}_y\mathsf{J}_x = i\hbar\mathsf{J}_z$, $\mathsf{J}_y\mathsf{J}_z - \mathsf{J}_z\mathsf{J}_y = i\hbar\mathsf{J}_x$, and $\mathsf{J}_z\mathsf{J}_x - \mathsf{J}_x\mathsf{J}_z = i\hbar\mathsf{J}_y$. Show that the Pauli matrices $\boldsymbol{\sigma}$ correspond to $\mathbf{S} = \hbar\boldsymbol{\sigma}/2$.
   Hint: By symmetry, it is sufficient to consider the x and y components of $\mathbf{S}$. Note that $\hbar$ is often omitted in practice, which amounts to measuring angular momenta in dimensionless units and magnetic moments in units of μ_B. In some notations, there is an additional factor -1 relating moment and momentum, due to the choice $\pm e$ for the negative electron charge.
8. ***Fermi level for noncubic macroscopic shapes.*** To derive k_F and E_F, we have assumed that the solid has the form of a cube of volume L^3. What would be different for a solid of spherical shape?
9. ***Bandwidth for semicircular DOS.*** A band is characterized by the second moment $z|t|$ and approximated by a semicircular or "Hubbard" density of states. Calculate the bandwidth.
10. ***Ferromagnetism of yttrium intermetallics.*** YFe_2 is ferromagnetic, but the equistructural alloy YNi_2 is paramagnetic. Why?
11. ***Ferromagnetism across the 3d series.*** Band-filling arguments predict antiferromagnetism for elements in the middle of the 3d series (Mn, Cr) but ferromagnetism for nearly empty and nearly full 3d bands. However, ferromagnetism is encountered in Fe, Co, and Ni but not in the early transition-metal elements, such as V and Ti. Why?

12. ***Real-space wave functions vs. spin functions.*** In the introductory part of this chapter, we have derived the exchange between spins in terms of real-space wave functions, as compared to spin functions. Why is this justified?
13. ***The zoo of holes around electrons in solids.*** Explain the differences between orthogonality, exchange, and correlation holes.
14. ****Exchange hole.*** Consider a $\uparrow\uparrow$ pair of free electrons and calculate the probability of finding them at $\mathbf{r}$ and $\mathbf{r}'$.
    Hint: Free electrons are described by wave functions $\exp(\mathrm{i}\mathbf{k}\cdot\mathbf{r})$. Assume that the two electrons have wave vectors $\mathbf{k}$ and $\mathbf{k}'$, construct an antisymmetric two-electron state from these two states, and evaluate the pair distribution function $g(\mathbf{r}-\mathbf{r}') \sim \Psi^*(\mathbf{r}, \mathbf{r}')\Psi(\mathbf{r}, \mathbf{r}')$.
    Answer: The pair distribution function $g(\mathbf{r}-\mathbf{r}') \sim 1 - \cos[(\mathbf{r}-\mathbf{r}')\cdot(\mathbf{k}-\mathbf{k}')]$ is zero for $\mathbf{r}=\mathbf{r}'$, as expected from the Pauli principle, but nonzero elsewhere. Integration over all k-states and taking into account that $\uparrow\downarrow$ pairs do not experience an exchange hole yields a function $g \approx 0$ for small $k_\mathrm{F}|\mathbf{r}-\mathbf{r}'|$ and $g \approx 1/2$ for large $k_\mathrm{F}|\mathbf{r}-\mathbf{r}'|$.
15. ***Distance dependence of hopping integrals.*** Show that hopping integrals exhibit an approximately exponential decay with increasing interatomic distance. What is the consequence for ferromagnetism?
16. ***Unitary transformations.*** Finding the energy levels of a Hamiltonian corresponds to a diagonalizing unitary transformation (rotation in Hilbert space). The columns of the corresponding unitary matrix U are the normalized eigenfunctions of the matrix. Sometimes, the eigenfunctions are only approximately known, and $\mathsf{U}^+\mathsf{H}\mathsf{U}$ doesn't yield a complete diagonalization. Use the following sets of eigenfunctions to rotate the 4×4 matrix (2.13) and discuss the corresponding simplifications of the matrix: $|\mathrm{s}\rangle$, $|\mathrm{s\,a}\rangle + |\mathrm{a\,s}\rangle$, $|\mathrm{s\,a}\rangle - |\mathrm{a\,s}\rangle$, $|\mathrm{a\,a}\rangle$ and $|LL\rangle + |RR\rangle$, $|LL\rangle - |RR\rangle$, $|LR\rangle + |LR\rangle$, $|LR\rangle - |RL\rangle$.
17. ***Pseudopotential.*** Calculate the wave function and electron density for a long-wavelength electron coupled to an s-type core state.
    Hint: Write $|\psi\rangle = |\mathbf{k}\rangle - |\phi\rangle\langle\phi|\mathbf{k}\rangle$, where $|\psi\rangle$ is the true wave function, $|\mathbf{k}\rangle \sim \exp(\mathrm{i}\mathbf{k}\cdot\mathbf{r}) \sim 1$, and $|\phi\rangle$ is the core wave function.
    Answer: In fair approximation, $\exp(\mathrm{i}\mathbf{k}\cdot\mathbf{r}) = 1$, so that $\psi(\mathrm{r}) \sim 1 - \phi(\mathrm{r}) \int \phi(\mathbf{r}')\mathrm{d}\mathbf{r}'$. Since $\psi(\mathbf{r})$ has a zero, the average electron density is reduced near the core. This "orthogonality hole" is a manifestation of the pseudopotential.
18. ***Hubbard exchange.*** Use the large-U limit of $J_\mathrm{eff} = J + U/4 - (U^2/16 + T^2)^{1/2}$ to discuss the exchange between two atoms. Compare the approximate solution with exact result for $J = 0.1\,\mathrm{eV}$, $U = 10\,\mathrm{eV}$, and $T = 1\,\mathrm{eV}$.
    Hint: Exploit the fact that $(1+x)^{1/2} \approx 1 + x/2$.
19. ***Eigenfunctions of spin.*** Determine the eigenfunctions of σ_x, σ_y, σ_z, and $\mathbf{e}\cdot\boldsymbol{\sigma}$, where $\mathbf{e} = \sin\theta\cos\phi\,\mathbf{e}_\mathrm{x} + \sin\theta\sin\phi\,\mathbf{e}_\mathrm{y} + \cos\theta\,\mathbf{e}_\mathrm{z}$.
    Hint: Start from $|\uparrow\rangle = (1, 0)$ and use the wave functions $|\downarrow\rangle = (0, 1)$, $\mathsf{U}(\phi, \theta)\,(1, 0)^\mathrm{T}$ and $\mathsf{U}(\phi, \theta)\,(0, 1)^\mathrm{T}$.
20. ****Magic numbers.*** Free electrons confined to nanoscale features such as clusters and nanodots form energy levels similar to particle-in-a-box states, and the filling of these levels leads to "magic-number" stability when the HOMO-LUMO splitting

between highest occupied and lowest unoccupied orbitals is large. Determine and discuss the four lowest-lying energy levels for particles of cubic shape.
Hint: The behavior of simple metals is well described by the free-electron or jellium model.

21. ****RKKY interaction.*** Derive the oscillatory distance dependence of the RKKY interaction by using second-order perturbation theory. What happens if one restricts the consideration to first-order perturbation theory?
Hint: Use the free-electron dispersion relation $E = \hbar^2 k^2 / 2m_e$, where $k \leq k_F$, and consult Section A2.5.

3

Models of magnetic anisotropy

The energy of a magnet depends on the direction of the magnetization with respect to the crystal axes. This important property, known as the *magnetic anisotropy*, is the origin of hysteresis and coercivity. Anisotropy-energy densities vary from less than $0.005\,\mathrm{MJ/m^3}$ in very soft magnets to more than $10\,\mathrm{MJ/m^3}$ in some rare-earth permanent magnets. Figure 3.1(a) shows a cylindrical magnet with the magnetization along a preferential or easy axis (top) and a magnetization direction with high anisotropy energy (bottom). Anisotropy modeling tackles a number of questions, such as the existence and orientation of easy axes, the symmetry and strength of the anisotropy, and their dependence on chemical composition and crystal structure. Magnetic anisotropy is intimately related to phenomena such as orbital-moment formation, magnetoelasticity, and magnetoresistance. The models considered in this chapter fall into three categories: phenomenological models, pair-interaction models, and anisotropy models involving spin-orbit coupling.

In most magnetic materials, the main source of anisotropy is *magnetocrystalline* anisotropy, which involves electrostatic crystal-field interaction and relativistic spin-orbit coupling (Bloch and Gentile 1931). This applies to both bulk and surface anisotropies, and the same mechanism explains the orbital moment and magnetoelasticity of most magnets, and the anisotropic magnetoresistance of metallic ferromagnets.

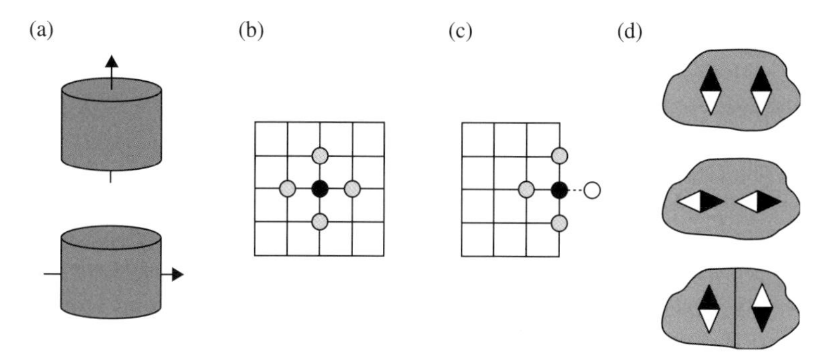

Fig. 3.1 Phenomenology of magnetic anisotropy and popular explanations: (a) magnetization directions with generally different energies, (b) and (c) anisotropy due to broken exchange bonds (a rather simplistic explanation) and (d) compass-needle analogy of shape anisotropy.

Typical anisotropy-energy densities (anisotropy constants) of iron-series transition-metal magnets are $0.05\,\mathrm{MJ/m^3}$ (0.5×10^6 ergs/cc) for bcc Fe and $0.5\,\mathrm{MJ/m^3}$ for hcp Co. Much higher room-temperature anisotropies, of order $10\,\mathrm{MJ/m^3}$, are obtained in rare-earth magnets. This reflects the large spin-orbit coupling of the rare-earth 4f electrons.

A popular but incorrect explanation of anisotropy is that removing atomic neighbors, or reducing the symmetry by lattice strain, weakens exchange bonds and leads to magnetic anisotropy. The argumentation is often furnished by referring to Néel's famous paper on surface anisotropy (1954). For example, it is argued that surfaces exhibit planar or perpendicular anisotropy because the exchange energy depends on whether the magnetization is in-plane or perpendicular, as illustrated in Figs 3.1(b) and (c). However, the Heisenberg Hamiltonian $J_{\mathrm{ij}}\,\mathbf{s}_{\mathrm{i}} \cdot \mathbf{s}_{\mathrm{j}}$ is *isotropic*, even if the exchange bonds $J_{\mathrm{ij}} = J(\mathbf{r}_{\mathrm{i}} - \mathbf{r}_{\mathrm{j}})$ are anisotropic. Only relative angles between neighboring spins matter, and Heisenberg exchange does not care in which direction the net magnetization points.

Magnetostatic interactions may yield some anisotropy, but only in a few systems are they the leading contribution. First, as emphasized long ago (Bloch and Gentile 1931), the dipolar anisotropy of cubic magnets is zero, in contrast to the experimentally observed finite anisotropy of cubic materials such as Fe and Ni. Second, from the compass-needle analogy shown in Fig. 3.1(d) it is straightforward to estimate that the corresponding shape-anisotropy energy densities are of the order of a few $0.1\,\mathrm{MJ/m^3}$. This is one to two orders of magnitude smaller than the anisotropy of rare-earth permanent magnets. Third, macroscopic magnetization states are generally nonuniform, and anisotropies obtained by comparing nonuniform magnetizations tend to be poor estimates. For example, in Fig. 3.1(c) the shape-anisotropy easy axis is horizontal, because the magnetostatic energy of the configuration in the middle is lower than that of the top configuration. However, physically realized configurations tend to be multidomain (bottom), and there is no longer a well-defined shape anisotropy.

3.1 Phenomenological models

Summary The dependence of the magnetic energy on the magnetization angle is usually parameterized in terms of anisotropy constants. The simplest phenomenological anisotropy model is lowest-order or second-order uniaxial anisotropy, $E_{\mathrm{a}}/V = K_1 \sin^2\theta$, where K_1 is the first uniaxial anisotropy constant. Most magnetic materials exhibit additional higher-order and, depending on their crystal symmetry, nonuniaxial contributions. These are often, but not always, small corrections to K_1. An important exception is cubic anisotropy, where the second-order terms are zero by symmetry and the leading term is fourth-order. Other parameterization tools are anisotropy coefficients, such as κ_2, and anisotropy fields, such as $2K_1/\mu_{\mathrm{o}}M_{\mathrm{s}}$.

Since the spontaneous magnetization $M_{\mathrm{s}} = |\mathbf{M}|$ is essentially fixed by intra- and interatomic exchange, we can express the anisotropy energy in terms of the magnetization angles ϕ and θ. Let us choose a coordinate frame by writing the magnetization as

$$\mathbf{M} = M_{\mathrm{s}}(\sin\theta\cos\phi\,\mathbf{e}_{\mathrm{x}} + \sin\theta\sin\phi\,\mathbf{e}_{\mathrm{y}} + \cos\theta\,\mathbf{e}_{\mathrm{z}}) \tag{3.1}$$

where the unit vectors $\mathbf{e}_x$, $\mathbf{e}_y$, and $\mathbf{e}_z$ may or may not correspond to the crystallographic a, b, and c axes. The simplest anisotropy-energy expression for a magnet of volume V is

$$E_a = K_1 V \sin^2\theta \tag{3.2}$$

Here V is the volume of the magnet and K_1 is the first- or second-order uniaxial anisotropy constant.

Equation (3.2) provides a simple but powerful parameterization of the magnetic anisotropy. This applies not only to typical hard magnets, where lowest-order uniaxial anisotropy is the leading anisotropy contribution, but also to magnets where the use of uniaxial anisotropy requires special care. Naturally, one limitation of (3.2) is the exclusion of higher-order terms, such as $\sin^4\theta$. This means that $K_1 = 0$ for cubic magnets, and the use of uniaxial expressions for cubic materials needs further justification. Another challenge is the generally unknown symmetry axis. In magnets with uniaxial crystal structure, the symmetry axis ($\theta = 0$) corresponds to the crystallographic c-axis, whereas $\theta = \pi/2$ denotes the basal or a–b plane. An example is Nd-Fe-B permanent magnets, which tend to have shapes similar to Fig. 3.1(a). In small particles with cubic crystal structure but elongated shape, the symmetry axis is parallel to the long or "shape" axis of the particle. Polycrystalline magnets have a spatially varying symmetry axis $\mathbf{n}(\mathrm{r})$, and it is convenient to replace $K_1 \sin^2\theta$ by $-K_1(\mathbf{n}\cdot\mathbf{M})^2/M_s^2$.

The physical interpretation of (1) depends on whether K_1 is positive or negative. For $K_1 > 0$, the energy has minima at $\theta = 0$ and $\theta = \pi$, so that the easy axis is parallel to the symmetry axis. This is known as *easy-axis* anisotropy. When K_1 is negative, the energy minimum is at $\theta = \pi/2$. In this regime, the magnetization is free to rotate in the basal plane, which is known as *easy-plane* anisotropy.

3.1.1 Uniaxial anisotropy

One way of generalizing (3.2) is to include higher-order powers of $\sin\theta$. The corresponding anisotropy-energy density is

$$\frac{E_a}{V} = K_1 \sin^2\theta + K_2 \sin^4\theta + K_3 \sin^6\theta \tag{3.3}$$

where K_2 and K_3 are the second and third uniaxial anisotropy constants, respectively. By definition, odd-order terms, such as $\sin\theta$ and $\sin^3\theta$, are not included in this expansion. Figure 3.2 shows some energy landscapes created by (3.2). The top row compares isotropic, easy-axis, and easy-plane magnets, whereas the bottom row illustrates the influence of fourth- and sixth-order terms.

Equation (3.3) gives rise to two questions: How many terms are necessary to describe magnetic anisotropy, and what determines the relative magnitudes of the anisotropy constants? Experiment suggests that the magnitude of the individual terms strongly decreases with increasing order, but there are exceptions. A more specific question is whether there is a small parameter that makes higher-order terms negligible. In rare-earth intermetallics (Section 3.4.1), the small parameter is the ratio R_{4f}/a, where $R_{4f} \approx 0.5$ Å is the rare-earth 4f-shell radius and $a \approx 3$ Å is the interatomic

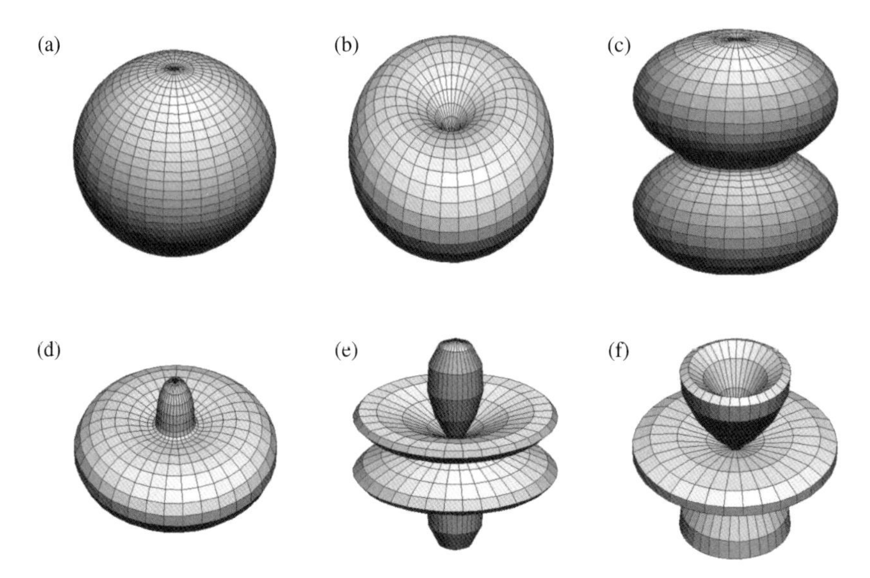

Fig. 3.2 Uniaxial anisotropy-energy landscapes: (a) isotropic, (b) easy axis, (c) easy plane, (d) easy cone, and (e–f) sixth-order landscapes.

distance. Significant higher-order contributions, as implied in the bottom row of Fig. 3.2, indicate the absence of a small expansion parameter is some systems or reflect competing lowest-order anisotropy contributions. For example, different sublattices may yield anisotropy contributions of opposite sign, and K_1 may change sign as a function of temperature (Section 5.5).

The inclusion of higher-order terms explains various experimental findings. Restoring the consideration to the K_1-term (lowest-order uniaxial anisotropy) yields two possible equilibrium states, namely easy-axis, as in Fig. 3.3(b), and easy-plane, as in Fig. 3.3(c). Higher-order uniaxial anisotropies give rise to spin structures with angles $0 < \theta < \pi/2$. A well-known example is *easy-cone* magnetism, which occurs for $K_1 < 0$ and $K_2 > -K_1/2$. Figure 3.3(d) illustrates this regime, where the equilibrium magnetization forms an angle $\theta_c = \arcsin(|K_1|/2K_2)$ with the c-axis. Fourth-order anisotropies tend to be smaller than second-order anisotropies by one or two orders of magnitude, so that easy-cone magnetism is limited to specific temperatures and composition windows.

3.1.2 Second-order anisotropy of general symmetry

Up until now, we have restricted ourselves to uniaxial anisotropy. In many cases, this is a good approximation, but there are exceptions. First, the crystal symmetry may be nonuniaxial, that is, cubic, orthorhombic, monoclinic, or triclinic. Second, there may be competing uniaxial anisotropies of different orientations. For example, the axis of revolution of an ellipsoidal particle may be unrelated to the crystalline easy axis, and the sum of magnetocrystalline and shape anisotropy can no longer be written in form of eqs. (3.2) or (3.3).

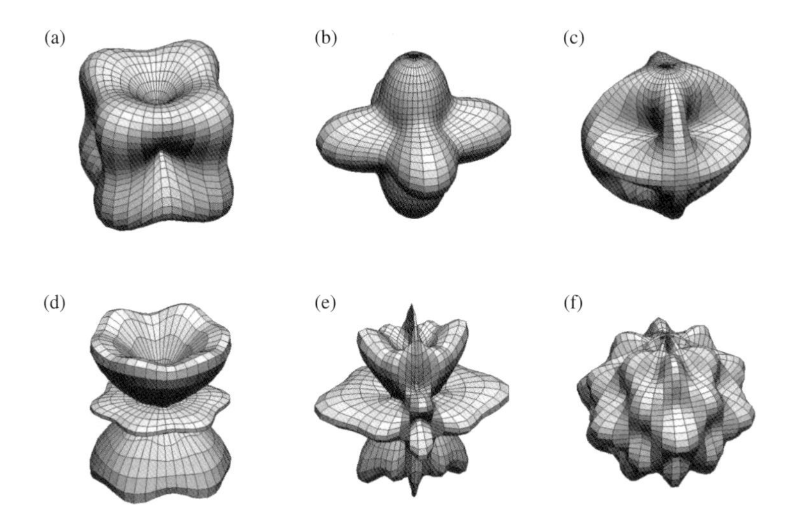

Fig. 3.3 Energy surfaces for higher-order anisotropies: (a–c) cubic magnets and (d–f) magnets with low symmetry.

Let us for the moment ignore fourth- and higher-order terms. The most general second-order anisotropy energy density can be written as

$$\frac{E_a}{V} = K_1 \sin^2 \theta + K_1' \sin^2 \theta \cos(2\phi) \tag{3.4}$$

This equation describes solids with low crystal symmetry, such as orthorhombic, monoclinic, and triclinic crystals. It must also be used for ellipsoids with three unequal principal axes, randomly shaped magnets, and for many surfaces, such as bcc (011). The first anisotropy constants K_1 and K_1' are generally of comparable magnitude, so that the fitting of experimental data to uniaxial anisotropies may give errors comparable to the magnitude of K_1.

It may be convenient to rotate the coordinate frame so that $K_1 > K_1' > 0$. Then the z-axis is then the global easy axis and K_1' describes the secondary in-plane anisotropy, the y-axis being the in-plane easy axis. The in-plane part of the anisotropy can be observed, for example, by using a magnetic field to confine the magnetization in the a–b plane. When $K_1 < 0$ for any choice of axes, the magnetization lies in the basal plane, as in Fig. 3.2(c), and the K_1'-term selects a favorable in-plane direction.

A general expression is obtained by writing the magnetization as $\mathbf{M} = M_s \mathbf{s}$, so that $|\mathbf{s}| = 1$ and s_x, s_y, and s_z can be considered as direction cosines

$$\mathbf{M} = M_s(s_x \mathbf{e}_x + s_y \mathbf{e}_y + s_z \mathbf{e}_z)$$

The second-order anisotropy energy density is then

$$\frac{E_a}{V} = -K_{xx}s_x^2 - K_{yy}s_y^2 - K_{zz}s_z^2 - 2K_{xy}s_x s_y - 2K_{xz}s_x s_z - 2K_{yz}s_y s_z \tag{3.5}$$

This equation contains six anisotropy constants, but three of them are necessary to fix the coordinate frame. A fourth constant determines the isotropic energy zero, but does not affect the anisotropy. The latter is seen by putting $s_z^2 = 1 - s_x^2 - s_y^2$ into (3.5) and redefining K_{xx} and K_{yy}. Extracting the anisotropy constants and identifying the angles θ and ϕ requires principal-axis transformation. In matrix form, $E_a/V = \mathbf{s} \cdot \underline{\underline{K}} \cdot \mathbf{s}$, and diagonalization of $\underline{\underline{K}}$ yields three eigenvalues $K_I \geq K_{II} \geq K_{III}$ and $K_1 = -(K_I + \frac{1}{2}K_{II} + \frac{1}{2}K_{III})$ and $K_1' = \frac{1}{2}K_{III} - \frac{1}{2}K_{II}$.

3.1.3 Higher-order anisotropies of nonuniaxial symmetry

The anisotropy contributions of (3.4) are higher-order but uniaxial. Figure 3.3 illustrates how the energy landscapes change on adding nonuniaxial fourth- and sixth-order terms. Examples (e) and (f) describe rather exotic magnets with very low symmetry, but energy landscapes of the types (a–d) are frequently encountered in practice. Anisotropies of eighth or higher order are also possible, but they are typically very small (Section 3.3).

Here we restrict ourselves to some common types of fourth- and sixth-order anisotropy. The fourth-order expression

$$\frac{E_a}{V} = K_1 \sin^2\theta + K_2 \sin^4\theta + K_2' \sin^4\theta \cos 4\phi \tag{3.6}$$

provides a convenient description of magnets with tetragonal crystal structure, but it can also be used to describe hexagonal, rhombohedral, and cubic crystals. However, hexagonal and rhombohedral crystals are characterized by $K_2' = 0$, so that (3.6) degenerates into a sum the second- and fourth-order uniaxial anisotropies. In tetragonal magnets, both K_2 and K_2' are fourth-order and generally nonzero, though often smaller than K_1. In cubic crystals, K_1, K_2, and K_2' are all fourth-order, and only two of the three constants are independent. The smallness of the anisotropy of cubic magnets is the main reason for preferring hexagonal, tetragonal, or rhombohedral structures as permanent magnets and recording materials.

With increasing order, the anisotropy-energy densities become quite complicated, even for high symmetries. For example, generalizing (3.6) to sixth order yields two different expressions. For tetragonal symmetry

$$\frac{E_a}{V} = K_1 \sin^2\theta + K_2 \sin^4\theta + K_2' \sin^4\theta \cos 4\phi + K_3 \sin^6\theta + K_3'' \sin^6\theta \cos 4\phi \tag{3.7a}$$

whereas hexagonal and rhombohedral magnets are described by

$$\frac{E_a}{V} = K_1 \sin^2\theta + K_2 \sin^4\theta + K_3 \sin^6\theta + K_3''' \sin^6\theta \cos 6\phi \tag{3.7b}$$

Equation 3.7(a) can also be used to describe cubic magnets, but this approach is cumbersome and rarely used. Even more complicated expression are obtained for more complicated symmetries, such as monoclinic.

3.1.4 Cubic anisotropy

For cubic crystals, the leading terms of the expansion of the anisotropy-energy density are

$$\frac{E_a}{V} = K_1(s_x^2 s_y^2 + s_y^2 s_z^2 + s_x^2 s_z^2) + K_2^{(c)} s_x^2 s_y^2 s_z^2 \tag{3.8}$$

Typical energy surfaces for cubic magnets are shown in Fig. 3.3(a–c). $K_1 > 0$ yields an easy magnetization directions along the (001)-type cube edges. This regime, shown in Fig. 3.3(a), is called *iron-type anisotropy*. For $K_1 < 0$ the preferential magnetization directions are parallel to the (111)-type cube diagonals. The energy landscape is illustrated in Fig. 3.3(b) and known as *nickel-type* anisotropy. $K_2^{(c)} > 0$ favors easy axes along the cube diagonals, too, but Fig. 3.3(c) shows that the symmetry of this contribution is different from Ni-type anisotropy.

Comparing eqs. (3.8) and (3.7b) we convince ourselves that the cubic and tetragonal anisotropy constants K_1 have the same value. However, this is accidental, and the conversion of the other cubic and tetragonal anisotropy constants is nontrivial. Explicitly,

$$K_2 = -\frac{7}{8}K_1 + \frac{1}{8}K_2^{(c)} \tag{3.9a}$$

$$K_2'' = -\frac{1}{8}K_1 - \frac{1}{8}K_2^{(c)} \tag{3.9b}$$

$$K_3 = -\frac{1}{8}K_2^{(c)} \tag{3.9c}$$

$$K_3'' = \frac{1}{8}K_2^{(c)} \tag{3.9d}$$

There is, usually, no need to use these conversions, but they are useful when considering cubic magnets that are strained parallel to the cube axes. The corresponding magnetoelastic anisotropy (Section 3.5.1) may be much higher than the original cubic anisotropy. For example, the magnetic hardness of steel magnets is due to interstitial carbon atoms, which yield a tetragonal lattice distortion in form of a martensitic phase transition. Similar sets of conversion rules exist for other pairs of symmetries, such as cubic and orthorhombic.

3.1.5 Anisotropy coefficients

Equation (3.9) indicates that the conversion rules between anisotropy constants are rather complicated. This reflects a far-reaching inadequacy of anisotropy constants, namely the nonorthogonality of the underlying functions $E_a(\theta, \phi)$. In particular, anisotropy constants mix anisotropy contributions of different order, which complicates the mathematical handling and physical interpretation of the anisotropy constants. A good example is cubic anisotropy, where K_1 looks like a second-order anisotropy constant but is, in fact, fourth-order.

A better description is provided by orthogonal functions. A natural choice is spherical harmonics, which are both complete and orthonormal. The corresponding constants are known as *anisotropy coefficients* and denoted by their order. For uniaxial anisotropy, (3.3),

$$\frac{E_a}{V} = \frac{\kappa_2}{2}(3\cos^2\theta - 1) + \frac{\kappa_4}{8}(35\cos^4\theta - 30\cos^2\theta + 3)$$
$$+ \frac{\kappa_6}{16}(231\cos^6\theta - 315\cos^4\theta - 105\cos^2\theta - 5) \tag{3.10}$$

The relation between the uniaxial anisotropy constants and the uniaxial anisotropy coefficients is $K_1 = -3\kappa_2/2 - 5\kappa_4 - 21\kappa_6/2$, $K_2 = 35\kappa_4/8 + 189\kappa_6/8$, and $K_3 = -231\kappa_6/16$. Nonuniaxial anisotropy coefficients carry two indices. For example, κ_{22} describes K_1' contributions. Equation (3.10) may look slightly more complicated than (3.3), but it has a number of advantages. Anisotropy coefficients are easier to access by experiment, easier to calculate, and correspond to a given order of magnetocrystalline anisotropy. For example, an accurate determination of K_1 requires the knowledge of higher-order terms such as κ_4 and κ_6, which can be quite a challenge. By contrast, the κ_m are independent of $\kappa_{\mathrm{m}'}$ where $\mathrm{m}' > \mathrm{m}$.

The preferential use of anisotropy constants has practical reasons. For second-order uniaxial anisotropy, κ_2 differs from K_1 by a factor of $-3/2$, which complicates the notation without yielding any specific benefit. Another point is that that measurements and theoretical calculations of anisotropy-energy densities are sometimes restricted to directions parallel and perpendicular to the c-axis, and the difference is then equated to K_1.

3.1.6 Anisotropy fields

Dimensional analysis shows that $K_1/\mu_\mathrm{o}M_\mathrm{s}$ has the dimension of a magnetic field. This is very convenient, because it makes it possible to compare the effect of the anisotropies with fields such as the applied field and the coercivity H_c. It is customary to define the corresponding *anisotropy field* of uniaxial magnets with second-order anisotropy as

$$H_\mathrm{a} = \frac{2K_1}{\mu_\mathrm{o}M_\mathrm{s}} \tag{3.11}$$

As we will see in Section 4.1, the factor 2 allows a direct comparison of H_a with H_c. The inclusion of higher-order anisotropies gives rise to different non-equivalent definitions of the anisotropy field. For example, using (3.3) and comparing the energies for $\theta = 0$ and $\theta = 90°$ leads to $H_\mathrm{a} = 2(K_1 + K_2 + K_3)/\mu_\mathrm{o}M_\mathrm{s}$. The initial slope of the perpendicular magnetization curves yields $H_\mathrm{a} = 2(K_1 + K_2 + K_3)/\mu_\mathrm{o}M_\mathrm{s}$, whereas the ideal coercivity of uniaxial magnets is not affected by K_2 and K_3, so that (3.11) remains valid.

In cubic magnets, the anisotropy field for iron-type anisotropy ($K_1 > 0$) is described by (3.11), whereas nickel-type anisotropy ($K_1 < 0$) leads to $H_\mathrm{a} = -4K_1/3\mu_\mathrm{o}M_\mathrm{s}$. The different expression for Ni-type anisotropy means that the energy landscape in the vicinity of the (111) direction in Fig. 3.3(b) is quantitatively different from that in the (001) direction in Fig. 3.3(a).

3.2 Models of pair anisotropy

Summary Magnetostatic dipole interactions are one source of magnetic anisotropy, although typical magnetostatic anisotropies are much smaller than the leading magnetocrystalline anisotropy contribution of electronic origin. Aside from a lattice contribution to the magnetocrystalline anisotropy, magnetostatic anisotropy manifests itself as macroscopic shape anisotropy. Due to magnetization inhomogeneities, shape anisotropy is limited to very small length scales. An atomic pair-anisotropy model is the Néel model, where the anisotropy is parameterized in terms of bond directions. The Néel model yields the correct symmetry but fails to reproduce the single-ion origin of the anisotropy of most bulk materials and surfaces.

Figure 3.1(d) illustrates that magnetostatic dipole interactions may produce magnetic anisotropy, even if the mechanism shown in the figure is not very realistic. There are two types of magnetostatic anisotropy contributions. First, pairs of magnetic atoms in noncubic structures yield a magnetostatic contribution to the crystalline anisotropy. Second, the long-range character of magnetostatic interactions means that the magnetic energy depends on macroscopic orientation of the magnetization. The second contribution is known as *shape anisotropy*.

3.2.1 Dipolar interactions and shape anisotropy

Dipolar and shape anisotropies are based on the magnetostatic interaction energy of a pair of magnetic dipoles $\mathbf{m}_{i/j}$ located at $\mathbf{r}_{i/j}$:

$$E_{ms}(i,j) = -\frac{1}{4\pi\mu_o}\frac{3\,\mathbf{m}_i\cdot\mathbf{R}_{ij}\,\mathbf{m}_j\cdot\mathbf{R}_{ij} - \mathbf{m}_i\cdot\mathbf{m}_j R_{ij}^2}{R_{ij}^5} \tag{3.12}$$

where $\mathbf{R}_{ij} = \mathbf{r}_i - \mathbf{r}_j$. In principle, this equation can be used to calculate the magnetostatic contribution to the anisotropy energy of any system, but the summation $\Sigma_{i>j} E_{ms}(i,j)$ over pairs of atomic dipoles is very cumbersome. It is convenient to replace the summations by integrals, $\Sigma_i \ldots \mathbf{m}_i = \int \ldots \mathbf{M}(\mathbf{r})\,\mathrm{d}V$, and to rewrite the total magnetostatic energy as $E_{ms} = \frac{1}{2}\mu_o \int \mathbf{M}(\mathbf{r})\cdot\mathbf{H}_D(\mathbf{r})\,\mathrm{d}V$, where H_D is the demagnetizing or self-interaction field. Some aspects of the involved mathematics are discussed in the Appendix.

There is no simple general expression for E_{ms}, but for homogeneously magnetized ellipsoids, the calculation is relatively straightforward and yields

$$E_{ms} = \frac{\mu_o}{2} M_s^2\,(D_x\,s_x^2 + D_y\,s_y^2 + D_z\,s_z^2)\,V \tag{3.13}$$

Here x, y, and z refer to the principal axes of the ellipsoid, and the demagnetizing factors D_x, D_y and D_z describe the demagnetizing field in the respective principal directions. For example, $H_{D,x} = -D_x M_x$. This equation is a special case of eq. (3.5) and yields, in general, two anisotropy constants K_1 and K_1', as in (3.4).

Let us now restrict ourselves to the uniaxial case of *ellipsoids of revolution*, where $D_x = D_y$. Furthermore, $D_x + D_y + D_z = 1$, so that $D = D_z = 1 - 2D_x$, and $s_x^2 + s_y^2 + s_z^2 = 1$. Putting these expressions into (3.11) and extracting the anisotropy yields

$$K_{sh} = \frac{\mu_o}{2}(1 - 3D)M_s^2 \tag{3.14}$$

where K_{sh} is the shape anisotropy contribution to K_1. Figure 3.4 summarizes the physics behind the calculation of D. In a real magnet, Fig. 3.4(a), the atomic dipoles yield both microscopic and macroscopic contributions. Modeling the magnet as a homogeneously magnetized ellipsoid of revolution, Fig. 3.4(b), ensures that the demagnetizing field is homogeneous, so that $\mathbf{H}_D = -D\mathbf{M}$ and $\int \mathbf{M}\cdot\mathbf{H}_D\,\mathrm{d}V = \mathbf{M}\cdot\mathbf{H}_D V$.

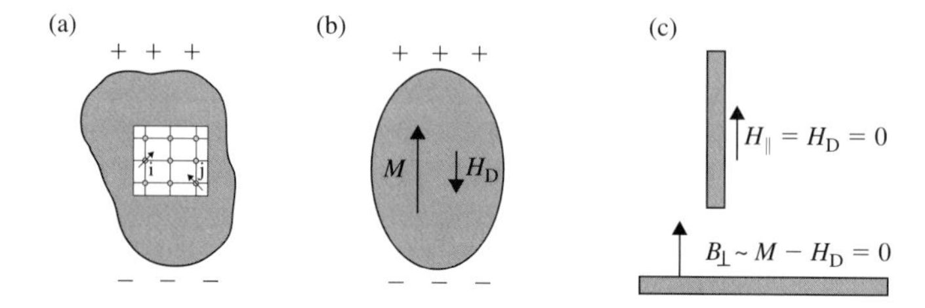

Fig. 3.4 Magnetostatic contributions to the anisotropy: (a) atomic picture, (b) macroscopic picture, and (c) considerations determining the demagnetizing field.

3.2.2 Demagnetizing factors

For ellipsoids of revolution, it is sufficient to consider the demagnetizing factor $D = D_z$ parallel to the axis of revolution. The demagnetizing factor in the basal plane then follows from $D_x + D_y + D_z = 1$ and is equal to $\frac{1}{2}(1 - D)$. For spheres, $D = 1/3$ by symmetry, and both K_{sh} and the corresponding anisotropy field are zero. This must be compared to the negative demagnetizing field, $H_D = -M/3$. In two other limits, shown in Fig. 3.4(c), the demagnetizing factors are obtained by analyzing Maxwell's equations. First, in the absence of macroscopic currents, $\nabla \times \mathbf{H} = 0$ and the field component $H_\parallel$ parallel to any surface is continuous. For strongly elongated or needle-shaped ellipsoids (top of figure), this means that H_D is equal to the field outside the magnet. However, this field is zero, because any finite field would yield an unphysical divergence of the total magnetostatic energy $\mu_o \int \mathbf{H}^2 \, dV/2$. This yields $H_D = 0$ and $D = 0$ for needle-shaped magnets. Second, $\nabla \cdot \mathbf{B} = 0$ means that the normal flux component $B_\perp$ is continuous at any surface. For strongly oblate or plate-like magnets, bottom of Fig. 3.4(c), this means that $B_\perp = \mu_o(M + H_D)$ inside the magnet is equal to $B_\perp = \mu_o H$ outside the magnet. To ensure a proper convergence of the total magnetostatic energy, we must put $H = 0$, so that $H_D = -M$ and $D = 1$.

Demagnetizing factors for general ellipsoids were discussed by Osborn (1945). For prolate and oblate ellipsoids of revolution with intermediate aspect ratios $\kappa_o = R_z/R_x$

$$D = \frac{1}{\kappa_o^2 - 1}\left(\frac{\kappa_o}{\sqrt{\kappa_o^2 - 1}} \operatorname{arcosh} \kappa_o - 1\right) \tag{3.15a}$$

and

$$D = \frac{1}{1 - \kappa_o^2}\left(1 - \frac{\kappa_o}{\sqrt{1 - \kappa_o^2}} \arccos \kappa_o\right) \tag{3.15b}$$

respectively. As discussed above, spherical particles exhibit $\kappa_o = 1$ and $D = 1/3$. In the limits of needle-shaped ($\kappa_o \gg 1$) and plate-like ($\kappa_o \ll 1$) magnets, the respective expressions reduce to

$$D = \frac{1}{\kappa_o^2}(\ln \kappa_o - 0.307) \quad \text{and} \quad D = 1 - \frac{\pi}{2}\kappa_o \tag{3.16}$$

Of course, these equations reproduce $D = 0$ for long cylinders and $D = 1$ for plates or thin films. Two intermediate demagnetizing factors are 0.527 and 0.174 for aspect ratios of 0.5 and 2.0, respectively.

The shape anisotropy associated with the demagnetizing factor must not be confused with the demagnetizing-field corrections to the hysteresis loop (Fig. 4.2). The latter is a macroscopic effect unrelated to magnetic anisotropy. For example, (3.13–14) are based on the assumption of a parallel spin orientation throughout the magnet (coherent rotation). In Section 4.3 we will see that this requirement is not realized in macroscopic magnets, and the corresponding field corrections have a magnitude comparable to DM.

3.2.3 Applicability of the shape-anisotropy model

Shape anisotropy originates from macroscopic charges at the magnet's surface. Atomic aspects of magnetostatic dipole interactions, as illustrated in Fig. 3.4(a), are ignored but can be incorporated very easily into the magnetocrystalline anisotropy. For example, some magnets have a layered atomic structure, and having the magnetization in the layer plane may be magnetostatically favorable. In most cases, this magnetostatic contribution is a small correction to the leading electronic contributions.

A popular but unphysical interpretation is to identify the shape-anisotropy field $2K_{\mathrm{sh}}/\mu_{\mathrm{o}}M_{\mathrm{s}}$ with the demagnetizing field H_{D} caused by the poles on the magnet's surface. In fact, the demagnetizing field is always negative, $H_{\mathrm{D}} = -DM$, where $D > 0$ is the shape-dependent demagnetizing factor. Arbitrarily "correcting" this result by subtracting demagnetizing fields for different directions doesn't provide a solution, because the obtained anisotropy energies are incorrect, except for spheres and thin films.

A severe condition is the requirement of a uniform magnetization state. Magnetostatic surface charges cost dipolar energy, and there is a general trend towards the formation of inhomogeneous magnetization states, as illustrated in the bottom part of Fig. 3.1(d). These inhomogeneous magnetization states may have the character of equilibrium domains, or they may appear as inhomogeneous modes during magnetization reversal. Since the derivation of (3.12) relies on a homogeneous magnetization, there is no well-defined shape anisotropy in magnets with inhomogeneous magnetization. The magnetic energy of a magnet with arbitrary spin structure is likely to reflect phenomena such as domain wall motion and local magnetization rotations. By definition, this has nothing to do with magnetic anisotropy, but is a micromagnetic phenomenon.

The trend towards domain formation competes with the interatomic exchange, which favors parallel spin orientation. Since interatomic exchange is a short-range phenomenon, the shape-anisotropy model works best for *small* magnetic particles. Examples are powders of elongated nanoparticles, such as Fe, and alnico-type permanent magnets, where long needles of $Fe_{65}Co_{35}$ are embedded in a nonmagnetic Al-Ni matrix. To realize a useful degree of shape anisotropy, the Fe-Co needles must be very thin, less than 100 nm in diameter. Magnetostatic interactions are also important in magnetic nanostructures, such as nanowires (Sellmyer, Zheng, and Skomski 2001). We will discuss the involved length scales in Chapter 4.

3.2.4 The Néel model

A famous pair-interaction model was developed by Néel (1954). Aside from an isotropic zero-point energy, the anisotropy energy per pair of magnetic atoms is

$$E(i,j) = L(r)\left(\cos^2\psi_{\mathrm{ij}} - \frac{1}{3}\right) + (\mathrm{h.o.}) \tag{3.17}$$

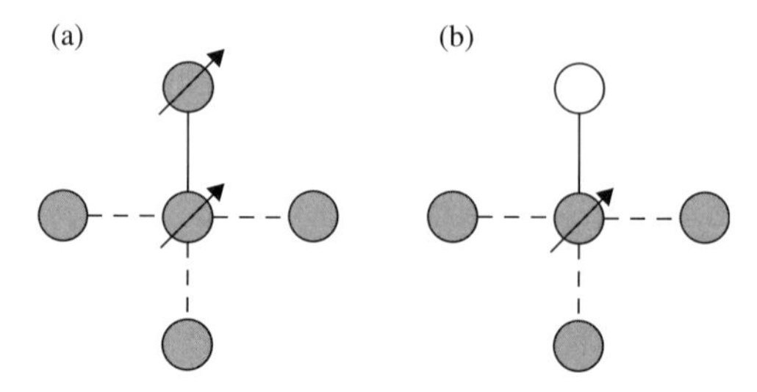

Fig. 3.5 Models of magnetic anisotropy: (a) Néel model and (b) single-ion model. Both models reproduce the correct symmetry, but (b) is physically more adequate for most systems.

where ψ_{ij} is the angle between the bond axis and the magnetization, $\cos\psi_{ij} = \mathbf{M}\cdot\mathbf{R}_{ij}/M_s R$. This expression includes but is not restricted to magnetostatic dipole interactions. A specific advantage is the use of orthogonal functions (Legendre polynomials), very similar to the parameterization in terms of anisotropy coefficients (Section 3.1.5).

The Néel model provides an adequate description of the symmetry of the magnetic anisotropy, including surface effects such as that indicated in Fig. 3.1(c), and has been used to describe thin films and surfaces (Chuang, Ballantine, and O'Handley 1994, Millev, Skomski, and Kirschner 1998). However, this does not mean that Néel's pair approach provides a physical understanding of anisotropy. Néel himself was well aware of the phenomenological nature of his expansion, even if he was partly motivated by magnetostatic dipole terms. There are also electronic terms of pair character, but the main contribution to the magnetocrystalline anisotropy of most bulk magnets and surfaces is single-ion anisotropy.

The difference is seen by considering the effect of *nonmagnetic* atomic neighbors. Figure 3.5(a) illustrates that the Néel model requires pairs of interacting magnetic atoms. By contrast, single-ion anisotropy is caused by an environment that is not necessarily magnetic. The principal failure of the Néel model is seen by comparing rare-earth transition-metal intermetallics Sm_2Fe_{17} and $Sm_2Fe_{17}N_3$. The interstitial nitrogen is nonmagnetic but acts as an electronegative crystal-field charge, changing the room-temperature anisotropy from relatively soft, $K_1 = -0.8\,\mathrm{MJ/m^3}$, to very hard, $K_1 = 8.6\,\mathrm{MJ/m^3}$ (Skomski and Coey 1999). Since both the single-ion and Néel models reproduce the correct symmetry, there is no true advantage in using the Néel model unless imposed by the underlying physics.

3.3 Spin-orbit coupling and crystal-field interaction

Summary Spin-orbit coupling and crystal-field interaction are key requirements for magnetocrystalline anisotropy. Spin-orbit coupling is a higher-order term in the relativistic Pauli expansion in terms of the small parameter v/c, where v is the velocity of the electrons. Magnetic interactions, such as spin-orbit coupling, tend to be much smaller than the leading electrostatic and exchange interactions, but the high effective nuclear charge of inner electrons in rare-earth atoms enhances the spin-orbit coupling.

In solids, the spin-orbit coupling competes against the crystal-field splitting, which favors the suppression (quenching) of the orbital moment. Quenched orbitals have a standing-wave character and adapt more easily to the crystal field than unquenched or running-wave orbitals. The outcome of the competition between spin-orbit coupling and crystal-field interaction determines the degree of quenching and the magnitude of the magnetic anisotropy. 3d electrons tend to undergo strong quenching. For example, iron has a magnetization of about 2.2 μ_B, but only about 5% of this moment are of orbital origin. The opposite is true for the 4f electrons in rare-earth ions, which are close to the nucleus and therefore combine a weak crystal-field interaction with a strong spin-orbit coupling.

The leading electronic contribution to the magnetic anisotropy involves two mechanisms, spin-orbit coupling and crystal-field interaction. This section starts with a brief discussion of the relativistic origin of magnetism. Magnetic energies, including spin-orbit coupling, are obtained as small corrections to the leading electrostatic energies. We then analyze the electrostatic energy contributions, by modeling the atomic wave functions as hydrogen-like orbitals embedded in a crystal field, and investigate how spin-orbit coupling modifies the orbitals.

3.3.1 Relativistic origin of magnetism

The relativistic nature of matter means that all basic equations can be written in a space-time symmetric form. An example is the propagation of light, $x^2 + y^2 + z^2 - c^2t^2 = 0$, where $\mathbf{r}$ and t can be considered as a four-dimensional vector. Electromagnetism is described by a four-vector containing the vector potential $\mathbf{A}$ and the scalar potential ϕ. The source of the potential is a four-vector containing the current density $\mathbf{j}$ and the charge ρ, and the four Maxwell equations can be combined into a set of two four-dimensional differential equations.

Nonrelativistic equations can be used if the velocity v of the electron is much smaller than the speed of light. For example, expanding the relativistic energy

$$E = m_e c^2 \sqrt{1 + \frac{v^2}{c^2}} \tag{3.18a}$$

into powers of the small parameter v/c yields

$$E = m_e c^2 + \frac{1}{2} m_e v^2 - \frac{1}{8} m_e v^4 \tag{3.19}$$

The first term on the right-hand side is the rest energy of the electron, $m_e c^2 = 0.511\,\text{MeV}$, the second term is the nonrelativistic kinetic energy, and the third term is the lowest-order relativistic correction to the kinetic energy. The second term is also a good estimate for the electrostatic energy of electrons in atoms, where kinetic and potential energies are of comparable magnitude.

The relativistic character of (3.18) becomes explicit on substituting $v = p/m_e$, so that

$$m_e^2 c^4 = (E^2 - c^2 p^2) \tag{3.18b}$$

In this equation, $\mathbf{p}$ and E form the energy-momentum four-vector. To account for the quantum-mechanical nature of atomic matter, we must convert (3.18b) to a linear wave equation known as the *Dirac equation*. This is possible, because $\mathbf{p}$ and E have a four-vector analog in the respective operators $-\mathrm{i}\hbar\nabla = -\mathrm{i}\hbar\,\partial/\partial\mathbf{r}$ and $\hbar\,\mathrm{i}\partial/\partial\mathrm{t}$. To derive the wave equation, it is convenient to write (3.18b) as $c^2p^2 = (E + m_\mathrm{e}c^2)(E - m_\mathrm{e}c^2)$. Let us, for the moment, ignore the vector character of $\mathbf{p}$, so that the above equation has two roots $cp = E \pm m_\mathrm{e}c^2$. The roots correspond to electrons ($E \approx m_\mathrm{e}c^2$) and positrons ($E \approx -m_\mathrm{e}c^2$).

Unfortunately, the three-dimensional equivalent of $cp = E \pm m_\mathrm{e}c^2$ doesn't make sense, because $c\mathbf{p} = E \pm m_\mathrm{e}c^2$ equates a vector with a scalar. To solve the problem, we exploit an important property of the Pauli matrices, namely that $(\boldsymbol{\sigma}\cdot\mathbf{a})^2 = \mathbf{a}^2$ for any vector $\mathbf{a}$. This yields $c\boldsymbol{\sigma}\cdot\mathbf{p} = E \pm m_\mathrm{e}c^2$. Since the Pauli matrices are 2×2 matrices, the resulting wave function has two components. As we can guess from the involvement of the Pauli matrices, the two components denote the *spin* of the electron. This shows that the spin is a natural consequence of the four-dimensional character of relativistic space. There is neither need nor justification to assume a rotation of the electron about its own axis!

Magnetic interactions are obtained by incorporating the vector potential, as one adds the electric potential ϕ to the energy E. The solution of the Dirac equation, outlined in the Appendix, exhibits two features. First, there is generally some admixture of antimatter or positronic features to the electron wave function. This requires a careful book-keeping of terms for each power of v/c, and the resulting wave equation has the character of a perturbation series. Second, $\mathbf{p} = -\mathrm{i}\hbar\nabla$ gives rise to derivatives of the vector potential, which have the character of magnetic fields. This means that Zeeman interactions are automatically included in the expansion. Restricting ourselves to the most important terms, we obtain the Pauli expansion

$$E\psi = m_\mathrm{e}c^2\psi - \frac{\hbar^2}{2m_\mathrm{e}}\nabla^2\psi + V(\mathrm{r})\psi + \mu_\mathrm{B}\,\mathbf{H}\cdot(\mathbf{L} + 2\mathbf{S})\psi + \lambda\,\mathbf{L}\cdot\mathbf{S}\psi \tag{3.20}$$

where $\mathbf{L}$ and $\mathbf{S}$ are one-electron orbital-moment and spin operators, respectively, and λ is the spin-orbit coupling constant. Note that this familiar L-S form of the spin-orbit coupling is limited to spherical potentials V.

It is illustrative to compare this expression with (3.17). For hydrogen-like atoms, the velocity of the electrons $v \approx Z\alpha c$, where Z is the effective nuclear charge and $\alpha = \mathrm{e}^2/4\pi\varepsilon_\mathrm{o}\hbar c \approx 1/137$ is Sommerfeld's fine-structure constant. This yields

$$E = m_\mathrm{e}c^2 + \frac{1}{2}m_\mathrm{e}Z^2\alpha^2c^2 - \frac{1}{8}m_\mathrm{e}Z^4\alpha^4c^2 \tag{3.21}$$

Analyzing the order of magnitudes of the terms in (3.20), we obtain one term of order $\alpha^0(m_\mathrm{e}c^2)$, two terms of order α^2 (the kinetic and potential energies), and two terms of order α^4 (the magnetostatic interaction and the spin-orbit coupling). This indicates that magnetic interactions are much smaller than electrostatic interactions. For $Z = 1$, the electrostatic and magnetostatic (or Zeeman) energies scale as $\alpha^2m_\mathrm{e}c^2/2 = 13.6\,\mathrm{eV}$ and $\alpha^4m_\mathrm{e}c^2/8 = 0.18\,\mathrm{meV}$, respectively. One example is the Pauli paramagnetism of simple metals, where the field H competes against the kinetic energy. The corresponding susceptibility is of the order of α^2, and very high fields are necessary to produce a substantial spin polarization. Another example is the Zeeman interaction of a spin moment with an external magnetic field, which is epitomized by the Bohr magneton, $\mu_B = 0.672\,\text{K}$ per tesla. By contrast, exchange corresponds to temperatures (Curie temperatures) of the order of 1000 K. Spin-orbit coupling suffers from the same fundamental limitation, but by cleverly exploiting the large Z of inner-shell electrons, it is possible to obtain high anisotropies in rare-earth intermetallics and other materials.

3.3.2 Hydrogen-like atomic wave functions

Since magnetostatic and spin-orbit interactions are small corrections to the leading electrostatic terms, we start with a discussion of the electrostatic potential $V(r)$. This part of the argumentation is closely related to the treatment of interatomic hopping and exchange (Chapter 2). It is convenient to separate the spherical atomic potential $V_o(\text{r})$ and to write

$$V(\mathbf{r}) = V_o(\mathbf{r}) + V_{CF}(\mathbf{r}) \tag{3.22}$$

where $V_{CF}(\mathbf{r})$ is the *crystal-field* potential. The same division has been used to derive hopping integrals, and it can be shown that hopping and direct crystal-field interactions have a very similar effect on the magnetic anisotropy. To model the crystal-field interaction, we take hydrogen-like atomic wave functions and put these orbitals into the crystal field.

3.3.3 Crystal-field interaction

The minima and maxima of the 3d wave functions translate into an aspherical charge distribution, as shown in Fig. 3.6(c). Putting these charge clouds into the crystal field $V_{CF}(r)$ means that different orbitals may have different electrostatic energies. The crystal-field splitting of the energy levels is outlined in Fig. 3.7, and Fig. 3.8 shows the physical origin of the splitting. In free space, the $|xy\rangle$ and $|x^2 - y^2\rangle$ orbitals have the same energy, but putting the electrons in a crystalline environment yields energy changes. The reason for the higher energy of the $|xy\rangle$ orbital is the repulsive electrostatic interaction between the negative charges (dark regions). The $|x^2 - y^2\rangle$ orbital has the lower energy, because the negative crystal-field charges are in directions where the electron density is zero.

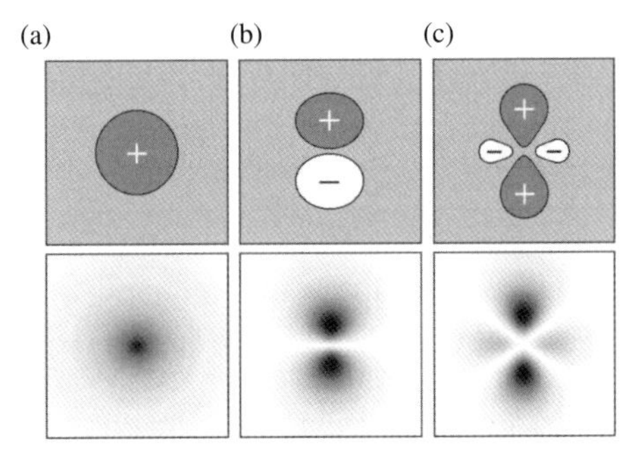

Fig. 3.6 Some hydrogen-like wave functions: (a) $|1\text{s}\rangle$, (b) $|2\text{p(z)}\rangle$ and (c) $|3\text{d}(\text{z}^2)\rangle$. The top and bottom rows show the signs of the wave functions and the charge densities, respectively. The $|3\text{d}(\text{z}^2)\rangle$ orbital (c) has the quantum numbers $n = 3$, $l = 2$, and $m = 0$.

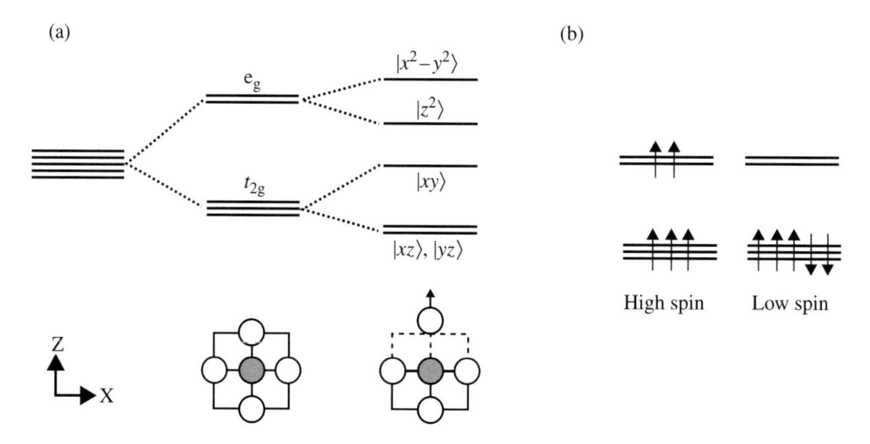

Fig. 3.7 Crystal-field splitting: (a) $e_g - t_{2g}$ splitting in an octahedral environment (center) and splitting in a tetragonal field (right) and (b) high-spin low-spin transition for Mn^{2+} in an octahedral field.

Crystal-field interactions are often visualized by energy-level diagrams, such as that shown in Fig. 3.7. In many cubic compounds, the crystal field is octahedral. For example, in rock salt oxides such as NiO and CoO, the six nearest neighbors are on the x, y, and z axes. With the partial exception of hydrogen, neighboring atoms acts as *negative* crystal-field charges, so that the lobes of the 3d orbitals don't want to point towards neighboring atoms. By inspection of Fig. 2.17, we see that the $|xy\rangle$, $|xz\rangle$, and $|yz\rangle$ orbitals are *between* the x, y, and z axes. This lowers their energy compared to the $|x^2 - y^2\rangle$ and $|z^2\rangle$ orbitals and gives rise to the famous $e_g - t_{2g}$ splitting into an e_g dublet and a t_{2g} triplet, Fig. 3.7(a). Replacing the octrahedral environment by a tetrahedral or cubal environment reverses the sign of the splitting. The four or eight crystal-field charges are now on cube diagonals, that is, between the x, y, and z axes, and the energy of the triplet is higher than that of the dublet. The $e_g - t_{2g}$ splitting governs the electron occupancy in cubic oxides, although hybridization effects (broadening of levels into bands) are often important.

In noncubic materials, the situation is more complicated. Figure 3.7(a) illustrates this point by showing the tetragonal splitting of $e_g - t_{2g}$ levels. Frequent reasons for tetragonal crystal fields are the creation of a surface (figure), chemical substitutions leading to layered compounds, and mechanical strain. One consequence of large crystal-field splitting is high-spin low-spin transitions, as introduced in Section 2.2.4 and illustrated in Fig. 3.7(b). Note that the specific high-spin low-spin transition shown in the figure is linked to the rather weak $e_g - t_{2g}$ splitting in cubic crystals. Second-order crystal-field splittings, such as the tetragonal splitting shown in Fig. 3.8, are larger than typical fourth-order or cubic splittings by a factor of order $R_{3d}^2/D^2 \sim 5$ (Section 3.4.2), where R^{3d} is the 3d shell radius and D is the interatomic distance (Bethe 1929). This enhances the trend towards high-spin low-spin transitions so long as unoccupied $\downarrow$ orbital are available.

In practice, there is some interatomic hopping, and the energy levels broaden into bands (Section 2.4). In some oxides, the bandwidth is smaller than the crystal-field

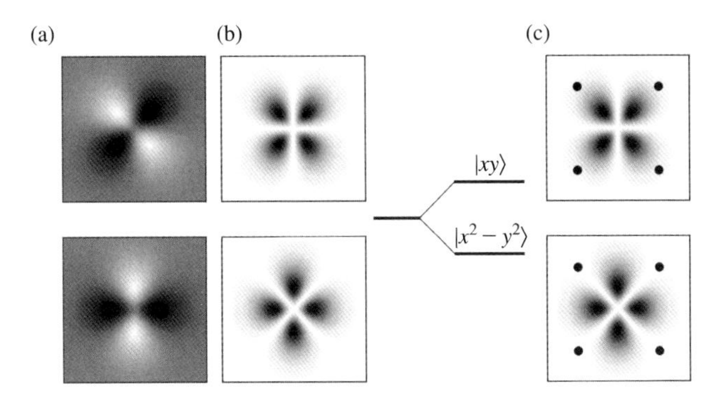

Fig. 3.8 Crystal-field interaction: (a) quenched 3d wave functions, (b) charge density, and (c) embedding in a crystal represented by four negatively charged atoms (black dots).

splitting, so that electronic structure is essentially given by the crystal-field splitting. However, many oxides are intermediate cases (Mattheis 1972), and in metals, the crystal field yields relatively small corrections to the band splitting.

3.3.4 Quenching

Let us now consider a far-reaching question. Spin-orbit coupling requires an orbital moment, but experiment shows that the magnetic moment of Fe, Co, and Ni originates nearly exclusively from the *spin* of the 3d electrons. This is striking, because the magnetic quantum number m_z is supposed to describe the orbital moment. To check this point from a basic quantum-mechanical point of view, let us calculate the orbital moment $\langle L_z \rangle \sim -i\hbar \int \psi^* \partial\psi/\partial\phi \, d\phi$ for the wave functions of (2.30). The integral is *zero* for all orbitals, in agreement with experiment but at unease with the orbital-moment interpretation of the magnetic quantum number. How can we explain this disagreement?

The answer lies in the orthogonality of the set of 3d wave functions (2.30). Any orthogonalized combination of these functions provides an equally valid mathematical description but may contain different physics. For example, combining a wave function of orbital moment $+m_z$ with a wave function of orbital moment $-m_z$ may yield two wave function with zero orbital moment. Let us consider the functions $|xy\rangle$ and $|x^2 - y^2\rangle$. Exploiting the relation $\exp(\pm ix) = \cos(x) \pm i\sin(x)$, we construct

$$|\pm 2\rangle = \sqrt{\frac{15}{32\pi}} \, R_{3d}(r) \, \sin^2\theta \, e^{\pm 2i\phi} \tag{3.23}$$

where $\langle L_z \rangle = \pm 2$. More generally, we have two sets of wave functions, namely real wave functions $|xy\rangle$, $|yz\rangle$, $|zx\rangle$, $|x^2 - y^2\rangle$ and $|z^2\rangle$, and complex wave functions $|0\rangle$, $|\pm 1\rangle$, $|\pm 2\rangle$. The sets of wave functions are linear combinations of each other, as in the above example, where $|\pm 2\rangle \sim |x^2 - y^2\rangle \pm |xy\rangle$. A similar distinction is known from other quantum-mechanical problems. For example, the wave function $\psi \sim \exp(ikx)$ describes a moving particle of momentum $p = \hbar k$, whereas $\psi \sim \cos(kx)$ is a standing wave. This analogy is the key to the understanding of the orbital moment. The $|xy\rangle$ state shown in Fig. 3.8(c) is the ground state, because the electrons oscillate in the energy valleys between the hills (black dots). Moving the electrons uphill, by occupying

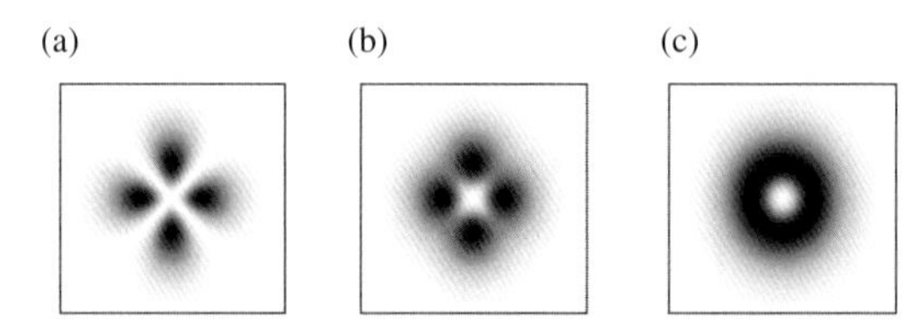

Fig. 3.9 Quenching and electron density: (a) quenched, (b) partially quenched, and (c) unquenched. Pictorially, (a–b) mean that the 3d electrons oscillate in the valleys of the crystal-field potential.

the $|x^2 - y^2\rangle$ state, costs crystal-field energy. The same is true for an orbital motion, where the path is that of a wanderer and the average energy is intermediate between valleys and hills.

Figure 3.9 shows the electron densities for quenched, partially quenched, and unquenched $|\pm 2\rangle$ orbitals. So, what determines the degree of quenching? Crystal-field interaction favors quenched orbitals but competes against spin-orbit coupling. According to (3.18), the presence of a spin moment makes it favorable to develop an orbital moment of appropriate sign. The stronger the spin-orbit coupling, the more pronounced the circular-current or "running-wave" character of the orbital and the higher the orbital moment.

3.3.5 Spin-orbit coupling

Spin-orbit coupling is an important term in the relativistic Pauli expansion, (3.18). Its strength is parameterized by the one-electron spin-orbit coupling λ. In a simple but essentially valid classical picture, spin-orbit coupling is the dipole interaction between the spin of an electron and the magnetic field created by the electron's own orbital motion. In (3.18), this aspect of spin-orbit coupling is contained in the rather specific $\mathbf{L} \cdot \mathbf{S}$ form of the coupling. Basically, the $\mathbf{L}$ refers to the orbital moment associated with the spherical atomic potential V_o. However, as emphasized by Jones and March (1973), the sphericity of the potential isn't a necessary condition, and any potential gradient ∇V gives rise to a spin-orbit coupling $\boldsymbol{\sigma} \cdot (\nabla V \times \mathbf{p})$. Examples are the Rashba splitting in two-dimensional electron gases (Bychkov and Rashba 1984) and a small interstitial anisotropy contribution in layered magnets (Skomski 1996a).

For spherical potentials, spin-orbit coupling constant is equal to the quantum-mechanical average $\lambda = Ze^2\langle 1/r^3\rangle/8\pi\varepsilon_\mathrm{o} m_\mathrm{e}^2 c^2$. The $1/r^3$ dependence indicates that spin-orbit coupling is large for inner electrons but small for conduction electrons and macroscopic currents, similar to Section 1.2. For hydrogen-like orbitals, the evaluation of $\langle 1/r^3\rangle$ yields

$$\lambda = \frac{m_\mathrm{e}}{2}\, Z^4\, \alpha^4\, c^2\, \frac{1}{n^3 l(l+\frac{1}{2})(l+1)} \tag{3.24}$$

This equation reveals the strong influence of the effective charge. The heavier the element and the closer the electrons to the nucleus, the higher Z and λ. Furthermore, being close to the nucleus helps to reduce the crystal field interaction, which competes against the spin-orbit coupling. Iron-series 3d electron and rare-earth 4f electrons have respective spin-orbit couplings of about 0.05 eV and 0.2 eV.

3.4 The single-ion model of magnetic anisotropy

Summary Single-ion anisotropy is usually much stronger than pair anisotropy and combines electrostatic crystal-field and relativistic spin-orbit interactions. The spin-orbit coupling creates current loops that interact with the anisotropic crystal field. Rare-earth anisotropy involves unquenched wave functions, and the magnetic anisotropy energy is approximately equal to the electrostatic energy of rigid 4f ions in the crystal field. This is exploited in models such as the point-charge model and the superposition model. In 3d magnets, the orbital moments are largely quenched, and spin-orbit coupling is a small perturbation to the leading crystal-field and hopping terms. Heavy transition-metal atoms combine strong crystal-field interaction with strong spin-orbit coupling and are intermediate between 3d and 4f atoms.

Single-ion anisotropy is the leading anisotropy mechanism in most magnetic materials. In a nutshell, it means that the rotation of atomic moments (spins) modifies the charge cloud of the electrons and thereby changes the electrostatic energy of the electron. The interaction between the spin and the charge cloud is mediated by spin-orbit coupling, whereas the electrostatic energy is that of the atomic orbitals in the crystal field. This important connection was first pointed out by Bloch and Gentile (1931). Alternatively, spin-orbit coupling creates orbital currents, which interact with the anisotropic crystalline environment. The details of the mechanism depend on the degree of quenching.

The main distinction is between rare-earth 4f and iron series 3d magnets. Rare-earth moments are largely unquenched, and the spin-orbit coupling ensures that the 4f charge cloud is rigidly coupled to the spin. By contrast, 3d moments are often largely quenched, and the shape and orientation of the 3d charge clouds is determined by the crystal field. Spin-orbit coupling is a small correction, and both the orbital moment and the anisotropy are obtained by perturbation theory. These two limits can be quantified in terms of typical interaction energies. Iron-series 3d electrons, which are relatively extended, have crystal field-energies of the order of 1 eV, as compared to spin-orbit couplings of about 0.05 eV. By contrast, rare-earth 4f electrons are close to the nucleus, largely screened from the crystal field, and characterized by spin-orbit couplings of about 0.2 eV and crystal-field interactions of the order of 0.01 eV. The magnetism caused by 4d, 5d, and 5f electrons is intermediate, characterized by spin-orbit and crystal-field interactions that are both very strong.

3.4.1 Rare-earth anisotropy

The room-temperature anisotropy of some rare-earth transition-metal intermetallics is significantly higher than that of any other magnetic materials. Examples of K_1 are 17.0 MJ/m^3 in $SmCo_5$, 4.9 MJ/m^3 in $Nd_2Fe_{14}B$, and 8.6 MJ/m^3 in $Sm_2Fe_{17}N_3$. By comparison, bcc iron has $K_1 = 0.05$ MJ/m^3. In the rare-earth intermetallics, most of the anisotropy comes from the rare-earth, in spite of small numbers of rare-earth

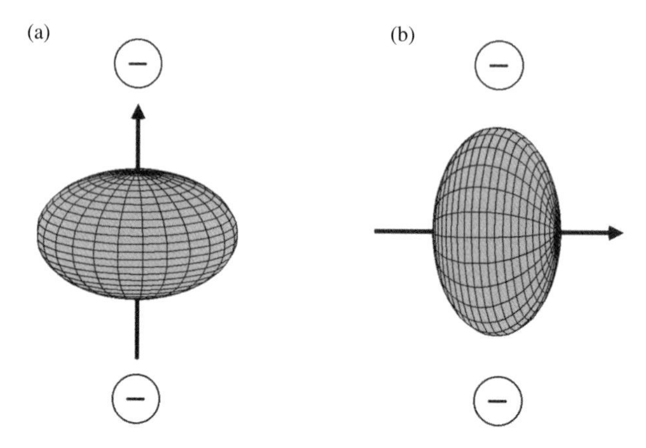

Fig. 3.10 Single-ion anisotropy of a Nd^{3+} ion in an axial crystal field: (a) magnetization along the easy direction and (b) magnetization along a hard direction. The Nd moment (arrow) is rigidly coupled to the oblate 4f charge cloud, and the easy axis is determined by the electrostatic interaction of the charge cloud with the negative crystal-field charges.

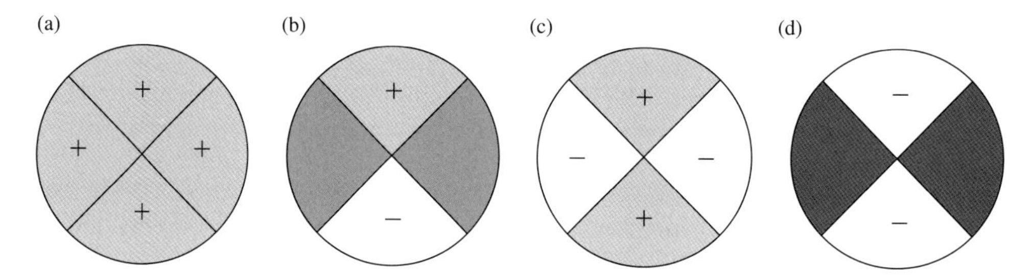

Fig. 3.11 Electrostatic multipole moments: (a) electric charge or monopole, (b) electrostatic dipole moment, and (c–d) typical quadrupole moments. Rare-earth 4f shells are of the type (d). For the derivation of multipole moments, see Section A.4.2.

atoms per formula unit. Figure 3.10 illustrates the physics of the rare-earth anisotropy. The dominance of the rare-earth spin-orbit coupling ensures an unquenched orbital moment, and the rotational symmetry of the Nd^{3+} charge clouds is an equivalent of the unquenched character of the 3d wave function shown in Fig. 3.9(c). Furthermore, the crystal field acting on the 4f electrons is largely screened by rare-earth 5d and 6s electrons (not shown in Fig. 3.10). On changing the spin direction, the orbital moment and the charge cloud follow the spin, because the crystal field is too weak to break the spin-orbit coupling. However, changing the spin direction modifies the crystal-field energy. In Fig. 3.10, the energy of configuration (a) is lower than that of (b), because the negative 4f charges and the negative crystal-field charges repel each other. This repulsion is the source of the magnetocrystalline anisotropy of rare-earth magnets.

Quantifying the rare-earth anisotropy involves two tasks: (i) the determination of the shape of the rare-earth ions and (ii) the analysis of the crystal-field interaction. This will be done in the framework of the point-charge model of crystal-field interaction, which is a special case of the superposition model. For simplicity, we focus

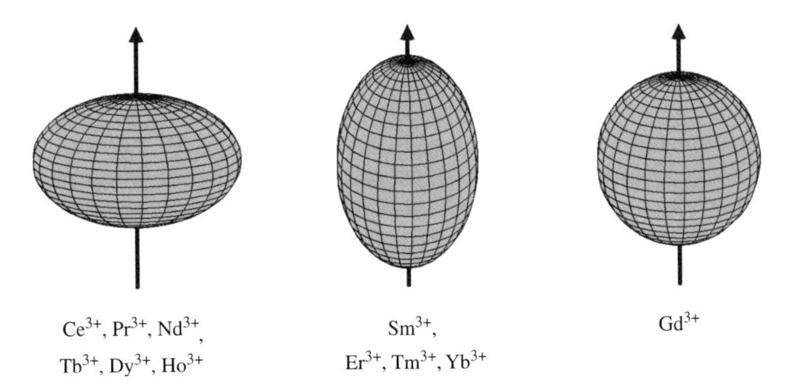

Fig. 3.12 Charge distribution of the 4f electron clouds of tripositive rare-earth ions. There are three types: oblate or pancake-like, prolate or cigar-like, and spherical.

on the lowest-order uniaxial anisotropy constant K_1. Higher-order contributions have been discussed, for example, by Herbst (1991) and Skomski and Coey (1999).

The prolate (cigar-like) or oblate (pancake-like) shape of the rare-earth 4f shells is described by the quadrupole moment of the 4f charge cloud. The idea is to represent the electron distribution $n_{4f}(\mathbf{r})$ as a multipole expansion. Figure 3.11 illustrates the physical meaning of some low-order multipole moments. The zeroth moment, $Q_0 = \int n_{4f}(\mathbf{r})\,dV$, is equal to the total number of 4f electrons, or to the (negative) total charge of the 4f shell. However, this contribution is isotropic and does not influence the magnetic anisotropy. The electrostatic dipole moment $Q_1 = \int n_{4f}(\mathbf{r}) \cos\theta' r\,dV$ is zero by the symmetry, in analogy to Fig. 3.6(c). Here we have used θ' to denote the angle between the 4f symmetry axis and the 4f charge element, so that $n_{4f}(\mathbf{r}) = n_{4f}(r, \theta')$. The lowest nonvanishing anisotropic moment, the *quadrupole moment*,

$$Q_2 = \int n_{4f}(\mathbf{r})(3\cos^2\theta' - 1)\, r^2\, dV \tag{3.25}$$

is responsible for the crystal-field interactions shown in Fig. 3.10. The sign of Q_2 decides whether the 4f electron cloud is prolate ($Q_2 > 0$) or oblate ($Q_2 < 0$). In turn, projected onto the quadrupole moment, the 4f charge density is

$$n_{4f}(\mathrm{r}) = \frac{5Q_2}{16\pi\langle r^2\rangle_{4f}}(3\cos^2\theta' - 1) f(r) \tag{3.26}$$

In the equation, $f(r) = R_{4f}(r)$ is the radial part of the 4f wave function.

The quadrupole moments are specific properties of the Hund's-rules rare-earth ions. Explicitly,

$$Q_2 = \alpha_J \langle r^2\rangle_{4f}\, (2J^2 - J) \tag{3.27}$$

where $\langle r^2\rangle_{4f}$ is the squared 4f-shell radius and $\alpha_J = \theta_2$ is the second-order Stevens coefficient (Hutchings 1964). Figure 3.12 illustrates the distinction between negative, positive, and zero quadrupole moments, whereas Table 3.1 specifies the quadrupole moment and some other Hund's-rules properties across the 4f series (Freeman and Watson 1962). Note that the shapes of Fig. 3.12 correpond to second-order anisotropies

Table 3.1 Ground states of rare-earth 4f ions (R^{3+}). The arrows indicate the Hund's-rules f-shell filling from $m_z = 3$ (left) to $m_z = -3$ (right). For the g-factors, see Table 2.1.

Ion	R^{3+}	3	2	1	0	−1	−2	−3	S	L	J	Q_2	Q_2/a_o^2
$4f^1$	Ce^{3+}	↑	–	–	–	–	–	–	1/2	3	5/2	⊖	−0.686
$4f^2$	Pr^{3+}	↑	↑	–	–	–	–	–	1	5	4	⊖	−0.639
$4f^3$	Nd^{3+}	↑	↑	↑	–	–	–	–	3/2	6	9/2	⊖	−0.232
$4f^4$	Pm^{3+}	↑	↑	↑	↑	–	–	–	2	6	4	θ	0.202
$4f^5$	Sm^{3+}	↑	↑	↑	↑	↑	–	–	5/2	5	5/2	θ	0.364
$4f^6$	Eu^{3+}	↑	↑	↑	↑	↑	↑	–	3	3	0	θ	–
$4f^7$	Gd^{3+}	↑	↑	↑	↑	↑	↑	↑	7/2	0	7/2	⊖	0.000
$4f^8$	Tb^{3+}	↑↓	↑	↑	↑	↑	↑	↑	3	3	6	⊖	−0.505
$4f^9$	Dy^{3+}	↑↓	↑↓	↑	↑	↑	↑	↑	5/2	5	15/2	⊖	−0.484
$4f^{10}$	Ho^{3+}	↑↓	↑↓	↑↓	↑	↑	↑	↑	2	6	8	⊖	−0.185
$4f^{11}$	Er^{3+}	↑↓	↑↓	↑↓	↑↓	↑	↑	↑	3/2	6	15/2	θ	0.178
$4f^{12}$	Tm^{3+}	↑↓	↑↓	↑↓	↑↓	↑↓	↑	↑	1	5	6	θ	0.427
$4f^{13}$	Yb^{3+}	↑↓	↑↓	↑↓	↑↓	↑↓	↑↓	↑	1/2	3	7/2	θ	0.409
$4f^{14}$	La^{3+}	↑↓	↑↓	↑↓	↑↓	↑↓	↑↓	↑↓	0	0	0	⊖	0.000

and do not yield fourth-order anisotropy (cubic anisotropy). To describe higher-order anisotropies, one must consider hexadecapole deviations from the quadrupole shapes, described by the Stephens coefficients $\beta_J = \theta_4$ (electrostatic hexadecapole) and $\gamma_J = \theta_6$ (hexacontatetrapole or 64-pole moments)

The prolate or oblate shape of the 4f charge clouds is a consequence of Hund's second rule. In each half-series, the filling of the level starts with states having a large m_z. Figure 3.7 shows that the orbital with the largest m_z are oblate rather than prolate and lie in the $x - y$ basal plane. Physically, this amounts to a pronounced in-plane orbital or "circular-current" motion of the electrons. When the oblate orbitals with positive m_z are used up, the filling continues with the prolate orbitals, making the total shape prolate at some point. The orbitals added last are oblate, with $m_z < 0$, and ensure that each half shell is spherical. As a consequence, tripositive ions of the first three lanthanides of each half-shell, Ce, Pr, Nd and Tb, Dy, Ho, exhibit oblate 4f charge distributions, whereas the 4f orbitals of Pm, Sm, Eu and Er, Tm, Yb are prolate. Spherical shells (La^{3+} and Gd^{3+}) do not contribute to the anisotropy. La^{3+} is actually nonmagnetic, whereas Gd^{3+} combines zero magnetocrystalline anisotropy of spin-orbit origin with a large spin-only moment. This means that Gd^{3+} doesn't exhibit magnetocrystalline anisotropy of relativistic origin, but there may be magnetostatic dipole anisotropy. (In a strict sense, crystal-field interactions, including hybridization, yield some admixture of excited multiplets and a small residual 4f anisotropy.)

The *anisotropy energy* is equal to the electrostatic energy of the 4f charge distribution $n_{4f}(\mathbf{r})$ in the crystal field

$$E_a(\theta, \phi) = -\frac{e}{4\pi\varepsilon_0} \int \frac{n_{4f}(r, \theta'; \phi, \theta)\, \rho(\mathbf{R})}{|\mathbf{R} - \mathbf{r}|} \mathrm{d}\mathbf{r}\, \mathrm{d}\mathbf{R} \qquad (3.28a)$$

where $\rho(\mathbf{R})$ is the charge density of the crystal-field charges. This equation can also be written as

$$E_{\mathrm{a}}(\theta, \phi) = \int \mathrm{n}_{4\mathrm{f}}(r, \theta'; \phi, \theta)\, V_{\mathrm{CF}}(r)\, \mathrm{d}V \qquad (3.28\mathrm{b})$$

where $\mathrm{d}V = \mathrm{d}\mathbf{R}$. A straightforward way to determine the magnetocrystalline anisotropy is to numerically evaluate this integral for different magnetization directions θ and ϕ. However, this is a rather cumbersome, and the question arises how to parameterize the crystal field. In the *point-charge model*, the crystal field is approximated by electrostatic charges $\rho(\mathbf{r}) = Q_{\mathrm{i}}\delta(\mathbf{r} - \mathbf{R}_{\mathrm{i}})$. In the model, introduced by Bethe (1929) to explain atomic energy levels in solids, Q_{i} and $\mathbf{R}_{\mathrm{i}}$ are the charge and the position of the crystal-field creating atom, respectively. Figure 3.13 illustrates the idea for cubic and tetragonal symmetries.

In a given crystal-field environment, the sign of the rare-earth anisotropy depends on whether the ion is prolate or oblate. Replacing the prolate ion in Fig. 3.13 by an oblate ion reverses the sign of the anisotropy, making (c) the easy-axis configuration and changing the anisotropy of (b) to easy-plane, because an oblate charge cloud would be repelled by the essentially planar distribution of the negative crystal-field charges. This effect has been used to create and optimize permanent magnets. For example, in the $R_2Fe_{14}B$ series, the crystal field is of the type of Fig. 3.13(c). Sm^{3+} ions are prolate and, consequently, $Sm_2Fe_{14}B$ exhibits easy-plane anisotropy and is magnetically rather soft. The opposite is true for Nd^{3+}, and $Nd_2Fe_{14}B$ is the most powerful permanent magnet developed so far (Herbst 1991, Coey 1996). Note that the crystal-field creating charges are not necessarily magnetic. A good example is interstitial nitrogen in Sm_2Fe_{17}, which changes the anisotropy from easy-plane to easy-axis. In the corresponding interstitial compound $Sm_2Fe_{17}N_3$, the negatively charged nitrogen atoms surround the Sm^{3+} ions in the a–b plane (Skomski and Coey 1999).

3.4.2 Point-charge model

To derive explicit expressions for the anisotropy constants, it is necessary to quantify the crystal-field interaction. In the point-charge model, the crystal field is a sum of atomic contributions $Q_{\mathrm{i}}(\mathbf{r}_{\mathrm{i}})$. This model was introduced by Bethe (1929) to explain optical spectra of solids. It is convenient to expand the crystal-field potential

$$V_{\mathrm{CF}}(\mathbf{r}) = -\frac{\mathrm{e}}{4\pi\varepsilon_{\mathrm{o}}} \sum_{\mathrm{i}=1}^{\mathrm{N}} \frac{Q_{\mathrm{i}}}{|\mathbf{R}_{\mathrm{i}} - \mathbf{r}|} \qquad (3.29)$$

into powers of $\mathbf{r}$. The expansion is physically meaningful, because the radius of the 4f charge clouds, about 0.5 Å, is smaller than $R_{\mathrm{i}} \geq 3$ Å. Creating a cubic crystal field by putting six charges Q on the axes ($x = \pm a,\ y = \pm a,\ z = \pm a$) yields

$$V_{\mathrm{CF}}(\mathbf{r}) = \frac{35 Q_{\mathrm{e}}}{8\pi\varepsilon_{\mathrm{o}} a^5}(x^2y^2 + y^2z^2 + z^2x^2) \qquad (3.30)$$

A tetragonal distortion of this environment, so that $a = b \neq c$, yields in lowest order

$$V_{\mathrm{CF}}(\mathbf{r}) = \frac{Q\mathrm{e}}{4\pi\varepsilon_{\mathrm{o}}}\left(\frac{1}{a^3} - \frac{1}{c^3}\right)(3z^2 - r^2) \qquad (3.31)$$

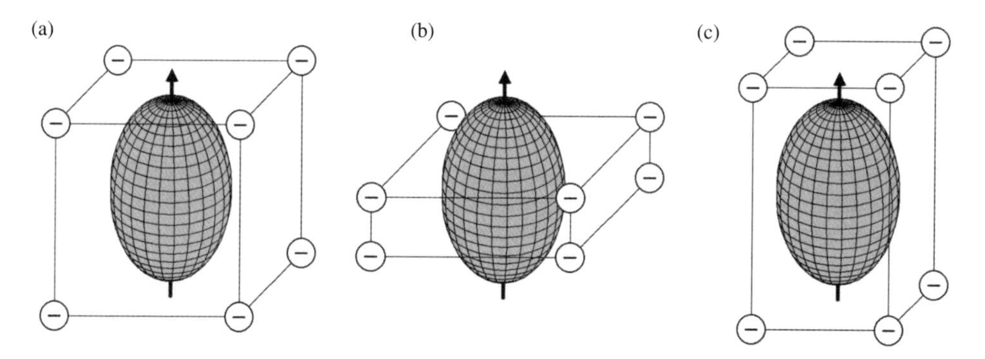

Fig. 3.13 Point-charge model, as applied to the crystal-field interaction of a prolate ion, such as Sm^{3+}: (a) cubic crystal field and (b–c) tetragonal crystal fields of opposite crystal-field parameter A_2^0. For the shown prolate ion, (b) and (c) are easy-axis and easy-plane, respectively. For instance, in (c), the negative 4f charges at the ends of the ion are repelled by the negative crystal-field charges, which would force the magnetization (arrow) from the *c*-axis direction into the *a*–*b* plane.

Figure 3.13 illustrates the tetragonal distortion of the cubic crystal field for a cubal environment (8 nearest neighbors).

The x, y, and z dependent functions in (3.30–31) are, essentially, spherical harmonics, in close analogy to Section 3.1. The terms that depend on the crystal-field creating charges only are conveniently collected in the form of *crystal-field parameters.* For example,

$$V_{\mathrm{CF}}(\mathbf{r}) = A_2^0(3z^2 - r^2) \tag{3.32}$$

where $A_2^0 = Q_e(1/a^3 - 1/c^2)/4\pi\varepsilon_o$ is the second-order uniaxial crystal-field parameter for the considered environment. For a given series of isostructural compounds, the crystal-field parameters A_2^0 are therefore largely independent of the rare-earth. Examples are $300\,\mathrm{K/a_o^2}$ for $R_2Fe_{14}B$, $34\,\mathrm{K/a_o^2}$ for R_2Fe_{17}, and $-358\,\mathrm{K/a_o^2}$ for $R_2Fe_{17}N_3$. Since the crystal-field expansion amounts to a decomposition of the crystal-field potential into orthogonal contributions, similar relations are obtained for higher-order and non-uniaxial crystal fields. It can be shown that n-th order point-charge crystal fields scale as $(a_o/R)^{n+1}$. This is one reason for the relative smallness of higher-order rare-earth anisotropy contributions.

Putting eqs. (3.32) and (3.26) into (3.28b) and evaluating the energy as a function of the magnetization angle θ yields

$$E_a = \frac{1}{2}Q_2A_2^0(3\cos^2\theta - 1) \tag{3.33}$$

and the lowest-order uniaxial anisotropy constant

$$K_1 = -\frac{3}{2V_R}Q_2A_2^0 \tag{3.34}$$

Here V_R is the crystal volume per rare-earth atom. Equation (3.34) solves the anisotropy problem by expressing K_1 in terms of the shape of the 4f shell, described by Q_2, and the crystal environment, described by A_2^0. The easy-axis anisotropy is achieved by using oblate ions, such as Nd^{3+}, on sites where the crystal-field parameter A_2^0 is positive, and prolate ions, such as Sm^{3+}, in crystalline environments where A_2^0 is negative.

3.4.3 The superposition model

A specific and somewhat complicated feature of the derivation of (3.33) is the integration over the angle θ' denoting the position of the crystal-field charges relative to the magnetization direction. Direct integration of (3.28b) is an option, but an easier way is to use the *superposition model*of crystal-field interaction (Newman and Ng 1989). The model is based on two basic assumptions. First, the total crystal field is a sum of atomic crystal-field contributions. Second, the charge density of each crystal-field creating atom or "ligand" is axially symmetric about the line from the rare earth to the ligand. The point charge model satisfies these two criteria, and we can write the total crystal field as a sum of *intrinsic* crystal-field contributions. For example,

$$A_2^0 = \sum_{\mathrm{i}} A_2'(R_{\mathrm{i}}) \frac{1}{2}(3\cos^2\Theta_{\mathrm{i}} - 1) \tag{3.35}$$

where the summation includes all neighbors and $\mathrm{R_i}$ is the distance between the rare-earth atom and the i-th ligand. The angle Θ_{i} is the coordination angle of the i-th ligand. For example, in the respective Figs 3.10(a) and (b), $\Theta = 0$ and $\Theta = 90°$.

The intrinsic crystal-field parameter A_2' is obtained most conveniently by putting $\Theta = 0$, that is, by considering atoms in axial coordination, as in Fig. 3.10(a). Starting from eqs. (3.29) and (3.31), we exploit that $z = r = R$ and obtain

$$A_2'(\mathrm{r}) = -\frac{\mathrm{e}Q}{4\pi\varepsilon_{\mathrm{o}}} \frac{1}{2R^3} \tag{3.36}$$

Typical experimental crystal-field charges Q are *negative*, which means that the ligands' electron shells are more important than their nuclei. An exception is hydrogen, which may give rise to zero or positive (protonic) crystal-field charges.

The point-charge model was originally designed to describe insulators, where the assumption of electrostatic point charges is, to some extent, meaningful. Crystal-field charges in metals are strongly screened by conduction electrons. Describing the conduction electrons as a free-electron gas yields (Skomski 1994)

$$\mathrm{A}_2'(\mathrm{r}) = -\frac{\mathrm{e}Q}{4\pi\varepsilon_{\mathrm{o}}} \frac{\mathrm{e}^{-qR}}{2R^3} \left(1 + qR + \frac{1}{3}q^2R^2\right) \tag{3.37}$$

where $q \approx 2.3/\text{Å}$ is an inverse Thomas-Fermi screening length. This reduces the crystal-field charges to a reasonable order of magnitude. More generally, the superposition model is relatively insensitive to the physical origin of the crystal field. For example, the electrostatic interaction of (3.29) may be replaced by hybridization contributions (Ballhausen 1962).

3.4.4 Transition-metal anisotropy

The calculation of the rare-earth anisotropy was simplified by the dominance of the spin-orbit coupling. The rare-earth wave functions are unquenched, similar to Fig. 3.9(c), so that electrostatic energy of the 4f charge clouds is equal to the anisotropy energy. By contrast, in *iron-series 3d* metals and nonmetals, the spin-orbit coupling is a small correction to the crystal-field. If the spin-orbit coupling were zero, the orbits would be fully quenched, corresponding to zero anisotropy. In reality, the quenching is incomplete, establishing a picture somewhere between Fig. 3.9(a) and 3.9(b). The residual orbital moment leads to some anisotropy in iron-series magnets. Typically, this 3d anisotropy is much smaller than rare-earth anisotropies, but some noncubic or strained cubic 3d magnets exhibit room-temperature anisotropies higher than $1\,\mathrm{MJ/m^3}$. A well-known example is YCo_5 ($5\,\mathrm{MJ/m^3}$) which historically anticipates rare-earth permanent magnets, especially the isostructural $SmCo_5$ ($14\,\mathrm{MJ/m^3}$). The origin of palladium-series (4d), platinum-series (5d), and actinide (5f) anisotropies is similar to that of iron-series (3d) magnets but characterized by stronger spin-orbit coupling. High anisotropies are common in these series, especially in actinides. These anisotropies have been exploited for many decades in $L1_0$ magnets such as PtCo (A.5.1). Another example is the "giant anisotropy" in thin-film structures such as Pt-Co (Gambardella *et al.* 2003), where Co atoms create polarization clouds in the Pt environment. This leads to some anisotropy per Pt atom and therefore to a large anisotropy per 3d atom, similar to the creation of giant moments in dilute alloys, Section 2.4.3.

In terms of the d-electron wave functions $|\Psi_1\rangle = |xy\rangle$ and $|\Psi_2\rangle = |x^2 - y^2\rangle$, the Hamiltonian is

$$E_{\mathrm{ik}} = \begin{pmatrix} A & 0 \\ 0 & -A \end{pmatrix} + 2\lambda\cos\theta \begin{pmatrix} 0 & \mathrm{i} \\ -\mathrm{i} & 0 \end{pmatrix} \tag{3.38}$$

where $2A$ is the crystal-field splitting, λ is the spin-orbit coupling constant, and θ is the angle between spin direction and z-axis. In the Hamiltonian, the spin-orbit interactions is a perturbation that couples the two crystal-field levels. The energy levels (eigenvalues) the Hamiltonian are

$$E_1 = +\sqrt{A^2 + 4\lambda^2\cos^2\theta} \tag{3.39a}$$

$$E_0 = -\sqrt{A^2 + 4\lambda^2\cos^2\theta} \tag{3.39b}$$

Next we assume that the lower-lying level E_0 is occupied by one electron and the other level is empty, so that the energy of the system is equal to $E_0(\theta)$. The corresponding anisotropy constant is estimated by comparing the energies for $\theta = 0$ and $\theta = 90°$

$$K_1 = \frac{1}{V_{\mathrm{T}}}\left(\sqrt{A^2 + 4\lambda^2} - A\right) \tag{3.40}$$

where V_{T} is the crystal volume per 3d atom. Expanding K_1 for small spin-orbit coupling yields $K_1 = 2\lambda^2/AV_{\mathrm{T}}$. This result is specific to the present model, but the λ^2/A

dependence is quite general and describes second-order 3d anisotropies obtained by perturbation theory. When $A \sim O$ in (3.30–40), as it happens in some special crystal-field environments, then K_1 may approach or exceed $|M\rangle/\mathrm{m}^3$ for 3d magnets.

In 3d oxides, the level splitting $2A$ is provided by the electrostatic crystal field (Bloch and Gentile 1931), but in 3d, 4d, and 5d metals it reflects interatomic hopping. For some k-vectors, the splitting may be very low, amounting to a disproportionately strong anisotropy contribution. However, a condition is that the Fermi level lies between E_0 and E_1. If both levels are occupied, the total energy $E_0 + E_1$ does not depend on θ, and the corresponding K_1 contribution is zero. The lowest-order expression for *cubic* anisotropy scales as λ^4/A^3. This explains the low cubic anisotropy of bcc iron ($0.05\,\mathrm{MJ/m^3}$) and Ni ($-0.005\,\mathrm{MJ/m^3}$), as compared to that of hexagonal Co ($0.5\,\mathrm{MJ/m^3}$). To obtain quantitative itinerant anisotropy constant, it is necessary to perform numerical calculations (Brooks 1940, Daalderop, Kelly, and Schuurmans 1990, Trygg *et al.* 1995, and Johnson *et al.* 1996).

When the magnetization is parallel to the z-axis ($\theta = 0$), the wave function belonging to E_0 is

$$|\psi\rangle = \cos\frac{\chi}{2}\,|x^2 - y^2\rangle + \mathrm{i}\sin\frac{\chi}{2}|xy\rangle \tag{3.41}$$

where $\chi = \mathrm{arccot}(A/2\lambda)$. The admixture of $|xy\rangle$ character gives rise to the orbital moment

$$m_\mathrm{L} = \frac{4\lambda}{\sqrt{A^2 + 4\lambda^2}}\,\mu_\mathrm{B} \tag{3.42}$$

The equation describes the degree of quenching. In the absence of spin-orbit coupling, $m_\mathrm{L} = 0$, $\lambda \gg A$ yields $m_\mathrm{L} \approx 2$.

An approach to the treatment of arbitrary crystals is perturbation theory. In second-order, the anisotropy energy is obtained from

$$E_\mathrm{a} = E_\mathrm{o} - \lambda^2\Sigma_\mathrm{m}\frac{\langle\Psi_\mathrm{o}|\mathbf{L}\cdot\mathbf{S}|\Psi_\mathrm{m}\rangle\langle\Psi_\mathrm{m}|\mathbf{L}\cdot\mathbf{S}|\Psi_\mathrm{o}\rangle}{E_\mathrm{m} - E_\mathrm{o}} \tag{3.43}$$

where $|\Psi_\mathrm{o}\rangle$ and $|\Psi_\mathrm{m}\rangle$ are unperturbed (many-electron) wave functions. Summation over all orbital degrees of freedom yields the *spin Hamiltonian*

$$\mathbf{H} = -\lambda^2\mathbf{S}\cdot\underline{\underline{\Lambda}}\cdot\mathbf{S} \tag{3.44}$$

$$\text{where } \underline{\underline{\Lambda}} = \Sigma_\mathrm{m}\frac{\langle\Psi_\mathrm{o}|\mathbf{L}|\Psi_\mathrm{m}\rangle\langle\Psi_\mathrm{m}|\mathbf{L}|\Psi_\mathrm{o}\rangle}{E_\mathrm{m} - E_\mathrm{o}} \tag{3.45}$$

The matrix is, essentially, equivalent to (3.5). In the uniaxial limit, $\Lambda_\mathrm{xx} = \Lambda_\mathrm{yy}$, and $K_1 = \lambda^2(\Lambda_\mathrm{zz} - \Lambda_\mathrm{xx})$.

Numerical calculations of the anisotropy start from (3.44) or from more general energy expressions. Perturbative electronic-structure calculations imply the small parameter λ/W, where W is the bandwidth (Brooks 1940, Daalderop, Kelly, and Schuurmans 1990). However, as mentioned above, certain k-points tend to give disproportionately strong anisotropy contributions and negatively affect the accuracy. This is of particular importance in low-dimensional systems, such as thin films and wires. Furthermore, the net anisotropy is a sum of relatively large positive (easy-axis) and negative (easy-plane) contributions. Minor errors may drastically change the K_1 prediction or even change the sign of K_1. In addition, there are physically significant K_1 oscillations as a function of the d-band filling. These oscillations are akin to the sign changes of the rare-earth quadrupole moments, Table 3.1. An approximate equation is $K_1 \sim (5 - \mathrm{n})(10 - \mathrm{n})(15 - 2\mathrm{n})$, where n is the number of d electrons, but individual levels have a strong impact and yield less smooth and strongly crystal-structure dependent functions $K_1(\mathrm{n})$.

The situation is particularly complex for heavy transition-metal atoms. This refers to 4d or palladium-series, 5d or platinum series, and 5f or actinide series. The bandwidths in alloys containing heavy transition metals are roughly comparable to those of 3d elements and alloys, but the spin-orbit couplings in 4d, 5d, and 5f elements are of the order 0.5 to 1 eV, as compared to $\lambda = 0.05\,\mathrm{eV}$ for 3d metals. This leads to much higher orbital moments, and perturbation theory is no longer meaningful. A further complication is that the ferromagnetism of many heavy transition-metal atoms relies on the vicinity of 3d atoms. The 3d atoms spin-polarize the local density of states of the heavy atoms, and the anisotropy is a secondary consequence of this spin polarization.

3.5 Other anisotropies

Summary Magnetoelastic anisotropy is caused by mechanical strain and tends to yield substantial anisotropy contributions in soft magnets. Physically, it is equivalent to magnetocrystalline anisotropy, because a strained cubic lattice can be considered as an unstrained lattice with reduced symmetry. Surface and interface anisotropies are of magnetocrystalline origin, too. A key feature is that their strengths and symmetries depend on the indexing of the surfaces. Other anisotropies involving spin-orbit coupling are the Dzyaloshinski-Moriya interaction in magnets with very low symmetry and the anisotropic exchange, which must not be confused with anisotropic exchange bonds. The so-called unidirectional anisotropy in exchange-coupled magnets is a biasing effect that does not involve spin-orbit coupling.

3.5.1 Magnetoelasticity

Subjecting cubic magnets to uniaxial mechanical strain yields a uniaxial anisotropy contribution. This *magnetoelastic anisotropy* has long been exploited in iron-based magnets, such as carbon steels and related alloys (Fe-Cr, Fe-Co). It is also important in soft magnets, for example in permalloy-type magnets ($\mathrm{Fe_{100-x}Ni_x}$), where the cubic anisotropy is small and the magnetoelastic contribution easily dominates the total anisotropy. Magnetoelasticity is closely related to magnetostriction, where a rotation of the magnetization direction creates a mechanical strain.

The main source of magnetoelasticity is magnetocrystalline single-ion anisotropy. For example, the crystalline configurations shown in Fig. 3.13(b) and (c) may be considered as strained cubic environments. Uniaxial magnetoelasticity is described by

$$\frac{E_{\mathrm{ME}}}{V} = -\frac{\lambda_{\mathrm{S}} E}{2}(3\cos^2\theta - 1)\varepsilon + \frac{E}{2}\varepsilon^2 - \varepsilon\sigma \tag{3.46}$$

where σ is the uniaxial stress, $\varepsilon = \Delta 1/l$ denotes the elongation along the stress axis, E is Young's modulus, and θ is the angle between the magnetization and strain axes. The strength of the magnetoelastic coupling is described by the saturation magnetostriction λ_s. Typical values for λ_s, averaged over all cubic crystal directions and measured in parts per million, are -7 for Fe, 40 for Fe_3O_4, and 1800 for $TbFe_2$.

Putting $\sigma = 0$ and $\theta = 0$ and minimizing the magnetoelastic energy with respect to ε yields the elongation $\varepsilon = \lambda_s$. This means that λ_s is the spontaneous magnetostriction in the magnetization direction. A magnet that has a spherical shape in the paramagnetic state becomes a prolate ferromagnet when $\lambda_s > 0$ but an oblate ferromagnet when $\lambda_s < 0$. A simple explanation is provided by Fig. 3.13. If the crystal-field charges in (a) were free, then the electrostatic quadrupole repulsion between the 4f charge cloud and cubic lattice would yield a distortion as in (b). Since λ_s is very small in most compounds, moderate stress $\sigma = E\varepsilon$ outweighs the spontaneous magnetostriction. This yields the magnetoelastic anisotropy energy density

$$\frac{E_a}{V} = -\frac{\lambda_s \sigma}{2}(3\cos^2\theta - 1) \tag{3.47}$$

and the magnetoelastic K_1 contribution $3\lambda_s\sigma/2$. Magnetoelastic anisotropy can be quite large, but the practical challenge is to create a big strain.

A crystal-field phenomenon occurring in highly symmetric crystals and requiring a degenerate ground state is the *Jahn-Teller* effect. Jahn-Teller ions can lower their energy by spontaneously distorting the surrounding lattice. This energy gain is small but proportional to ε, as compared to the elastic energy, which is proportional to ε^2. Minimizing the total energy results in a finite lattice distortion (see e.g. Blundell 2001). However, aside from the crystal-field analogy, the Jahn-Teller effect is unrelated to magnetocrystalline anisotropy.

3.5.2 Anisotropic exchange

Heisenberg exchange $J\mathbf{S} \cdot \mathbf{S}'$ is isotropic, because it depends on the relative orientations of $\mathbf{S}$ and $\mathbf{S}'$ but is independent of the angle between magnetization and crystal axes. The isotropy is a consequence of the electrostatic origin of the exchange. *Anisotropic* exchange means that the energy depends not only on the relative orientation of the interacting spin but also on the crystalline orientation of the spins. An extreme example is the Ising exchange, $J\ S_z\ S'_z$, where only the z-component of the spin contributes. As we will discuss in Chapter 5, the main purpose and advantage of the Ising model is the description of finite-temperature magnetization effects. Applied to *real* magnets, it amounts to the unreasonable assumption of infinite ratio of anisotropy to exchange. There are a few systems with artificially suppressed interatomic exchange (de Jongh and Miedema 1975, Sellmyer and Nafis 1986), but for most materials, the Ising model is a completely inadequate starting point.

Writing the exchange as $J_{xx}S_xS'_x + J_{yy}S_yS'_y + J_{zz}S_zS'_z$, we find that typical exchange anisotropies are $|J_{xx} - J_{zz}|/J_{zz}$ and $|J_{yy} - J_{zz}|/J_{zz}$ are very small. The same is true for intra-atomic exchange. The effect is due to spin-orbit coupling and observed, for example, as a small dependence of the spontaneous magnetization M_s on the magnetization direction. In hexagonal Co, M_s decreases by about 0.5% on turning the magnetization from the easy magnetization direction (c-axis) into the $a-b$ basal plane (Wijn 1991). In rare-earth magnets, the effect is not much bigger, because $K_1 V_R$ tends to be much smaller than $J \sim k_B T_c$.

The small relativistic anisotropy of the Heisenberg exchange must not be confused with the generally much stronger *bond anisotropy.* For example, many intermetallic compounds and artificial structures have a layered structure, with strong intralayer exchange but much weaker interlayer exchange. This leads to a real-space anisotropy of the exchange stiffness but does not mean that turning the spin system from in-plane to perpendicular changes the energy of the magnet.

By definition, the anisotropy energy remains unchanged on reversing the magnetization direction, $E_a(\mathbf{M}) = E_a(-\mathbf{M})$. In a strict sense, this means that there are no odd-order anisotropy contributions. However, some systems exhibit odd-order energy contributions that are traditionally classified as anisotropies. Exchange bias due to the coupling to an antiferromagnetic phase yields a uni*directional* anisotropy observed as a shift of the hysteresis loop. This effect was discovered by Meiklejohn and Bean (1956), who investigated Co nanoparticles surrounded by cobaltous oxide (CoO). The sign of the shift is determined by the antiferromagnetic sublattice that interacts more strongly ith the ferromagnetic Co phase.

An unconventional anisotropy contribution of relativistic exchange interactions is the *Dzyaloshinski-Moriya* or DM interaction $\mathsf{H}_{DM} = -\frac{1}{2}\Sigma_{ij}\mathbf{D}_{ij} \cdot \mathbf{S}_i \times \mathbf{S}_j$, where the vector $\mathbf{D}_{ij} = -\mathbf{D}_{ji}$ reflects the local environment of the magnetic atoms. Net DM interactions require local environments with sufficiently low symmetry (absence of inversion symmetry). Physically, the orbit of an electron and therefore its crystal-field interaction depend on the spin direction, and electrons on sites without inversion symmetry can minimize the crystal-field energy by forming a slightly noncollinear spin structure. Phenomenologically, the interaction favors noncollinear spin states, because parallel spins $\mathbf{S}_i$ and $\mathbf{S}_j$ mean that $\mathsf{H}_{DM} = 0$.

DM interactions occur in some crystalline materials, such as α-Fe_2O_3 (hematite), in amorphous magnets (Moorjani and Coey 1984), spin glasses (Fischer and Hertz 1991), and in magnetic nanostructures (Sandratskii 1998, Skomski 2003). The resulting canting is small, because it competes against the leading Heisenberg exchange, but it is relatively easily observed in "weakly ferromagnetic" antiferromagnets such as hematite (α-Fe_2O_3), where there is no ferromagnetic background. These noncollinear states must not be confused with micromagnetic spin structures, such as domains and domain walls (Section 4.2), although both phenomena involve spin-orbit coupling. They are also different from noncollinear structures caused by competing interatomic exchange (Section 2.3.3).

3.5.3 Models of surface anisotropy

An anisotropy contribution of particular importance in magnetic thin films and nanostructures is surface and interface anisotropy. A necessary condition is a reduced symmetry at the surface or interface. However, this condition is not sufficient, and contradictory to widespread belief there is no such thing as a normal anisotropy that automatically appears at a surface. As any other anisotropy, surface anisotropy obeys the laws of crystal-field theory, and the strength and symmetry of the anisotropy is determined by the interplay between crystal-field interaction and spin-orbit coupling.

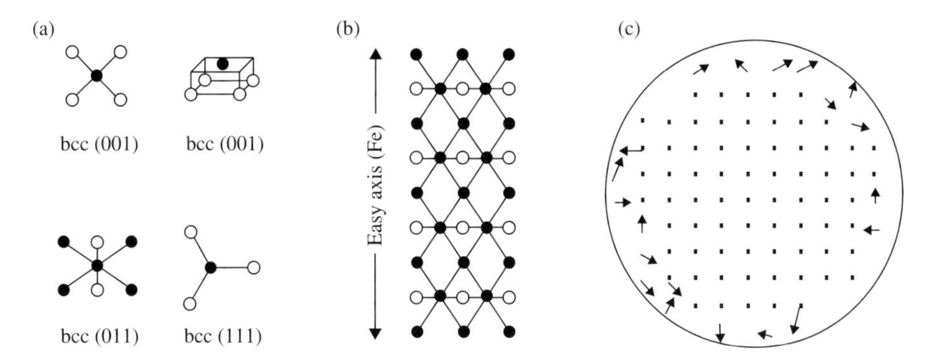

Fig. 3.14 Surface anisotropy and symmetry: (a) some bcc surfaces, (b) a nanowire with a pronounced easy axis along the nanowires, and (c) surface anisotropy of an irregular nanoparticle. Small white circles denote subsurface atoms.

Broken exchange bonds may indicate or represent crystal-field changes, but they do not contribute to the magnetic anisotropy. The Néel model (Section 3.2) does not provide an explanation, because it parameterizes the anisotropy in terms of relatively weak pair interactions of magnetostatic origin, rather than considering single-ion crystal-field interactions.

Figure 3.14 illustrates the atomic origin of surface anisotropy. Both the direction and the strength of the anisotropy depend on the indexing of the surface. For example, in Fig. 3.14(a), only the bcc (011) surface supports a strong second-order in-plane anisotropy (Sander *et al.* 1996). The corresponding in-plane anisotropies of the (001) and (111) surfaces are fourth- and sixth-order, respectively. In appropriately structured nanoparticles, the second-order contributions yields a preferred magnetization direction, as in Fig. 3.14(b). The easy axis follows from the (slightly) oblate character of the Fe 3d clouds and the negative crystal-field charge of the Fe atoms. In many structures, the prolaticity changes sign between Fe and Co, so that these findings cannot be generalized to other elements. For example, a prolate 3d cloud would yield an easy-axis contribution *normal* to the film plane, with corresponding changes in Fig. 3.14(b).

Starting with Gay and Richter (1986), first-principle surface and interface calculations have been performed for a variety of systems. such as multilayers (Johnson *et al.* 1996), and the same applies to magnetic clusters (Wang, Wu, and Fieeman 1993), interfaces (Victora and McLaren 1993), and nanowires (Komelj *et al.* 2002, Eisenbach *et al.* 2002). These calculations confirm and specify the findings summarized in Figs 3.14(a–b). Depending on the indexing or the surface and on the considered element, the preferential magnetization axis is in-plane or normal, and K_1' is generally nonzero.

An important point is that surface anisotropies easily dominate the bulk anisotropy of *cubic* materials. Fourth-order bulk anisotropies are about two orders of magnitude smaller than second-order anisotropies. Due to the comparatively large number N_s of surface atoms of small particles or clusters, the surface contribution dominates the bulk anisotropy in particles smaller than about 3 nm, even if one takes into account that the net surface anisotropy is not necessarily linear in N_s but tends to scale as

$N_s^{1/2}$ due to random-anisotropy effects. The point is that typical nanoparticles contain surface patches with many different indexings, and the corresponding anisotropy contributions are likely to compensate each other. The resulting net anisotropy is generally biaxial, involving both K_1 and K_1', and there is generally no physical justification for considering nanoparticles as uniaxial magnets.

The competition of a variety of bulk and surfaces anisotropies of different orders and symmetries may give rise to spin-reorientation transitions as a function of film thickness or temperature. Typical surface-anisotropy contributions are limited to surface and subsurface atoms, with very small contributions from atoms deeper in the bulk. Applied to thin films, this has given rise to the phenomenological anisotropy expression

$$K_1 = \frac{K_s}{t} + K_o \tag{3.48}$$

where t is the film thickness, K_s is the surface anisotropy, and K_o includes the bulk magnetocrystalline anisotropy $K_1(\text{bulk})$ and the shape anisotropy $-DM_s$. For 3d magnets, a typical order of magnitude is $K_s = 0.5\,\text{mJ/m}^2$. When K_o and K_s favor in-plane and perpendicular anisotropy, respectively, then there is a spin-reorientation transition at the thickness $K_s/|K_o|$.

Exercises

1. ***S* = 0 *and* *S* = 1/2 *anisotropy.*** Show that quenched ions with $S = 0$ and $S = 1/2$ have zero anisotropy.
2. ***Energy of s electrons.*** Compare the kinetic and potential energies of hydrogen 1s electrons.
3. ***Quenching and symmetry.*** Confirm or refute the argument that the reduced symmetry at the surface enhances the orbital moment (and the anisotropy), as contrasted to the quenching of the orbital moment by the cubic bulk crystal field.
4. ***Lowest-order uniaxial anisotropy.*** Show that $-K_1(\mathbf{n}\cdot\mathbf{M})^2/M_s^2$ is essentially equivalent to $K_1 V \sin^2\theta$. What is the advantage of the first expression?
5. ***Spin and orbital motion.*** Show that the spin is unrelated to the real-space motion of the electron.
   Answer: Let us take into account that $\mu_B = eRv/2$ and assume a "classical" electron radius R obtained from the electrostatic field energy $m_e c^2 = \frac{1}{2}\varepsilon_o \int_{r>R} E^2\,dV$, where $E = e^2/4\pi\varepsilon_o r^2$. The calculation yields the physically unreasonable result $v > c$. Incidentally, electrons are point-like, which makes things even worse.
6. ***Iron-series transition-metal anisotropy.*** Equation (3.38) contains 2λ rather than λ. Why?
   Answer: The involved wave functions have $L_z = 2$.
7. ***Cubic and tetragonal anisotropy constants.*** Show that the tetragonal K_1 is equal to the cubic K_1 but $K_2 \neq K_2^{(c)}$.
8. ***Quenched and unquenched.*** Expand the unquenched wave function $|\pm 2\rangle$ into $|x^2 - y^2\rangle$ and $|xy\rangle$.
9. ***Spin-orbit coupling of core electrons.*** The spin-orbit coupling of 1s, 2s, and 2p electrons in very heavy elements is much stronger than that of rare-earth 4f electrons. Why is the anisotropy contribution of these electrons negligible?

10. ***Anisotropy and symmetry.*** Convince yourself that the top and bottom atoms in Fig. 3.10 yield the same anisotropy contribution, in spite of being located at different ends of the magnetic dipole.
11. ***Cubal crystal field.*** Use the point-charge model to calculate the crystal-field potential for an atom in a cubal environment with eight nearest neighbors, as in Fig. 3.13(a). How could the crystal field be used to determine the magnetic anisotropy?
Hint: Expand $V_{\mathrm{CF}} = -\Sigma_{\mathrm{i}}(\mathrm{e}Q/4\pi\varepsilon_{\mathrm{o}}|\mathbf{r} - \mathbf{R}_{\mathrm{i}}|)$, into powers of x, y, and z.
Answer: $V_{\mathrm{CF}} = 35\mathrm{e}Q/18\pi\varepsilon_{\mathrm{o}}(x^4 + y^4 + z^4)$. To calculate the anisotropy energy, one must evaluate the crystal-field energy $\int \psi^*(\mathbf{r})\,\psi(\mathbf{r})\,V(\mathbf{r})\,\mathrm{d}\mathbf{r}$ as a function of the spin direction. For 3d atoms, this requires an explicit calculation of the wave function (Section 3.3.3), whereas 4f ions have rigid charge distributions $\rho(\mathbf{r}) = -\mathrm{e}\int \psi^*(\mathbf{r})\,\psi(\mathbf{r})\,\mathrm{d}\mathbf{r}$ and are therefore easier to handle. Howver, the interaction is of the hexadecapole type and deviates from simple oblate or prolate charge distributions considered in this chapter.
12. ***Crystal-field expressions.*** Some authors use the cubic crystal-field expression $x^2y^2 + x^2z^2 + y^2z^2$ rather than $x^4 + y^4 + z^4$. Is just justified?
13. ***Pseudocubic anisotropy.*** So-called pseudocubic intermetallics, such as slightly strained PtCo, have $a = b = c$ but exhibit a pronounced tetragonal anisotropy. How is this possible?
14. ***Strength of dipolar anisotropy.*** Estimate the order of magnitude of typical dipolar anisotropies by considering two atomic point dipoles. Is this result representative of macroscopic magnets?
15. ***Limitations of uniaxial anisotropy.*** Consider the two uniaxial anisotropy constants K_1 and K_2. In hexagonal and rhombohedral crystals, these constants provide an adequate description of second- and fourth-order anisotropies, but this is not the case for tetragonal magnets. Why?
16. ***Anisotropy of isostructural rare-earth intermetallics.*** A Nd-containing intemetallic has an anisotropy $K_1 = 5\,\mathrm{MJ/m^3}$. Estimate the anisotropy of the isostructural Sm compound.
17. ***Dirac equation and Pauli matrices.*** One way of deriving the Dirac equation exploits that $(\mathbf{a}\cdot\sigma)(\mathbf{a}\cdot\sigma) = a^2$, where $\mathbf{a}$ is any three-dimensional vector and $\boldsymbol{\sigma} = (\sigma_{\mathrm{x}}, \sigma_{\mathrm{y}}, \sigma_{\mathrm{z}})$ is the vector of the Pauli matrices. Prove this relation for arbitrary vectors $\mathbf{a}$.
18. ****Easy-cone phase diagrams.*** Draw the easy-cone phase diagram for $K_2 = 0$ and $K_3 > 0$. Compare the phase diagram with the more frequently considered case $K_2 > 0$ and $K_3 = 0$, where the phases are marked in the $K_1 - K_2$ plane.
Hint: Draw the anisotropy energy as a function of θ and then identify the minima of the energy.
19. ****Mixed lowest-order anisotropy.*** Calculate K_1 and K_1' for a "cigar-shaped" (prolate) magnetic particle with magnetocrystalline anisotropy. The strengths of the shape and magnetocrystalline anisotropies are 0.5 and 1.0 $\mathrm{MJ/m^3}$, respectively, and the angle between the particle's axis of revolution and the crystallographic c-axis is 60°.
Hint: The solution of this problem amounts to an eigenvalue analysis. The first step is to write down and appropriately simplify the total anisotropy energy.

4 Micromagnetic models

Hysteresis is a key feature of magnetic materials, and its prediction from atomic or intrinsic parameters such as M_s and K_1 is a major challenge in magnetism. Figure 4.1 compares two types of permanent magnets, a cumbersome nineteenth-century horseshoe magnet and a compact permanent magnet, as introduced in recent decades. In addition to the convenient shape, permanent magnets have become much smaller—10 grams of Nd-Fe-B now to replace 1 kg of carbon steel! This progress has two legs: improved intrinsic properties and their realization in the hysteresis loop.

The determination of hysteresis loops from local quantities such as $K_1(\mathbf{r})$ and $M_s(\mathbf{r})$ is the subject of a branch of magnetism known as *micromagnetics*. Its scope also includes phenomena such as magnetic domains and domain walls. Hysteresis has been known for a long time, but modern micromagnetism starts with the paper by Landau and Lifshitz (1935), who put Bloch's earlier ideas (1932) onto a sound physical basis. The prefix "micro" refers to small continuum entities having sizes of at least a few interatomic distances (Brown 1963a). The term is somewhat unfortunate, because most micromagnetic features are nanoscale, realized on length scales between 1 nm and 1 μm (see e.g. Skomski 2003).

As a complex nonlinear, nonequilibrium, and nonlocal phenomenon, hysteresis is caused by energy barriers associated with the magnetic anisotropy, Figs 3.2–3. The magnetization state is captured in a local energy minimum, and an additional magnetic field is necessary to cause the magnetization to jump into another local minimum. However, in most cases, there is no simple relation between hysteretic properties and the anisotropy constants. The reason is the *extrinsic* dependence of the hysteresis on

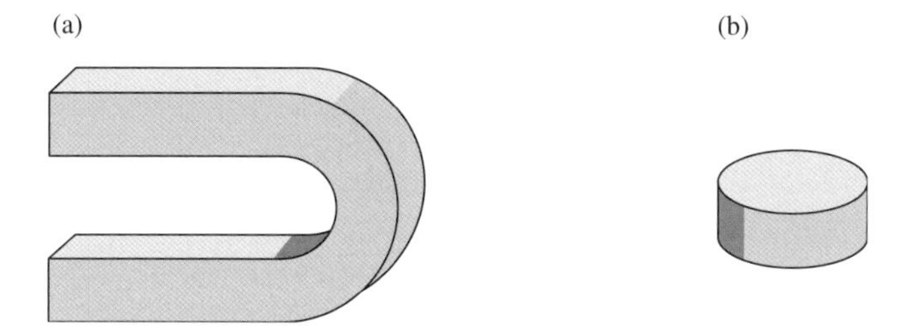

Fig. 4.1 Shape of permanent magnets: (a) horseshoe magnet and (b) compact rare-earth magnet. In addition to the improved shape, rare-earth magnets are much smaller than steel magnets.

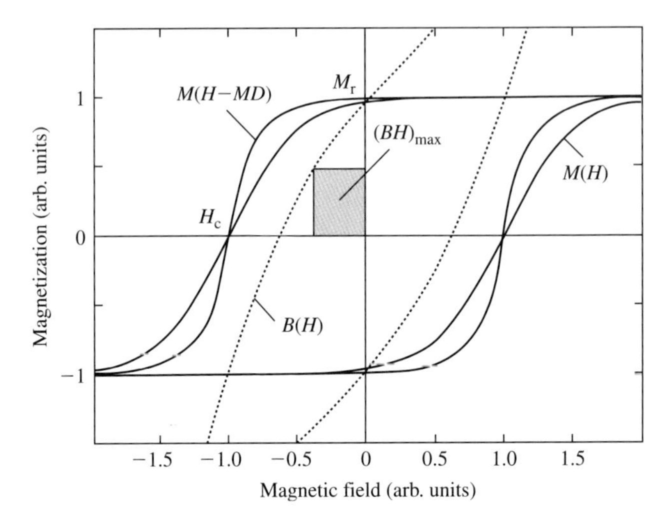

Fig. 4.2 Hysteresis loops: $M-H$ loops without and with demagnetizing-field correction (solid lines) and $B-H$ loop (dotted line). The energy product (gray area) describes a magnet's ability to store magnetostatic energy in free space.

real-structure features such as metallurgical and chemical inhomogeneities, as contrasted to the atomic character of intrinsic properties. A well-known example is the low coercivity of as-cast permanent-magnet alloys. To enhance the coercivity, it is necessary to subject the alloy to microstructural improvements such as sintering. This treatment has very little effect on the average values of the intrinsic parameters but yields major coercivity changes, typically by an order of magnitude.

Let us start with an introduction to the *phenomenology of hysteresis.* Important hysteretic or extrinsic properties are the coercivity H_c, the remanent magnetization or remanence M_r, and the energy product $(BH)_{max}$. In addition, there are parameters describing the loop shape, such as the slope or "micromagnetic susceptibility" dM/dH, which is often measured at H_c. There are, in fact, many types of hysteresis loops. Figure 4.2 shows three examples, all referring to a single magnetic sample. The classification of hysteresis loops reflects the plotted field and magnetization variables, the field range, and the directions of field and magnetization. Furthermore, it is necessary to consider shape differences, such as that between Fig. 4.1(a) and (b). Figure 4.2 shows various types of *major hysteresis loops*, where the field varies between $-\infty$ and $+\infty$.

Most frequently considered are ordinary $M-H$ loops, where the magnetization M is plotted as a function of the external field H. These loops are also known as extrinsic loops. By contrast, intrinsic loops display M as a function of the internal field $H - DM$, where the demagnetizing factor D (Section 3.2.2) describes the shape of the magnet. The name "intrinsic" is somewhat misleading, because intrinsic hysteresis is an extrinsic or real-structure phenomenon, except that the macroscopic sample shape is taken into account.

A second type of loop is the $B-H$ loop, where the magnetic flux density $B = \mu_o(M + H)$ is plotted as a function of the magnetic field. In practice, the $B-H$ loop

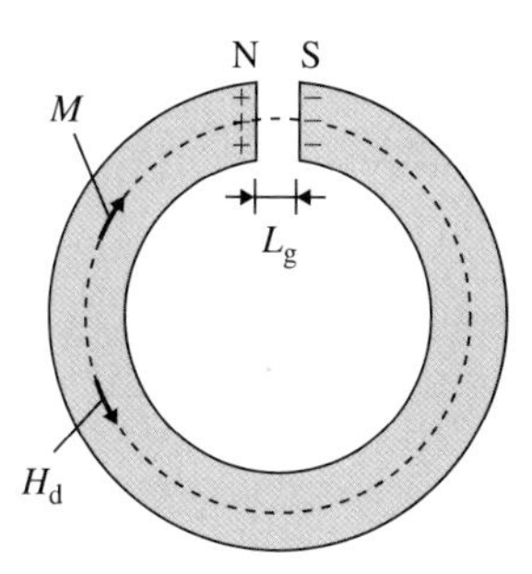

Fig. 4.3 Magnetic toroid. The gap width L_g determines the demagnetizing field. The flux and field lines in the toroid are circular, and the bold arrows show the direction of the magnetization.

serves to determine the energy product *energy product* $(BH)_{max}$, which is twice the maximum magnetostatic energy stored in free space by a permanent magnet of unit volume (gray area in Fig. 4.2). The energy product depends on the field in the magnet and, therefore, on the magnet's demagnetizing field. This is rationalized most easily by considering the magnetic toroid of Fig. 4.3. The basic assumption of the model is to confine the magnetic field H in free space to the gap (see exercise on flux leakage).

To calculate the fields and magnetic energies, we exploit $\nabla \times \mathbf{H} = 0$ and $\nabla \cdot \mathbf{B} = 0$, as in Section 3.2. The first equation amounts to $\oint \mathbf{H} \cdot d\mathbf{L} = 0$ and yields $H_D L_m + H L_g = 0$, where $L_m + L_g = L$ is equal to the contour length $2\pi R$ of the toroid. The second equation means that $B = M + H$ does not change at the interface, so that $H = M - H_D$. From these two equations we obtain $D = L_g/L$ and $H = (1 - D)M$. For narrow gaps, $D \approx 0$ and $H \approx M$. This field is quite large, up to 2.15 T for Fe and 2.43 T for $Fe_{65}Co_{35}$, but the requirement of a narrow gap means that huge quantities of magnetic materials must be used. This is the idea behind Fig. 4.1. To specify this finding, we calculate the magnetostatic energy E in the gap,

$$\frac{\mu_o}{2} \int_g H^2 \, dV = \frac{\mu_o}{2} (1 - D)^2 \, M^2 \, V_g \tag{4.1}$$

where V_g is the gap volume, and convince ourselves that E is equal to the integral $\frac{1}{2} \int_m |BH| \, dV$ over the magnetic material, where $|BH| = \mu_o D(1 - D)M^2$. Minimizing this expression with respect to D yields $D = 1/2$, corresponding to an ellipsoid of revolution with a slightly oblate shape ($R_z/R_x = 0.55$). The corresponding energy product $(BH)_{max} = \mu_o M^2/4$ is the maximum energy product one can get from a magnet of magnetization M. In steel magnets, this high energy product cannot be realized, because the magnetization collapses when DM reaches H_c. In other words, the coercivity limits the area of the $(BH)_{max}$ rectangle in Fig. 4.2, and magnets with small H_c require small D, corresponding to bulky shapes.

The energy product of carbon steel is of the order of 1 kJ/m^3 but has increased by two orders of magnitude since the late nineteenth-century (Section A.5). This development is based on the exploitation of the magnetocrystalline anisotropy of magnetic materials with noncubic crystal structure. It started with hexagonal ferrites in the 1950s and has continued with the exploitation of the large spin-orbit coupling of rare-earth permanent magnets since the 1970s and 1980s. At present, $Nd_2Fe_{14}B$ magnets with energy products exceeding 440 kJ/m^3 have been produced in industrial laboratories in Europe and Japan.

Minor loops are obtained for finite maximum fields and lie inside the major loop (see below, Fig. 4.14). Recoil loops are minor loops where the field varies between $-H$ and 0. They are an example of loops without inversion symmetry, as are loops shifted by a bias field. After thermal demagnetization, or heating beyond T_c, the application of a magnetic field yields the initial or virgin curve, which is used in the investigation of magnetization processes and coercivity mechanisms.

Most hysteresis loops refer to magnetizations and fields parallel to the symmetry or c-axis of the magnet. Sometimes, field and magnetization form a nonzero angle with the symmetry axis, giving rise to an *angular dependence* of the hysteresis. For example, the analysis of thin-film hysteresis often involves the comparison of perpendicular loops, where the field is normal to the film plane, with one or more in-plane loops. Hysteresis loops measured in different directions are useful to determine anisotropy constants and to gauge polycrystallinity. Less common measurements probe the magnetization in a direction different from the field direction, and there are phenomena such as rotational hysteresis, where the direction of the field is changed, rather than its magnitude.

4.1 Stoner-Wohlfarth model

Summary A simple but powerful micromagnetic model is the Stoner-Wohlfarth or coherent rotation model. It assumes a rigid exchange coupling between the atomic spins in a ferromagnetic body and reproduces the exact micromagnetic behavior in the limit of very small particles. This includes structurally inhomogeneous particles and features such as a grain boundaries, but the length-scale requirements are quite stringent. The model is a useful starting point for the discussion of the angular dependence of magnetization curves and predicts spin-reorientation transitions, for example from easy-axis to easy-cone magnetism. A key prediction of the Stoner-Wohlfarth model is that the coercivity is equal to the anisotropy field. This is rarely observed, due to the size of particles encountered in practice. In large particles, magnetostatic interactions lead to incoherent magnetic reversal, even in the absence of morphological inhomogeneities and in single-domain particles.

A very popular and highly instructive—though sometimes overstretched—model of hysteresis is the *Stoner-Wohlfarth* model (1948). The basic assumption of the original model is a constant magnetization throughout the magnet. As a consequence, the exchange energy remains unchanged during magnetization reversal, and the energy of the particle is essentially equal to the anisotropy energy. The Stoner-Wohlfarth model, also known as the uniform-rotation or coherent-rotation model, provides an adequate description of non- and weakly interacting small particles, where the interatomic exchange is sufficiently strong to ensure parallel spin alignment on a local scale.

The model is not limited to ferromagnets but is easily generalized to ferrimagnets and other spin structures. This includes noncollinear spin structures caused by competing exchange. In fact, the main criterion is the *smallness* of the particles. In very small particles, the exchange wins, the magnetization projections $\mathbf{M}(\mathbf{R}_i) \cdot \mathbf{M}(\mathbf{R}_j)$ are fixed, and anisotropic and magnetostatic effects can be treated by zeroth-order perturbation theory, that is, by rotating the spin system as a whole. However, the smallness criterion is nontrivial, as epitomized by the distinction between coherent rotation and single-domain magnetism (Section 4.2).

4.1.1 Aligned Stoner-Wohlfarth particles

Let us start with the simple but instructive case of *aligned* uniaxial ellipsoids of revolution, where the c-axis is parallel to the external field $\mathbf{H} = H\mathbf{e}_z$. The total energy density is

$$\frac{E}{V} = K_1 \sin^2\theta + (1 - 3D)M_s^2 \sin^2\theta - \mu_o M_s H \cos\theta \tag{4.2}$$

where θ is the angle between $\boldsymbol{M}$ and $\mathbf{e}_z$. Figure 4.4 shows this energy landscape for several field values. In small reverse fields (top), there are two minima, at $\theta = 0$ and $\theta = \pm\pi$. Depending on the sample past, the magnetization remains in the $\uparrow$ state ($\theta = 0$) or in the $\downarrow$ state ($\theta = \pm\pi$). *This is the origin of hysteresis.* With increasing field magnitude, one of the minima becomes shallow and finally vanishes at the coercivity. Mathematically, the micromagnetic instability has the character of a catastrophe, meaning that a small change in a control parameter leads to macroscopic changes (Pinto 1987).

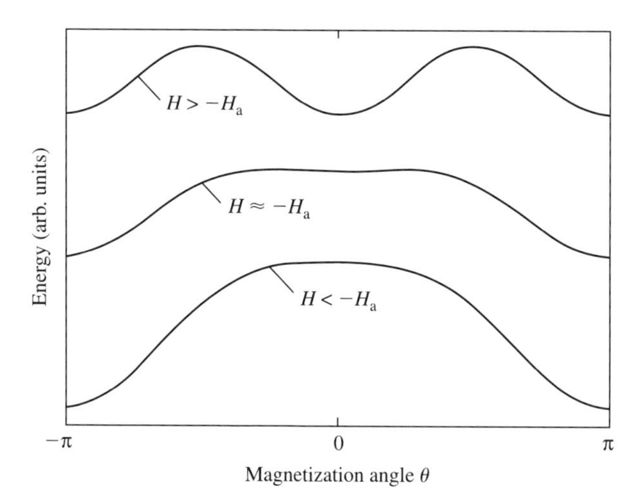

Fig. 4.4 Energy landscape and coercivity for the aligned Stoner-Wohlfarth model (4.2). The curves show minima at $\theta = 0$ ($\uparrow$) and $\theta = \pi$ ($\downarrow$) of equal energy at $H = 0$ (top), approach to coercivity (middle), and absence of the $\uparrow$ minimum in reverse field larger than the coercivity (bottom). At the coercivity, $H = -H_a$, the $\uparrow$ state becomes unstable and the systems switches into the $\downarrow$ state.

To determine the coercivity, we start from the ↑ state at $\theta = 0$ and analyze the stability of (4.2) for small magnetization angles θ. Taking into account that $\sin(x) \approx x$ and $\cos(x) \approx 1 - x^2/2$, we obtain

$$\frac{E}{V} = \left(K_1 + \frac{\mu_o}{4}(1 - 3D)M_s^2 + \frac{\mu_o}{2} M_s H\right)\theta^2 \tag{4.3}$$

When the parenthesized term is positive, $\theta = 0$ corresponds to an energy minimum. At coercivity, the term becomes zero, indicating a transition from a stable energy minimum to an unstable maximum (center and bottom graphs in Fig. 4.4). The corresponding coercivity is

$$H_c = \frac{2K_1}{\mu_o M_s} + \frac{1}{2}(1 - 3D)M_s \tag{4.4}$$

This important expression is also known as the Stoner-Wohlfarth *nucleation* field H_N, because it is obtained by the stability analysis of the fully magnetized state. However, $H_c = H_N$ is limited to nucleation models, because domain-wall pinning (Section 4.3.2) may lead to $H_c > H_N$.

From (4.4) we see that the Stoner-Wohlfarth coercivity is equal to the anisotropy field $2K_1/\mu_o M_s$ plus the shape-anisotropy field $\frac{1}{2}(1 - 3D)M_s$. The latter is different from the demagnetizing field $-DM_s$, indicating that the demagnetizing field is more than just an additional external-field contribution. For example, the demagnetizing field in a sphere ($D = 1/3$) is $-M_s/3$, as contrasted to the absence of shape anisotropy in spheres. We will return to this problem in Section 4.3.1.

4.1.2 Angular dependence

Very often, the magnetic field forms an angle θ_o with the symmetry axis of the magnet. For example, the magnetization of a magnetic thin film may be measured in the film plane and perpendicular to the plane. Another example is ensembles of randomly oriented nanoparticles (Section 7.4.3). Incorporating the shape anisotropy of (4.2) into an effective anisotropy, we obtain

$$\frac{E}{V} = K \sin^2\theta - \mu_o M_s H \cos(\theta - \Theta) \tag{4.5}$$

where $K = K_1 + \mu_o M_s^2(1 - 3D)/4$. From this equation, the loops are obtained by stability analysis, $\partial E/\partial\theta = 0$ and $\partial^2 E/\partial\theta^2 > 0$. Figure 4.5 shows the magnetization in field direction for various angles Θ.

As a function of Θ, the nucleation field H_N decreases, reaches a minimum at 45°, and then increases again. The nucleation field $H_N(\Theta)$ is symmetric about $\Theta = 45°$, as illustrated by the dashed lines connecting the 30° and 60° curves in Fig. 4.5. However, for angles $\Theta > 45°$, the coercivity is smaller than the nucleation field. For example, the 60° curve in Fig. 4.5 crosses the $M = 0$ line *before* the nucleation event. In a less common definition (Givord and Rossignol 1996), coercivity is actually given by the maximum of $\mathrm{d}M/\mathrm{d}H$. In this definition, $H_c = H_N$ for both aligned and misaligned Stoner-Wohlfarth particles, but not necessarily for other magnets.

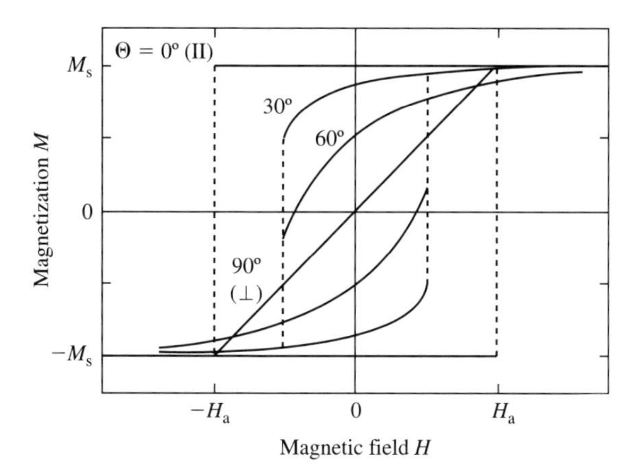

Fig. 4.5 Angular dependence of the Stoner-Wohlfarth model.

Two interesting limits are $\Theta = 0$ (parallel loop) and $\Theta = 90°$ (perpendicular loop). For $\Theta = 0$, the model predicts a square loop whose coercivity is equal to the anisotropy field $2K/\mu_o M_s$. For $\Theta = 90°$, the loop degenerates into a straight line of zero energy product and slope $\mathrm{d}M/\mathrm{d}H = 2K/\mu_o M_s^2$. This slope is fairly easy to measure and frequently used to estimate the anisotropy constants of magnetic materials. More generally, adding the fourth-order uniaxial anisotropy term $K_2 \sin^4 \theta$ to (4.2), exploiting that $\cos(\theta - 90°) = \sin\theta$, and putting $\partial E/\partial\theta = 0$ we obtain the equation of state

$$2K \sin\theta + 4K_2 \sin^3\theta - \mu_o M_s H = 0 \tag{4.6}$$

Introducing the magnetization component $M_x = M_s \sin\theta$ and plotting H/M_x as a function of M_x^2 yields a straight line whose slope and intercept correspond to K and K_2, respectively. This procedure is known as the Sucksmith-Thompson method.

4.1.3 Spin reorientations and other first-order transitions

Uniaxial magnets with positive anisotropy constant K_1 and zero higher-order anisotropy constants are known as *easy-axis magnets*, because the preferential zero-field magnetization direction is parallel to the magnets' symmetry axis, as in Fig. 3.2(b). Changing the sign of K_1 to negative yields *easy-plane magnetism*, Fig. 3.2(c). In ideal easy-plane magnets, the magnetization is free to rotate in the basal plane. However, real magnets tend to exhibit some basal-plane anisotropy. Examples of easy-axis and easy-plane magnets are hexagonal Co and $Sm_2Fe_{14}B$, respectively (Section A.5). In cubic magnets, a negative K_1 means easy-axis magnetism along the cube diagonals, Fig. 3.3(b), rather than easy-plane magnetism.

Adding fourth-order uniaxial anisotropy, the K_2 term in (3.6), gives rise to a variety of zero-field spin configurations When both K_1 and K_2 are positive, one encounters easy-axis anisotropy ($\theta = 0$), whereas $K_1 < 0$ and $K_2 < 0$ yield easy-plane anisotropy (easy-plane anisotropy, $\theta = \pi/2$). *Easy-cone magnetism*, Fig. 3.2(d), occurs if the $K_1 < 0$ and $K_2 > -\frac{1}{2}K_1$ are satisfied simultaneously. Furthermore, there are regions where easy-axis and easy-plane magnetism coexist. The phase diagram Fig. 4.6 shows

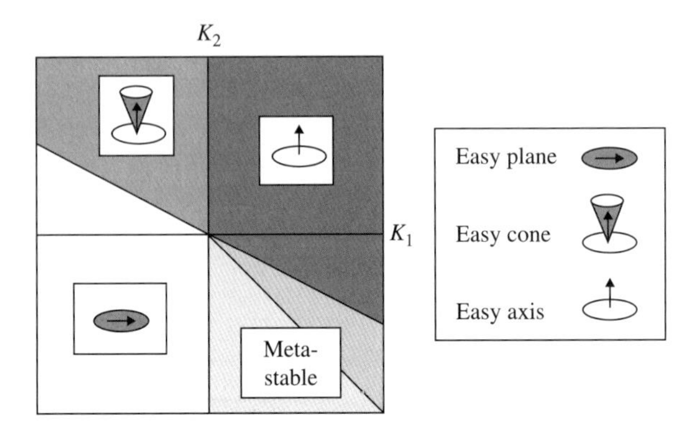

Fig. 4.6 Easy-cone magnetism in the K_1–K_2 plane.

the spin structure in the K_1–K_2 plane. The tilt angle between the z-axis and the easy magnetization direction is given by

$$\theta_c = \arcsin \sqrt{\frac{|K_1|}{2K_2}} \tag{4.7}$$

Since the temperature dependences of K_1 and K_2 are generally different—K_2 is often negligible at high temperatures (Section 5.5)—the preferential magnetization direction may change upon heating. This is an example of a *spin-reorientation* transition (SRT). A similar effect occurs in some magnetic thin films with competing bulk and surface anisotropies (Section 3.5.3), where the SRT can be triggered by changing the temperature or varying the film thickness, for example across a wedge-shaped film.

Spin-reorientation transitions and their field-dependent analog, first-order magnetization processes (FOMPs), must not be confused with metamagnetic phase transitions, such as high-spin low-spin transitions. Both involve first-order magnetization changes as a function of parameters such as magnetic field and temperature, but SRTs reflect competing anisotropies, whereas metamagnetic phase transitions are due to exchange. SRTs and FOMPs occur in many rare-earth intermetallics (Coey 1996).

A phenomenon involving both exchange and anisotropy is the spin-flop transition in antiferromagnets. Putting an isotropic antiferromagnet (AFM) into a magnetic field yields a canted spin-structure, as in Fig. 2.19(c). A characteristic feature of this spin configuration is that the sublattice magnetizations are nearly, though not completely, perpendicular to the field. When an AFM with uniaxial anisotropy is subjected to a magnetic field parallel to easy axis, then the configuration of Fig. 2.19(c) costs a substantial amount of anisotropy energy. For this reason, the magnet remains in its easy-axis state with zero net magnetization until the field is sufficiently strong to compensate the anisotropy effect. The sublattice magnetization angles and the net magnetization then jump to finite values. Figure 4.7 illustrates thin *spin-flop transition*.

The spin-flop field H^* at which the transition occurs is obtained by adding a second-order uniaxial anisotropy to (2.40). In the parallel case, nothing happens until the external field switches the sublattice magnetizations in directions approximately

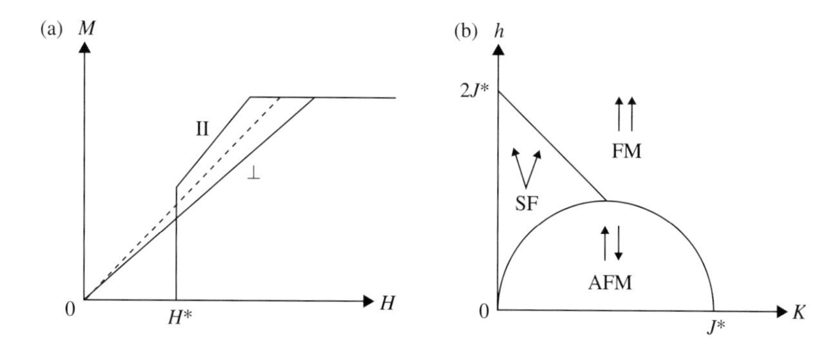

Fig. 4.7 Spin-flop transition in antiferromagnets with uniaxial anisotropy: (a) $M-H$ curve and (b) phase diagram. The symbols $\|$ and $\perp$ in (a) refer to angle between field and easy axis, and dashed line is the isotropic limit of Section 2.40.

perpendicular to the easy axis. The corresponding spin-flop field is, in appropriately chosen units,

$$H^* = 2\sqrt{K(J^* - K)} \tag{4.8}$$

and can be used to determine the magnetic anisotropy of antiferromagnets, because J^* is essentially given by the ordering temperature (Néel temperature).

A spin-flop transition where the net magnetization jumps directly from zero to saturation is known as a spin-flip transition. In the $M-H$ diagram, a spin-flip transition is characterized by the absence of the finite-slope part of the $\|$ curve in Fig. 4.7(a). Examples of AFMs with spin-flop and spin-flip transitions are MnF_2 and $FeCl_2$, respectively. A related transition is the Morin transition in hematite (α-Fe_2O_3). The transition occurs at about 260 K and is accompanied by a change of the easy-axis direction by about 90°. The difference is the additional involvement of the anisotropic DM exchange (Section 3.5.2).

4.1.4 Limitations of the Stoner-Wohlfarth model

The Stoner-Wohlfarth model is a very simple approach but a useful starting point for the description of weakly interacting ensembles of small particles. The key assumption of the original Stoner-Wohlfarth model is a uniform or coherent magnetization throughout the magnet. The exchange punishes magnetization inhomogeneities $\nabla\mathbf{M}$ (Section A.2.4) but competes against magnetostatic flux closure and structural imperfections, which both favor $\nabla\mathbf{M} \neq 0$. In macroscopic magnets, the last two contributions often dominate, and the Stoner-Wohlfarth predictions are poor. Figure 4.8 compares Stoner-Wohlfarth loops with typical experimental findings. We see that the Stoner-Wohlfarth model overestimates both the coercivity and the loop squareness. Even in perfected permanent magnets, the coercivities are only 10% to 40% of the Stoner-Wohlfarth prediction. In a slightly different context (Section 4.3.1), this disagreement is known as Brown's coercivity paradox.

As mentioned above, one may define the Stoner-Wohlfarth model by rigid angles between the moments $\mathbf{m}_\text{i}$ and $\mathbf{m}_\text{j}$ of atoms located at $\mathbf{R}_\text{i}$ and $\mathbf{R}_\text{j}$. Physically, the interatomic exchange ensures that the direction cosines $\mathbf{m}_\text{i} \cdot \mathbf{m}_\text{j} \sim \mathbf{M}(\mathbf{R}_\text{i}) \cdot \mathbf{M}(\mathbf{R}_\text{j})$

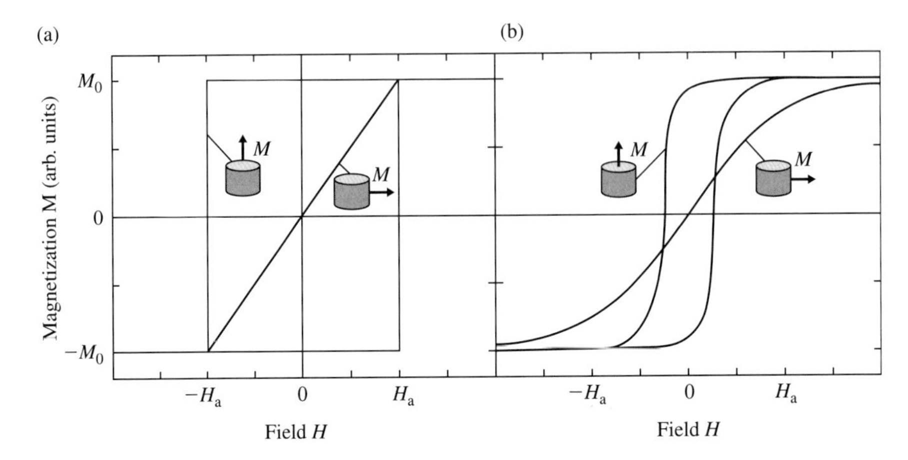

Fig. 4.8 Hysteresis loops: (a) Stoner-Wohlfarth predictions and (b) typical loops encountered in practice. Both the coercivity and the loop shape are poorly reproduced by the Stoner-Wohlfarth model.

remain unchanged during magnetization reversal and that the magnetic particle acts as a giant though not necessarily ferromagnetic "macrospin". In this regime, the external field and the anisotropy are small corrections to the leading interatomic exchange, and the only magnetization parameters are the magnetization angles ϕ and θ. The regime is realized for small particles and can be used to describe, for example, hysteresis loops and spin-orientation transitions. It does *not* include spin-flop and other metamagnetic transitions, which involve not only θ and ϕ but also additional degrees of freedom, such as the net magnetization.

The uncritical use of the Stoner-Wohlfarth model is an example of poor phenomenological modeling. Since the coercivity is equal to the anisotropy field $2K/\mu_o M_s$, it yields unphysically low anisotropies for most materials. The rationale behind the approach is to interpret the obtained anisotropy as some volume-averaged value $\langle K \rangle$, but experimental coercivities reflect local minima of $K(\mathbf{r})$ rather than $\langle K \rangle$. In practice, coherent rotation is realized in noninteracting or weakly interacting particles having radii smaller than about 10 nm. On this length scale, the model is actually quite robust against imperfections, such as grain boundaries. It can also be used as a starting point for the description of needle-shaped nanoscale particles, such Fe-Co precipitates in alnico-type magnets, so long as the needles are sufficiently thin. In contrast to widespread belief, the single-domain character of magnetic particles is not sufficient to ensure coherent rotation. As we will analyze below, the term "elongated single-domain" or ESD particles confuses coherent rotation with the absence of magnetic domains (Section 4.2.5).

4.2 Hysteresis

Summary The local magnetization $\mathbf{M}(\mathbf{r})$ is determined by the competition between interatomic exchange, anisotropy, Zeeman energy, and magnetostatic self-interaction. The energy contributions establish a complicated nonlinear and nonlocal problem, involving metastable magnetic energy minima and leading to a history dependence known as hysteresis. In some cases, it is possible to linearize the micromagnetic equations. In the Stoner-Wohlfarth model, the hysteresis is independent of the exchange, which is assumed to be sufficiently strong to ensure a rigid coupling between the spins. In reality, this is rarely the case, although exchange favors relatively smooth magnetization variations. Magnetostatic contributions dominate in macroscopic magnets, where they lead to the formation of magnetic domains, separated by domain walls. The wall width $\delta_B = \pi(A/K_1)^{1/2}$ reflects competition between exchange and anisotropy, but there are no comparable simple relations for the domain size.

Hysteresis is caused by local rotations of the magnetization vector $\mathbf{M}(\mathbf{r})$, traced as a function of the applied field $\mathbf{H}$. A simple example is the Stoner-Wohlfarth model (4.2), where one considers a single magnetization angle θ. Compared to magnetization rotations, field-dependent changes of the spontaneous magnetization $M_s(\mathbf{r}) = |\mathbf{M}(\mathbf{r})|$ are negligible. This is because typical micromagnetic energies of order 0.0001 eV per atom cannot compete against electronic energies of order 1 eV per atom. We also disregard the anisotropy of M_s, which arises from spin-orbit coupling and is about 0.5% for hcp Co. More generally, on a length scale of a few interatomic distances, the spin structure is of intrinsic origin (Section 2.3.3 and Section 2.4.3) and unrelated to hysteresis. Thermal equilibrium is established rapidly in this intrinsic regime, and quantities such as M_s and K_1 are determined by the local crystalline environment and used as micromagnetic parameters. However, ferromagnetic exchange is unable to ensure parallel spin alignment on length scales larger than a few nanometers. This is the micromagnetic domain considered in this chapter.

It is important to note that Maxwell's equations are constraints rather than solutions to micromagnetic problems. For example, Maxwell's equations predict the magnetic fields $\mathbf{H}$ and $\mathbf{B}$ from the magnetization $\mathbf{M}$ but do not explain the origin of the magnetization. In this section, we consider fundamental features of micromagnetism. The findings of this section are the basis for numerical simulations (Schrefl, Fidler,and Kronmüller 1994) and for the specific models considered in Section 4.3–4.

4.2.1 Micromagnetic free energy

The magnetization $\mathbf{M}(\mathbf{r})$ is obtained from the micromagnetic (free) energy E or F. The *free* character of the energy means that the involved materials parameters are temperature-dependent. For example, the room-temperature magnetization of most materials is lower than the zero-temperature magnetization, $M_s(\mathrm{RT}) < M_s(0)$, because thermal excitations yield a temporary reduction of the local moment. Since these processes are very fast, we can use thermal equilibrium averages determined from atomic partition functions (Section 5.1–3). However, this equilibrium character does not carry over to micromagnetic phenomena, and the dynamics of micromagnetic magnetization rotations (Section 6.4) is very different from the temperature-dependent change of the spontaneous magnetization.

Micromagnetic processes are generally multidimensional, and one-dimensional energy landscapes, such as that of Fig. 4.4, are rarely encountered in practice. Treating the magnetization as a classical vector, as is usually done in micromagnetism, yields $2N$ degrees of freedom ϕ_i and $\theta_i (i = 1 \ldots N)$. In reality, the strong interatomic exchange fixes the relative spin directions of neighboring atoms, and the number of relevant degrees of freedom is much smaller than $2N$. This is the basis for treating micromagnetism on a continuum level. For example, ferromagnetic exchange J_{ij} between neighboring atoms is replaced by the *exchange stiffness* $A \approx J/a$, where a is the interatomic distance.

The formulation of micromagnetism as a continuum theory has a long history (Bloch 1932, Becker and Döring 1939, Brown 1963a). Here we focus on uniaxial magnets, where

$$E = \int \left\{ A \left[\nabla \left(\frac{\mathbf{M}}{M_s} \right) \right]^2 - K_1 \frac{(\mathbf{n} \cdot \mathbf{M})^2}{M_s^2} - \mu_o \mathbf{M} \cdot \mathbf{H} - \frac{\mu_o}{2} \mathbf{M} \cdot \mathbf{H}_d(\mathbf{M}) \right\} dV \tag{4.9}$$

The terms in the integrand are the interatomic exchange (A), the second-order uniaxial anisotropy (K_1), the Zeeman interaction (H), and the magnetostatic self-interaction (H_d). The anisotropy term, where $\mathbf{n}$ is the unit vector of the easy-axis direction, is easily generalized to cubic and other symmetries (Section 3.1). The exchange term, where the ∇ punishes deviations from ferromagnetic spin alignment, and the self-interaction term will be discussed in some more detail below. All parameters entering (4.9) are local and temperature-dependent, determined by real-structure features such as local chemistry, crystal structure, and crystallite orientation. Of particular practical importance is the strong dependence of $K_1(\mathbf{r})$ on chemical composition and temperature.

4.2.2 *Magnetostatic self-interaction

The long-range character of the magnetostatic self-interaction adds considerable complexity to micromagnetics. There are two equivalent ways of looking at the magnetostatic self-interaction energy $E_{ms} = -\mu_o \int \mathbf{M} \cdot \mathbf{H}_d \, dV$. In the first approach, we divide our magnet into small dipoles, evaluate the magnetic field $\mathbf{H}_i$ created at $\mathbf{r}_i$ by the j-th dipole, and evaluate the total energy $\frac{1}{2} \sum_{ij} \mu_o \mathbf{m}_i \cdot \mathbf{H}_i(\mathbf{r}_j)$, where the factor $\frac{1}{2}$ accounts for the double-counting of dipolar interaction bonds ij and ji. In the second approach, we obtain the self-interaction field by solve Maxwell's equations for a given magnetization $\mathbf{M}(\mathbf{r})$. Both methods are equivalent, because the dipole fields used in the first approach are solutions of Maxwell's equations. The self-interaction field $\mathbf{H}_d$ is

$$\mathbf{H}_d(\mathbf{r}) = \frac{1}{4\pi} \int \frac{3(\mathbf{r} - \mathbf{r}')(\mathbf{r} - \mathbf{r}') \cdot \mathbf{M}(\mathbf{r}') - |\mathbf{r} - \mathbf{r}'|^2 \, \mathbf{M}(\mathbf{r}')}{|\mathbf{r} - \mathbf{r}'|^5} \, dV' \tag{4.10}$$

In ellipsoids of revolution homogeneously magnetized along the axis of symmetry, the self-interaction field is equal to the demagnetizing field, $\mathbf{H}_d = -D\mathbf{M}$ (Section 3.2.2). The self-interaction energy $E_{ms} = -\frac{1}{2}\mu_o \int \mathbf{M} \cdot \mathbf{H}_D \, dV$ is then equal to $\mu_o D M^2/2$.

One way to deal with complicated magnets is to exploit the fact that $(3\mathbf{r}\,\mathbf{r} - r^2)/r^5 = -\nabla(\mathbf{r}/r^3)$ and $\nabla \cdot (a\,\mathbf{b}) = a \nabla \cdot \mathbf{b} + \nabla a \cdot \mathbf{b}$. This makes it possible to write (4.10) in terms of the magnetic charge density $\rho_M = -\nabla \cdot \mathbf{M}$ (Section A.4.2). In close analogy to electrostatics, the self-interaction energy then assumes the form

$$E_{ms} = \frac{\mu_o}{4\pi} \int \frac{\rho_M(\mathbf{r})\, \rho_M(\mathbf{r}')}{|\mathbf{r} - \mathbf{r}'|} \, dV' \tag{4.11}$$

Homogeneously magnetized bodies of arbitrary shape have their magnetic charges at the surface. This leads to a relatively high magnetostatic energy. This energy is reduced by flux closure and domain formation but paid by an increase in exchange and anisotropy energies (Section 4.2.4). In addition, most magnets are structurally inhomogeneous, so that $\nabla \cdot \mathbf{M} \neq 0$ inside the magnet. Fourier transformation of $\nabla \cdot (\mathbf{M} + \mathbf{H}) = 0$ and $\mathbf{H} = -\nabla \phi_\mathrm{M}$ (Section A4.2) yields the demagnetizing-field components $\mathbf{H}_\mathrm{k} = -\mathbf{k}\,\mathbf{k} \cdot \mathbf{M}_\mathrm{k}/k^2$. For many calculations, this transformation is of little use, because few magnetic systems are periodic, but the wave-vector dependence of the magnetic properties can be investigated by neutron scattering (Weismüller *et al.* 2001).

It is important to note that self-interaction fields (demagnetizing fields) cannot be regarded as a local correction to the external field. This is seen, for example, from the factor 1/2 in (4.9). In isolated Stoner-Wohlfarth particles, E_ms assumes the form of an anisotropy term (shape anisotropy), rather than adding to the external field, but in general neither the shape anisotropy picture nor the external-field picture is correct. An exception is weakly interacting ensembles of nonequivalent small particles, where the interparticle interactions can be described by an interaction field (Section 7.4.4).

4.2.3 *Exchange stiffness

In (4.9), the exchange appears in form of the *exchange stiffness A*. This quantity is also important in other areas of magnetism, such as spin waves (Section 6.1.3). On a continuum level, the exchange energy $E_\mathrm{ex} = -\frac{1}{2}\sum_\mathrm{ij} J_\mathrm{ij}\,\mathbf{s}_\mathrm{i} \cdot \mathbf{s}_\mathrm{j}$ assumes the form

$$E_\mathrm{ex} = \int A(\nabla \mathbf{s})^2 \,\mathrm{d}V \tag{4.12}$$

where $\mathbf{s} = \mathbf{M}/M_\mathrm{s}$. This expression (Kittel 1949, Brown 1963a) ensures the ferromagnetic character of the exchange by punishing magnetization inhomogeneities $\nabla\mathbf{s}$.

To derive A from the interatomic exchange constants J_ij, we consider an arbitrary but smooth magnetization inhomogeneity, calculate the exchange energy, and compare the result with (4.12). A convenient choice is $\mathbf{s}(\mathbf{R}_\mathrm{i}) = \cos(\theta_\mathrm{i})\,\mathbf{e}_\mathrm{z} + \sin(\theta_\mathrm{i})\,\mathbf{e}_\mathrm{y}$, where $\theta_\mathrm{i} = \theta(x_\mathrm{i})$. In other words, we confine the magnetization vector to the y–z plane and assume that the magnetization angle θ depends on x only. The exchange energy can be written as

$$-\frac{1}{2}\sum_\mathrm{ij} J_\mathrm{ij}\cos(\theta_\mathrm{i} - \theta_\mathrm{j}) \approx -\frac{1}{2}\sum_\mathrm{ij} J_\mathrm{ij}\left(1 - \frac{1}{2}(\theta_\mathrm{i} - \theta_\mathrm{j})^2\right) \tag{4.13}$$

where we have taken into account that the angle $\theta_\mathrm{i} - \theta_\mathrm{j}$ between neighboring spins is small. Using the expansion $\theta_\mathrm{j} = \theta_\mathrm{i} + \nabla_\mathrm{x}\theta \cdot (x_\mathrm{j} - x_\mathrm{i})$, we obtain

$$E_\mathrm{ex} \approx \frac{1}{4}\sum_\mathrm{ij} J_\mathrm{ij}(x_\mathrm{j} - x_\mathrm{i})^2\,(\nabla_\mathrm{x}\theta)^2 \tag{4.14}$$

where we have ignored the physically uninteresting zero-point energy $-\frac{1}{2}\sum_\mathrm{ij} J_\mathrm{ij}$. Next, we replace the double summation by an integral. The sum over the first variable (i or j) is essentially a volume integral, whereas the second sum is limited to nearest neighbors.

Let us consider z nearest neighbors, described by $J_{\mathrm{ij}} = J$ and $\mathbf{r}_{\mathrm{i}} - \mathbf{r}_{\mathrm{j}} = \mathbf{D}_{\mathrm{o}}$. Unless the distribution of nearest neighbors is anisotropic, we can replace the term $(x_{\mathrm{j}} - x_{\mathrm{i}})^2$ in (4.14) by $D_{\mathrm{o}}^2/3$. The energy now becomes $E_{\mathrm{ex}} \approx \int J\, D_{\mathrm{o}}^2\, (\nabla_{\mathrm{x}}\theta)^2\, \mathrm{d}V/6V_{\mathrm{o}}$, where V_{o} is the crystal volume per magnetic atom. For the assumed magnetization inhomogeneity, (4.12) yields $E_{\mathrm{ex}} = \int A(\nabla_{\mathrm{x}}\theta)^2\, \mathrm{d}V$, so that

$$A = \frac{z\, J\, D_{\mathrm{o}}^2}{12\, V_{\mathrm{o}}} \tag{4.15}$$

Of course, (4.15) is linked to the definition of J in terms of the pair energy $-J_{\mathrm{ij}}\mathbf{s}_{\mathrm{i}} \cdot \mathbf{s}_{\mathrm{j}}$, where $|\mathbf{s}_{\mathrm{i}}| = 1$. Different definitions yield different expressions for A as a function of J.

For simple-cubic lattices (lattice parameter a and $z = 6$), we find $D_{\mathrm{o}} = a$, $V_{\mathrm{o}} = a^3$, and $A = J/2a$. Typical experimental exchange stiffnesses of the order of 10 pJ/m are consistent with the microscopic parameters $J \sim 10^{-20}$ J and $a \approx 2.5$ Å. Note that the main structural quantity in (4.15) is the interatomic distance D_{o}. Expressions such as $A = J/2a$ are somewhat unfortunate, because they associate the exchange stiffness A with the lattice constant a. Complicated intermetallic alloys tend have lattice constants much larger than 2.5 Å, but this is not accompanied but a proportionate decrease in A.

Due to its continuum character, the concept of exchange stiffness becomes questionable for large wave vectors $\mathbf{k} \sim \nabla$, but even on an atomic scale, the relative errors may be smaller than 20%. Another limitation is the assumed isotropy of the exchange, which leads to the replacement of $(x_{\mathrm{j}} - x_{\mathrm{i}})^2$ by $D_{\mathrm{o}}^2/3$. This procedure is exact for cubic magnets, where the x, y, and z components yield equal contributions to $(\mathbf{r}_{\mathrm{i}} - \mathbf{r}_{\mathrm{j}})^2$. It is also a reasonable approximation for dense-packed noncubic magnets, such as hcp Co. However, the method cannot be used for strong bond anisotropy, as encountered for example in layered compounds. In this case, A must be replaced by the 3×3 exchange-stiffness tensor $A_{\mu\nu}$, corresponding to different exchange stiffnesses in different directions.

The derivation of (4.15) is based on (4.14) and therefore limited to short-range interactions. Long-range interactions, such as RKKY interactions, cause the nearest-neighbor sum in (4.14) to diverge. This is of some practical interest, because exchange interactions in metals have oscillating tails (Section 5.3.5). The origin of the divergence is the approximation (4.13), and performing the summation directly over the cosine function removes the divergence (Skomski *et al.* 2005).

4.2.4 Linearized micromagnetic equations

The local magnetization $\mathbf{M}(\mathbf{r})$ is obtained by analyzing the local minima of the free-energy functional (4.9), similar to the Stoner-Wohlfarth example of Fig. 4.4. However, the analysis of nonlinear energy landscapes has remained a demanding challenge, especially in multidimensional phase spaces. In some cases, it is useful or sufficient to consider harmonic energies, which correspond to linear equations of state. In the Stoner-Wohlfarth analogy, this approximation corresponds to the transition from (4.2) to the quadratic equation (4.3). Figure 4.9 shows that the harmonic approximation can be used to determine the stability of a spin configuration but is unable to predict alternative energy minima.

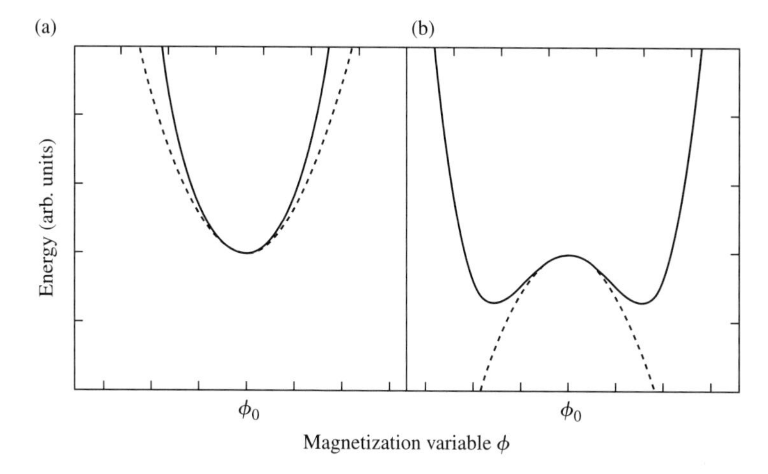

Fig. 4.9 Harmonic approximation: (a) stable spin configuration and (b) unstable configuration. In the approximation, nonlinear energies (solid lines) are approximated by a quadratic functions (dashed lines).

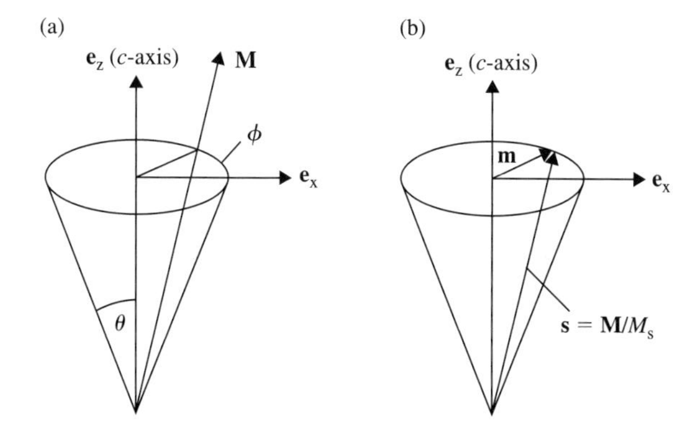

Fig. 4.10 Magnetization direction: (a) in terms of ϕ and θ and (b) in terms of m.

To realize the harmonic approximation in (4.9), consider small perpendicular deviations $\mathbf{m}$ and $\mathbf{a}$ from the complete saturation and perfect crystalline alignment, respectively. The meaning of these two quantities is illustrated in Fig. 4.10. In terms of $\mathbf{m}$ and $\mathbf{n}$, we can write

$$\mathbf{M}(\mathbf{r}) = M_s \left(\sqrt{1 - m^2(\mathbf{r})}\, \mathbf{e}_z + \mathbf{m}(\mathbf{r}) \right) \tag{4.16a}$$

and

$$\mathbf{n}(\mathbf{r}) = \sqrt{1 - a^2(\mathbf{r})}\, \mathbf{e}_z + \mathbf{a}(\mathbf{r}) \tag{4.16b}$$

Ignoring cubic and higher-order terms, the magnetization becomes

$$\mathbf{M}(\mathbf{r}) = M_s \left(1 - \frac{m(\mathbf{r})^2}{2}\right) \mathbf{e}_z + M_s\, \mathbf{m}(\mathbf{r}) \tag{4.17a}$$

and the easy-axis vector

$$\mathbf{n}(\mathbf{r}) = \left(1 - \frac{a(\mathbf{r})^2}{2}\right) \mathbf{e}_z + \mathbf{a}(\mathbf{r}) \tag{4.17b}$$

Note that magnets with partial crystalline misalignment, $|\mathbf{a}| > 0$, are also known as *textured* magnets (Müller *et al.* 1994). They are intermediate between aligned magnets ($\mathbf{n} = \mathbf{e}_z$) and isotropic magnets, where $\mathbf{n}$ is random with $\langle \mathbf{n} \rangle = 0$. Texture is usually treated on a Stoner-Wohlfarth level, although the corresponding small-grain criterion is often stringent, especially in the limit of well-aligned magnets (small texture).

Aside from a physically unimportant zero-point energy, series expansion of (4.9) yields, for $\mathbf{H} = H\mathbf{e}_z$,

$$E = \int \left[A(\nabla \mathbf{m})^2 + K_{\text{eff}}(\mathbf{m} - \mathbf{a})^2 + \frac{1}{2}\left(\mu_o\, M_s\, H_{\text{eff}}\, \mathbf{m}^2\right)\right] \mathrm{d}V \tag{4.18}$$

Here we have incorporated the magnetostatic self-interaction into $K_{\text{eff}} = K_1 + K_{\text{sh}}$ and $H_{\text{eff}} = H + H_{\text{loc}}$. This is exact for a variety of problems, such as nucleation modes in perfect ellipsoids of revolution, but a rather crude approximation from a general point of view.

The energy $E = \int \eta\, \mathrm{d}V$ is minimized by putting $\delta F/\delta \mathbf{m}(\mathrm{r}) = 0$, where $\delta E/\delta \mathbf{m}(\mathrm{r}) = -\nabla(\partial\eta/\partial\nabla\mathbf{m}(\mathrm{r})) + \partial\eta/\partial\mathbf{m}(\mathrm{r})$ is the functional derivative (Section A.2.4). The result is

$$-\nabla(A\nabla\mathbf{m}) + \left(K_{\text{eff}} + \frac{\mu_o}{2} M_s\, H_{\text{eff}}\right) \mathbf{m} = K_{\text{eff}}\, \mathbf{a}(\mathbf{r}) \tag{4.19}$$

As in the original equations, A, M_s, K_{eff}, and H_{eff} are local parameters, for example $K_{\text{eff}}(\mathbf{r})$. However, since $A \approx 10\,\text{pJ/m}$ for a wide range of materials, A is nearly constant in many cases, and $\nabla(A\nabla\mathbf{m}) \approx A\,\nabla^2\mathbf{m}$. One exception is grain boundaries, which may exhibit a significantly reduced local exchange (Section 4.4).

Equation (4.19) means that the easy-axis disorder $\mathbf{a}(\mathbf{r})$ acts as an inhomogeneity. However, the magnetic stability is determined by the left-hand side of (4.19). The nucleation field H_N and the nucleation mode $\mathbf{m}(\mathbf{r})$ are obtained by eigenmode analysis of the operator acting on $\mathbf{m}$. In Section 4.3, we will consider a number of examples.

4.2.5 Micromagnetic scaling

Dimensional analysis of the free energy (4.9) yields a variety of micromagnetic length scales. There are three basic quantities, namely the exchange stiffness A, measured in J/m, the anisotropy K_1, measured in J/m^3, and the magnetostatic self-energy $\mu_o M_s^2$, also measured in J/m^3. External fields are often comparable to M_s, so that the self-energy is also a crude estimate of the Zeeman energy.

There are two fundamental lengths. The *wall-width parameter*

$$\delta_o = \sqrt{\frac{A}{K_1}} \tag{4.20}$$

determines the thickness of the domain wall separating magnetic domains of different magnetization directions and the spatial response of the magnetization to local perturbations. It varies from about one nanometer in extremely hard materials to several hundred nanometers in very soft materials. The parameter δ_o is sometimes regarded as an exchange length, determining the effective range of exchange interactions. If this were a valid consideration, then ideally soft materials, characterized by $K_1 = 0$ and $\delta_o = \infty$, would realize exchange coupling on a macroscopic scale. However, this is not observed and at odds with a refined micromagnetic analysis.

The second length is the (proper) *exchange length*

$$l_o = \sqrt{\frac{A}{\mu_o M_s^2}} \tag{4.21}$$

It describes the competition between interatomic exchange and magnetostatic self-interaction and determines, for example, the transition from coherent rotation to curling (Section 4.3.1) and the grain size below which mixtures of two magnetic phases with different anisotropies yields single-phase hysteresis loops (Fig. 4.14). Since typical ferromagnets have magnetizations of the order of 1 T and exchange stiffnesses of the order of 10 pJ/m, l_o is between 1 and 2 nm for a broad range of materials. In practice, most lengths derived from l_o carry a factor of the 5 (Section 4.3), so that experimental exchange-length scales are close to 10 nm. From an atomic point of view, the order of magnitude of l_o is $a_o/\alpha = 7.52$ nm, where a_o is the Bohr length and $\alpha \approx 1/137$ is Sommerfeld's fine-structure constant (Skomski, Oepen, and Kirschner 1998). Physically, the involvement of the fine-structure constant reflects the higher-order relativistic character of the anisotropy. Combinations of l_o and δ_o yield other lengths, such as the critical single-domain radius (4.27).

4.2.6 Domains and domain walls

The magnetostatic self-interaction (4.11) favors *magnetic domains* with partial or complete flux closure. Figure 4.11 shows various domain configurations. Historically, the concept of domains was introduced to explain why two pieces of soft iron do not attract each other. In 1907, Weiss postulated that ferromagnetism on a local scale, created by mean-field interactions, is accompanied by a loss of net magnetization due to domain formation. Bloch (1932) introduced the concept of domain walls, and the first quantitative calculations by Landau and Lifshitz (1935) are now regarded as the starting point of modern domain theory. More theoretical developments in the mid twentieth century are the analysis of pinning mechanisms (Becker and Döring 1939, Kersten 1943) and nucleation modes (Brown 1963a).

The first experimental verification of domains was due to Barkhausen (1919), who measured magnetizations jumps associated with domain-wall motion. The first direct domain observations were made by Hámos and Thiessen in 1931 and, later the same year, Bitter. Since then, many papers, books, and reviews have been published on

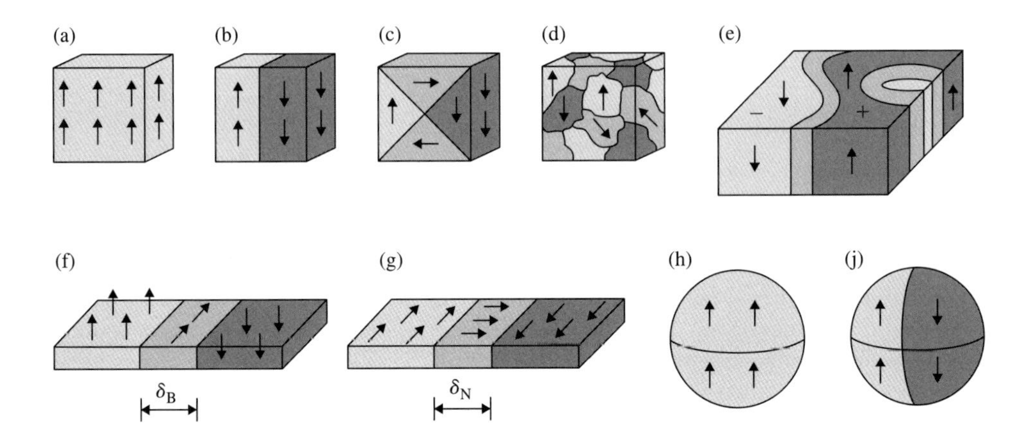

Fig. 4.11 Micromagnetic spin configurations: (a) single-domain state, as observed in very small particles, (b) two-domain configuration, as encountered in fairly small particles with uniaxial anisotropy, (c) flux-closure in cubic magnets, (d) complicated domain structure in a polycrystalline magnet, (e) Bloch wall in a thin film with perpendicular anisotropy, (f) Néel wall in a thin film with in-plane anisotropy, (h) single-domain particle and (j) two-domain particle.

experimental and general aspects of magnetic domains. Examples are Kittel (1949), Bozorth (1951), Kooy and Enz (1960), Craik and Tebble (1961), Chikazumi (1964), Oepen and Kirschner (1989), and Hubert and Schäfer (1998). Recent theoretical developments will be discussed below, in connection with specific models and phenomena.

Magnetic domains are separated by comparatively thin, though not atomically sharp, domain walls (Bloch 1932). Incidentally, Bloch's original estimate of the domain-wall width was based on the incorrect assumption that $M_s = 0$ inside the wall. In 1932, the role of magnetization rotations was not yet established, and the modern theory of Bloch walls starts with Landau and Lifshitz (1935). The idea is to start from (4.9), to put $H = 0$, and to assume that $H_d = 0$. It can be shown that a magnetic field yields domain-wall motion rather than domain-wall widening or narrowing, and that domain walls in bulk magnets are free of magnetic poles ($H_d = 0$). The only remaining energy contributions are then the exchange energy (A) and the anisotropy energy (K_1). The only length that can be formed from these two parameters is the wall-width parameter $\delta_o = \sqrt{A/K_1}$, and a refined calculation yield the *Bloch-wall width*

$$\delta_B = \pi\sqrt{\frac{A}{K_1}} \tag{4.22}$$

We see that the anisotropy (K_1) favors narrow domain walls. Inside the domains, the magnetization lies in a favorable direction, but in the wall, the magnetization is no longer parallel to an easy axis. This costs anisotropy energy and causes the wall to narrow. On the other hand, narrow walls correspond to large magnetization gradients and are unfavorable from the point of view of exchange (A).

To derive (4.22), we start from (4.9) and rewrite $\mathbf{M}$ in term of the magnetization angles ϕ and θ. For a Bloch wall in the y–z plane, $\mathbf{M} = M_s \cos\theta\, \mathbf{e}_z + M_s \sin\theta\, \mathbf{e}_y$, and (4.9) yields the energy per wall area L^2:

$$\frac{E}{L^2} = \int \left(A \left(\frac{\partial \theta}{\partial x} \right)^2 + E_K(\theta) \right) \mathrm{d}x \tag{4.23}$$

where $E_K(\theta)$ is the anisotropy-energy density. Minimizing this energy with respect to $\theta(x)$ (Section A.2.4) yields the Euler equation $-2A\,\partial^2\theta/\partial x^2 + \partial E_K(\theta)/\partial\theta = 0$. We now multiply the Euler equation by $\partial\theta/\partial x$, exploit $\partial(\partial\theta/\partial x)^2/\partial x = 2\,\partial\theta/\partial x\,\partial^2\theta/\partial x^2$, and integrate over x. The integration is trivial, of the type $\int \partial f/\partial x\,\mathrm{d}x = f$, and yields

$$\frac{\mathrm{d}\theta}{\mathrm{d}x} = \pm\sqrt{\frac{E_K(\theta)}{A}} \tag{4.24}$$

The solution of (4.24) is $x = \pm\int (A/E_K(\theta))^{1/2}\,\mathrm{d}\theta$. In the simplest case, $E_K(\theta) = K_1 \sin^2\theta$. Using $M_z = M_s \cos\theta$, we obtain $\mathrm{d}M_z/(M_s^2 - M_z^2) = \pm(K_1/A)\,\mathrm{d}x$ and

$$M_z(x) = -M_s \tanh\left(x\sqrt{\frac{K_1}{A}} \right) \tag{4.25}$$

Figure 4.12 shows M_z as a function of $x/(K_1 A)^{1/2} = x/\delta_o$. Defining the domain-wall width δ_w in terms of $\mathrm{d}M_z/\mathrm{d}x$ at $M_z = 0$ (dashed line) yields $d_w = 2\delta_o$. However, it is common to define the wall width by considering $\theta(x)$ rather than $M_z(x)$, and from $\mathrm{d}\theta/\mathrm{d}x$ at $\theta = \pi/2$ one obtains the "conventional" Bloch-wall width $\delta_B = \pi\delta_o$, that is, equation (4.22). The energy of a 180° Bloch wall is obtained by putting (4.25) into (4.23). The result is

$$\gamma = 4\sqrt{A K_1} \tag{4.26}$$

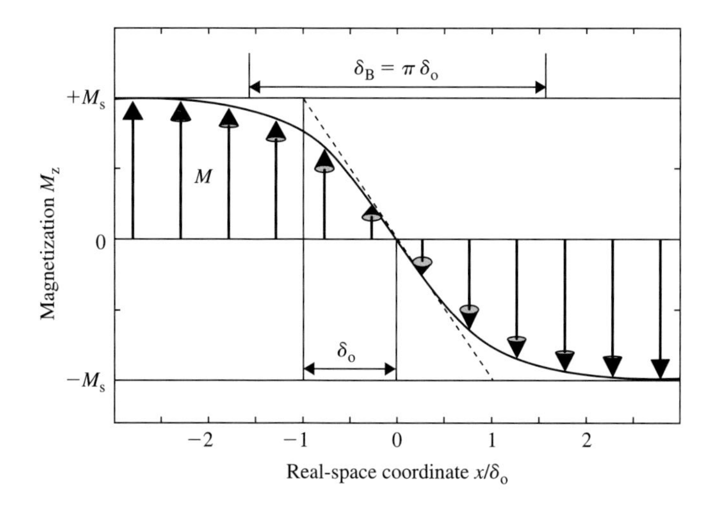

Fig. 4.12 Fine structure of a 180° Bloch wall in the bulk. The wall is located in the y–z plane, and in the center of the wall ($x = 0$) the magnetization vector is parallel to the y axis (perpendicular to the paper plane).

Typical domain-wall widths are 5 nm and 100 nm for hard and soft magnetic materials, respectively. Domain-wall energies range from about $0.1\,\mathrm{mJ/m^2}$ for soft materials to about $50\,\mathrm{mJ/m^3}$ in very hard materials.

Figure 4.11 indicates that there are several types of domain walls, such as 180° Bloch walls (b and f), 90° walls (c), and Néel walls (g). Equation (4.22) refers to 180° Bloch walls in uniaxial bulk magnets. As mentioned, Bloch walls do not have any magnetostatic energy, except at the surface of the magnet. This is because $\nabla \cdot \mathbf{M} = 0$ for $\mathbf{M} = M_\mathrm{s} \cos\theta\, \mathbf{e}_\mathrm{z} + M_\mathrm{s} \sin\theta\, \mathbf{e}_\mathrm{y}$. However, other walls may have considerable magnetostatic self-energies. An example is the Néel wall between 180° domains in the plane of very thin films, Fig. 4.11(g). This spin configuration occurs very frequently in soft-magnetic thin films, where the demagnetizing field confines the magnetization to the film plane. The magnetization of a Néel wall remains in the film plane, which gives rise to magnetic charges in the film, at the right and left ends of the wall shown in (g). The transition to a Bloch wall, realized by switching the magnetization to a perpendicular direction, would cost even more magnetostatic energy, because it creates surface charges all across the wall area. For soft-magnetic thin films with in-plane magnetization, such as Fig. 4.11(g), the transition from Bloch walls in thick films to Néel walls in thin films occurs at a thickness that is some multiple of the exchange length l_o.

Domain formation is magnetostatically favorable but costs domain-wall energy. A simple criterion for the existence of equilibrium domains is obtained by comparing the energies of the two configurations Fig. 4.11(h–j). The wall energy in (j) is $\gamma\pi R^2$, whereas the gain in magnetostatic energy is roughly half the single-domain energy, that is, $\mu_\mathrm{o} M_\mathrm{s}^2 V/12$. Domain formation in spheres is therefore favorable for particles whose radius exceeds a *critical single-domain radius*

$$R_\mathrm{sd} \approx \frac{36\sqrt{AK_1}}{\mu_\mathrm{o} M_\mathrm{s}^2} \tag{4.27}$$

This value varies between a few nm in soft magnets and about $1\,\mu\mathrm{m}$ in very hard magnets.

Since the derivation of (4.27) involves the comparison of competing ground-state energies rather the analysis of hills and valleys in the energy landscape, single-domain behavior is *unrelated to hysteresis*. Experimentally, the single-domain character of a particle is accessible after thermal demagnetization, that is, by heating above the Curie temperature and subsequent cooling. In micromagnetism, this is known as the initial or virgin state. The application of an external field, as required for hysteresis, may create nonuniform magnetization states in single-domain particles (Section 4.3.1) or remove domain walls in multidomain particles. A good example is submicron single-domain particles of permanent-magnet materials such as $BaFe_{12}O_{19}$ and $SmCo_5$, whose magnetization reversal is almost inevitably nonuniform (Skomski and Coey 1999). The equating of single-domain magnetism and uniform rotation, as epitomized by the unfortunate term "elongated single-domain particle" (ESD), has its origin in the focus on soft and semi-hard magnets in the first half of the twentieth century, where the difference between uniform rotation and single-domain behavior is less striking.

Domain sizes are generally more difficult to predict than domain-wall width. One aspect of the problem is the large number of competing domain structures, Fig. 4.11 (b–e) and (f), and Fig. 4.13. A second aspect is the difficult-to-treat long-range character of the magnetostatic interactions. For example, the size of domains such as Fig. 4.11(e) exhibits a square-root dependence on the film thickness b. This is because both the top and the bottom of the film contain magnetic charges which, according to (4.11), interact over long distances.

A crude estimate of the domain size is obtained by assuming stripe domains of length L and width w in a relatively thick film of area L^2. The domain size is then obtained by comparing the self-interaction energy of order $\mu_o M_s^2 L^2 b(w/b)$ with the wall energy $\gamma bL(L/w)$. Here the factor w/b is a crude estimate for the demagnetizing factor of the domain, corresponding to an ellipsoid of axes $R_x = w$, $R_y = \infty$, and $R_z = b$. We see that narrow domain walls are magnetostatically favorable but cost wall energy. Comparison of the two energies yields

$$w \sim \sqrt{\frac{\gamma b}{\mu_o M_s^2}} \tag{4.28}$$

This simple calculation indicates that domain sizes (and critical single-domain sizes) strongly depend on anisotropy and sample geometry. In thin films, this dependence is sometimes exponential (Málek and Kamberský 1958, Skomski, Oepen, and Kirschner 1998), and features such as stray fields created by domain walls play an important role (Hubert and Schäfer 1998, Kuch *et al.* 2003) As in other areas of magnetism, numerical micromagnetic methods have become a valuable tool in the investigation of domain structures and magnetization processes (Schrefl, Fidler, and Kronmüller 1994).

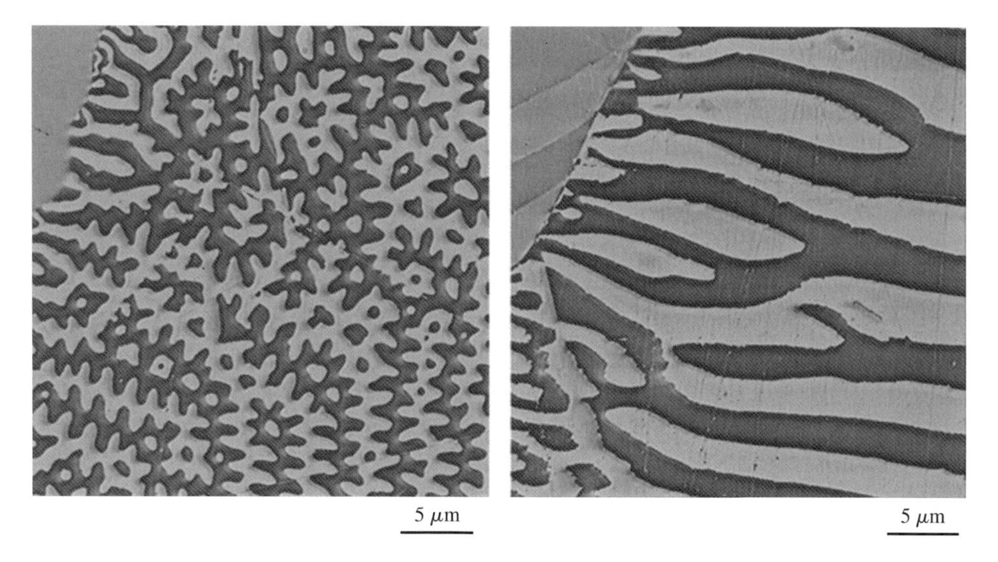

Fig. 4.13 Spike domains in $Nd_2Fe_{14}B$ magnet, as probed by magnetic force microscopy (MFM). The top view (left) and the side view (right) corresponds to the schematic domain structure of Fig. 4.11(e). *Courtesy S.-H. Liou.*

When the radius of soft-magnetic nanoparticles is smaller than δ_o, then the "wall" covers the whole particle and $\nabla\mathbf{M} \sim M_s/R$. The magnetostatic and exchange energies scale as $\mu_o D M_s^2 R^3$ and AR, respectively, and a strongly nonuniform ground state is realized when R is much larger than $(A/\mu_o D M_s^2)^{1/2}$. In other words, the critical radius R_F for the onset of a flux-closure or "vortex" state scales as $l_o/D^{1/2}$. In plate-like soft-magnetic thin-film dots with in-plane magnetization, the demagnetizing factor $D \sim b/R$, so that $R_F \sim b^{-1/2}$ (Skomski *et al.* 2004b). On somewhat larger submicron length scales, for example in lithographically patterned thin films, flux-closure configurations are common, too, but the domain walls keep their individuality (Hirohata *et al.* 2000).

4.3 Coercivity

Summary As other extrinsic or hysteretic properties, coercivity is strongly real-structure dependent. Aside from coherent rotation, important coercivity mechanisms are curling, localized nucleation, and domain-wall pinning. Nucleation refers to the onset of magnetization reversal and determines the coercivity in nearly defect-free magnets. With increasing size, the nucleation mechanism in perfect ellipsoids of revolution changes from coherent rotation to curling. The curling mode costs some exchange energy but is magnetostatically favorable due to vortex-like flux closure. However, both coherent rotation and curling greatly overestimate the coercivity of most magnetic materials. This disagreement, known as Brown's paradox, is solved by considering localized nucleation due to imperfections. Micromagnetic localization costs exchange energy, too, but is favorable from the point of view of anisotropy, because it exploits local anisotropy minima. The transition from coherent rotation to curling or localized nucleation is unrelated to the single-domain character of the magnet, and magnetization reversal in single-domain particles is not necessarily coherent. *Pinning* means that the motion of domain wall is impeded by imperfections. It determines the coercivity in strongly inhomogeneous magnets. Some pinning mechanisms are Kersten pinning, Gaunt-Friedel pinning and weak pinning. The above considered micromagnetic models must be distinguished from phenomenological models and methods, such as Preisach models and remanence plots.

The most intriguing aspect of hysteresis is the coercive force or *coercivity*. It describes the stability of the remanent state and gives rise to the classification of magnets into hard magnetic materials (permanent magnets), semihard materials (storage media), and soft-magnetic materials. Coercivity goes back to the first half of the twentieth century, to the publications by Kondorski (1937), Becker and Döring (1939), and Stoner and Wohlfarth (1948). The physical origin of the so-called "static" coercivity, as treated in this chapter, is metastable energy minima that vanish on applying a reverse magnetic field, very similar to Fig. 4.4. In addition to the static coercivity, there are small time-dependent corrections to the hysteresis. For convenience, they will be discussed in the context of dynamic magnetization effects (Section 6.4).

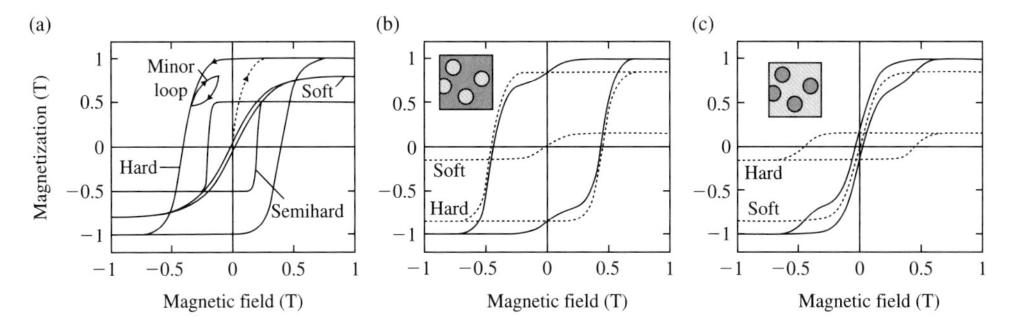

Fig. 4.14 Typical hysteresis loops: (a) hard, semihard, and soft magnets, (b) inflected loop (85% hard and 15% soft phases), and (c) wasp-shaped loop (15% hard and 85% soft phases). In (b) and (c), dark and bright regions correspond to hard and soft regions, respectively. Also shown in (a) are a virgin curve (dotted line) and a minor loop.

Figure 4.14 shows some types of hysteresis loops encountered in practice. The main difference between permanent or hard magnets and soft magnets is the magnitude of the coercivity, ranging from less than $10\,\mu$T (soft) to more than 1 T (hard). The two-phase loops shown in Fig. 4.2(b–c) are superpositions of single-phase hysteresis loops and correspond to macroscopic phase mixtures. In nanocomposites, exchange interactions between the phases tend to smooth the loops, and the magnetic phase analysis of Fig. 4.2 must be replaced by micromagnetic calculations.

Here we are concerned with the coercivity of macroscopically homogeneous magnets, ignoring the technologically important but scientifically rather boring superposition effects shown in Fig. 4.14(b–c). A major challenge in the understanding of coercivity is that experimental coercivities are much smaller than predicted from the Stoner-Wohlfarth theory. For example, $Nd_2Fe_{14}B$ has an anisotropy field of about 8 T, but as-cast $Nd_2Fe_{14}B$ magnets are quite soft, and sophisticated processing techniques are necessary to achieve coercivities of order 2.5 T. This finding is known as *Brown's paradox* and is solved by taking into account the real structure of the magnet, such as defects, crystalline texture, and grain boundaries.

There are two basic coercivity mechanisms, namely *nucleation* and *pinning*. Nucleation refers to the stability of the fully magnetized state in a reverse field. A trivial example of a nucleation model is the Stoner-Wohlfarth model, as epitomized by Fig. 4.4. Pinning means that the coercivity is due to the interaction of domain wall with real-structure features such as defects. An example is iron, where reverse domains nucleate very easily but defects may create coercivity by impeding the motion of the domain walls. For example, the coercivity of technical iron doubles by adding 0.01 wt.% nitrogen (Kersten 1943). Figure 4.15 shows a typical reversal process in a magnetic particle.

The key challenge in coercivity modeling is to determine H_c from the magnet's real structure. This aim is complicated by the fact that one set of magnetic data may be reproduced by different micromagnetic theories. For example, there is no simple way to tell whether a given major loop is nucleation- or pinning-controlled. Some guidance is provided by the virgin curve after thermal demagnetization, that is, by the dotted

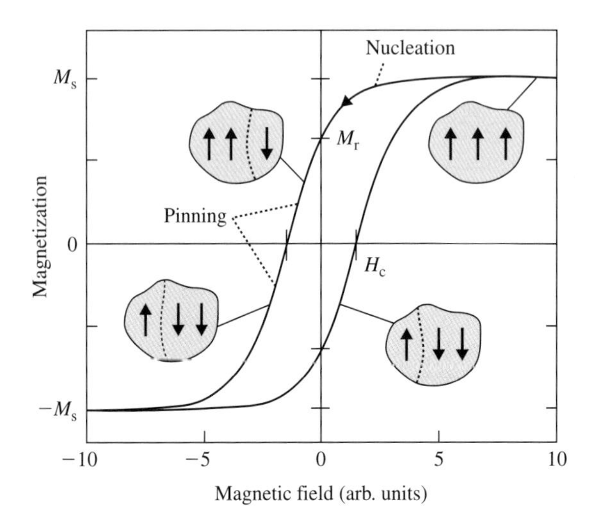

Fig. 4.15 Origin of hysteresis. The schematic figure shows one typical coercivity mechanism in ferromagnetic particles. In this example, the fully magnetized magnetization state (top right) is rather unstable, so that the coercivity is due to domain-wall pinning. Since the particle is free of bulk defects, the pinning of the wall occurs at surface irregularities.

line in Fig. 4.14(a). In particles having radii larger than R_{SD}, the thermally demagnetized equilibrium state contains magnetic domains. Pinning-controlled magnets, characterized by difficult-to-move domain walls, exhibit a small initial susceptibility $\chi = \mathrm{d}M/\mathrm{d}H$, whereas the absence of pinning centers in nucleation-controlled magnets gives rise to mobile domain walls and a large initial susceptibilities.

4.3.1 Nucleation

Nucleation means that a magnetization state becomes unstable at the *nucleation field* $H_z = -H_N$. A simple example is coherent rotation in spherical particles, where $H_N = H_a$ (Section 4.1.1). In contrast to localized nucleation processes in phase-transition kinetics, such as the formation of liquid droplets from the gas phase, micromagnetic nucleation may or may not be localized. Delocalized nucleation means that the nucleation mode extends over the whole magnetic body. For example, coherent rotation is delocalized, because $\mathbf{M}(\mathbf{r}) = \mathbf{M}(\theta, \phi)$ throughout the magnet.

In nearly perfect magnets, H_N is a good estimate for the coercivity, $H_c \approx H_N$. In other words, it is sufficient to consider the *onset* of magnetization reversal, as contrasted to phenomena such as domain-wall motion. The nucleation mode depends on the real structure of the magnet. Nucleation in very small particles is delocalized but not necessarily coherent. Localized nucleation is always incoherent (nonuniform) and frequently encountered in practice.

To determine the nucleation field, we perform a stability analysis similar to that in Figs 4.4 and 4.9, except that our phase space is multidimensional (continuous). Nucleation occurs when the curvature of the dashed line in Fig. 4.9 reaches zero. Let us start from (3.19) and make the model assumption that $\nabla A = 0$ (homogeneous exchange) and $\mathbf{a} = 0$ (perfect crystalline alignment). The equation of state is then

$$-A\nabla^2\mathbf{m} + \left(K(\mathbf{r}) + \frac{\mu_o}{2} M_s H\right)\mathbf{m} = 0 \tag{4.29}$$

The nucleation instability is obtained by eigenmode analysis, as outlined in Section A.2.2. Each eigenmode corresponds to a curvature of type shown in Fig. 4.9, and the stability is determined by the lowest curvature or eigenvalue. The eigenfunction $\mathbf{m}(\mathbf{r})$ that corresponds to the nucleation field is known as the *nucleation mode*.

Assuming that $K(\mathbf{r}) = const.$ and putting the coherent mode $\mathbf{m}(\mathbf{r}) = \mathbf{m}_o$ into (4.29) yields $K + \mu_o H M_s/2 = 0$, or $H_N = 2K/\mu_o M_s$. This is, of course, the familiar Stoner-Wohlfarth expression for coherent rotation. Inhomogeneities, such as anisotropy inhomogeneities $K(\mathbf{r})$, mean that $\mathbf{m}(\mathbf{r}) = \mathbf{m}_o$ is no longer an eigenmode of (4.29) but acquires the character of a trial function, similar to the variational approach in quantum mechanics (Section A.3). This effective-anisotropy approximation, based on $K_{\text{eff}} = \langle K(\mathbf{r})\rangle$, is a micromagnetic analog to the virtual-crystal approximation in band-structure calculations (Section 7.1.1). The corresponding nucleation field $2\langle K\rangle/\mu_o M_s$ *overestimates* the coercivity, because regions with small $K(\mathbf{r})$ contribute disproportionately strongly to the nucleation field. For example, in a composite containing hard and soft regions, the nucleation starts in a soft region, irrespective of the average anisotropy.

Equation (4.29) considers magnetostatic self-interactions by incorporation into K and H. This procedure is exact for coherent rotation but not necessarily for other reversal modes. For example, flux closure is magnetostatically favorable and may therefore facilitate magnetization reversal. This has indeed be shown for the *curling mode*

$$\mathbf{m}(\phi, z, r) = m_o(z, r)\,(\cos\phi\,\mathbf{e}_y - \sin\phi\,\mathbf{e}_x) \tag{4.30}$$

This mode is nonuniform ($\nabla\mathbf{m} \neq 0$) and costs exchange energy, but the flux closure implied by (4.30) lowers the magnetostatic energy. Figure 4.16 shows curling modes for several geometries. The determination of the magnetic energy is moderately complicated, but the essence of the calculation is seen by comparing the coherent-rotation and curling modes in spheres, Fig. 4.16(d–e). Coherent rotation (d) leaves the magnetostatic self-energy unchanged, because the reduced magnetization component in z direction is compensated by a magnetization component in the x–y plane. In (4.4), this corresponds to $D = 1/3$ and $\frac{1}{2}(1-3D) = 0$. Curling (e) implies a reduction of the z component, very similar to coherent rotation, but this reduction is *not* accompanied by the creation of a net magnetization in the x–y plane. In fact, the in-plane magnetization component (4.30) has no poles (magnetic charges) and does not contribute to the magnetostatic energy. The corresponding change in magnetostatic energy is proportional to $-D$, as compared to $\frac{1}{2}(1 - 3D)$ for coherent rotation.

The calculation of the exchange energy requires the knowledge of the function $m_o(z, r)$, such as Bessel functions for cylinders and tubes. An order-of-magnitude estimate is obtained by taking into account that $A\nabla^2 \approx A/R^2$, where R is the radius of the ellipsoid of revolution. An explicit calculation for the ellipsoid of revolution yields the curling nucleation field

$$H_N = \frac{2K_1}{\mu_o M_s} - DM_s + \frac{c(D)A}{\mu_o M_s R^2} \tag{4.31}$$

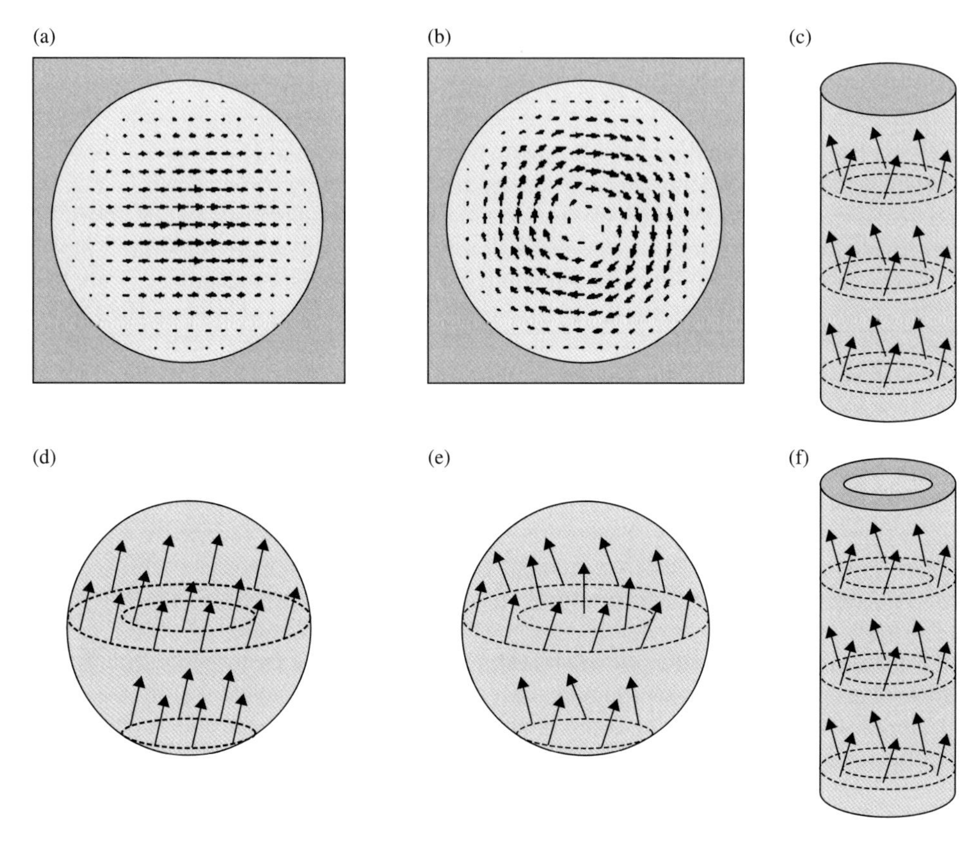

Fig. 4.16 Delocalized nucleation modes: (a) quasicoherent rotation or bulging, (b) clamped curling, (c) curling in a cylinder, (d) coherent rotation in a sphere, (e) curling in a sphere, and (f) curling in a nanotube. The modes (a) and (b) are shown as top views on **m** in the equator plane and describe soft inclusions in a very hard matrix.

where the values of c are 8.666 for spheres ($D = 1/3$) and 6.678 for needles ($D = 0$). For details about the modes in Fig. 4.16, see Brown 1963a and Aharoni 1996 (c–e) and Skomski, Liu, and Sellmyer 1999a (a–b), and Sui *et al.* 2004 (f).

Figure 4.17 compares the curling nucleation field (4.31) with the Stoner-Wohlfarth prediction (4.4). For large particles, the gain in magnetostatic energy overcompensates the increase in exchange energy, and curling is more favorable than coherent rotation. In spheres, the corresponding transition or coherence radius $R_{\text{coh}} = 5.099\, l_{\text{o}}$, whereas in wires $R_{\text{coh}} = 3.655\, l_{\text{o}}$. Since R_{coh} reflects the competition between magnetostatic and exchange energies, it is *independent* of the anisotropy constant K_1. Any change in the uniaxial anisotropy shifts both the coherent-rotation line (left) and the curling line (right) by the same amount and leaves R_{coh} unaffected. Typical coherence radii are of the order of 10 nm for a broad range of ferromagnetic materials. Experimental coercivities are often much smaller than the predictions of (4.4) and (4.31). This is *Brown's paradox*. Very small particles have a low coercivity due to superparamagnetic excitations (Section 6.4.6), but otherwise Brown's paradox is explained by localized

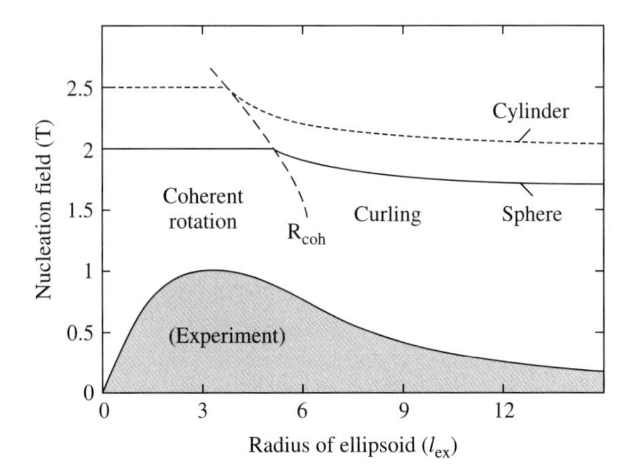

Fig. 4.17 Delocalized nucleation in a sphere (solid line) and a cylinder (dotted line). The dashed line separating coherent rotation and curling is a guide for the eye.

nucleation. Statistically, big particles are more likely to contain imperfections harmful to coercivity, so that the coercivity tends to decrease with increasing particle radii $R \gg l_{ex}$.

The coherent-rotation and curling modes in homogeneous ellipsoids of revolution are *exact* solutions of the nucleation problem. This must be contrasted to the critical single-domain radius, which depends on the choice of the domain-wall geometry and also on the only approximately known magnetostatic energy. It is also important to keep in mind that both (4.4) and (4.31) differ from the naïve demagnetizing-field correction, which yields $H_N = 2K_1/\mu_o M_s - DM_s$.

Real magnets tend to contain chemical or microstructural imperfections, and their shapes often deviate from the required ellipsoidal geometry. The coherent and curling modes are then no longer eigenfunctions of (4.29), and the nucleation field decreases. In fact, a single submicron imperfection may control the magnetization reversal of a macroscopic volume, which is known as the Barkhausen effect (1919). In permanent magnets, inhomogeneities larger than a few nm are very harmful to coercivity. Industrial nucleation-type magnets ($Nd_2Fe_{14}B$, $BaFe_{12}O_{19}$, $SmCo_5$) require sophisticated annealing or liquid-phase sintering to reduce the number of morphological inhomogeneities.

In terms of (4.29), imperfections lead to a local reduction of the anisotropy $K(\mathbf{r})$. If $K(\mathbf{r})$ were the only consideration, then H_N would be given by the global minimum of $K(\mathbf{r})$. However, the corresponding magnetization inhomogeneity is punished by an increase in exchange energy. To describe the nucleation of reverse domains near imperfections, we exploit that linearized micromagnetic equation (4.29) is reminiscent of the Schrödinger equation

$$-\frac{\hbar^2}{2m}\nabla^2\psi + (V(\mathbf{r}) - E)\psi = 0 \tag{4.32}$$

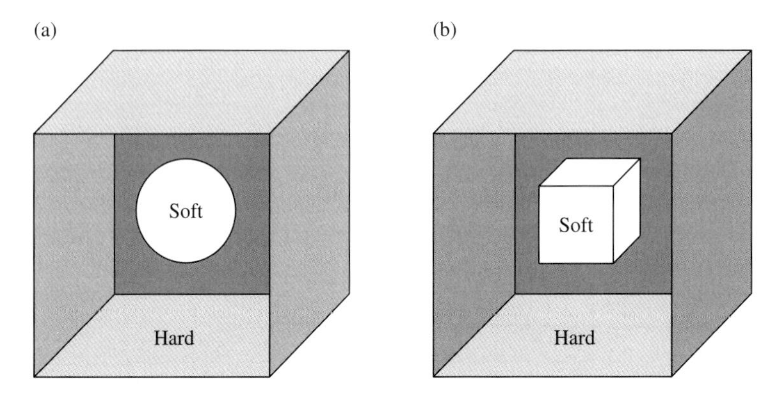

Fig. 4.18 Soft-magnetic inclusions ($K = 0$) in a hard matrix ($K = \infty$).

for an electron in a potential $V(\mathbf{r})$. In this analogy, the nucleation mode and the nucleation field are analogous to the ground-state energy and wave function, respectively. This makes it possible to use ideas known from quantum mechanics to solve micromagnetic nucleation problems.

A simple example is a soft-magnetic inclusion ($K = 0$) embedded in a very hard matrix ($K = \infty$). Figure 4.18 shows two geometries. Let us consider a soft inclusion of cubic shape (b) and of volume L^3. The corresponding eigenfunctions are particle-in-a-box states, and the ground state is

$$\Psi(x, y, z) = N \sin\left(\frac{\pi x}{K}\right) \sin\left(\frac{\pi y}{L}\right) \sin\left(\frac{\pi z}{L}\right) \tag{4.33}$$

Here N is the normalization factor. The ground-state energy $3\pi^2\hbar^2/2m_e L^2$ corresponds to the nucleation field

$$H_N = \frac{6\pi^2 A}{\mu_o M_s L^2} \tag{4.34a}$$

where M_s is the spontaneous magnetization of the soft phase. This result can also be written as

$$H_N = H_a \frac{3\,\delta_h^2\, M_h}{L^2\, M_s} \tag{4.34b}$$

where the index h and $H_a = 2K_h/\mu_o M_h$ refer to the *hard* phase, respectively. Typically, magnetizations M_h and M_s of comparable magnitude, so that (4.34) predicts a nucleation-field reduction of the order of δ_h^2/L^2. Taking $\delta_h = 5\,\text{nm}$ and $L = 25\,\text{nm}$, and $M_h = M_s$ yields $H_N = 0.12\,H_a$, in agreement with typical experimental values.

In a nutshell, equation (4.34) solves Brown's paradox by considering local inhomogeneities. Of course, the cubic shape, as well as the approximate incorporation of the magnetostatic interaction into $K(\mathbf{r})$ limit the applicability of (4.34), but the model provides a physically reasonable semiquantitative description of nucleation processes in real materials. Refined models have been used to calculate nucleation fields for a variety of geometries, including planar defects and multilayers (Kronmüller 1987, Nieber and Kronmüller 1989), spherical inhomogeneities (Skomski and Coey 1993), core-shell nanoparticles (Skomski *et al.* 1999a), and nanowires (Skomski *et al.* 2000). We will return to these calculations in Section 7.4. In addition, it is now possible to reproduce nucleation fields and complete hysteresis loops by micromagnetic simulations (Schrefl, Fidler, and Kroumüller 1994).

A striking feature of (4.34) is the prediction of a diverging coercivity for very small inclusion sizes L. This is unphysical, because the coercivity of a hard-soft composite cannot be higher than that of the hard phase. In reality, the anisotropy field H_a of the hard phase provides a cut-off to the coercivity. Putting $H_N = H_a$ in (4.34) we see that this is the case for $L \approx \sqrt{3}\,\delta_h$, that is, for box dimensions comparable to the Bloch-wall width δ_h of the hard phase. In the quantum-mechanical analogy, the energy of an electron confined to a box of volume L^3 increases with decreasing L, and when the energy of the electron is larger than the potential energy of the box, the electron becomes delocalized. The same delocalization is encountered in micromagnetics as illustrated in Fig. 4.19 for the physically reasonable case of finite anisotropy K_h. For large inclusions, Fig. 4.19(a), the nucleation mode is localized and well approximated by (4.34). Small inclusions, Fig. 4.19(b), cause the tails of $m(\mathbf{r})$ to extend far into the hard phase. The mode becomes delocalized, and the reversal eventually degenerates into coherent rotation or curling, with $H_N \approx 2K_h/\mu_o M_h$.

4.3.2 Pinning

If localized nucleation leads to complete magnetization reversal, then the coercivity H_c is equal to the nucleation field H_N. However, this process requires the growth of the reverse domain, and the corresponding domain-wall motion may be impeded by real-structure imperfections (Becker and Döring 1939, Kersten 1943) This mechanism is known as *domain-wall pinning*. It is the main source of hysteresis and coercivity in

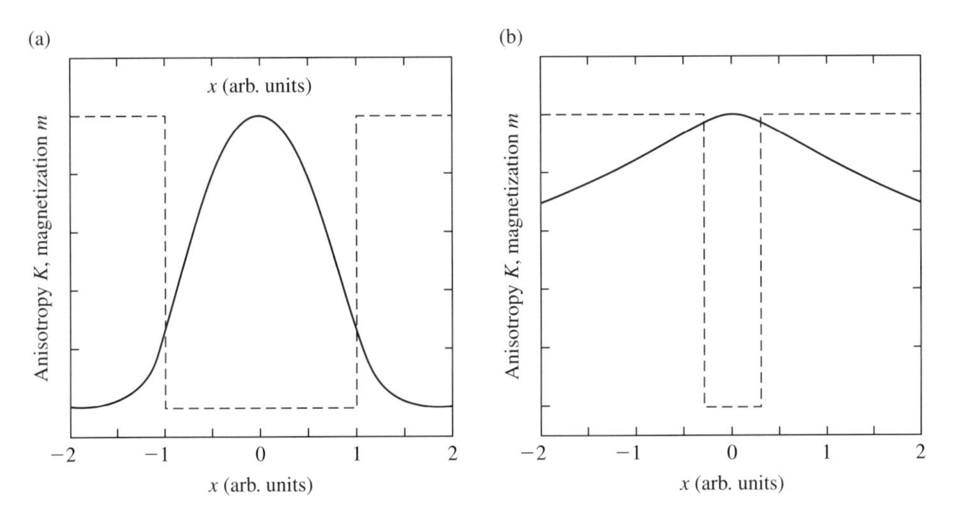

Fig. 4.19 Nucleation modes in inhomogeneous magnets: (a) well-localized and (b) largely delocalized.

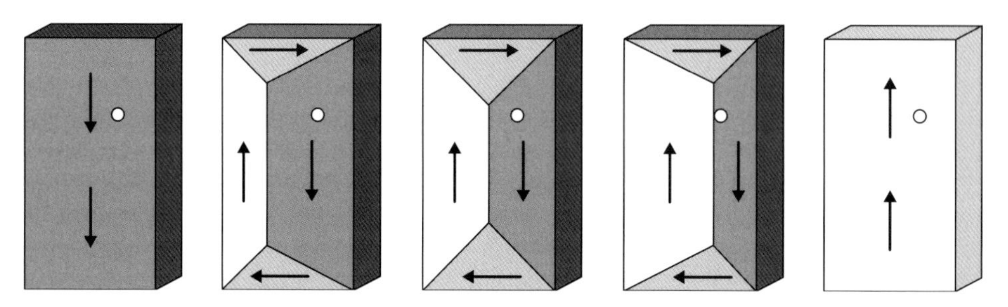

Fig. 4.20 Pinning in a submillimeter piece of iron containing a small inhomogeneity. The external magnetic field points in the ↑ direction and increases from left to right.

strongly imhomogeneous magnets and yields coercivities $H_c > H_N$. Since the domain-wall energy is proportional to the square root of K_1, domain-wall pinning often occurs in regions where K_1 is small due to chemical disorder. Simply speaking, the domain wall is trapped in a region with low domain-wall energy, and the coercivity is given by the pinning or "depinning" field H_p necessary to push the wall over the pinning-energy barrier. Figure 4.20 shows the example of a small piece of soft-magnetic material.

The are several pinning mechanisms, including strong pinning (Kersten and Gaunt-Friedel pinning) and weak pinning. Here we focus on *strong* domain-wall pinning, where the domain wall is pinned by individual imperfections (pinning centers), as in Fig. 4.20. Weak pinning is caused by a large number of very small pinning centers, such as atomic defects, and involves structural averages over a distance of order δ_B. In Section 7.4.3, we will discuss an explicit example, namely the coercivity of random-anisotropy magnets.

A very simple pinning model considers the energy $\gamma(x)$ of a planar domain wall as a function of the wall position x. In practice, $\gamma = 4(AK_1)^{1/2}$ is often determined by the local chemistry, which determines $K_1(x)$. Figure 4.21 shows an example where the wall energy of the defect is *higher* than that of the main phase. This regime is known as repulsive pinning and means the penetration of the wall into the defect costs energy. By contrast, attractive pinning amounts to the capturing of a wall in a region of low wall energy. In both regimes, the coercivity is given by analyzing the magnetic energy $E(x)$ as a function of the wall position x. For a wall of area L^2, the total wall energy

$$E(x) = \gamma(x)L^2 - 2\mu_o M_s H x L^2 \tag{4.35}$$

where $\gamma(x)$ depends, via K_1, on the microchemistry. The right-hand term in this equation is the Zeeman energy, and the factor 2 means that the magnetization changes from $-M_s$ to $+M_s$.

For any given field, the wall position x is obtained by putting $\mathrm{d}E/\mathrm{d}x = 0$, that is, $H = (\mathrm{d}\gamma/\mathrm{d}x)/2\mu_o M_s$. The derivative $\mathrm{d}\gamma/\mathrm{d}x$ is a function of x but reaches some maximum in the vicinity of the defect, where the slope is largest (dashed line in Fig. 4.21). Any further increment in H pushes the wall over the pinning defect, so

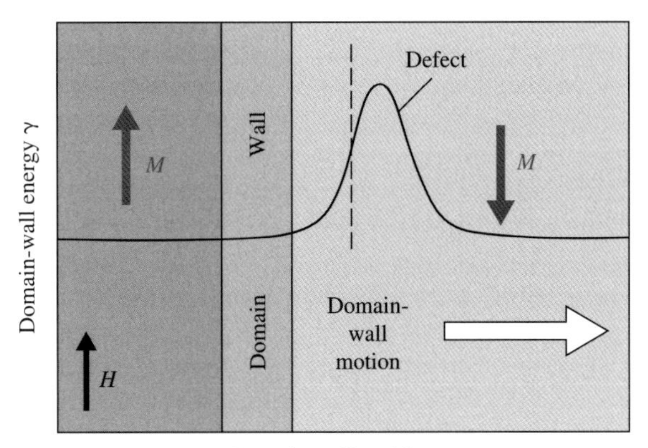

Fig. 4.21 Kersten pinning of a planar domain wall. The dashed line show the wall position where $\mathrm{d}\gamma/\mathrm{d}z$ and the pinning strength are largest.

that

$$H_\mathrm{p} = \frac{1}{2\mu_\mathrm{o} M_\mathrm{s}} \max\left(\frac{\mathrm{d}\gamma}{\mathrm{d}x}\right) \tag{4.36}$$

In general, the wall energy γ depends on the ratio of defect size to domain-wall width. A frequently considered anisotropy profile is rectangular, $K_1(x) = K_\mathrm{o}$ outside the defect and $K_1(x) = K_\mathrm{o} + \Delta K$ inside a defect of thickness b. For small ΔK and b, lowest-order perturbation theory yields the pinning coercivity (Becker and Döring 1939)

$$H_\mathrm{p} = H_\mathrm{a} \frac{\pi b}{3\sqrt{3}\,\delta_\mathrm{B}} \frac{|\Delta K|}{K_\mathrm{o}} \tag{4.37}$$

For small inhomogeneities of arbitrary profile (Skomski *et al.* 2004a):

$$E(x) = \int K_1(x)\left(1 - \tanh^2\left(\frac{\pi}{\delta_\mathrm{B}}(\xi - x)\right)\right) \mathrm{d}\xi - 2\mu_\mathrm{o} M_\mathrm{s} H x L^2 \tag{4.38}$$

This equation shows that the wall energy is, in fact, a convolution of the wall profile, as epitomized by δ_B, and the anisotropy profile $K_1(x)$. Pinning is most effective when the size b of the pinning centers is comparable to the domain-wall width δ_B. Very small effects lack pinning strength, as described by (4.37), whereas extended defects yield no additional pinning strength but tend to reduce the gradient $\mathrm{d}\gamma/\mathrm{d}x$. Refined calculations include features such as the spatial variation of the exchange stiffness (Givord and Rossignol 1996) and pinning at surface defects (Hubert und Schäfer 1998).

Equations (4.36–38) describe the pinning of a domain wall by a planar defect of thickness b. The planar character of the defect is epitomized by the *pinning force* $p = L^2\,\mathrm{d}\gamma(x)/\mathrm{d}x$, which has the dimension of a force ($1\,\mathrm{N} = 10^5\,\mathrm{dyn}$). In terms of the pinning force, $H_\mathrm{p} = p/2\mu_\mathrm{o} M_\mathrm{s} L^2$. The planar model is a reasonable approximation for features such as extended grain boundaries, but most strong pinning centers have compact shapes, characterized by diameters b. For example, pinning centers in bulk materials can often be described as spherical inclusions (Kersten 1943). The finite pinning area of the inclusions leads to a factor b^2/L^2 in the pinning force, which is now proportional to $b^2 \mathrm{d}\gamma/\mathrm{d}x$. Between the pinning centers, the wall is curved, and the curvature radius depends on $H = H_\mathrm{c}$. In good approximation, one can treat the wall pieces as spherical caps, and the curvature radius is obtained by minimizing the energy

$$E(R) = 4\pi\gamma R^2 - 2\mu_\mathrm{o} M_\mathrm{s} H \left(\frac{4\pi}{3}\right) R^3 \tag{4.39}$$

This expression is reminiscent of droplet nucleation from vapor, where the surface energy is analogous to the domain-wall energy. However, (4.39) describes the propagation of a domain wall in a homogeneous magnet *after* nucleation, as contrasted to the physically very different nucleation (Section 4.3.1) in an inhomogeneous magnet. Minimization of (4.39) yields the curvature radius $R = \gamma/\mu_\mathrm{o} M_\mathrm{s} H$.

It is straightforward to show that (4.36) is essentially *independent* of the wall curvature, so that that $H_\mathrm{p} = p/2\mu_\mathrm{o} M_\mathrm{s} L^2$ remains valid. This is because the wall position x and the curvature radius R are, in lowest order, decoupled (see exercise on domain-wall curvature). This mechanism, illustrated in Fig. 4.22(a) is known as *Kersten* pinning (1943). However, the curvature is not the only consideration. As shown in Fig. 4.22(b–c), the distance ξ between centers actually involved in the pinning may be *larger* than the average distance L between defects. This saves domain-wall energy, especially in the limit of high domain-wall energies. The calculation of the corresponding Gaunt-Friedel pinning field H_p requires the self-consistent determination of the correlation length ξ from H_p (Gaunt 1986).

First, we take into account the fact that the Barkhausen volume covered by each domain-wall jump is about L^3, so that the enhanced ξ must be compensated by a reduced jumping distance Δx, $L^3 \approx \xi^2 \Delta x$. The pinning field is now equal $p/2\mu_\mathrm{o} M_\mathrm{s} \xi^2$, whereas the curvature radius $R = \gamma/\mu_\mathrm{o} M_\mathrm{s} H$ remains unchanged. Together with the height $\Delta x \sim \xi^2/R$ of the spherical cap, we must solve four equations for four unknowns, namely $\Delta\xi$, x, H_p, and R. The calculation is trivial and yields $H_\mathrm{p} \sim p^2/\gamma M_\mathrm{s} L^3$. In

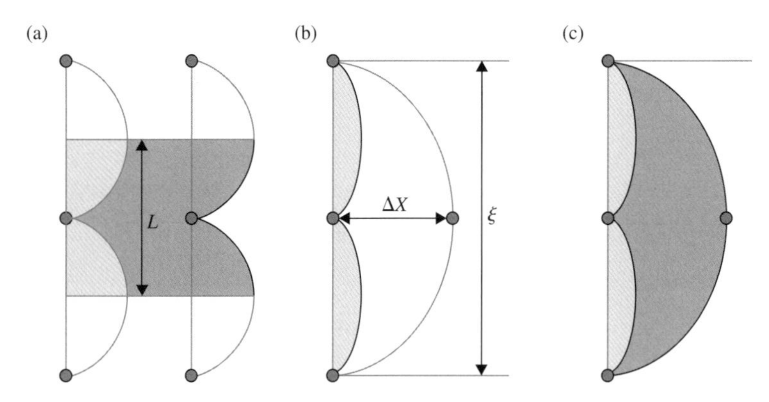

Fig. 4.22 Strong domain-wall curvature: (a) Kersten pinning and (b–c) Gaunt-Friedel pinning.

other words, the pinning coercivity is *quadratic* in the pinning force. In magnetic thin films, an analogous calculation yields $H_p \sim p^{3/2}$.

The range of pinning mechanisms is mirrored by the materials that exhibit pinning-controlled coercivity. The coercivity of iron and steel magnets reflects pinning at magnetoelastic inhomogeneities and martensitic lattice distortions due to interstitial carbon. In soft magnets, the reduction of domain-wall pinning is a challenge met by using materials with low intrinsic anisotropy, few defects, and low magnetostriction. Alternatively, one can use random-anisotropy magnets (Section 7.4.3). Pinning coercivity in industrial Sm-Co permanent magnets involves Sm_2Co_{17} crystallites having a size of about 100 nm are surrounded by a thin $SmCo_5$ grain-boundary phase. Additives such as Cu and Zr help to establish the microstructure and to tune the properties of the phases (Kumar 1988). Other pinning mechanisms exist in "superferro-magnets", where nanoscale grains are exchange-coupled by a matrix. (Zhou *et al.* 2005a), and in thin films with constant surface anisotropy but varying thickness (Sander *et al.* 1996).

The pinning energy is, in fair approximation, proportional to the anisotropy difference, and by changing the chemical composition or the temperature it is possible to adjust the anisotropy and to tune the pinning behavior. One example is Sm-Co-Cu-Ti permanent-magnet alloys for high-temperature applications, with coercivities more than 1.2 T at 500°C (Zhou *et al.* 2000). In Chapter 6 we will see that the temperature dependence of the coercivity is dominated by the intrinsic temperature dependence of the anisotropy. Thermally activated jumps over energy barriers yield only small sweep-rate corrections to the coercivity.

4.3.3 Phenomenological coercivity modeling

Aside from microscopic coercivity models, such as those discussed in the previous two subsections, there exist various phenomenological models. The scientific scope of most phenomenological models is limited, but they are useful in engineering, where the physics behind the coercivity is of secondary importance. Some techniques are Preisach modeling, first-order reversal curve analysis (FORC), Barkhausen analysis, and micromagnetic mean-field modeling (Bertotti 1998, Della Torre 1999).

A simple phenomenological model is the superposition model, which exists in many variants. Any hysteresis-loop branch can be represented by a normalized switching-field distribution $P_{SF}(H)$

$$M(H) = -M_s + 2M_s \int_{-\infty}^{H} P_{SF}(h)\, dh \tag{4.40}$$

The distribution P_{SF} is closely related to the micromagnetic susceptibility, $\chi = dM/dH$, or $\chi = 2M_s P_{SF}(H)$. To exclude reversible magnetization processes, it is customary to restrict the consideration to the irreversible part χ_{irr} of the susceptibility, which is obtained by a minor-loop analysis. Agreement with experimental loops may also be achieved by considering superpositions of hyperbolic tangents $\tanh((H - H_c)/\Delta H)$. Two or three functions of this type are often sufficient to reproduce a hysteresis loop nicely. For example, Fig. 4.14(b–c) uses two hyperbolic functions. The physical disadvantage of this magnetic phase analysis is the neglect of magnetic interactions, which are swept under the rug by assuming a superposition of individual loops. In a broader sense, this approach includes techniques that average over ensembles of small magnetic particles or crystallites. One example is the single-point detection or SPD method by Asti and Rinaldi (1974), whose aim is to determine anisotropy constants rather than explaining micromagnetics.

Various phenomenological models take into account magnetic interactions in an approximate way, often considering or exploiting minor hysteresis loops. Examples are the Jiles-Atherton model (1986), the Preisach model (1935), FORC analysis (Davies *et al.* 2005), models focusing on texture (Jahn, Schumann, and Christoph 1985), and approaches based on Wohlfarth's remanence relation (Wohlfarth 1958). The last category includes Henkel, delta-M, and delta-H plots, where recoil loops are analyzed. The Jiles-Atherton model assumes domain-wall pinning on a phenomenological basis and yields hysteresis loops as a function of the anhysteretic (single-valued) magnetization curve $M_{\rm an}(H)$. The total magnetization is the sum of irreversible and reversible magnetization contributions, $M = M_{\rm irr} + M_{\rm rev}$, where $M_{\rm irr} = M_{\rm an} - M_\Delta$, $M_{\rm rev} = cM_\Delta$, and

$$\frac{\mathrm{d}M_\Delta}{\mathrm{d}H} + \frac{M_\Delta}{k - \alpha\, M_\Delta} = \chi_{\rm an}(H) \qquad (4.41)$$

The parameters in these equations describe domain-wall pinning (k), magnetostatic interactions (a), and reversible domain-wall motion (c). The nonlinear differential equation (4.41) yields M_Δ (and M) from the phenomenological source term $\chi_{\rm an} = \mathrm{d}M_{\rm an}/\mathrm{d}H$.

The *Preisach model* (1935) is based on the assumption of hysteresis quanta or hysterons with rectangular but not necessarily symmetric hysteresis loops. The hysterons, also referred to as "hysteresis particles", are mathematical constructions and generally unrelated to structural and magnetic features such as grains and nucleation modes. Due to its input-output character, the Preisach model is essentially a mathematical model, closer to nonlinear system theory than to magnetism. FORCs are obtained by applying a large positive field to saturate the magnet, reducing H to the recoil field $H_{\rm r}$, and finally increasing the field to $H_{\rm a}$, where $M_{\rm z} = M_{\rm a}$ The distribution of first-order reversal curves (FORC) is then obtained by sampling all points in the $H_{\rm r}$–$H_{\rm a}$ plane or, equivalently, by considering $\partial^2 M_{\rm a}/\partial H_{\rm r}\partial H_{\rm a}$.

The inner-loop methods discussed in this paragraph can be used to discuss *weak* interactions between small particles qualitatively, as in magnetic rocks. The idea is that different parts of the systems create positive (exchange) or negative (dipolar) interaction fields, which affect the inner hysteresis loops. However, interaction-field models are of the mean-field type and suffer from a number of shortcomings. In (4.41), the mean-field character is manifest from the term containing k and a, which is similar to the susceptibility of the Stoner model. In the Preisach model, a hysteron switches if the total field (external field plus interaction field) exceeds the switching field of the particle. It has been known for a long time that mean-field models poorly describe and often overestimate the coercivity of strongly interacting grains or particles (Callen, Liu, and Cullen 1977). A good example is the curling mode (Fig. 4.16), which cannot be mapped onto an interaction field. As we will discuss in Section 7.4.4, strongly interacting particles behave cooperatively, reminiscent of macrospins with strong interatomic exchange, and there is no point in adding the huge internal interaction field to the coercivity.

A phenomenological expression with a sound micromagnetic basis is the so-called Kronmüller equation (1987)

$$H_c = \alpha_K \frac{2K_1}{\mu_o M_s} - D_{eff} M_s \tag{4.42}$$

where α_K is a complex parameter and D_{eff} a magnetostatic interaction parameter. The parameter α_K may be derived for many nucleation and pinning models and varies between about 1% in as-cast alloys and more that 40% in magnets whose production involves highly sophisticated processing techniques.

4.4 Grain-boundary models

Summary The spin structure of magnets is modified by imperfections such as grain boundaries and nanojunctions. On a continuum level, grain boundaries are modeled by taking into account the appropriate boundary conditions. Even for well-localized and weak imperfections, the magnetization perturbation extends several nanometers into the adjacent ferromagnetic regions. Micromagnetic problems with atomic resolution can, in principle, be calculated from first principles, but the large number of affected atoms and the involved small energies make these calculations very difficult. In fact, models with atomic resolution tend to yield rather small corrections to the continuum results. At granular interfaces, both the reduced grain-boundary exchange and grain misalignment contribute to the perturbation of the spin structure. Changes in the interatomic exchange yield large magnetization gradients, whereas anisotropy changes at grain boundaries of hard-magnetic materials are much less effective in perturbing the spin structure.

The spin structure at grain boundaries and geometrical constraints is important in various areas of magnetism, because it affects hysteresis and magnetoresistance. This includes permanent magnetism, magnetic recording, soft magnetism, and spin electronics. Micromagnetic problems are usually solved on a continuum level. For example, the magnetization M_s considered in micromagnetism is generally *averaged* over a few interatomic distances and can be regarded as a temperature-dependent materials constant (micromagnetic parameter). Narrow-wall phenomena, which have been studied for example in rare-earth cobalt permanent magnets (Hilzinger und Kronmüller 1975) and at grain boundaries (Skomski 2001, 2003), involve individual atoms and atomic planes and lead to comparatively small corrections to the extrinsic behavior. In cases where atomic-scale effects are important, such as $L1_0$ magnets, multiscale modeling is a valuable option (Garcia-Sanchez *et al.* 2005). However, here we focus on analytical models of grain-boundary exchange.

4.4.1 Boundary conditions

As outlined in Section 4.2.3, the local magnetization $\mathbf{M}(\mathbf{r})$ is obtained by finding the local minima of the micromagnetic free energy. In the linear approximation, we can use the equation of state (4.19), which considers a small perpendicular magnetization component $\mathbf{m}(\mathbf{r})$. This introduces quantitative errors, such as the factor $\pi/2$ mapping

$m = 1$ onto $\theta = \pi/2$, but leaves the physical picture largely unaffected. Our strategy is to solve the linearized micromagnetic equations at features such as grain boundaries and to determine the spin structure, $\mathbf{m}(\mathbf{r})$. This yields magnetic properties such as the magnetization $\mathbf{M}(\mathbf{r}, H)$ and the effective intergranular exchange.

The term $\nabla(A\nabla\mathbf{m})$ in (4.19) is important when A changes across the material (Skomski and Coey 1993). A relatively simple limit is sharp grain boundaries between adjacent homogeneous phases I and II. Inside each phase $\nabla(A\nabla\mathbf{m}) = A\nabla^2\mathbf{m}$, but at the interface, the $\nabla(A\nabla\mathbf{m})$ term yields the Erdmann-Weierstrass boundary conditions $\mathbf{m}_\mathrm{I} = \mathbf{m}_\mathrm{II}$ and

$$\left(A(x)\frac{\partial \mathbf{m}}{\partial x}\right)_\mathrm{I} = \left(A(x)\frac{\partial \mathbf{m}}{\partial x}\right)_\mathrm{II} \tag{4.43}$$

Here we have assumed that the grain boundary is in the $y{-}z$ plane. Figure 4.23 illustrates the physical meaning of the boundary condition. For $A_\mathrm{I} = A_\mathrm{II}$, (4.43) reduces to $(\partial m/\partial x)_\mathrm{I} = (\partial m/\partial x)_\mathrm{II}$, that is, the slope of the magnetization is continuous at the interface. Since many ferromagnetic materials have an exchange stiffness of about 10 pJ/m, as compared to anisotropy constants varying over several orders of magnitudes, $A_\mathrm{I} = A_\mathrm{II}$ is often a good approximation. Figure 4.23(a) shows the example of an interface between hard and soft phases. It is interesting to note that the slope is continuous at the interface but the curvature change sign. If a soft phase is in contact with a hard phase of infinite anisotropy, the magnetization is clamped, that is, $\mathbf{m}(\mathrm{r}) = 0$ at the interface.

Figure 4.23(b) shows that a jump in $A(x)$ changes the slope of the perpendicular magnetization component $m(x)$ but leaves the magnetization continuous. However, reduced exchange in a thin grain-boundary region yields a *quasi-discontinuity* of the magnetization, as shown in Fig. 4.23(c). This discontinuity is unrelated to the hard or soft character of the involved phases and therefore qualitatively different from Fig. 4.23(a).

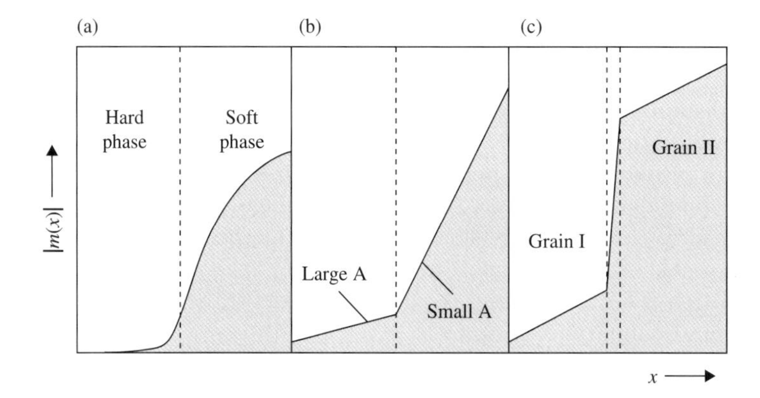

Fig. 4.23 Spin structure in the vicinity of grain boundaries: (a) hard-soft interface with common exchange stiffness A, (b) two ferromagnetic phases with different A, and (c) quasi-discontinuity of the magnetization due to strongly reduced grain-boundary exchange. The dashed line shows the grain-boundary plane.

4.4.2 Spin structure at grain boundaries

Let us start with the determination of the relative strength of the discontinuity in Fig. 4.23(c) for two weakly misaligned semi-infinite grains (index I/II) in the absence of a magnetic field. In each phase, and in the grain-boundary region, (4.19) reduces to

$$-A\nabla^2\mathbf{m} + K_1\mathbf{m} = K_1\mathbf{a} \tag{4.44}$$

where $\mathbf{a} = \mathbf{a}_{I/II}$ describes the misalignment of the easy axis. We choose our coordinate frame so that the grain boundary is in the y–z plane and $\mathbf{m} = m(x)\,\mathbf{e}_y$. For simplicity, we assume that the grain boundary is soft, $K_1 = 0$, so that there is no need to fix $\mathbf{a}$ for the grain-boundary region. However, it is straightforward to show that finite anisotropy changes in the grain-boundary regions have very little effect on the magnetization (see exercise on grain-boundary anisotropy).

For the considered one-dimensional geometry, the magnetization at $x = \pm\infty$ is parallel to the local easy axis, $m_{I/II} = a_{I/II}$. Elsewhere in the two grains, the solution of (4.44) is exponential, $m_{I/II}(x) - a_{I/II} = c_\pm \exp(\pm x/\delta_o)$. The two constants $c_\pm$ are determined by the boundary conditions at $x = \pm t/2$, where t is the thickness of the grain-boundary region. Figure 4.24 shows the magnetization component m across the grain boundary. The magnetization exhibits a quasi-discontinuity $\Delta = |m(t/2) - m(-t/2)|/|a_I - a_{II}|$ of relative magnitude

$$\Delta = \frac{1}{1 + \dfrac{2A'\delta_o}{At}} \tag{4.45}$$

where $A' < A$ is the exchange stiffness in the grain boundary (Skomski 2001 and 2003). When A and A' are comparable, the denominator is determined by the relative large ratio δ_o/t and the quasi-discontinuity Δ is very small. In the limit of strongly reduced

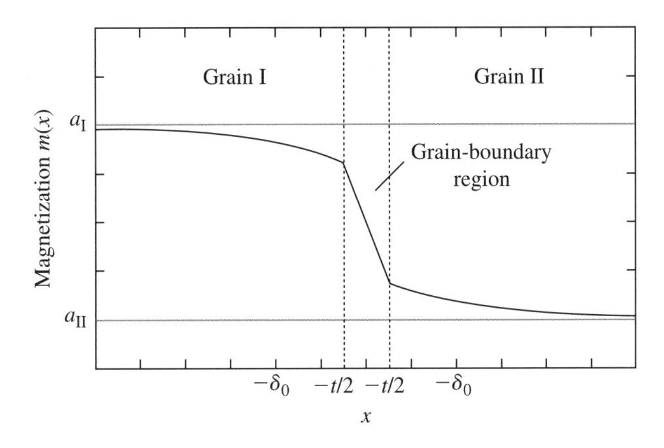

Fig. 4.24 Spin structure at a grain boundary with reduced exchange. The exponentially decaying tails are analogous to the "tunneling" of micromagnetic modes in Fig. 4.19.

grain-boundary exchange, $A' \ll A$, the quasi-discontinuity $\Delta \approx 1$. However, this limit is difficult to achieve in practice, because there is always some residual exchange due to RKKY-type interactions and interatomic hopping (tunneling).

A general effect of grain boundaries and other defects is a magnetic perturbation of range δ_o, typically a few nanometers. Figure 4.24 shows that this perturbation has the character of an exponentially decaying tail. Aside from magnetostatic corrections, the decay length is independent of the strength of the perturbation—only the *magnitude* of the magnetization perturbation depends on the strength of the inhomogeneity. In addition, the details of the magnetization tail depend on the geometry and dimensionality of the problem. In one-dimensional systems, including planar grain boundaries, the decay is exponential, whereas three-dimensional perturbations are described by $K_{1/2}(r/\delta_o)$, where $K_{1/2}(\xi) \sim \exp(-\xi)/\xi$ is a spherical Bessel function. This situation is reminiscent of small spherical soft inclusions in a matrix of large but finite anisotropy, where the nucleation mode is centered around the inclusion but exhibits a tail extending into the hard matrix phase.

The magnetization distribution $m(\mathbf{r})$ costs exchange and anisotropy energy. Due to the magnetization tails, the energy is not confined to the grain boundary but partially stored in the grains. The energy scales as $J_{\text{eff}}(a_{\text{I}} - a_{\text{II}})^2$, where J_{eff} is the effective intergranular exchange. Integration of (4.18) over x yields

$$J_{\text{eff}} \approx L^2 \sqrt{AK_1} \frac{1}{1 + \dfrac{At}{2A'\delta_o}} \tag{4.46}$$

where L^2 is the interface area (Skomski 2003). This exchange is much smaller than the "naïve" grain boundary exchange, which is equal to the interatomic exchange J times the number of surface atoms per unit area. It is also smaller than the $1/t$ exchange obtained by confining the magnetization inhomogeneity to the grain boundary. This is because the system lowers its energy by developing magnetization tails.

As mentioned, atomic-scale *exchange* inhomogeneities have a particularly strong effect on the spin configuration. For example, a fictitious grain-boundary layer with zero exchange would completely decouple the adjacent grains. By comparison, anisotropy changes average over a length scale of a few nanometers, and a single atomic layer with zero anisotropy has a relatively small effect on the micromagnetic behavior.

4.4.3 Models with atomic resolution

The continuum approximation breaks down when the thickness t of the boundary region becomes comparable to the interatomic distances. In a *layer-resolved analysis*, (4.9) must be replaced by the discrete expression

$$E = L^2 \sum_{\text{n}=-\infty}^{+\infty} \left(J_{\text{n,\,n+1}} \frac{(\mathbf{M}_\text{n} - \mathbf{M}_{\text{n}+1})^2}{M_\text{s}^2} - K_1\, t_\text{o} \frac{(\mathbf{n}_\text{n} \cdot \mathbf{M}_\text{n})^2}{M_\text{s}^2} - \mu_\text{o} \mathbf{M}_\text{n} \cdot \mathbf{H}\, t_\text{o} \right) \tag{4.47}$$

where L^2 is the interface area, $J_{\text{n,\,n+1}} \approx A(\mathbf{r})t_o$ is the interlayer exchange coupling between the n-th and $(n+1)$-th layer, and each layer has a thickness t_o. For $H = 0$,

(4.18–19) become

$$E = L^2 \sum_{n=-\infty}^{+\infty} (J_{\mathrm{n,n+1}}(m_\mathrm{n} - m_\mathrm{n+1})^2 - K_1\, t_\mathrm{o}(m_\mathrm{n} - a_\mathrm{n})^2) \tag{4.48}$$

and

$$J_{\mathrm{n,n+1}}(m_\mathrm{n} - m_\mathrm{n+1}) + K_1\, t_\mathrm{o} m_\mathrm{n} = K\, t_\mathrm{o} a_\mathrm{n} \tag{4.49}$$

In the simplest case, the grain boundary is described by an interface exchange $J_{01} = J'$ smaller than the bulk exchange $J_{\mathrm{n,n+1}} = J$.

As in the continuum case, the solution of the equation of state (4.49) is exponential, $m_{\mathrm{I/II}}(\mathrm{n}) - a_{\mathrm{I/II}} \sim \exp(\pm x/\lambda)$, but the decay length $\lambda = 1/\mathrm{arcosh}(1 + K_1 t_\mathrm{o}/2J)$ is slightly different from δ_o. In addition, intrinsic parameters, such as the magnetization and anisotropy, may be different in layers close to the interface, so that the use of bulk parameters is a relatively crude approximation. However, for most micromagnetic problems, the continuum approximation works very well, and corrections do not exceed a few percent. This is because the wall-width parameter tends to be much larger than the interatomic distance. One exception is the case of extremely hard materials, such as $SmCo_5$, where the domain walls are very narrow, $\delta_\mathrm{o} \approx 1.5\,\mathrm{nm}$, and the corrections are of the order of 10% (Hilzinger and Kronmüller 1975). In numerical magnetism, the use of site-resolved parameters from first-principle calculations is known as *multiscale modeling*. An example is the case of small grains of $L1_0$ materials such as FePt, where missing or changed atomic neighbors at the surface have a disproportionately strong effect on the magnetic anisotropy (Belashchenko and Antropov 2002, Garcia-Sanchez *et al.* 2005).

4.4.4 Nanojunctions

The spin structure in grain boundaries and nanojunctions is important in spin electronics, because it affects the magnetoresistance (Section 7.2.7). On a one-electron level, the scattering reflects the spin dependence of the exchange potential $V_\sigma(\mathbf{r}_\mathrm{i})$, so that the resistance is a functional of the local magnetization $\mathbf{M}(\mathbf{r})$. In particular, large magnetization gradients $\nabla\mathbf{M}(\mathbf{r})$ are expected to yield strong scattering contributions. Typical domain walls are smooth and extend over many interatomic distances, but grain-boundaries and nanojunctions may have regions with very large gradients. A crude measure to gauge the spin-dependent scattering ability of an interface is the integral $\int(\nabla\mathbf{M})^2\,\mathrm{d}x \approx M_\mathrm{s}^2 \int(\nabla\mathbf{m})^2\,\mathrm{d}x$.

Let us start by considering planar grain boundaries with strongly reduced grain-boundary exchange, $A' \ll A\delta_\mathrm{o}/t$. Using (4.45) we obtain $\Delta \approx 1$ and $\int(\nabla\mathbf{M})^2\,\mathrm{d}x \sim M_\mathrm{s}^2/t$. This is a considerable enhancement compared to Bloch-wall scattering, where $\int(\nabla\mathbf{m})^2\,\mathrm{d}x \approx 1/\delta_\mathrm{o}$, but this regime is difficult to realize in practice, because δ_o is usually much larger than t. In fact, for fixed A', the scattering is maximized by choosing an interface thickness of order $\delta_\mathrm{o}A'/A$ (Skomski 2001, 2003). Compared to Bloch walls, the corresponding scattering is enhanced by a factor A/A'. For example, taking $\delta_\mathrm{o} = 10\,\mathrm{nm}$ and $A' = 0.1\,A$ yields a maximum scattering for boundaries having a thickness of 1 nm. Note that A' is difficult to reduce to very low values, because

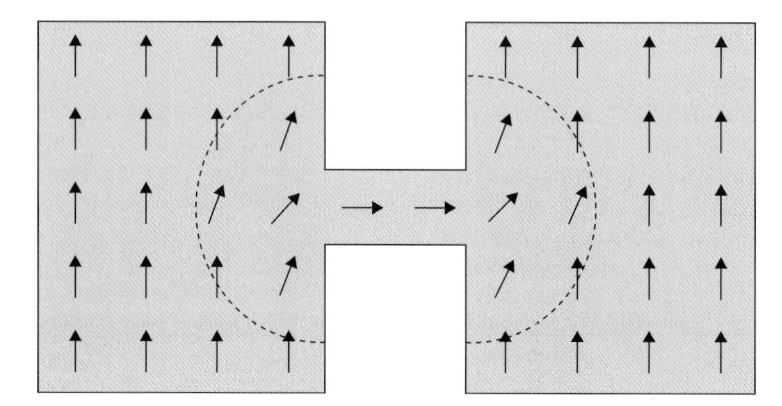

Fig. 4.25 Schematic spin structure in the vicinity of a nanojunction. The radius of the dashed hemispheres is about $\delta_o = (AK_1)^{1/2}$.

a complete exchange decoupling would suppress the current. In practice, there is also a substantial contribution due to anisotropic magneto-resistance.

In *nanojunctions*, such as that shown in Fig. 4.25, the magnetization gradient is maximized for junction diameters smaller than δ_o. In this regime, $\int(\nabla\mathbf{m})^2\,dx \approx 1/L$, where L is the length of the junction. It is therefore possible to enhance the scattering, reducing the length of the junction. However, a general feature of the magnetic response to the imperfections is the magnetization tail of length $\delta_o = \sqrt{K_1/A}$, as discussed in the previous subsection. The dashed line in Fig. 4.25 shows the extension of this tail in a nanojunction. Most of the scattering is usually realized in the junction, but with decreasing length L, the contribution of the tails increases. This reduces the electron scattering by the magnetic inhomogeneity and puts an upper limit to the reduction of L as a method to enhance the scattering (exercise on magnetization inhomogeneities at nanojunctions). Note that $1/\delta_o$ increases with anisotropy, so that the spin-dependent electron scattering is largest for hard-magnetic materials. However, hard magnets require high switching fields and are therefore difficult to handle in applications.

Exercises

1. ***Shape of permanent magnets.*** Show that the shape of a magnet with $D = 1/2$ is flat (oblate) rather than elongated (prolate).
   Answer: Spheres have $D = 1/3$, whereas magnets with $D < 1/3$ and $D > 1/3$ are prolate and oblate respectively. Note that $D = 1/2$ corresponds to the maximum energy product of hard magnets with ideal rectangular hysteresis loops.
2. ***Energy product of iron.*** Estimate the energy product of soft-magnetic iron.
3. ***Micromagnetic mean-field approaches.*** Mean-field models are of limited applicability in micromagnetism. Why?
4. ***Flux leakage from toroids.*** How can a magnetic toroid (Fig. 4.3) be redesigned to avoid to avoid flux leakage for large gap widths L_g?

5. ***Fourier transform of exchange.*** Consider a micromagnetic model where the exchange stiffness $A\nabla^2$ is replaced by the more general expression $J(\mathbf{r}-\mathbf{r}')$ and find the Fourier transform equivalent to $-Ak^2$.
6. ***Grain-boundary anisotropy and micromagnetic spin structure.*** Show that a moderate enhanced or reduced anisotropy in a grain boundary has virtually no effect on the spin structure $\mathbf{M}(\mathbf{r})$.
   Hint: Use a quantitative analysis based on the slope and curvature of $m(x)$.
7. ***Hysteresis of Ising and Heisenberg models.*** Consider isolated spins with $\mathrm{S} = \frac{1}{2}$ and calculate the low-temperature coercivity for two cases: (a) Heisenberg spins and (b) Ising spins.
8. ***Dysprosium-containing permanent magnets.*** Many transition-metal-rich rare-earth intermetallics containing heavy rare earths, such as Dy, exhibit high Curie temperatures, anisotropies, and coercivities. Why are they not used as permanent magnets?
9. ***Coherence radii in ferrimagnets.*** Show that ferrimagnetic materials have relatively large coherence radii. Can the accompanying coercivity improvement be exploited in permanent magnets and recording media?
10. ***Nucleation field and K_2.*** Show that the nucleation field of a c-axis aligned uniaxial magnet is independent of K_2.
    Answer: In the vicinity of the nucleation field, where θ is small, the term $K_2 \sin^4\theta \sim K_2\theta^4$ can be ignored compared to $K_1 \sin^2\theta \sim K_1\theta^2$.
11. ***Coercivity of alnico.*** Estimate the coercivity of alnico-type permanent magnets.
    Answer: Alnico magnets consist of long needles of soft-magnetic Fe-Co embedded in a nonmagnetic Ni-Al matrix. Using the equation for the curling nucleation field and taking $D = 0$, $A = 10\,\mathrm{pJ/m}$ ($10^{-11}\,\mathrm{J/m}$), $\mu_o M_s = 2.43\,\mathrm{T}$, and $R = 20\,\mathrm{nm}$ we obtain $\mu_o H_c = 0.088\,\mathrm{T}$, which is a typical result for alnico.
12. ***Energy product for model loop.*** Determine the energy product for a loop that is linear in the second quadrant, that is, a straight line connecting $M(0) = M_r$ and $M(-|H_c|) = 0$.
    Hint: Use $\mathrm{d}(BH)/\mathrm{d}H = 0$.
13. ***Energy product of different materials.*** Estimate the maximum energy product for the following three materials: (a) $\mu_o M_s = 1\,\mathrm{T}$; $\mu_o H_c = 1\,\mathrm{mT}$, (b) $\mu_o M_s = 0.3\,\mathrm{T}$; $\mu_o H_c = 0.1\,\mathrm{T}$, (c) $\mu_o M_s = 1.5\,\mathrm{T}$; $\mu_o H_c = 1.5\,\mathrm{T}$.
    Hint: Assume (a) rectangular loops and (b) linear $M(H)$ curves in the second quandrant which yields altogether six energy-product values.
14. ****Nucleation field for a spherical inclusion.*** Calculate the nulceation field for a soft spherical inclusion of radius R in a very hard magnet. Express the solution (a) as a function of the exchange stiffness of the soft phase and (b) as a function of the Bloch-wall width of the hard phase. Discuss the nucleation field in the limit of very small inclusions.
    Hint: Ignore the magnetostatic self-interaction, which is of of secondary importance in very hard magnets. The calculation involves the spherical Bessel function $j_0(x) \sim \sin(x)/x$.
15. ****Kersten pinning and domain-wall curvature.*** Show that the domain-wall curvature leaves the Kersten pinning field essentially unchanged.

16. ***Gaunt-Friedel pinning in two dimensions.*** Derive the Gaunt-Friedel pinning field H_p and the corresponding correlation length ξ for two-dimensional magnets and compare the result with Gaunt's three-dimensional calculation.
17. ***Demagnetizing factors of embedded particles.*** Determine the effective demagnetizing factor for magnetic particles of demagnetizing factor D_P embedded in a nonmagnetic matrix of demagnetizing factor D_o.
18. ***Magnetization inhomogeneities at nanojunctions.*** Determine the integral $\int(\nabla\mathbf{M})^2\,\mathrm{d}V$ for a small cylindrical nanojunction (or pinhole) of radius R, length L, and exchange stiffness A'. Maximize the integral as a function of R and L.
Hint: Determine the spin structure by minimizing the total magnetic energy—junction plus adjacent material—and estimating (rather than calculating) the energy contributions.

5

Finite-temperature magnetism

Magnetic properties are usually temperature-dependent. For example, a large zero-temperature moment does not necessarily translate into room-temperature ferromagnetism, because thermal excitations adversely affect magnetic order. How can we explain the vanishing of the magnetization at a sharp Curie temperature T_c, and what determines the temperature dependence of intrinsic properties such as magnetization and anisotropy? Figure 5.1 illustrates that the thermal excitations primarily affect the *direction* of the local magnetization, yielding the "net" or spontaneous magnetization M_s as a thermal average. In a classical picture, thermal excitations randomize the magnetization angles θ and ϕ, whereas a simple quantum-mechanical interpretation is the involvement of spin states $S_z < S$ (Fig. 2.13). Compared to the direction of the moment, the *magnitude* of the moment remains largely unchanged. This is because intra-atomic exchange is typically of the order of 1 eV, whereas the total interatomic exchange per atom does not exceed about 0.1 eV. A notable exception is very weak itinerant ferromagnets, such as $ZrZn_2$, where both inter- and intra-atomic exchange are small (Section 5.2.5).

This chapter deals with *equilibrium* models of finite-temperature magnetism. Intrinsic properties are realized on an atomic scale, characterized by very fast equilibration times, and usually well described by equilibrium models. Emphasis is on equilibrium at and above room temperature, whereas low-temperature excitations such as spin waves will be treated in Chapter 6, in connection with magnetization dynamics. Finite-temperature magnetism amounts to embedding the magnet in a heat bath of temperature T. In equilibrium, the physical nature of the heat bath is of secondary

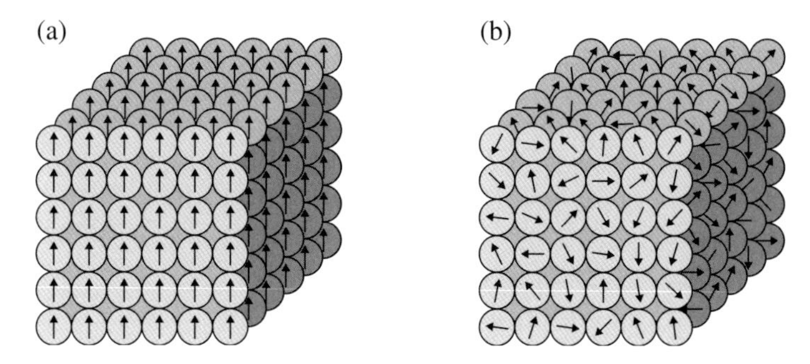

Fig. 5.1 Spin structure of a ferromagnet: (a) zero temperature and (b) finite temperature.

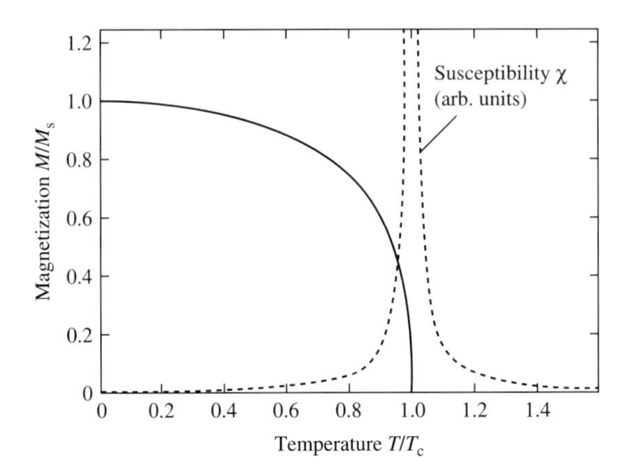

Fig. 5.2 Temperature dependence of the spontaneous magnetization (solid line) and susceptibility (dashed line). Both curves refer to ferromagnets in the absence of an external magnetic field.

importance. For example, it does not matter whether the thermal equilibrium is realized by phonons (lattice distortions) or conduction electrons. This makes it possible to use standard tools of statistical mechanics, such as the partition function Z. Sections 5.1–2 provide a brief introduction to these methods.

One key feature of finite-temperature magnetism is the existence of a sharp Curie temperature T_c, as contrasted to a smooth decay of the spontaneous magnetization $M_s(T)$. Figure 5.2 shows the temperature dependence of spontaneous magnetization and the zero-field susceptibility $\chi = \mathrm{d}M/\mathrm{d}H$ for a simple ferromagnet. Both the magnetization (solid line) and the susceptibility (dashed line) are singular at the Curie point. This leads to a number of questions. What is the origin of the singularity, and what determines the magnitude of the Curie temperature? How does the transition depend on factors such as the spatial dimensionality of the magnet and the type of interatomic exchange? What phase transitions exist in other magnets, such as antiferromagnets?

In the nineteenth century, attempts were made to explain T_c by magnetostatic interactions, but the smallness of the Bohr magneton, $\mu_B/k_B = 0.672\,\mathrm{K/T}$, means that magnetostatic fields in solids are unable to explain ferromagnetic order above about 1 K. The key to the understanding of the Curie transition is the involvement of the interatomic exchange, $J \sim 0.01\,\mathrm{eV}$, or $J/k_B \sim 100\,\mathrm{K}$ (Section 1.4). An intriguing feature of the Curie transition is that the divergence of the susceptibility at T_c is accompanied by a divergence of the spatial fluctuations. These *critical fluctuations* complicate the modeling of the critical point, because they mean that a large number of spins must be taken into account (Section 5.3–4). They also interfere with deviations from perfect crystal periodicity, such as surfaces and nanoscale features (Section 7.4).

5.1 Basic statistical mechanics

Summary Finite-temperature equilibrium amounts to the minimization of the free energy $F = E - TS$, where the entropy S describes thermal disorder. Zero-temperature equilibrium means that only the lowest-lying state is occupied, but for nonzero temperatures $T > 0$, the interaction with the heat bath leads to the population of excited states. The probability of finding an equilibrium spin configuration (index μ) is given by the Boltzmann distribution $\exp(-E_\mu/k_B T)$. Thermal averages are conveniently obtained as derivatives of the partition function $Z = \Sigma_\mu \exp(-E_\mu/k_B T)$. The partition function leads to general relationships, such as $F = k_B T \ln Z$ and the fluctuation-response theorem relating real-space correlations to the susceptibility, and to model-specific predictions. The main challenge is the large number spin configurations, which increases exponentially with the size of the magnet. A topic of particular interest is phase transitions, especially the continuous (or second-order) phase transition at the critical or Curie temperature T_c. A simple phenomenological free-energy model is the Landau model, which treats the critical behavior on a mean-field level.

At zero temperature, equilibrium is realized by minimizing the total energy E, which is often equated with the internal energy U but generally includes the interaction with external fields and forces. At nonzero temperatures, the trend towards lower energy competes against thermal disorder, and equilibrium is realized by minimizing the *free energy* $F = E - TS$, where S is the entropy (not to be confused with the spin) and T is the temperature. Entropy means that the randomness of thermal excitations favors disordered states, and it can actually be considered as a measure of disorder. In a mechanical analogy, zero temperature equilibrium corresponds to a tiny steel ball moving towards the bottom of a bowl ($E = 0$), coming to a standstill after dissipating its kinetic energy. At nonzero temperatures, thermal excitation realizes a state that costs energy ($E > 0$) but is entropically favorable due to thermal disorder.

The entropy of magnetic systems is largely *configurational*, associated with the randomness of the atomic spins located at $\mathbf{R}_i$. There are other entropy contributions in magnetic solids, such as vibrational entropies and the entropy of conduction electrons, but these contributions are usually of secondary importance. In a classical picture, the configurational entropy of a magnet reflects the magnetization orientation $\mathbf{s}_i = \mathbf{M}(\mathbf{R}_i)/M_s$. At zero temperature, a small field aligns the spins ($\mathbf{s}_i = \mathbf{e}_z$), but thermal excitations randomize the spins so that the thermally averaged magnetization $\langle \mathbf{s}_i \rangle$ becomes smaller. It is convenient to calculate the free energy directly from the partition function Z (next subsection), so that there is usually no need for an explicit calculation of S. However, it is useful keep in mind that both F and Z contain both energetic and entropic contributions.

The randomization of atomic spins is exploited in *magnetic cooling*, where aligned spins randomize and the entropy change leads to heat extraction from the environment. To maximize the effect, one needs a large entropy change close to the application temperature. The maximum entropy change per spin of length N is $k_B \ln(2N + 1)$. This has led to a search for materials containing large spins (S = 7 for Gd) and forming suitable magnetic phases (Pecharsky, Gschneidner, and Pecharsky 2003). Since heat capacity and hysteresis losses of magnetic particles are approximately linear in N, the magnetocaloric performance of larger Stoner-Wohlfarth particles is poor. However, magnetic cooling based on particles with $N = 100 \ldots 1000$ may be realized in very convenient field and temperature regions, because their Curie susceptibility is large.

5.1.1 Probability and partition function

Figure 5.1 indicates that finite-temperature magnetism involves a large number of spin configurations (index μ). Each spin configuration has a probability p_μ, and the expectation value of any physical quantity A is obtained as the thermal average

$$\langle A \rangle = \Sigma_\mu A_\mu p_\mu \tag{5.1}$$

Panel 7 Entropy and probability

A simple derivation of the entropy is based on the consideration that E, F, and S are all extensive quantities. For interacting systems, this means that $S = S_1 + S_2$. A way of measuring disorder (S) is to count the states of a system. For example, in the absence of magnetic field, N noninteracting spins ($\uparrow$ and $\downarrow$) have $\Omega = 2^{\mathrm{N}}$ possible states with equal probability $p = 1/\Omega$. The larger the number of available states, the higher the entropy. Since Ω and p are multiplicative, as exemplified by $p = p_1 p_2$, the entropy must depend logarithmically on Ω and p, $S = k_\mathrm{B} \ln \Omega = -k_\mathrm{B} \ln p$. Here the Boltzmann constant $k_\mathrm{B} = 1.38 \times 10^{-23}$ J/K provides the conversion between energy and temperature units. If there is only one state ($\Omega = 1$), then $S = 0$, but otherwise the entropy is positive.
In noninteracting systems, the free energy $F = E - TS$ reduces to $F = -TS$, so that $F = -k_\mathrm{B} T \ln \Omega$. A simple and instructive example is a single atom in the gas phase (figure). In this case, the number of states is equal to the volume, $\Omega \sim V$, $S \sim k_\mathrm{B} T \ln V$, and the pressure $P = -\partial F/\partial V = k_\mathrm{B} T/V$. For N noninteracting atoms, this yields the ideal-gas law $PV = N k_\mathrm{B} T$. Other examples of noninteracting systems with $F = -TS$ are paramagnetic gases ($J_{\mathrm{ij}} = 0$) and rubber elasticity (free links between statistical polymer segments).

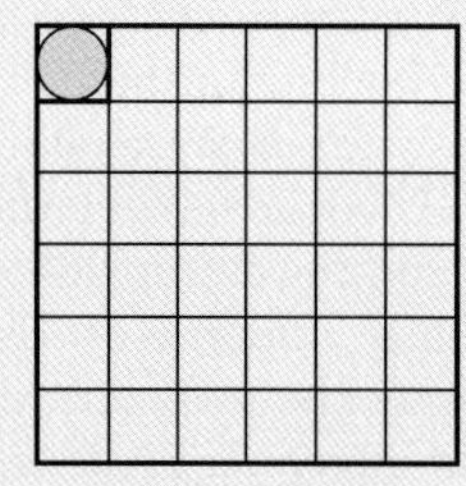
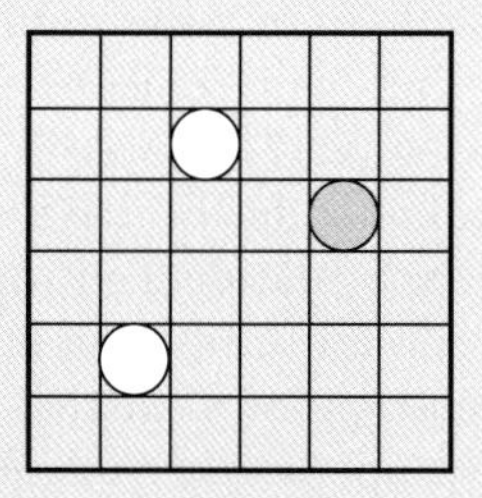

Volume and entropy: $S = 0$ for gas atom confined to a single cell (left) and $S \sim k_\mathrm{B} T \ln V$ for a gas atom in a finite volume (right).

In general, probabilities p_μ differ from each other, because they depend on the energy E_μ. The entropy is then obtained as an average, $S = -k_\mathrm{B} \sum_\mu p_\mu \ln p_\mu$, and with $E = \sum_\mu E_\mu p_\mu$, the free energy becomes $F = \Sigma_\mu E_\mu p_\mu + k_\mathrm{B} \sum_\mu p_\mu \ln p_\mu$. Minimization of the free energy, $\partial F/\partial p_\mu = 0$, yields the Boltzmann distribution $p_\mu = Z^{-1} \exp(-E_\mu/k_\mathrm{B} T)$. Here the *partition function* $Z = \Sigma_\mu \exp(-E_\mu/k_\mathrm{B} T)$ is obtained from the condition $\Sigma_\mu p_\mu = 1$. The partition function is a very useful tool. For example, it is straightforward to show that $F = -k_\mathrm{B} T \ln Z$.

Exercises

Calculate the entropy of an atom in a cube containing $6 \times 6 \times 6$ phase-space cells (s. figure).

In equilibrium, statistical mechanics reduces to the Boltzmann distribution

$$p_\mu = \frac{1}{Z} \exp\left(\frac{-E_\mu}{k_B T}\right) \tag{5.2}$$

where p_μ is the probability of finding the system in the μ-th state, E_μ is the energy of the state and the Boltzmann constant $k_B = 1.38 \times 10^{-23}$ J/K. The normalization constant

$$Z = \sum_\mu \exp\left(\frac{-E_\mu}{k_B T}\right) \tag{5.3}$$

is known as the *partition function*. It ensures not only that $\Sigma_\mu p_\mu = 1$ but is also an important tool for calculating thermodynamic properties. Note that the symbol Z stands for *Zustandssumme* (literally "sum over all states"), meaning that the calculation of thermodynamic properties essentially reduces to a straightforward though often lengthy sum.

It is convenient to treat (5.1–3) as the starting point for the description of finite-temperature phenomena, but Panel 7 shows how the Boltzmann distribution can be derived from the entropy. In practice, one wants to determine classical or quantum-mechanical averages

$$\langle A \rangle = \frac{\sum_\mu A_\mu \exp(-E_\mu k_B T)}{\sum_\mu \exp(-E_\mu / k_B T)} \tag{5.4}$$

The big challenge is to actually perform the summations over $\mu = 1 \ldots \Omega$, because the total number Ω of states increases exponentially with the size of the magnet. For example, in the Ising model (Chapter 1), there are only two spin states per atom, $\uparrow$ and $\downarrow$, so that $s_i = \pm 1$ and, for N spins, $\Omega = 2^N$. One tool is to exploit the fact that averages $\langle A \rangle$ can be expressed as functions of the partition function.

In magnetism, the summation in (5.3) includes all classical or quantum-mechanical spin states. In the quantum-mechanical case, the states μ are the eigenstates of the Hamiltonian, as determined from the Schrödinger equation. When one considers the interaction of spins with other degrees of freedom, such as lattice distortions (phonons), the summation includes those nonmagnetic degrees of freedom too.

5.1.2 *Fluctuations and response

In principle, the probabilities p_μ can be used to calculate thermally averaged quantities, such as magnetization and susceptibility. For simple systems, this is hardly a problem, although each quantity must be calculated separately. For big systems, this is both time-consuming and unnecessary, because thermal averages are easily determined from the partition function (5.3). In other words, a single exact or approximate summation (5.3) is sufficient, and there is no need to evaluate each average (5.4) separately. One example is the determination of the average energy

$$\langle E \rangle = \frac{1}{Z} \sum_\mu E_\mu \exp\left(\frac{-E_\mu}{k_B T}\right) \tag{5.5}$$

from (5.3). Comparing $\partial Z/\partial T = (1/k_B T^2) \sum_\mu E_\mu \exp(-E_\mu/k_B T)$ with (5.5) and exploiting that $d(\ln Z) = dZ/Z$ yields $\langle E \rangle = k_B T^2 \, \partial \ln Z/\partial T$. A similar calculation reproduces $S = -\partial F/\partial T$, where $F = -k_B T \ln Z$.

Very useful relations are obtained by specifying the energy levels E_μ. In particular, the magnetic field enters the Hamiltonian in the form of a Zeeman term, and the response of the magnet is described by derivatives of Z with respect to the magnetic field. Without loss of generality, we can restrict ourselves to the *Ising model*, introduced in Chapter 1 and defined by the Hamiltonian

$$\mathsf{H} = -\frac{1}{2} \sum_{ij} J_{ij} \, s_i \, s_j - \mu_o \mu_B \sum_i H_i s_i \tag{5.6}$$

Here $H_i = H_z(\mathbf{r}_i)$ is the magnetic field acting on the i-th atomic spin. In the corresponding partition function

$$Z = \sum_\mu \exp\left(\frac{-U_\mu + \sum_i h_i s_i}{k_B T}\right) \tag{5.7}$$

it is convenient to use the internal energy (exchange energy) $U_\mu = -\frac{1}{2} \sum_{ij} J_{ij} \, s_i \, s_j$ and the local field variable $h_i = \mu_o \mu_B H_i$.

Due to the involvement of the J_{ij}, there is no general solution for Z, but it is easy to derive thermal averages once Z is exactly or approximately known. Taking the derivative of (5.7) with respect to h_i yields the local magnetization

$$\langle s_i \rangle = \frac{k_B T}{Z} \frac{\partial Z}{\partial h_i} \tag{5.8}$$

and the quadratic average

$$\langle s_i \, s_j \rangle = \frac{{k_B}^2 T^2}{Z} \frac{\partial^2 Z}{\partial h_i \partial h_j} \tag{5.9}$$

The magnetic susceptibility, $\chi_{ij} = \partial \langle s_i \rangle / \partial H_i$ is proportional to $\partial \langle s_i \rangle / \partial h_i$ and obtained by taking the derivative of (5.8) with respect to h_j. The resulting $\partial Z/\partial h_i$ and $\partial^2 Z/\partial h_i \partial h_j$ terms are easily substituted from (5.8–9), so that

$$\chi_{ij} = \frac{\mu_o \mu_B}{k_B T} \left(\langle s_i \, s_j \rangle - \langle s_i \rangle \langle s_j \rangle \right) \tag{5.10}$$

This important equation is known as the *fluctuation-response theorem*. It relates the spatial fluctuations of the local magnetization, as described by the correlation function C_{ij} or

$$\langle (s_i - \langle s_i \rangle)(s_j - \langle s_j \rangle) \rangle = \langle s_i s_j \rangle - \langle s_i \rangle \langle s_j \rangle \tag{5.11}$$

to the magnet's equilibrium response, $C_{ij} \sim T \chi_{ij}$. Equation (5.10) shows that the linear response of a magnet to a small magnetic field reduces to the probing of fluctuations that are present even in the absence of the magnetic field. The larger the fluctuations, the higher the susceptibility. In turn, high susceptibilities indicate large fluctuations, as in the vicinity of the Curie point.

The fluctuation-response theorem is closely related to the fluctuation-*dissipation* theorem, which considers $\langle s_i(t)s_j(0)\rangle - \langle s_i(t)\rangle\langle s_j(0)\rangle$ and describes the equilibrium dynamics of the system. Subjecting a magnet to an external field, or compressing a gas with a piston, leads to a response that is determined by the system's fluctuations. Physically, the compression means that the gas molecules or atoms fluctuate away from the piston, so that the piston exploits a "temporary vacuum". In terms of the picture in Panel 7, this corresponds to going from right to left due to mechanical pressure or, equivalently, by waiting until the atom fluctuates to the top left corner.

The equivalence of fluctuations and response reflects a very general feature of equilibrium statistics, namely the equivalence of time and ensemble averages. In equilibrium, it does not matter whether thermal averages are obtained by considering snapshots of one magnet at different times or by simultaneously considering snapshots of several magnets. This is no longer true for nonequilibrium phenomena. For example, the mechanical susceptibility (compliance) of glasses is much smaller than predicted from the liquid-like structural or ensemble correlations, because the structure is frozen and the time-averages are solid-like. A related class of magnetic materials are *spin glasses* (Section 7.1.4).

5.1.3 Phase transitions

The spontaneous magnetization may exhibit a singular dependence on temperature or external magnetic field. This is an example of a *phase transition*. More generally, phase transitions are defined as singular changes of an order parameter (magnetization, fluid density) as a function of a control parameter (field, pressure, temperature). For example, infinitesimally small changes in pressure P or temperature T can lead to the condensation of water vapour. When order parameters, such as fluid density and magnetization, change discontinuously, the material undergoes a first-order phase transition (Fig. 5.3). An example is the boiling of water, which is accompanied by a density change $\Delta\rho > 0$. The temperature at which the order-parameter gap vanishes is called the *critical temperature*. At this point, the phase transition is referred to as

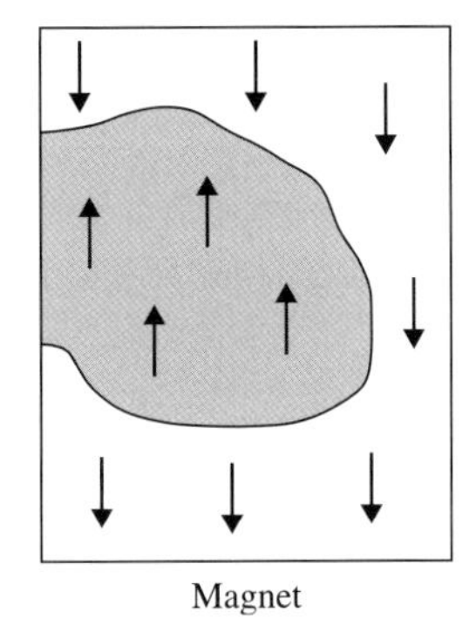

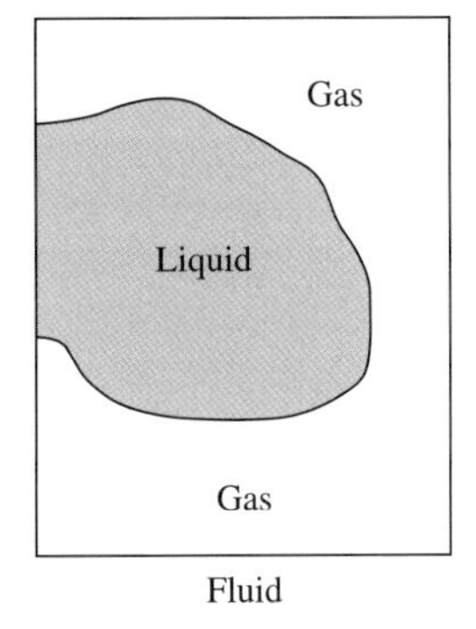

Fig. 5.3 Coexistence of phases below the critical temperature. The fractions of the phases are determined by external forces (pressure, magnetic field). In a magnet, the spin-up and spin-down regions must not be confused with domains.

a second-order or, more precisely, continuous phase transition. For water, the critical point ($T_c = 647\,\mathrm{K}$, $P_c = 22.1\,\mathrm{MPa}$) is characterized by $\Delta\rho = 0$, so that vapor and liquid are indistinguishable.

Phase transitions are caused by interatomic interactions. Without attractive interactions between gas molecules, there would be no transition to a liquid, and at low temperature one would obtain a very dense gas rather than a liquid. Similarly, without exchange, a paramagnetic gas is easily spin-polarized at low temperature but does not undergo a transition to ferromagnetism.

A very simple interpretation of magnetic phase transitions is based on the free energy $F = \langle E \rangle - TS$, where S is the entropy and $\langle E \rangle$ is the average energy. The paramagnetic phase has a relatively high energy, because the thermal randomization of the spins costs exchange energy. However, the entropy or "disorder" is stronger in the paramagnetic phase than that in the ferromagnetic phase. The term $-TS$ indicates that the relative contribution of the entropy increases with temperature, and above T_c the entropic term dominates and paramagnetism is more favorable than ferromagnetism. Of course, the problem remains actually to determine $\langle E \rangle$ and S, or alternatively, Z. We will see, for example, that Z depends on the size and dimensionality of the magnet, and that ferromagnetism is limited to infinite magnets. In finite magnets, thermal excitations cause the net moment to fluctuate between opposite directions, so that the thermally averaged magnetization is zero.

5.1.4 Landau theory

A simple phenomenological model of phase transitions was developed by Landau, who expanded the free energy F in terms of the average magnetization $\langle s \rangle$:

$$F = a_2 \langle s \rangle^2 + a_4 \langle s \rangle^4 - h \langle s \rangle \tag{5.12}$$

where a^2 and a^4 are phenomenological interaction parameters, and $h = g\mu_o\mu_B H$. The addition of a quartic term is necessary to ensure two (free) energy minima corresponding to ↑ and ↓ phases. Note that the expansion of the magnet's internal energy is limited to odd powers of $\langle s \rangle$, because positive and negative magnetization direction are equivalent. By comparison, the Zeeman term breaks this symmetry, because the external field discriminates between field directions.

To describe phase transitions, the parameter a_2 in (5.12) must change sign at T_c. Linearizing a_2 with respect to T then yields

$$F = a'(T - T_c)\langle s \rangle^2 + a^4 \langle s \rangle^4 - h \langle s \rangle \tag{5.13}$$

Figure 5.4 shows the free energy for different temperatures. Below T_c, the free energy has two minima, corresponding to ↑ and ↓ phases. This is known as *spontaneous symmetry breaking.* Figure 5.5 shows two magnetic phase diagrams derived from (5.13). The Curie transition means that both $T = T_c$ and $H = 0$ are satisfied. Arbitrarily small magnetic fields destroy the singularity of the phase transition and assimilate the $M(T)$ curve to a paramagnet (dashed line in Fig. 5.5). The Curie transition is an example of a second-order or continuous phase transition, characterized by $\Delta M = 0$, as contrasted to first-order transitions, where $\Delta M \neq 0$.

By phase mixing, it is possible to realize any magnetization between $-M_s$ and $+M_s$. The spontaneous magnetization is *not* related to the formation of macroscopic

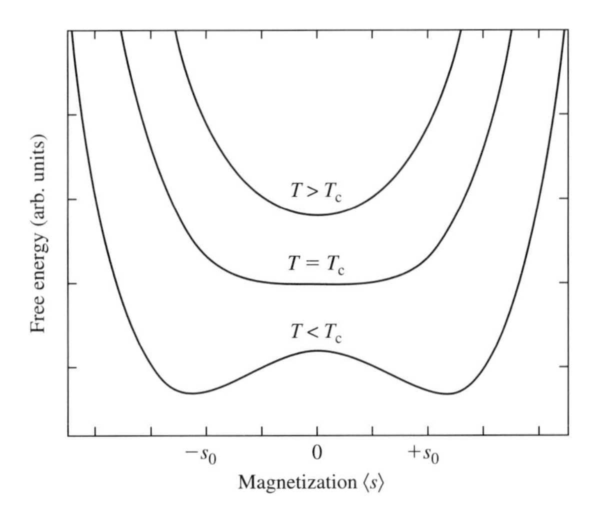

Fig. 5.4 Free energy in the vicinity of the critical point. Below T_c, there are two phases (spontaneous symmetry breaking).

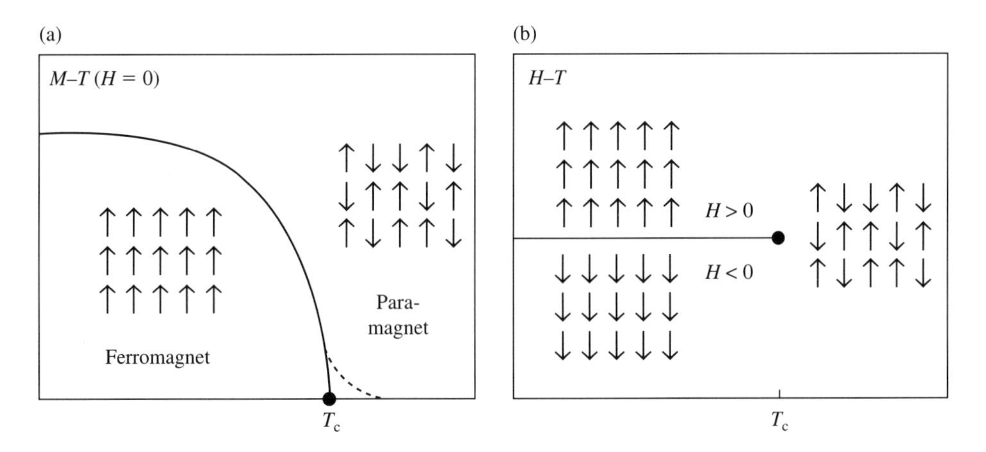

Fig. 5.5 Coexistence of different phase below and above the Curie temperature T_c: (a) schematic $M-T$ phase diagram and (b) $H-T$ phase diagram. Below T_c, phases with $M_z > 0$ and $M_z < 0$ may coexist. The M(T) curve in (a) is the spontaneous magnetization M_s = M(H = 0). At T_c, an arbitrary small field smoothes the phase-transition singularity (dashed line).

magnetic domains, because domain formation involves the magnetostatic self-interaction energy, which is not included in (5.13). This is seen for example, by comparing the domain-wall width δ_B (Section 4.2.5) with the thickness ξ of the phase boundaries shown in Fig. 5.5. At low temperatures, δ_B remains finite but ξ goes to zero. Another difference is that the magnetization in domain walls, $|\mathbf{M}(\mathbf{r})|$, is averaged over a few interatomic distances and equal to M_s, whereas phase boundaries exhibit $|\mathbf{M}(\mathbf{r})| < M_s$.

The Landau expansion (5.13) contains both equilibrium and nonequilibrium spin configurations. In equilibrium, $\partial F/\partial\langle s\rangle = 0$, so that

$$h = 2a'(T - T_c)\,\langle s\rangle + 4a_4\,\langle s\rangle^3 \tag{5.14}$$

Linearization of this equation with respect to $\langle s\rangle$ yields $\langle s\rangle = h/2a'(T - T_c)$, so that the isothermal zero-field susceptibility $\chi \sim \mathrm{d}\langle s\rangle/\mathrm{d}h$ scales as $1/(T - T_c)$. More generally, in the vicinity of T_c,

$$\chi \sim \frac{1}{|T - T_c|^\gamma} \tag{5.15}$$

where γ is an example of *critical exponent*. In the present model $\gamma = 1$. Critical exponents describe the susceptibility and other properties in the vicinity of the critical point, and their values depend on the models and approximations used. For example, the Landau model predicts $\gamma = 1$, but experiment and refined models yield different exponents for most systems. The $\gamma = 1$ susceptibility is also known the Curie-Weiss susceptibility. Figure 5.6 compares the susceptibility of a ferromagnet (solid line) with the paramagnetic $1/T$ Curie law (dashed line).

The spontaneous magnetization is defined by the absence of external magnetic fields and obtained by putting $h = 0$ in (5.14). The result is $\langle s\rangle \sim 1/\sqrt{T_c - T}$ or

$$\langle s\rangle \sim \frac{1}{|T_c - T|^\beta} \tag{5.16}$$

where the critical exponent β is equal to $\frac{1}{2}$ in the Landau model. This equation describes the spontaneous magnetization as the Curie temperature is approached, as

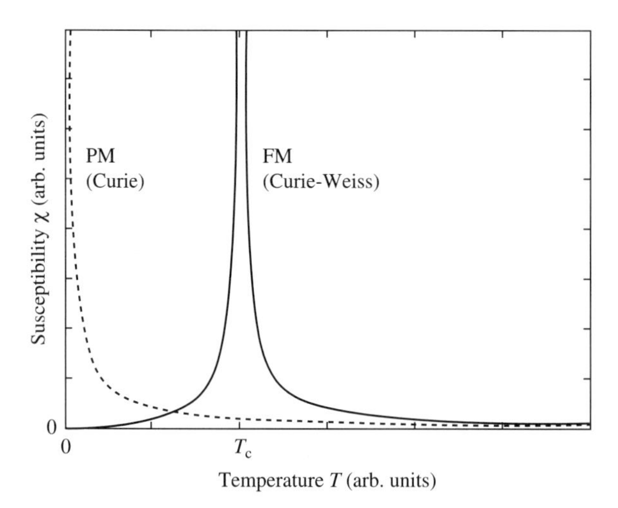

Fig. 5.6 Typical susceptibilities of ionic paramagnets (PM) and ferromagnets (FM). The strongly temperature-dependent Curie susceptibility of ionic paramagnets should not be confused with the small and largely temperature-independent Pauli susceptibility of metallic paramagnets.

shown in Fig. 5.5. As we will see below, $\beta = \frac{1}{2}$ is a rather crude approximation in many cases, and on approaching T_c, the magnetization change is more dramatic ($\beta < \frac{1}{2}$).

A third exponent, δ, describes the critical isotherm $T = T_c$. From (5.14) we find that

$$\langle s \rangle \sim \frac{1}{|H|^{1/\delta}} \tag{5.17}$$

and $\delta = 3$ for the Landau model. Another critical exponent, α, determines the temperature dependence of the heat capacity, and correlations at the critical point are described by

$$\langle s_i\, s_j \rangle - \langle s_i \rangle \langle s_j \rangle \sim \exp\left(\frac{-|\mathbf{R}_i - \mathbf{R}_j|}{\xi} \right) \tag{5.18}$$

where the correlation length

$$\xi \sim \frac{1}{|T_c - T|^{\nu}} \tag{5.19}$$

Using a spatially resolved extension of the Landau model (Section 5.3) yields $\nu = \frac{1}{2}$. Note that the different critical exponents are not independent but related to each other. For example, the fluctuation-response theorem (5.10) indicates an essential correspondence between correlation length and susceptibility. A detailed analysis of the partition function reveals that there are only two independent critical exponents (see e.g. Huang 1963, Yeomans 1992).

The phenomenological Landau model is closely related to various models and approximations of well-defined microscopic meaning. Examples are the van-der-Waals theory of the gas–liquid transition, the mean-field model, and the Bragg-Williams treatment of binary alloys.

5.2 Spin-Space modeling

Summary The modeling of atomic spins is a key aspect of finite-temperature magnetism. The simplest model is the Ising model, where each atom has two spin states $s_i = \pm 1$. The Ising model captures some essential features of magnetism but ignores the quantum-mechanical effects and amounts to the unphysical prediction of infinite magnetic anisotropy. An isotropic model is the Heisenberg model, which exists in form of classical and quantum-mechanical realizations. A generalization of Ising and Heisenberg models is the n-vector model, which also includes the classical limits of models such as the XY model. The quantum-mechanical Heisenberg model provides an adequate description of typical magnetic ions. In magnetic or exchange fields, the energy levels of the magnetic ions split into multiplets whose finite-temperature occupancy is described by Brillouin functions. A very complicated situation is encountered in itinerant magnets, such as Fe, Co, and Ni. The Stoner model greatly overestimates the Curie temperature, because thermal excitations create spin disorder and break the assumed Bloch symmetry of the wave functions. As a consequence, itinerant moments are fairly well-conserved at the Curie temperature and the spin structure resembles that of localized magnets. An exception are very weak itinerant ferromagnets, such as $ZrZn_2$, whose finite-temperature behaviour is determined by long-wavelength spin-fluctuations close to the Stoner limit.

A key aspect of magnetic modeling is the description of individual spins. The spins may be quantum mechanical or classical, isotropic or anisotropic, linked to individual atomic sites, or form a continuum. In this section, we introduce and discuss various spin models and investigate how the spin behave in a magnetic field. The models introduced in this section serve a the starting point for the treatment of interaction effects in Section 5.3–5.5.

5.2.1 Heisenberg models

A very important model is the *Heisenberg model* introduced in Section 2.1.4. It exists in the form of several quantum-mechanical and classical realizations. The quantum-mechanical spin-1/2 Heisenberg model is defined by

$$\mathsf{H} = -2 \sum_{i>j} J_{ij}\, \mathbf{s}_i \cdot \mathbf{s}_j - g\mu_o\mu_B \sum_i \mathbf{H}_i \cdot \mathbf{s}_i \tag{5.20}$$

where the spin operator is closely related to Pauli matrices (2.32), $\mathbf{s} = \frac{1}{2}\boldsymbol{\sigma}$. Ignoring the orbital-moment contribution, $g = 2$. The challenge posed by (5.20) is that the operators $\mathbf{s}_x$, $\mathbf{s}_y$, and $\mathbf{s}_z$ do not commute. The Heisenberg interaction $J\mathbf{s}_1 \cdot \mathbf{s}_2$ between two spins has been investigated in Section 2.1.4, whereas infinite Heisenberg systems will be discussed in the context of spin waves (Section 6.1).

Quantum-mechanical Heisenberg models of arbitrary spin S exhibit $2S + 1$ states per atom, as discussed in Section 2.2. Spin values $S > \frac{1}{2}$ are frequently encountered in magnetic oxides. In rare-earth ions, spin and orbital moments are coupled, and one must consider the total angular momentum rather than the spin, but the resulting interaction is of the Heisenberg type. Special cases are the above-mentioned $S = \frac{1}{2}$ Heisenberg model with two states (↑ and ↓) and the classical Heisenberg model, $S = \infty$, with a continuum of states. It is convenient to define the classical Heisenberg model as

$$\mathsf{H} = - \sum_{i>j} J_{ij}\, \mathbf{s}_i \cdot \mathbf{s}_j - g\mu_o\mu_B \sum_i \mathbf{H}_i \cdot \mathbf{s}_i \tag{5.21}$$

where the $\mathbf{s}_i$ are normalized classical spin vectors, $|\mathbf{s}_i| = 1$. This ensures that the energies of ferromagnetic and antiferromagnetic bonds are $\pm J$, as in the two-spin Heisenberg model of Section 2.1.4.

In both quantum-mechanical and classical models, the J_{ij} may be positive (ferromagnetism) or negative (antiferromagnetism). In complicated magnetic compounds and disordered magnets, the H_i and J_{ij} are site-specific, and one magnet may contains atoms with different spins s_i and coupling J_{ij}. This gives rise to complicated spin structures, such as ferrimagnetism, noncollinear structures, and spin-glass behavior. An example is the RKKY exchange introduced in Section 2.3.2, which amounts to exchange constants $J_{ij} = J(\mathbf{r}_i - \mathbf{r}_j)$. This equation also indicates that Heisenberg exchange is not restricted to nearest neighbors, although the interactions tend to rapidly decrease with distance. In the vicinity of the Curie transition, this range of interactions is crucial. Long-range interactions reproduce the qualitative behavior predicted by the Landau model, whereas short-range interactions yield essential corrections in low-dimensional magnets.

5.2.2 Ising, XY, and other n-vector models

The spin-$\frac{1}{2}$ Heisenberg Hamiltonian for a pair of spins may be written as $\mathsf{H} = -2J\mathbf{s}\cdot\mathbf{s}'$ or

$$\mathsf{H} = -2J(s_x \cdot s'_x + s_y \cdot s'_y + s_z \cdot s'_z) \tag{5.22}$$

This Hamiltonian is symmetric with respect to the three spin components s_x, s_y, and s_z. Generalizing (5.22) to arbitrary spin dimensions yields the n-component vector-spin model or *n-vector model*. For example, $n = 1$ yields the Ising model, with one spin component s_z. The spin dimensionality n must not be confused with the real-space dimensionality d of the lattice. For example, the square-lattice Ising model is characterized by $n = 1$ and $d = 2$.

The isotropy of the Heisenberg model, that is, the symmetry with respect to the n spin components s_x, s_y, and s_z, applies not only to bulk magnets with cubic crystal structure but also to magnets with anisotropic structures, such as thin films. In contrast to popular belief, features such as broken exchange bonds at surfaces are unable to introduce magnetic anisotropy. For example, in multilayers, the interlayer coupling tends to be weaker than the intralayer coupling, but coherently changing the magnetization from in-plane to perpendicular leaves the Heisenberg interaction unchanged.

The isotropy of the Heisenberg model is a consequence of the exchange interaction of Section 2.1.4 and means that there is no explicit spin dependence in the starting Hamiltonian. However, in the presence of spin-orbit coupling, the spins create orbital currents, which interact with both the lattice and with neighboring spins. Aside from giving rise to magnetocrystalline anisotropy (Section 3.4), the spin-orbit interaction creates some exchange anisotropy, for example

$$\mathsf{H} = -2J_{x-y}(s_x \cdot s'_x + s_y \cdot s'_y) - 2Js_z \cdot s'_z \tag{5.23}$$

in uniaxial magnets. The exchange anisotropy $|J_{x-y} - J|/J$ is usually very small, but it becomes important if J is small due to competing exchange contributions. For $|J_{x-y}| \gg |J|$, one can ignore J and obtains the XY or $n = 2$ vector-spin model with the pair interaction

$$\mathsf{H} = -2J_{x-y}(s_x \cdot s'_x + s_y \cdot s'_y) \tag{5.24}$$

In the opposite limit of very large J, one obtains the *Ising model* with the two-spin interaction

$$\mathsf{H} = -2Js_z \cdot s'_z \tag{5.25}$$

The spin-1/2 Ising model is the simplest model of finite-temperature magnetism. Its advantage is the restriction to two spin states per site, $s_i = \pm 1$, or $\uparrow$ and $\downarrow$, which are usually interpreted as the z components of the atomic moment. This greatly simplifies the calculation of the partition function, although the main problem, namely the exponential dependence on the number N of spins, remains. Nevertheless, for nearest-neighbor interactions, the ferromagnetic Ising model has been solved in one dimension (Ising 1925) and in two dimensions (Onsager 1944).

Since the exchange anisotropy tends to be small, (5.24–25) are very poor approximations for most magnetic materials. In particular, (5.24) and (5.25) should not be confused with magnets exhibiting easy-plane and easy-axis anisotropies, respectively, where the exchange is essentially isotropic. A variety of rather exotic compounds with XY or Ising character has been reviewed by de Jongh and Miedema (1975).

As mentioned above, the models (5.23–25) can be classified as n-vector models. According to the number n of spin components (spin dimensionality), we have $n = 1$ (Ising model), $n = 2$ (XY model), and $n = 3$ (Heisenberg model). Other n-vector models are the polymer model ($n = 0$) and the spherical model ($n = \infty$, Berlin and Kac 1952). The polymer model describes self-avoiding polymer chains in solution and is classified as an n-vector model because its partition function derives from the magnetic n-vector model as $n \to 0$ (de Gennes 1979).

The spin dimensionality n is unrelated to the real-space dimensionality d. For example, Ising models ($n = 1$) include Ising chains ($d = 1$), thin films ($d = 2$), and bulk magnets ($d = 3$). The n-vector model has been useful in the understanding of phase transitions, revealing essential differences between magnets with continuous symmetry ($n \geq 2$) and without continuous symmetry ($n = 1$). This is because the critical behavior near the Curie temperature reflects the nature of the symmetry rather than the strength of the anisotropy. Models that are hopelessly inadequate over a wide range of temperatures may therefore yield qualitatively correct predictions near T_c. For example, the Ising model may be derived by adding an infinite (and therefore pathological) anisotropy to the classical Heisenberg model, but it reproduces essential features of the critical behavior of anisotropic magnets.

5.2.3 *Other discrete and continuum spin models

There are various generalizations of the "ordinary" or $S = \frac{1}{2}$ Ising models. The $S = 1$ Ising model has three states with $s = -1$, 0, and +1. Its most general version has five phenomenological model parameters, as opposed to the two parameters J and H of the ordinary Ising model. It contains, for example, a biquadratic exchange proportional to $s^2 s'^2$. The q-state Potts model contains q states $s = 1, 2, \ldots, q$ per site, but the interaction energy is zero unless $s = s'$. The related p-state clock model (also known as the vector Potts model or the discrete XY model) is defined by the interaction

$$\mathsf{H} = -2J \cos(2\pi(s - s')/p) \tag{5.26}$$

where $s = 1, 2, \ldots, p$ (Yeomans 1992). Another version of the Potts models is the Ashkin-Teller model, describing interactions of four kinds of atoms on a lattice.

XY, Potts, and other models have been studied to improve the understanding of magnetic phase transitions and to describe specific, often nonmagnetic systems (Baxter 1982). For example, the two-dimensional XY model exhibits a Thouless-Kosterlitz transition to a low-temperature phase without long-range order but with weakly decaying (nonexponential) correlations. The transition involves bound pairs of vortices and is of interest in superconductivity. Another example is that of ice-type models describing the position of hydrogen atoms in crystals with hydrogen bonding. In ice, the oxygen atoms form a three-dimensional structure of coordination number four. Since the two-dimensional model is similar to but much easier to treat than the three-dimensional model, one replaces the real ice by a "square ice". The interactions of the ice-type models are electrostatic and have also been used to describe ferroelectrics.

So far, we have focused our attention on models where each atomic site carries one well-defined spin. This is important if one needs to distinguish between different types of lattices, such as square, triangular, honeycomb, simple cubic, face-centered cubic, and body-centered cubic. However, for many purposes it is sufficient to consider the "longitudinal" magnetization as a continuous quantity with an exchange proportional to the square gradient of the magnetization, similar to the perpendicular magnetization component $\mathbf{m}$ in micromagnetics (Section 4.2.4). In general, these models have the character of soft-spin models, that is, terms such as $(\mathbf{s}^2 - s_o^2)^2$ in the classical Hamiltonian ensure a finite magnitude of the magnetization $|\mathbf{s}|$. The resulting problem is nonlinear and difficult to treat. A special case is that of *Gaussian models*, where a harmonic approximation is used for the energy $E(\mathbf{s})$. One example is the Gaussian version of the spherical or $\mathrm{n} = \infty$ vector-spin model (Section 5.2.2).

5.2.4 Ionic excitations

Let us, for the moment, ignore the interactions and consider isolated spins in a magnetic field. In Section 2.2 we have seen that Heisenberg spins are characterized by $2J + 1$ energy levels. For example, ferric iron (Fe^{3+}) has the spin $J = S = 5/2$, and according to quantum mechanics it exhibits $2S+1 = 6$ spin orientations in an external magnetic field, namely $S_z = -5/2, -3/2, -1/2, 1/2, 3/2, 5/2$. In a magnetic field, these levels undergo an intramultiplet splitting according to the Zeeman energy

$$E = -g\mu_o\,\mu_B J_z H \tag{5.27}$$

The corresponding partition function

$$Z = \sum_{J_z=-J}^{J} \exp\left(\frac{g\mu_o\,\mu_B J_z H}{k_B T}\right) \tag{5.28}$$

has the character of a geometrical series

$$Z = \frac{\exp(g\mu_o\mu_B(J + \frac{1}{2} + H/k_B T) - \exp(-g\mu_o\mu_B(J + \frac{1}{2})H/k_B T)}{\exp(g\mu_o\mu_B J_z H/2k_B T) - \exp(-g\mu_o\mu_B J_z H/2k_B T)} \tag{5.29}$$

As outlined in Section 5.1, the thermally averaged moment $\langle m\rangle = g\mu_B\langle J_z\rangle$ is obtained by differentiation with respect to H, $\langle m\rangle = k_B T(\partial \ln Z/\partial H)/\mu_o$. The result is

$$\langle m\rangle = gJ\mu_B B_J\left(\frac{gJ\mu_B\mu_o H}{K_B T}\right) \tag{5.30}$$

where the *Brillouin functions* $B_J(x)$ are defined as

$$B_J(x) = \frac{2J+1}{2J}\coth\left(\frac{(2J+1)x}{2J}\right) - \frac{1}{2J}\coth\left(\frac{x}{2J}\right) \tag{5.31}$$

The functions are linear for small arguments and approach 1 for large arguments, corresponding to full spin alignment.

Special cases are the Langevin function $\mathsf{L}(x) = \coth x - 1/x$ in the classical limit ($J = \infty$) and the hyperbolic tangent $B_{1/2}(\mathrm{x}) = \tanh x$ in the limit $J = \frac{1}{2}$. The latter function is very similar to the Ising case, because the spin-$\frac{1}{2}$ Ising and Heisenberg models have two energy levels per spin, but the physics is different. In the Ising model, $s_\mathrm{x} = s_\mathrm{y} = 0$, whereas the Heisenberg spins exhibit $s_\mathrm{x} \neq 0$ and $s_\mathrm{y} \neq 0$ but $\langle s_\mathrm{x} \rangle = \langle s_\mathrm{y} \rangle = 0$.

Equation (5.30) describes the paramagnetism of magnetic ions. At high temperatures, $B_\mathrm{J}(\mathrm{x}) \approx (J+1)x/3J$ and the magnetization obeys *Curie's law*

$$\langle m \rangle = \frac{g^2 \mu_\mathrm{B} J(J+1)}{3 k_\mathrm{B} T} \mu_\mathrm{o} H \tag{5.32}$$

This equation defines the *Curie paramagnetism* of magnetic ions (dashed line in Fig. 5.6). The moment tends to small values at room temperature, but unlike the nearly temperature-independent Pauli paramagnetism (Section 2.4.3), the moment is strongly enhanced at low temperatures.

Equation (5.30) is important for the quantitative understanding of the magnetism of interacting spins, because interactions can often be approximated by interaction fields. Aside from the average $\langle J_\mathrm{z} \rangle$, we will need averages of powers of J_z, to determine the temperature dependence of the anisotropy (Section 5.5).

5.2.5 Spin fluctuations in itinerant magnets

So far, we have considered ionic models, where the magnetic moments are localized on individual atomic sites. This is a reasonable assumption for insulators and rare-earth metals but unrealistic for itinerant magnets, such as elemental iron. As we have seen in Section 2.4, itinerant magnetism involves interatomic hopping, and the eigenstates are well described by wave vectors **k**. It is therefore tempting to explain the Curie temperature of itinerant magnets by excitations between $\uparrow$ and $\downarrow$ states with well-defined wave vectors. Such excitations are known as *Stoner excitations*.

Unfortunately, the Stoner theory greatly overestimates the Curie temperature of Fe, Co, and Ni. Approximating the $\uparrow$ and $\downarrow$ densities of states by sharp peaks and assuming that the Curie temperature reflects thermally activated transitions between the peaks leads to $T_\mathrm{c} = \mathrm{mI}/4\mu_\mathrm{B} k_\mathrm{B}$. These Stoner temperatures are about five times larger than the observed Curie temperatures. The reason for this failure is the Bloch character of the electron states assumed in the Stoner theory. As exemplified by the tight-binding approximation (Section 2.4.2), the electrons are extended and described by wave vectors **k**, and the only way to reduce the magnetization is to transfer spins from $\mathbf{k}_\uparrow$ to $\mathbf{k}_\downarrow$ states. This amounts to the unphysical prediction that the moment and the spontaneous magnetization vanish simultaneously at the Stoner temperature. Alternatively, the Stoner theory ascribes the Curie temperature to a smearing of the Fermi surface, but we know that this thermal smearing is very small, $k_\mathrm{B} T \ll E_\mathrm{F}$.

In ferromagnets such as Fe and Co, the excitations responsible for the decrease of the spontaneous magnetization are localized, and the Curie temperature is much lower than the Stoner temperature. Figure 5.7 illustrates this point by comparing the zero-temperature spin structure of itinerant magnets (a) with the Stoner model (b) and

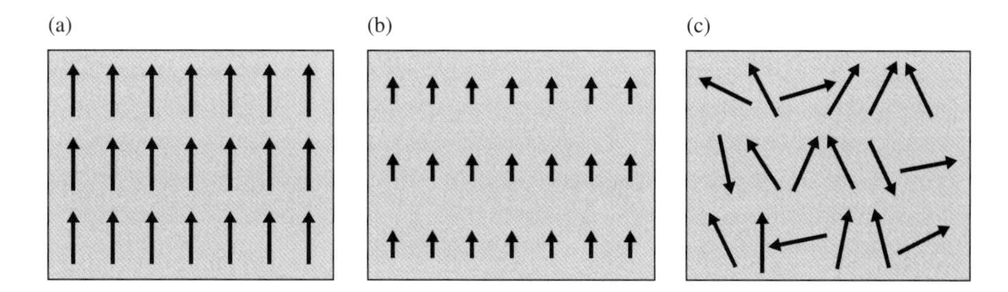

Fig. 5.7 Spin structure in itinerant magnets: (a) zero temperature, (b) Stoner picture, and (c) localized or Heisenberg picture. In (b) and (c), the temperature is just below T_c. In iron-series transition metals, the spin structure at T_c is close to (c), with little reduction in the magnitude of the moment.

with the localized picture (c). The difference between (b) and (c) is illustrated by a needle in a flame, so that the spontaneous magnetization collapses at the hot end. If the relevant excitations were of the Stoner type, the magnetization would decrease homogeneously at both the hot and cold ends of the needle, which is in contrast to experience.

Quantum-mechanically, the localization is similar to that observed in disordered magnets (Gubanov, Liechtenstein, and Postnikou 1992). In terms of Section 2.4, thermal activation leads to spin-disorder and yields a random spin-dependent potential $V_\sigma(\mathbf{r})$. This destroys the translational symmetry and means that the electron eigenfunctions are no longer of the Bloch type. Dynamically, the bandwidth W translates into a hopping time $\hbar/W$, so that electrons in the relatively narrow 3d bands are temporarily captured by individual atoms. Since the intra-atomic exchange of iron-series transition metals (about 1 eV) is much larger than the interatomic exchange (about 0.1 eV), atomic moments remain stable at and well above T_c, as illustrated in Fig. 5.7(c).

The degree of localization near T_c is, in general, intermediate between Fig. 5.7(b) and (c). The ferromagnetic elements Fe, Co, and to a lesser extent Ni, as well as most transition-metal alloys, are close to the localized picture of (c). An important exception is *very weak itinerant ferromagnets* such as $ZnZn_2$, which are barely ferromagnetic (Section 2.4) and where the vanishing of the spontaneous magnetization is accompanied by a nearly complete reduction of the magnetic moment, similar to Fig. 5.7(b).

To bridge the Stoner and Heisenberg limits, one must consider the wave-vector dependence of the moment. Putting $s = n_\uparrow - n_\downarrow$ in (2.52) and describing the interatomic interactions by a longitudinal exchange stiffness A_o yields the quasi-classical expression

$$\frac{\mathsf{H}}{V} = A_o\,(\nabla s)^2 + \frac{1}{4D_s(E_F)}s^2 - \mu_o\,\mu_B\,s\,H \tag{5.33}$$

In the itinerant limit of Fig. 5.7(b), we can ignore interactions such as the Heisenberg exchange J_{ij} and use $A_o = 1/48D_s(E_F)k_F^2$. Physically, moment inhomogeneities of wave vector k_o enhance $\nabla\psi(\mathbf{r})$ and yield an additional kinetic-energy term scaling as

k_o^2/k_F^2 (Qi, Skomski, and Coey 1994). Equation (5.33) corresponds to the wave-vector dependent susceptibility

$$\chi(k) = \frac{\chi_P}{1 - UD_s(E_F) + k^2/12k_F^2} \tag{5.34}$$

This equation describes not only the onset of ferromagnetism in very weak itinerant ferromagnets, which are close to Fig. 5.7(b), but also paramagnons in exchange-enhanced Pauli paramagnets such as Pd and Pt. Paramagnons are quasi-ferromagnetic spin fluctuations and occur because the Stoner criterion is nearly satisfied (Moriya 1985, Fulde 1991, and Mohn 2003).

Fourier transformation of (5.34) yields an exponentially decaying response to any point-like magnetic perturbation. The decay length increases with $UD(E_F)$ and diverges at the onset of long-range ferromagnetic order, where the Stoner criterion yields $\chi(0) = \infty$. In other words, any small local perturbation causes the system to become ferromagnetic. The exponential decay predicted by (5.34) must be contrasted to the oscillatory decay of the RKKY interaction (Section 2.3.2). In wave-vector space, the RKKY oscillations correspond to a specific nonanalytic feature of $\chi(k)$ at $k = 2k_F$ (Lindhard screening, Section 2.1.6), which is not reproduced by the harmonic approximation in the denominator of (5.34). Exchange-enhanced Pauli paramagnets combine a pronounced preasymptotic exponential decay with a small oscillatory tail. This is because $1-UD_s(E_F)$ is small and higher-order corrections in the denominator of (5.34) are less important. However, the exponential contribution decreases more rapidly than the oscillations, and the asymptotic behavior is determined by the small RKKY tail.

Mathematically, (5.33) is of the Landau-Ginzburg type. The harmonic energy (5.33) and its anharmonic generalizations are easily applied to a variety of zero- and finite-temperature phenomena in the bulk and at surfaces (see e.g. Mathon and Bergmann 1986, Miller and Dowben 1993, Skomski *et al.* 1998b). Basically, (5.33) must be minimized for the considered geometry, which is a straightforward problem, rather similar to the determination of nucleation modes. Some examples are mentioned in Section 7.4.

For the energy (5.33), the partition function is readily calculated (A2.4), but due to its harmonic character, it does not yield a Curie transition. As we seen in Section 5.1.4, phase transitions require higher-order energy contributions to $\mathsf{H}(s)$. Here we use what is known as the variational or Bogulyubov free energy, based on the relation

$$F \le F_o + \langle \mathsf{H} - \mathsf{H}_o \rangle_o \tag{5.35}$$

where H_o is any model Hamiltonian and the index o refers to H_o. If F contains a parameter λ, then minimization of $F_o + \langle \mathsf{H} - \mathsf{H}_o \rangle_o$ with respect to λ yields the best free energy estimate compatible with H_o. To make practical use of (5.35), one needs a *simple* Hamiltonian H_o. An example is the mean-field approximation (next section), which may be derived by putting $\mathsf{H}_o = -\lambda s$ (Yeomans 1992). To describe the spin fluctuations, we use a harmonic or Gaussian trial Hamiltonian $\mathsf{H}_o = \lambda s^2$. The approach, introduced by Murata and Doniach (1972), is known as self-consistent renormalization of spin fluctuations, mode-mode coupling, or classical functional-integral method. For quartic energies, minimizing the trial free energy with respect to λ yields the replacement $s^4 \to \zeta s^2$, where $\zeta = 3\langle s^2(T)\rangle_o$. The Curie temperature is proportional to $1/\zeta$, and self-consistency is ensured by putting $\langle s^2(T)\rangle_o = \langle s^2(T_c)\rangle_o$.

Very weak itinerant ferromagnets, which exhibit small zero-temperature moments of order $0.1\,\mu_B$, are well described by the Murata-Doniach theory. Unfortunately, it is not possible to apply the formalism to strong or nearly strong ferromagnets, such as Fe, Co, and Ni. Aside from the classical and continuum character of the model, there is the problem that the moments of Fe, Co, and Ni are comparable to the full spin polarization s_{max}. It has been attempted to generalize $s^4 \rightarrow \zeta s^2$ to arbitrary nonlinear functions, but the approach is semiconvergent, with results that are improved by the inclusion of the first terms in the series expansion but worsened by the inclusion of higher-order terms. This can be shown by considering arbitrary functions $\mathsf{H} = E(s)$. The calculation reveals that

$$\zeta \sim \int_{-\infty}^{+\infty} (s^2 - \langle s^2 \rangle_o)\, E(s) \exp\left(-\frac{s^2}{2\langle s^2 \rangle_o}\right) \mathrm{d}s \tag{5.36}$$

Since $E = \infty$ for $s > s_{max}$, this integral diverges, $\zeta = \infty$, and yields $T_c = 0$, in striking contrast to experiment. Pictorially, the assumed harmonic trial function λs^2 cannot be "squeezed" into a nonlinear potential that puts an upper limit to the spin (complete spin polarization).

5.3 Mean-field models

Summary Ferromagnetic order in most magenets is due to interatomic exchange. Mean-field models place individual atomic spins in an exchange field created by neighboring atoms. This maps the interaction problem onto the problem of noninteracting spins in a magnetic field, except that the field must be calculated self-consistently from the magnetization. Linearization of the mean-field equation yields the Curie temperature $T_c \sim zJ$, where z is the number of interacting neighbors and J is the interatomic exchange. Below T_c, the mean-field equations have two ferromagnetic solutions $\pm M_s(T)$. The mean-field model is easily generalized to two or more sublattices, where it yields complicated spin structures, such as ferrimagnets, antiferromagnets, and noncollinear magnets. In the most general case of N sublattices (or N non-equivalent atomic sites) it requires the solution of N coupled algebraic equations. The mean-field approximation provides a reasonable description of the finite-temperature magnetization for a broad range of models, including Ising and Heisenberg models. It also describes critical fluctuations, albeit on an dimensionality-independent Ornstein-Zernike level. Mean-field theory breaks down at very low temperatures, where the excitations have the character of spin waves, and near T_c, where critical fluctuations interfere. In particular, mean-field models tend to overestimate the Curie temperature. This failure is most pronounced in one-dimensional magnets where $T_c = 0$ but mean-field theory predicts $T_c > 0$.

Aside from the existence of atomic magnetic moments, ferromagnetism requires long-range magnetic order, typically created by interatomic exchange. To determine equilibrium properties such as the average spin $\langle s \rangle$, it is sufficient to know the partition

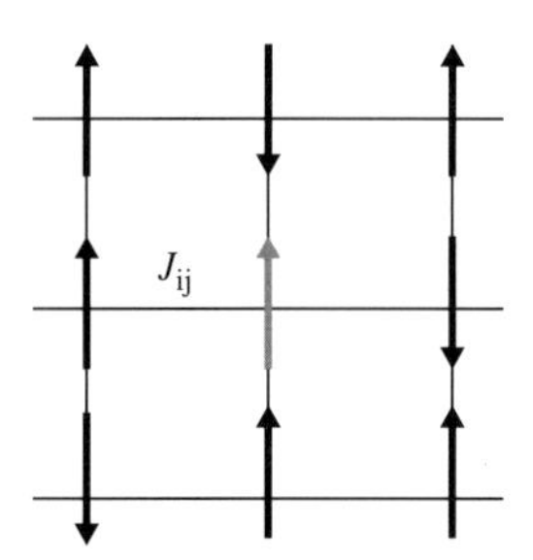

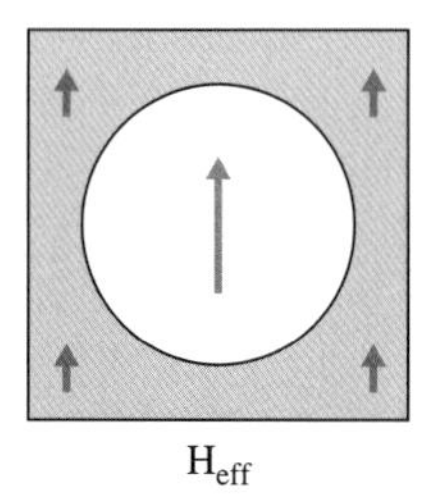

Fig. 5.8 Mean-field model: (a) magnet described by exchange interactions J_{ij} and (b) corresponding mean-field model. The mean field is proportional to $J_{ij}\langle s\rangle$ where $\langle s\rangle$ is the average spin, that is, the magnetization.

function (5.3), but the large number of magnetic degrees of freedom makes this task challenging. For example, Ising magnets exhibit 2^N different magnetization states, where N is the number of atoms. The idea of the mean-field model (Panel 3) is to replace the exchange interaction with individual neighbors (J_{ij}) by an interaction field, so that the partition function reduces to that of a single atom (Fig. 5.8). The field is known as the effective, molecular, or *mean* field H_{eff}.

The mean field is proportional to the average magnetization, $\mathbf{H}_{eff} = \lambda\langle\mathbf{S}\rangle$, where λ is the mean-field of molecular-field constant. It was introduced by Weiss (1907), who aimed at explaining ferromagnetism but was unaware of its physical origin. Experimental values of λ are very high, typically several hundred tesla. This is because the mean field is created by exchange, which is much stronger than magnetostatic interactions. Note that atomic mean field considered in this chapter must not be confused with the much smaller mean-field constants used to describe micromagnetic phenomena, such as interactions between magnetic nanoparticles (Chapters 4 and 6). The mean-field model can also be derived directly from microscopic Hamiltonians and has then the character of an approximation.

5.3.1 Mean-field Hamiltonians

Let us, for the moment, consider ions with quenched orbital moment, so that $\mathbf{J} = \mathbf{S}$. The corresponding mean-field Hamiltonian

$$\mathsf{H} = -2\mu_o\mu_B\,\mathbf{S}\cdot\mathbf{H}_{eff} \tag{5.37}$$

corresponds to a paramagnetic ion in a magnetic field, and it is straightforward to calculate the average spin or "magnetization" $\langle\mathbf{S}\rangle = f(\mathbf{H}_{eff})$. An example of the function f is the above-introduced Brillouin function, which describes paramagnetic Heisenberg spins. However, in contrast to paramagnets, the field is proportional to the magnetization, $\mathbf{H}_{eff} = \lambda\langle\mathbf{S}\rangle$, and we must find self-consistent solutions of the equation $\langle\mathbf{S}\rangle = f(\lambda\langle\mathbf{S}\rangle)$. We will do this for a number of mean-field models.

Quantum-mechanical effects are easy to treat on a mean-field level, because H and $\mathbf{S}$ have the same and commute. However, the construction of mean-field Hamiltonians requires some care. When $\mathbf{J}\neq\mathbf{S}$, as in rare-earth ions, we must replace $2\mathbf{S}$ by $g\mathbf{J}$. Furthermore, we must take into account that exchange interactions act on the spin $\mathbf{S}$, whereas external magnetic fields act on $\mathbf{L} + 2\mathbf{S} = g\mathbf{J}$. Subtracting $\mathbf{L} + \mathbf{S} = \mathbf{J}$ from this equation yields the spin projection $(g-1)\mathbf{J}$, meaning that $\mathbf{S}^2 = S(S+1)$ must be replaced by the *de Gennes* factor

$$(g-1)^2 \, \mathbf{J}^2 = (g-1)^2 \, J(J+1) \tag{5.38}$$

As we will see, this factor enters the expression for the Curie temperature. The de Gennes factor is largest for elements in the middle of the rare-earth series, explaining why Gd is the rare-earth element with the highest ordering temperature (T_c). However, we will not dwell on this point and focus on spin Hamiltonians, briefly mentioning rare-earth ions as we go along.

To derive the mean-field Hamiltonian from microscopic Hamiltonians, we must transform interactions of the type $\mathbf{s}\cdot\mathbf{s}'$ into noninteracting mean-field terms as outlined in panel 3. Let us start from the identity

$$\mathbf{s}\cdot\mathbf{s}' = \mathbf{s}\cdot\langle\mathbf{s}'\rangle + \langle\mathbf{s}\rangle\cdot\mathbf{s}' - \langle\mathbf{s}\rangle\cdot\langle\mathbf{s}'\rangle + C \tag{5.39}$$

where $C = (\mathbf{s} - \langle\mathbf{s}\rangle)\cdot(\mathbf{s}' - \langle\mathbf{s}'\rangle)$ describes spin correlations. The first two terms on the right-hand side of (5.39) have correct structure, because the spins $\mathbf{s}$ and $\mathbf{s}'$ interact with mean fields created by neighboring moments $\langle\mathbf{s}'\rangle$ and $\langle\mathbf{s}\rangle$, respectively. The third term, $\langle\mathbf{s}\rangle\cdot\langle\mathbf{s}'\rangle$, can safely be ignored, because it amounts to a physically irrelevant shift of the energy zero which doesn't affect the partition function. The key assumption of the mean-field model is the replacement of the correlation term C by its average $\langle C\rangle = \langle\mathbf{s}\cdot\mathbf{s}'\rangle - \langle\mathbf{s}\rangle\cdot\langle\mathbf{s}'\rangle$. As the third term, $\langle C\rangle$ merely changes the energy zero.

In practice, the mean field is obtained by summation over all neighbors (J_{ij}). For equivalent neighbors, there is no need to distinguish between $\mathbf{s}\cdot\langle\mathbf{s}'\rangle$ and $\langle\mathbf{s}\rangle\cdot\mathbf{s}'$, although the correct mean field expression differs by a factor two from the naïve replacement $\mathbf{s}\cdot\mathbf{s}' \rightarrow \mathbf{s}\cdot\langle\mathbf{s}'\rangle$. For nonequivalent neighbors, the mean fields acting on different atoms are generally different. We will exploit this important feature in the discussion of alloys and disordered magnets.

5.3.2 Basic mean-field predictions

There are several methods of self-consistently solving mean-field equations. Let us consider the nearest-neighbor spin-$\frac{1}{2}$ Ising model, which exhibits two levels per atom, corresponding to $\uparrow$ and $\downarrow$ states. The same level structure is encountered in the spin-$\frac{1}{2}$ Heisenberg model, making the mean-field versions of both models algebraically equivalent. As shown in Section 1.3, the paramagnetism of the model is described by $\langle s\rangle = \tanh(h/k_B T)$, where $h = \mu_o \mu_B H$. The effective field is equal to $h + zJ_o\langle s\rangle$, where J_o is the exchange between nearest neighbors and z is the number of nearest neighbors. Replacing the external field by the effective field yields the mean-field equation of state

$$\langle s\rangle = \tanh\frac{h + zJ_o\langle s\rangle}{k_B T} \tag{5.40}$$

Since the hyperbolic tangent is a nonlinear function, this equation may have several solutions. We will see that these roots describe ferromagnetic and paramagnetic states.

Figure 5.2 shows that the zero-field magnetization $\langle s\rangle$ reaches zero at T_c. This can be used to determine the Curie temperature from (5.40). Exploiting that $\tanh(x) = x$

for small arguments yields the linearized equation of state

$$(k_\mathrm{B}T - zJ_\mathrm{o})\langle s\rangle = h \tag{5.41}$$

For $h = 0$ and small $\langle s\rangle$, corresponding to the $T \approx T_\mathrm{c}$, the solution of this equation is $k_\mathrm{B}T = zJ_\mathrm{o}$, or

$$T_\mathrm{c} = \frac{zJ_\mathrm{o}}{k_\mathrm{B}} \tag{5.42}$$

This equation shows that the Curie temperature is, essentially, equal to the exchange energy.

Expanding (5.40) into powers of $\langle s\rangle$ reproduces the Landau theory, Section 5.1.4. For example, (5.41) predicts the susceptibility $\chi \approx \langle s\rangle/h$ to scale as $1/(T - T_\mathrm{c})$, that is, the critical exponent $\gamma = 1$. More generally, Landau and mean-field models exhibit the same critical exponents and are physically largely equivalent.

To calculate the spontaneous magnetization $\langle s\rangle$, we rewrite (5.40) as

$$T = \frac{h + zJ_\mathrm{o}\langle s\rangle}{k_\mathrm{B}\ \mathrm{atanh}\langle s\rangle} \tag{5.43}$$

This yields the temperature as function of $\langle s\rangle$, and by interchanging the T and $\langle s\rangle$ axes we obtain the familiar graph $\langle s\rangle(T)$. Figure 5.9 illustrates this simple procedure for $h = 0$. Below T_c, the paramagnetic states is unstable, as illustrated in Fig. 5.9(b).

Another method determines the magnetization by finding the intersection of the linear and nonlinear functions on the right- and left-hand sides of (5.40), respectively. Figure 5.10 illustrates how this graphical method works. Below T_c, there are three solutions, two ferromagnetic solutions $\langle s\rangle = \pm M/M_\mathrm{s}$ and one unstable paramagnetic solution $\langle s\rangle = 0$. Above T_c, there is only one solution, namely $\langle s\rangle = 0$. The graphical

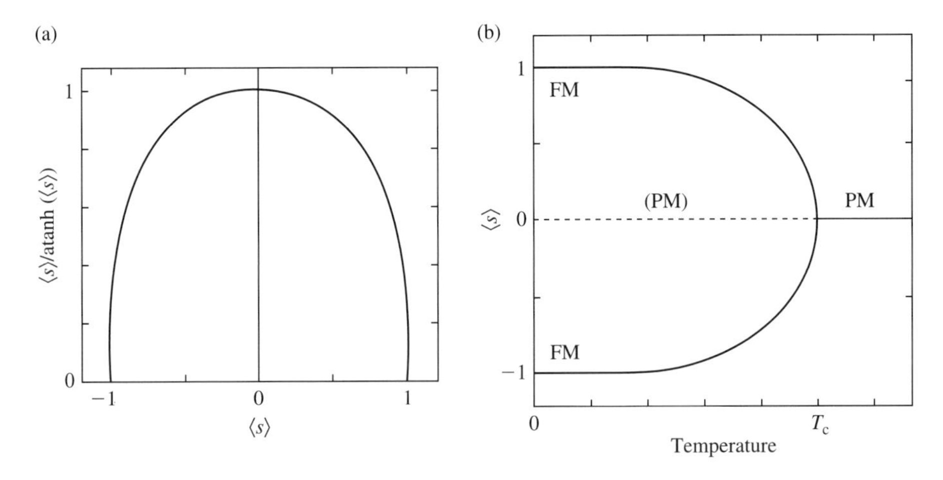

Fig. 5.9 Explicit zero-field solution of the mean-field equation (5.40): (a) temperature as a function of magnetization and (b) magnetization as a function of temperature. The lower branch of (b) is redundant and usually omitted.

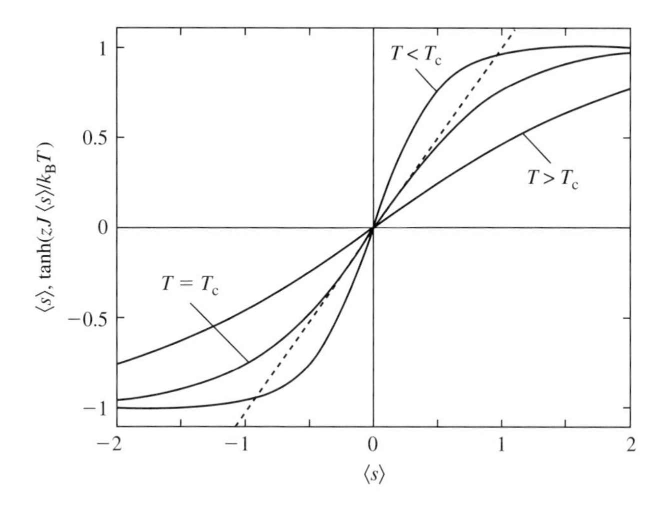

Fig. 5.10 Graphical solution of mean-field equations. The slope of the dashed lines equals the ratio of thermal energy, $k_B T$, to interatomic exchange, zJ_o.

method is less elegant than the inversion method shown in Fig. 5.10 but more practical unless the inverse function is explicitly known.

5.3.3 *Ornstein-Zernike correlations

On a mean-field level, the correlation length ξ and the corresponding exponent ν are obtained by the Ornstein-Zernike theory. The idea is to consider the linearized mean-field equation

$$k_B T \langle s_i \rangle = h_i + \sum_j J_{ij} \langle s_j \rangle \tag{5.44}$$

and to exploit that $\chi_{ik} = \partial \langle s_i \rangle / \partial h_k$. The fluctuation-response theorem (5.10) then yields the correlation function $C_{ij} = C(\mathbf{r}_i - \mathbf{r}_j)$ in matrix form:

$$C_{ij} = \frac{1}{\delta_{ij} - J_{ij}/k_B T} \tag{5.45}$$

where δ_{ij} is the Kronecker symbol (A.2.2). From this equation, the correlation function can be obtained as a series, based on $1/(1-x) = 1 + x + x^2 + \ldots$, but this method is very cumbersome.

To actually calculate C_{ij}, it is more convenient to diagonalize the matrix J_{ij}. In homogeneous magnets, the eigenmodes are diagonal in k-space and $C(\mathbf{k}) = 1/(1 - J_{\mathbf{k}}/k_B T)$. For hypercubic lattices ($z = 2d$), the relevant long-wavelength limit is, $J_{\mathbf{k}} = zJ - Ja^2k^2$ and

$$C(\mathbf{k}) = \frac{T}{T - T_c + Ja^2k^2/k_B} \tag{5.46}$$

Inverse Fourier transformation yields the leading correlation term $C(r) \sim \exp(-r/\xi)$ where $\xi \sim 1/(T - T_c)^{1/2}$, that is, $\nu = 1/2$. Alternatively, the same exponent is obtained by dimensional analysis of (5.46), exploiting that $k^2 \sim T - T_c$ and $\xi \sim 1/k$.

The Ornstein-Zernike approach has its historical origin in the statistical mechanics of gases and liquids. The theory, developed around 1914, focuses on the distinction between direct and total correlations (Goodstein 1975). Direct correlations correspond to short-range interactions, such as nearest-neighbor exchange J_{ij}, whereas the total correlation function (5.45–46) involves both direct correlations and correlations mediated by atoms other than i and j. This explains long-range magnetic order in terms of short-range exchange.

5.3.4 Magnetization and Curie temperature

Mean-field predictions depend on the underlying spin model. In the Heisenberg model, the magnetization is obtained from the Brillouin function B_S, as exemplified by $B_{1/2}(x) = \tanh(x)$. Figure 5.11 shows the spontaneous magnetization for some values of S. For any S, the approach to saturation is described by $M \sim \sqrt{T_c - T}$, corresponding to the mean-field critical exponent $\beta = \frac{1}{2}$. At low temperatures, $\mathrm{d}M/\mathrm{d}T = 0$, except for the classical limit $S = \infty$, where S_z/S forms a continuum.

The Curie temperature of the mean-field Heisenberg model is obtained by linearizing the Brillouin function. The result is

$$T_c = \frac{2S(S+1)}{3k_B} zJ_o \tag{5.47}$$

As above, z is the number of nearest neighbors. This important equation shows that the Curie temperature increases with the spin S, with the number z of interacting atomic neighbors, and with the exchange constant J_o.

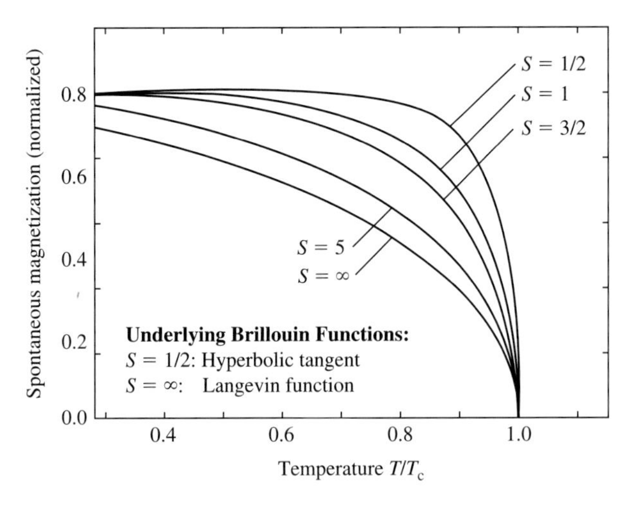

Fig. 5.11 Spontaneous magnetization of the mean-field Heisenberg model. The spin-$\frac{1}{2}$ Ising model exhibits the same temperature dependence as the spin-$\frac{1}{2}$ Heisenberg model.

Incidentally, for $S = \frac{1}{2}$, the Curie temperature $zJ_o/2k_B$ differs from (5.42) by a factor 2. To explain the difference, we recall our convention that ↑↓ and ↑↑ states have an energy-level splitting $\pm J_o$. This is trivial in the Ising model, where $s_{i/j} = \pm 1$ ensures $s_i s_j = \pm 1$. In the spin-$\frac{1}{2}$ Heisenberg model, $\mathbf{S}_i \cdot \mathbf{S}_j$ yields $-\frac{1}{4} \pm \frac{1}{2}$, so that we need a factor 2 to reproduce the level splitting ± 1. For Heisenberg spins $S > \frac{1}{2}$, the level splitting is $\pm(S^2 + S/2)$ and J_o can no longer be interpreted as an energy splitting $\pm J_o$ between FM and AFM spin configurations. Redefinition of J_o yields

$$T_c = \frac{(S+1)}{3(S+\frac{1}{2})k_B} zJ'_o \tag{5.48}$$

where $J'_o = (2S^2 + S)J_o$ ensures an FM-AFM energy-level splitting of $\pm J'_o$. Equation (5.48) is useful for discussing the transition to the classical limit $S = \infty$. In terms of J'_o, the classical Curie temperature is equal to $zJ'_o/3k_B$, as compared to $zJ'_o/2k_B$ for $S = \frac{1}{2}$.

Rare-earth magnetism is characterized by rigid spin-orbit coupling ($\mathbf{J} = \mathbf{S} + \mathbf{L}$), so that the level splitting of the rare-earth ions is $J_z = -J, \ldots J-1, J$. This corresponds to the de Gennes factor (Section 2.2.4) and yields

$$T_c = \frac{2(g-1)^2\, J(J+1)\, zJ_o}{3k_B} \tag{5.49}$$

Here the total moment (J) and the exchange (J_o) must not be confused. Partially quenched systems are intermediate between (5.48) and (5.49) and generally very difficult to treat. An approximate approach is to derive spin Hamiltonians by integrating over orbital degrees of freedom.

Note the orbital motion of the electrons ($\mathbf{L}$) depends not only on the spin (via spin-orbit coupling) but also on the crystalline environment. This gives rise to anisotropic interaction terms of the type $\cos(\theta_1)\cos(\theta_2)$, as contrasted to isotropic Heisenberg exchange $\cos(\theta_1 - \theta_2)$. Physically, the degree of quenching depends on the orientation of the orbital moment—it may be large along the c-axis of the crystal ($\mathbf{J} = \mathbf{S} + \mathbf{L}$), but small in the a–b plane ($\mathbf{J} = \mathbf{S}$), or vice versa.

5.3.5 *Mean-field Curie temperature of n-vector models

A very simple mean-field Curie temperature expression is obtained for the classical n-vector model. We consider the normalized magnetization $s_z = S_z/S$ and assume an FM-AFM splitting of $\pm J_o$, corresponding to a mean field $zJ_o\langle s_z\rangle$. In addition, we must take into account that the n-vector model has n spin components, such as s_x, s_y, and s_z in the Heisenberg model. The paramagnetic partition function $Z = \int_s \exp(hs_z/k_BT)\,\mathrm{d}\mathbf{s}$, where the integration is performed over all classical vectors $\mathbf{s}$ for which $\mathbf{s}^2 = 1$. This amounts to the integration over the surface ω_n of a n-dimensional unit sphere. The surface has an $n-1$ dimensional area ω_n, with the special cases $\omega_1 = 2$ (corresponding to $s_z = \pm 1$), $\omega_3 = 2\pi$ (circumference of circle), and $\omega_3 = 4\pi$ (surface of sphere).

To simplify the calculation, we exploit that $h = zJ_o\langle s_z\rangle$ is very small near T_c. This yields the expansion

$$\exp\frac{hs_z}{k_BT} = 1 + \frac{hs_z}{k_BT} + \frac{1}{2}\frac{h^2s_z^2}{2k_B^2T^2} + \cdots \tag{5.50}$$

and the partition function

$$Z = \omega_n\left(1 + \frac{h\langle s_z\rangle_s}{k_BT} + \frac{1}{2}\frac{h^2\langle s_z^2\rangle_s}{k_B^2\,T^2}\right) \tag{5.51}$$

where the averages $\langle\cdots\rangle_s$ refer to the whole unit sphere. By symmetry, $\langle s_z\rangle_s = 0$, whereas $\langle s_z^2\rangle_s = 1/n$, because the $s_z^2 + \cdots + s_n^2 = 1$ and the n spin components are equivalent on the unit sphere. This yields $Z = \omega_n(1 + h^2/2nk_B^2T^2)$, $\langle s_z\rangle = h/nk_BT$, and

$$T_c = \frac{1}{n}\frac{zJ_o}{k_B} \tag{5.52}$$

Special cases are the Ising model ($n = 1$) and the Heisenberg model ($n = 3$), corresponding to (5.42) and to the classical limit ($S = \infty$) of (5.48).

The *anisotropic* n-vector model is obtained by adding an anisotropy $-K_1s_z^2V_o$, where V_o is the volume per atom. The relevant lowest-order terms of the partition function are

$$Z = \omega_3\left(1 + \frac{K_1V_o\langle s_z^2\rangle_s}{k_B\,T} + \frac{1}{2}\frac{h^2\langle s_z^2\rangle_2}{k_B^2\,T^2} + \frac{1}{2}\frac{K_1V_oh^2\langle s_z^4\rangle_s}{k_B^3\,T^3}\right) \tag{5.53}$$

Exploiting $\langle s_z^2\rangle_s = 1/n$ and $\langle s_z^4\rangle_s = 3/n(n+2)$ we obtain the Curie temperature

$$T_c = \frac{1}{n}\frac{zJ_o}{k_B}\left(1 + \frac{2(n-1)K_1V_o}{(n+2)zJ_o}\right) \tag{5.54}$$

Typically, anisotropy energies per atom are much smaller than exchange energies, so that the anisotropy yields only small corrections to the mean-field Curie temperature. For the Ising model ($n = 1$), the correction is exactly zero, because its anisotropy is infinite by definition and any additional anisotropy leaves T_c unchanged. Comparing the Ising model ($T_c = zJ_o/k_B$) with the anisotropic Heisenberg model ($T_c \approx zJ_o/3k_B$) we find that K_1 must assume huge values to make a Heisenberg model Ising-like. It is therefore a poor approximation to describe typical anisotropic ferromagnets by the Ising model. An exception is the vicinity of the critical point, where symmetry is more important than the strength of the anisotropy (Section 5.4).

5.3.6 Two-sublattice magnetism

Aside from ferromagnetism, there are other types of magnetic order, such as antiferro- and ferrimagnetism. In Section 2.3.3 we have introduced these structures and briefly discussed their zero-temperature behavior, mentioning that they can be described in terms of magnetic sublattices. For simplicity, we focus on Ising magnets containing two sublattices A and B. The Hamiltonian (or energy) is

$$\mathsf{H} = -J_{AA}\,s_As_A - J_{AB}\,s_As_B - J_{BB}\,s_B\,s_B - h_As_A - h_Bs_B \tag{5.55}$$

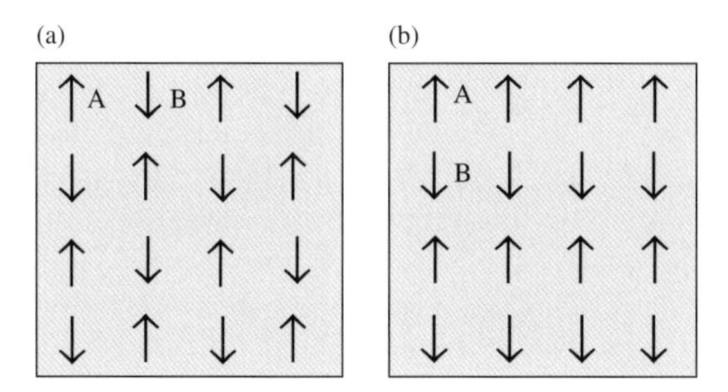

Fig. 5.12 Two antiferromagnetic spin structures on a square lattice: (a) checkerboard configuration and (b) stripe configuration.

where J_{AA} and J_{BB} are intrasublattice exchange constants and J_{AB} describes the intersublattice interactions. The three exchange constants are obtained by summation $\sum_j J_{ij}$ over all atomic sites that belong to the corresponding sublattice. In the simplest case, one restricts the exchange (and the summation) to nearest neighbors. Figure 5.12 shows two antiferromagnetic spin structures on a square lattice. In the checkerboard configuration (a), all nearest neighbors belong to different sublattices, so that $J_{AB} \neq 0$ but $J_{AA} = J_{BB} = 0$. By contrast, the stripe configuration (b) exhibits both intersublattice and intrasublattice exchange.

Let us assume that the spin configuration and the exchange constants J_{AA}, J_{BB}, and J_{AB} are known. The derivation of the mean-field Hamiltonian is straightforward and yields $\mathsf{H} = \mathsf{H}_A + \mathsf{H}_B$, where

$$\mathsf{H}_A = -(h_A + J_{AA}\langle s_A\rangle + J_{AB}\langle s_B\rangle)s_A \tag{5.56a}$$

$$\mathsf{H}_B = -(h_B + J_{AB}\langle s_A\rangle + J_{BB}\langle s_B\rangle)s_B \tag{5.56b}$$

These equations show that there are two effective fields, one for each sublattice, and that each effective field contains both intra- and intersublattice contributions. The equation of state is similar to (5.40), except that s_A and s_B must be treated separately:

$$\langle s_A\rangle = \tanh \frac{h_A + J_{AA}\langle s_A\rangle + J_{AB}\langle s_B\rangle}{k_B T} \tag{5.57a}$$

$$\langle s_B\rangle = \tanh \frac{h_B + J_{AB}\langle s_A\rangle + J_{BB}\langle s_B\rangle}{k_B T} \tag{5.57b}$$

Here h_A and h_B are the average magnetic fields acting on the sublattices. In practice, it is fair to assume that $h_A = h_B = h$.

In a few cases, it is possible to find exact solutions of these mean-field equations. An example is the checkerboard antiferromagnet shown in Fig. 5.11(a), where $J_{AB} < 0$ and $J_{AA} = J_{BB} = 0$. For $h = 0$, symmetry suggests $\langle s_B\rangle = -\langle s_A\rangle$, and exploiting $\tanh(-x) = -\tanh(x)$ we obtain $\langle s_A\rangle = \tanh(|J_{AB}|\langle s_A\rangle/k_B T)$. This is the mean-field equation for a ferromagnet, so that the sublattice magnetization $\langle s_A\rangle$ has the temperature dependence of the spin-$\frac{1}{2}$ Ising model (Fig. 5.9). However, the net magnetization $\langle s\rangle = \langle s_B\rangle + \langle s_A\rangle$ is zero at any temperature, as expected for an antiferromagnet. Sublattice magnetizations can be probed, for example, by neutron scattering, because

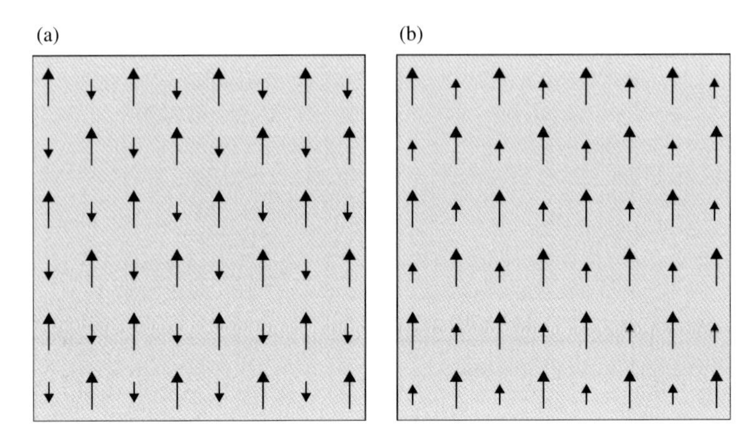

Fig. 5.13 Two-sublattice magnetism: (a) ferrimagnet and (b) two-sublattice ferromagnet.

neutrons carry a spin moment. The temperature at which the sublattice magnetizations vanish is known as the Néel temperature. In the present case, $T_N = -J_{AB}/k_B T$. Antiferromagnetism is a common phenomenon, especially in oxides. Examples are FeO with $T_N = 198\,K$, CoO with $T_N = 291\,K$, and MnF_2 with $T_N = 67\,K$.

A more complex situation is encountered if the sublattices are associated with different kinds of atoms, as contrasted to the spontaneous sublattice formation in Fig. 5.12. For example, rare-earth transition-metal intermetallics contain two or more sublattices with parallel or antiparallel magnetizations and different moments per atoms. Two-sublattice magnets with parallel and antiparallel moments are known as two-sublattice ferromagnets (FM) and ferrimagnets (FI) respectively. Figure 5.13 illustrates the two spin structures. Examples are the rare-earth transition-metallics $SmCo_5$ (FM) and $DyCo_5$ (FI) and the oxides Fe_3O_4 and $BaFe_{12}O_{19}$ (both FI). Ferrimagnetic moments are relatively small, for example 17.1 and 11.3 μ_B per formula unit for $Gd_2Fe_{14}B$ and $Dy_2Fe_{14}B$, respectively, as compared to 37.7 μ_B for the ferromagnet $Nd_2Fe_{14}B$. This is one reason for using Nd in 2:14:1-type permanent magnets.

To describe ferrimagnets and two-sublattice ferromagnets, we must consider arbitrary exchange constants J_{AA}, J_{BB}, and J_{AB}. In general, $J_{AB} \neq J_{BA}$, but since this merely changes J_{AB}^2 in the equations below to $J_{AB}J_{BA}$, we assume that $J_{AB} = J_{BA}$. Linearization of (5.57) yields the zero-field equations

$$k_B T\langle s_A\rangle = h_A + J_{AA}\,\langle s_A\rangle + J_{AB}\,\langle s_B\rangle \tag{5.58a}$$

$$k_B T\langle s_B\rangle = h_B + J_{AB}\,\langle s_A\rangle + J_{BB}\,\langle s_B\rangle \tag{5.58b}$$

Putting $h_B = h_B = 0$ and substituting $\langle s_B\rangle$ from (5.58a) into (5.58b) yields a quadratic equation for T whose larger root is the Curie temperature T_c. Alternatively, (5.58) establishes an eigenvalue problem with two eigenvalues obtained by diagonalizing the 2×2 interaction matrix, which amounts to the solution of a quadratic secular equation. The result is

$$T_c = \frac{1}{2k_B}(J_{AA} + J_{BB}) + \frac{1}{2k_B}\sqrt{(J_{AA} - J_{BB})^2 + 4J_{AB}^2} \tag{5.59}$$

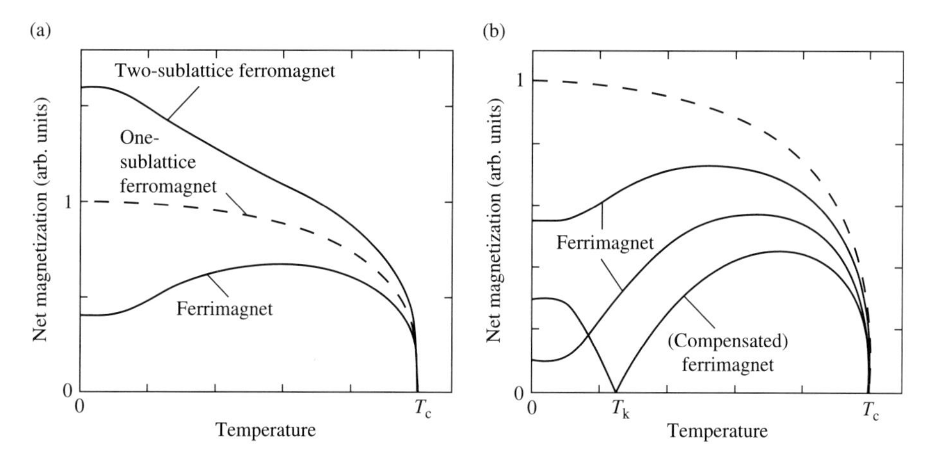

Fig. 5.14 Temperature dependence of the magnetization of ferro- and ferrimagnets: (a) comparison of ferro- and ferrimagnets and (b) compensation in ferrimagnets. The respective scenarios (a) and (b) are frequently encountered in transition-metal-rich rare-earth intermetallics and oxides (garnets).

The second root T^* of the secular equation has no meaning, because $\langle s_A \rangle$ and $\langle s_B \rangle$ are generally large and the linearized equation of state (5.58) is no longer valid below T_c. However, the paramagnetic response depends on T^*, especially in antiferromagnets. This is because the external field interacts with the mode $\langle s_A \rangle + \langle s_B \rangle$, as compared to the mode $\langle s_A \rangle - \langle s_B \rangle$ responsible for the Néel transition. As a consequence, paramagnetic or Curie-Weiss ordering temperatures θ are often larger than Néel temperatures.

Coupled mean-field equations such as (5.57) have been used to describe a variety of systems, most notably oxides (Smart 1966, Binek 2003) and intermetallics (Coey 1996). Figure 5.14 shows typical magnetization curves. In transition-metal-rich rare-earth intermetallics, such as $SmCo_5$, $DyCo_5$, and $Nd_2Fe_{14}B$, the relatively small rare-earth magnetization couples parallel (light rare earths) or antiparallel (heavy rare earths) to the transition-metal magnetization. The coupling between the rare-earth sublattice (B) and the transition-metal sublattice (A) is relatively weak, $J_{AB} \ll J_{AA}$, and the rare-earth intrasublattice exchange J_{BB} is negligible. The corresponding Curie temperature, $T_c = J_{AA}(1 + J_{AB}^2/J_{AA}^2)/k_B$, is enhanced for rare earths in the middle of the series, because J_{AB}^2 is proportional to the de Gennes factor. The weak exchange coupling of the rare-earth atoms is also responsible for the pronounced temperature dependence of the rare-earth sublattice magnetization, that is, the deviation from the dashed line in Fig. 5.14(a), which is reminiscent of a paramagnet in an external magnetic field.

The magnetization depends not only on the exchange but also on the number of atoms per sublattice and on the moments per atom. When ferrimagnetic sublattices have comparable magnitudes, the different temperature dependences of sublattice magnetizations may give rise to *compensation*. This effect, illustrated in Fig. 5.14(b), is particularly common in iron garnets such as $Gd_3Fe_5O_{12}$. Partial replacement of the Gd by a nonmagnetic rare earth, such as Y, reduces the compensation temperature and yields, eventually, a ferrimagnet without compensation temperature.

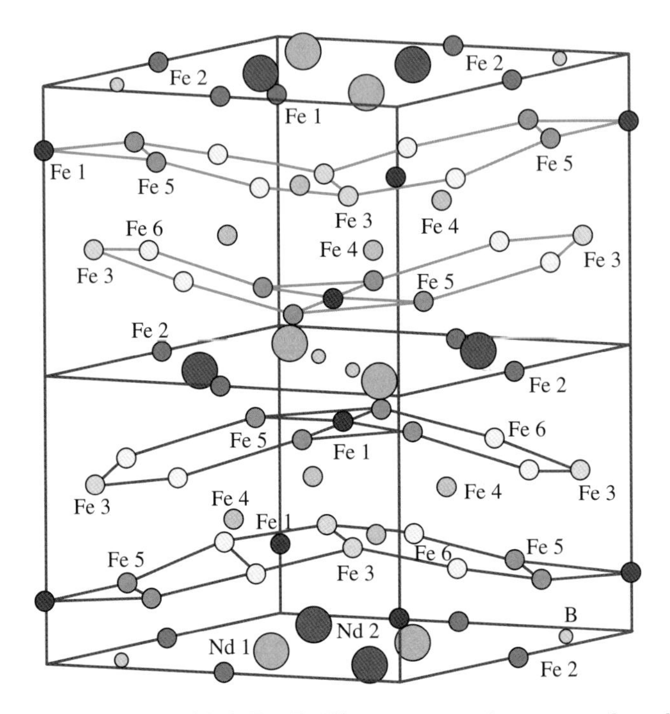

Fig. 5.15 Crystal structure of $Nd_2Fe_{14}B$. The symmetry is tetragonal, and the structure forms for a variety of $R_2T_{14}Z$ compositions, where R is a rare earth (big atoms), T is Fe or Co (intermediate size) and Z is boron or carbon (small atoms).

Typical rare-earth transition-metal intermetallics (RE-TM), contain more than two sublattices. Figure 5.15 hows the crystal structure of the $Nd_2Fe_{14}B$, which forms the main phase of neodymium-iron-boron permanent magnets. The main distinction is that between rare-earth and transition-metal atoms, but equation (5.57) is easily generalized to N sublattices. The mean-field Curie temperature is then obtained as the largest eigenvalue of the $N \times N$ matrix containing the intra- and intersublattice exchange constants. However, the basic distinction between rare-earth and transition-metal sites remains unchanged.

5.3.7 Merits and limitations of mean-field models

The mean-field model combines fascinating simplicity with powerful and transparent predictions. It captures the physical origin of magnetic order, namely interatomic exchange, and is widely used in magnetism, both as an explicit model and in the form of implicit model assumptions. In many cases, mean-field analysis is the first step towards the understanding of the investigated system, and in some cases it is the only feasible method. The mean-field model provides a reasonable overall description of the magnetization but breaks down at low temperatures, where M_s is reduced by spin waves, and close to T_c, where long-range critical fluctuations interfere.

At low temperatures, the mean-field version of the quantum-mechanical Heisenberg model predicts an exponential approach to zero-temperature saturation, $M_o - M_s \sim \exp(\Delta E/k_B T)$. In reality, spin waves yield a power law approach, $M_o - M_s \sim T^A$.

A well-known example is Bloch's law, where $A = 3/2$. Spin waves exhibit both equilibrium and nonequilibrium features and will be treated in Section 6.1.

Near the critical point, there are both quantitative and qualitative deviations from the mean-field predictions. Physically, the mean-field embedding of spins in an average environment is not necessarily realistic, because the correlation length diverges near the critical point. This is seen most easily by considering the paramagnetic phase, where the average magnetization is zero but spins may be located in correlated regions with predominant $\uparrow$ or $\downarrow$ character.

A widespread feature of mean-field models is an overestimation of the trend towards ferromagnetism. A good example is the one-dimensional Ising model, where the mean-field prediction $T_c = 2zJ/k_B$ is in striking contrast to the exact solution $T_c = 0$. The same trend is encountered in quantum-mechanical mean-field models such as the Stoner model (Section 2.4.3), which may incorrectly predict ferromagnetism because the introduction of average occupation numbers $\langle n_i \rangle$ yields a crude distinction between ferro- and paramagnetism but is unable to trace complicated spin structures, such as fluctuating ferromagnetic spin blocks.

A specific problem is the dependence of the Curie temperature on the size and dimensionality of the magnets. In the mean-field approximation, the Curie temperature depends on the number z of neighbors but is independent of the real-space dimensionality of the magnet. In reality, the Curie temperature is strongly dimensionality-dependent, irrespective of z, and there is no ferromagnetism in finite magnets, where mean-field models predict a nonzero Curie temperature. This is because thermally activated magnetization reversal destroys the statistically averaged or spontaneous magnetization of finite-size magnets. The thermodynamic limit of infinite ferromagnets is nonergodic, that is, thermally activated reversed domains collapse after a finite time and never lead to the magnetization reversal of the whole magnet. In finite magnets, there exist phenomena reminiscent of but not equivalent to phase transitions.

5.4 Critical behavior

Summary The behavior of the magnetization near the critical point is largely determined by long-range thermodynamic fluctuations. Mean-field models provide a poor description of this regime, because they assume that the interatomic exchange translates into a local exchange field. Critical behavior is described by scaling laws such as $M_s \sim (T_c - T)^\beta$, $\chi \sim |T - T_c|^{-\gamma}$, and $\xi \sim |T - T_c|^{-\nu}$, where the exponents generally differ from the mean-field predictions. In addition, critical fluctuations tend to reduce the Curie temperature. The deviations from mean-field behavior depend on the spatial dimensionality d of the magnet and on the considered model (spin dimensionality n and long- or short-range character of the exchange). However, critical exponents are independent of structural details such as the number z of nearest neighbors. By gauging the role of fluctuations as a function of d, one finds that mean field exponents are essentially exact for $d \leq 4$ (Ginzburg criterion). There are very few exact solutions in two or more dimensions, most notably the Onsager solution for the two-dimensional Ising model. A famous approach to the treatment critical fluctuations is renormalization-group analysis, where one exploits that the correlation length diverges at the critical point. The idea is to iteratively rescale the size of the magnet, $\xi' = \xi/b$, and to exploit the fact that $\xi' = \xi/b = \infty$ at the critical point.

At the Curie temperature, the spontaneous magnetization goes to zero, as contrasted to an asymptotic decrease. This is striking, because the partition function $Z = \sum_\mu \exp(-E_\mu/k_\mathrm{B}T)$ is smooth at all temperatures, as is the magnetization $M \sim \partial \ln Z/\partial H$. What is the origin of the sharp phase transition at T_c, and how can we explain the singularity of the magnetization at the Curie point? So far, we have restricted ourselves to a phenomenological discussion of phase transitions and to the mean-field model, where the existence of a spontaneous magnetization is assumed rather than derived. To provide an in-depth explanation of the Curie transition, we must analyze the partition function.

As remarked by Lee and Yang (1952), smooth partition functions may give rise to singularities in the thermodynamic limit of *infinite* crystals. This is seen most easily by considering the complex plane. The function exp(x) has no real zero, but $\exp(x) + \exp(-x) = 2\cos(\mathrm{i}x)$ indicates that there are roots elsewhere in the complex plane. The number of roots increases with the size of the magnet, and some roots may approach and finally touch the real axis, thereby establishing a phase transition. An alternative interpretation of the thermodynamic limit is that thermal fluctuations in the paramagnetic phase, $T > T_\mathrm{c}$, give rise to regions with predominant ↑ or ↓ character. The size of these correlated regions increases as the temperature approaches T_c. To distinguish between finite-size fluctuations and true ferromagnetism, it is therefore necessary to consider infinite magnets. This aspect of ferromagnetism is ignored in the mean-field model, where all spins experience the same local environment.

Finite-temperature critical behavior, as discussed in this section, must be distinguished from zero-temperature *quantum-critical behavior*. For example, we will see that thermal fluctuations ($T > 0$) destroy long-range ferromagnetic order in typical one-dimensional magnets, but at T = 0, quantum fluctuations tend to have a similar effect (Schofield 1999). The quantum critical point is tuned by control parameters—such as magnetic field, lattice constant and dopant concentration—and the finite-temperature response of the materials obeys scaling laws reminiscent of those discussed below.

5.4.1 One-dimensional models

To analyze the failure of the mean-field approximation and to investigate the nature of critical fluctuations, we consider the one-dimensional Ising model (1925). In the absence of an external magnetic field, the energy (or Hamiltonian) of the Ising chain

$$E = -\sum_\mathrm{i} J s_\mathrm{i}\, s_\mathrm{i+1} \tag{5.60}$$

where J is the nearest-neighbor exchange and $i = 1 \ldots N$ extends to $N = \infty$. Since each spin has two neighbors, the mean-field Curie temperature $T_\mathrm{c} = 2J/k_\mathrm{B}$. To find the exact Curie temperature, we must evaluate the partition function

$$Z = \sum_{\{\mathrm{s}\}} \exp\left(\frac{\sum_\mathrm{i} J s_\mathrm{i}\, s_\mathrm{i+1}}{k_\mathrm{B}T}\right) \tag{5.61}$$

where the summation includes all 2^N spin configurations. This is done most conveniently by introducing bond variables $\tau_i = s_i\, s_{i+1} = \pm 1$ rather than $s_i = \pm 1$, so that

$$Z = \sum_{\{\tau\}} \exp\left(\frac{\sum_i J\tau_i}{k_B T}\right) \tag{5.62}$$

This is actually the partition function for a paramagnetic gas where the magnetic field has been replaced by the exchange. For paramagnets, $T_c = 0$, and the same is true for (5.61–62).

A simple interpretation of the paramagnetism of the Ising chain is that a single broken bond, $\tau_i = -1$, destroys the magnetization of the whole chain. The probability that a given bond is broken, $\exp(-J/k_B T)$, is very small at low temperature, but this is compensated by the large number N of bonds. The correlation length, that is, the length of the ↑ and ↓ blocks between broken bonds, increases exponentially with decreasing temperature but remains finite at any nonzero temperature, so that $T_c = 0$.

The present argumentation applies to any one-dimensional system, because the energy to break a bond is always finite irrespective of the anisotropy. This includes nanostructures (Shen *et al.* 1997), although the energy is proportional to the cross section of the wire and may be so high that the magnet behaves like a bulk magnet. A generalization to two or more dimensions is not possible, because the spin variables can no longer be mapped onto bond variables. For example, on a 2×2 square (loop with four spins), there exists the spin configuration ↑↑↓↑ (on spin down), but the bond configuration $+ + - +$ (one broken bond) is unphysical, because spin switching in closed loops creates *pairs* of broken bonds.

5.4.2 Superparamagnetic clusters

Criticality is closely related to the thermodynamic limit, $N \to \infty$. The partition function of finite-size magnets is non-singular, and there is no Curie transition at $T_c > 0$ and $H = 0$. In fact, finite-size magnets exhibit a smooth decrease of finite-field magnetization with increasing temperature, and the spontaneous magnetization ($H = 0$) vanishes. Dynamically, there may be ferromagnetic correlations inside the magnet, but the net moment fluctuates between ↑ and ↓ moment orientations. This is known as *ergodicity*, meaning that the phase space is fully explored. Infinite ferromagnets are nonergodic, because thermally excitated ↓ regions in a ↑ ferromagnet are unstable and collapse after some time, never growing to infinite size.

It is instructive to see how the magnetization of small clusters depends on the cluster size (number N of atoms) and on the magnetic field. Here we focus on the equilibrium magnetization; the dynamics of magnetic clusters will be discussed in Section 6.4. Let us consider Ising-spin clusters ($S = \frac{1}{2}$) with nearest-neighbor exchange. The most trivial case is an isolated spin without exchange coupling ($N = 1$). We have solved this problem in Section 1.3, but is convenient to consider the partition function

$$Z = \exp\left(\frac{h}{k_B T}\right) + \exp\left(\frac{-h}{k_B T}\right) \tag{5.63a}$$

where $h = \mu_o \mu_B H$. The derivative of Z with respect to h yields the magnetization

$$\langle s \rangle = \frac{\exp\left(\frac{h}{k_B T}\right) - \exp\left(\frac{-h}{k_B T}\right)}{\exp\left(\frac{h}{k_B T}\right) + \exp\left(\frac{-h}{k_B T}\right)} \tag{5.64a}$$

that is, $\langle s \rangle = \tanh(h/k_B T)$. We see that $\langle s \rangle = 0$ for $h = 0$, indicating that the spontaneous magnetization vanishes.

The simplest nontrivial case is that of dumbbell spins ($N = 2$), with the four spin configurations $\uparrow\uparrow$, $\uparrow\downarrow$, $\downarrow\uparrow$, and $\downarrow\downarrow$. The partition function is

$$Z = \exp\left(\frac{2h + J}{k_B T}\right) + 2\exp\left(\frac{-2J}{k_B T}\right) + \exp\left(\frac{-2h + J}{k_B T}\right) \tag{5.63b}$$

and the magnetization

$$\langle s \rangle = \frac{\exp\left(\frac{(2h + J)}{k_B T}\right) - \exp\left(\frac{(-2h + J)}{k_B T}\right)}{\exp\left(\frac{2h + J}{k_B T}\right) + 2\exp\left(\frac{-2J}{k_B T}\right) + \exp\left(\frac{-2h + J}{k_B T}\right)} \tag{5.64b}$$

Since $\uparrow\downarrow$ and $\downarrow\uparrow$ are degenerate, there are only three terms in (5.63b). A square containing $N = 4$ atoms has $2^4 = 16$ different spin configurations, and

$$Z = \exp\left(\frac{4J + 4h}{k_B T}\right) + 4\exp\left(\frac{2h}{k_B T}\right) + 2\exp\left(\frac{4J}{k_B T}\right) + 4$$
$$+ 4\exp\left(\frac{-2h}{k_B T}\right) + \exp\left(\frac{4J - 4h}{k_B T}\right) \tag{5.65c}$$

With increasing size, the partition function becomes rapidly more complicated, and for large systems the determination of the partition function is cumbersome.

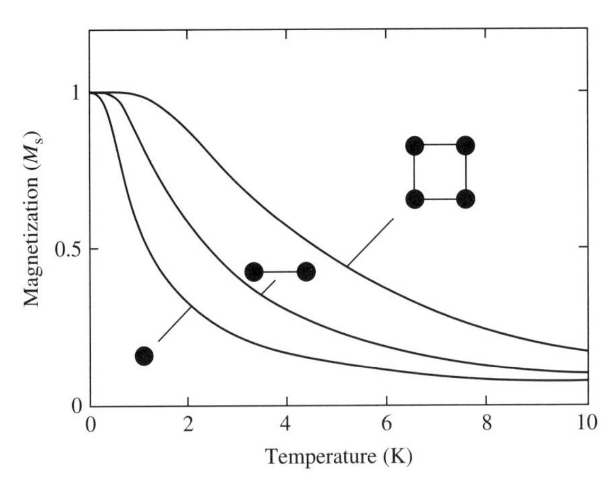

Fig. 5.16 Magnetization of small Ising spin clusters. The applied field is 10 T [100 kOe], and the strength of the exchange $Jk_B = 5$ K.

Figure 5.16 shows the temperature dependence of the magnetization for small Ising clusters. As expected for finite systems, the temperature dependence is smooth, without phase-transition singularity. In zero field, $\langle s(T)\rangle = 0$, corresponding to $T_c = 0$. It is interesting to note that the magnetization breaks down at low temperatures, in spite of the relatively large field, $\mu_o H = 10$ T. This is because $h = \mu_o \mu_B H$ contains the small Bohr magneton. We have encountered the same phenomenon in the context of Curie and Curie-Weiss paramagnetism (Section 5.6).

In the limit of strong exchange coupling ($J = \infty$), only the fully aligned ↑ and ↓ configurations contribute to the partition function, so that

$$Z = \exp\left(\frac{Nh}{k_B T}\right) + \exp\left(\frac{-Nh}{k_B T}\right) \tag{5.65}$$

The corresponding magnetization $\langle s\rangle = \tanh(Nh/k_B T)$ is that of a single paramagnetic spin, except that the effect of the field is enhanced by a factor N. Alternatively, in a given field, the temperature range is enhanced from 0.672 K/T to $N \cdot 0.672$ K/T. This is one aspect of *superparamagnetism* and explains why small magnetic nanoparticles ($N \sim 1000$) are easily spin-polarized at room temperature. Superparamagnetic behavior is restricted to temperatures $T \ll J/k_B$, because the simple partition function (5.65) relies on full ↑ or ↓ spin alignment throughout the cluster or particle. In fact, nanoparticles are often Heisenberg- rather than Ising-like, and the use of Langevin functions is more adequate than the hyperbolic tangent.

Nanoparticles containing hundreds or thousands of atoms are intermediate between the superparamagnetic small clusters (Section 6.4.6) and true ferromagnets. In a strict sense, they are nonferromagnetic, but in practice, it is often difficult to distinguish between finite-size magnetism (Fig. 5.16) and ferromagnetism smoothed by a small magnetic field (dashed line in Fig. 5.5). Note that experiments dealing with critical phenomena require high temperature and field resolutions ($\delta T \ll T$ and $\delta H \approx 0$), especially when attempting to trace finite-size effects in particles larger than a few interatomic distances. We will return to this point in the discussion of magnetic nanostructures (Section 7.4.2).

5.4.3 *Ginzburg criterion

The mean-field Curie temperature $T_c = 2J/z$ of the one-dimensional Ising ferromagnet is at striking odds with the exact result $T_c = 0$, because the mean-field model ignores magnetization fluctuations. Physically, different parts of the magnet are subject to different interaction fields, even in structurally homogeneous magnets. How can we estimate the strength of the fluctuations and their influence on the mean-field predictions?

Let us start from the fluctuation-response theorem (Section 5.1.2) and take into account that the total susceptibility $\chi = (\mu_B/V_o)\sum_j \chi_{ij}$, where V_o is the volume per spin. This yields

$$\chi = \frac{\mu_o \mu_B^2}{k_B T V_o} \sum_j (\langle s_i\, s_j\rangle - \langle s_i\rangle\langle s_j\rangle) \tag{5.66}$$

The magnetization $m = (\mu_B/NV_o)\sum_i s_i$, where N is the number of spins and $i = 1 \ldots N$, exhibits fluctuations

$$\langle m^2 \rangle - \langle m \rangle^2 = \frac{\mu_B^2}{N^2 V_o^2} \sum_{ij} (\langle s_i\, s_j \rangle - \langle s_i \rangle \langle s_j \rangle) \tag{5.67}$$

Putting (5.67) into (5.66) yields the following version of the fluctuation-response theorem:

$$\langle m^2 \rangle - \langle m \rangle^2 = \frac{k_B T \chi}{\mu_o\, N V_o} \tag{5.68}$$

This equation relates magnetization fluctuations in a volume NV_o to the susceptibility.

We are interested in critical fluctuations and consider a volume ξ^d, where ξ is the correlation length. Mean-field theory breaks down when $\langle m^2 \rangle - \langle m \rangle^2$ is comparable to or larger than $\langle m \rangle^2$. Equation (5.68) now assumes the form $\langle m \rangle^2 = \chi k_B T/\mu_o \xi^d$, and exploiting the mean-field scaling relations $\langle m \rangle \sim (T_c - T)^{1/2}$, $\chi \sim 1/|T - T_c|$, and $\xi \sim |T - T_c|^{-1/2}$ we find that mean-field theory is applicable for small $|T - T_c|^{d/2-2}$. This important condition is known as the *Ginzburg criterion*. It means that mean-field theory breaks down on approaching the critical point in less than four dimensions ($d < 4$). In more than four dimensions, the mean-field exponents remain applicable, whereas $d = 4$ is a borderline case (upper marginal or critical dimension).

5.4.4 Fluctuations and criticality

The essential failure of the mean-field predictions in less than four dimensions indicates that the partition function cannot be reduced to that of paramagnetic spins. A somewhat better approach is the Oguchi model, where paramagnetic spins are replaced by selfconsistently embedded small clusters. For two-spin clusters, the Curie temperature is

$$T_c = \frac{(z-1)J}{k_B} \left(1 + \tanh \frac{J}{k_B T_c}\right) \tag{5.69}$$

Compared to the mean-field result zJ/k_B, this equation yields a moderate reduction of T_c. The trend is correct, but for $z = 2$, as appropriate for the one-dimensional nearest-neighbor Ising model, T_c remains nonzero. This is because a two-spin cluster (dumbbell) embedded in a mean field environment is not able to properly distinguish, for example, between a nanowire environment ($d = 1$) and a bulk magnet ($d = 3$).

The number of spins that must be considered is determined by the correlation length ξ. Figure 5.17 illustrates the physical meaning of ξ for the square-lattice Ising model. Physically, the correlation length describes the thickness of the interface between ↑ and ↓ regions ($T < T_c$) or the size of the zero-field ↑ and ↓ fluctuations ($T > T_c$). Near T_c, where ξ is very large, the consideration of a finite volume is insufficient. For example, the $T \approx T_c$ configuration in Fig. 5.17 may show a smooth interface between ↑ and ↓ phases ($T < T_c$) or parts of ↑ and ↓ regions in the paramagnetic phase ($T > T_c$).

Aside from one-dimensional models, few exact results are known. An example is Onsager's famous solution of the two-dimensional Ising model (Onsager 1944, Yeomans

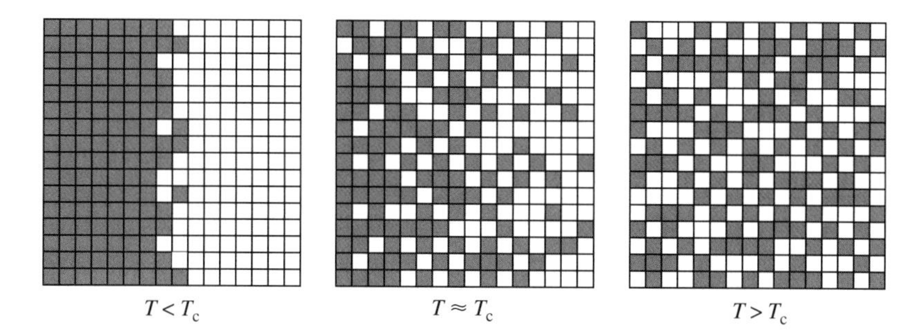

Fig. 5.17 Correlations in a two-dimensional Ising magnet. Dark and bright regions refer to ↑ and ↓ states, respectively. In the left and right figures, the correlation length ξ is comparable to the interatomic distance a. For the shown configuration at $T \approx T_c$, $\xi \sim 10a$.

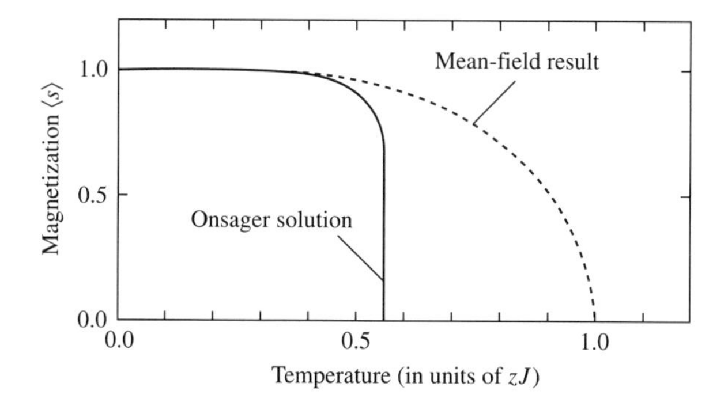

Fig. 5.18 Spontaneous magnetization of the square-lattice Ising model: mean-field approximation (dashed line) and Onsager solution (solid line).

1992). For the square lattice with nearest-neighbor exchange, the magnetization is

$$\langle s \rangle = \sqrt[8]{1 - \sinh^4(2J/k_B T)} \tag{5.70}$$

Figure 5.18 compares this result with the corresponding mean-field prediction. The Curie temperature is nonzero but significantly smaller than the mean-field prediction. More generally, magnets without continuous symmetry, such as Ising magnets and anisotropic Heisenberg magnets, exhibit ferromagnetism in two or more dimensions. Two-dimensional magnets *with* continuous symmetry, such as XY and isotropic Heisenberg magnets, do not exhibit long-range ferromagnetic order but are close to the onset of ferromagnetism. Adding a small anisotropy, $K_1 \ll J/a^3$, breaks the continuous symmetry and tips the scale towards ferromagnetism.

In addition to the reduced Curie temperature, the *temperature dependence* of the magnetization is qualitatively different from the mean-field prediction. In the vicinity of the Curie temperature, the magnetization scales as $(T_c - T)^{1/8}$, that is, the critical exponent $\beta = 1/8$. This is close to a step function ($\beta = 0$) and visibly different from the $\beta = 1/2$ mean-field behavior. Deviations from mean-field critical exponents are a general feature of low-dimensional magnets.

Critical exponents depend on the real-space dimensionality d and the spin dimension n but are independent of geometrical details of the lattice, such as the number z of nearest neighbors. This is in striking contrast to the mean-field model, where z is the only consideration. The reason is the involvement of long-range fluctuations, which leads to an averaging over atomic neighborhoods. Table 5.1 compares the critical exponents for several models, obtained by various methods. It is convenient to compare the Curie temperature with the mean-field prediction T_c(MF), that is, to consider the ratio $T_c/T_c(\mathrm{MF}) = nk_\mathrm{B}T_c/zJ_o$. Table 5.2 shows some values for Ising models ($n = 1$). For Heisenberg magnets (n = 3), the ratios are somewhat smaller, namely 0.610, 0.659, and 0.695 for simple cubic, bcc, and fcc magnets, respectively.

In several cases, the mean-field model works well. First, mean-field predictions such as T_c tend to improve with increasing number z of interacting neighbors. As a rule, this does *not* affect the critical exponents, because the exponents describe the magnets' behavior on macroscopic length scales and abstract from the atomic neighborhood. For example, bcc crystals ($z = 8$) and fcc crystals ($z = 12$) yield *exactly* the same critical exponents. An exception to the rule are models with infinite-range exchange interactions, $J(\mathbf{R}_i - \mathbf{R}_j) = J_o$ for $|\mathbf{R}_i - \mathbf{R}_j| = \infty$. In this case, z covers the whole magnet, and it is no longer possible to distinguish long-range fluctuations from interacting neighborhoods.

Table 5.1 Critical exponents for various nearest-neighbor n-vector models. In less than four dimensions, the spherical model ($n = \infty$) exhibits $\beta = 1/2, \gamma 2/(d-2)$, $\delta = (d+2)/(d-2)$, and $\nu = 1/(d-2)$.

d	n	Model	β	γ	δ	ν
$d = 2$	$n = 1$	Ising	1/8	7/4	15	1
$d = 3$	$n = 0$	polymer	0.302	1.16	4.85	0.588
$d = 3$	$n = 1$	Ising	0.324	1.24	4.82	0.630
$d = 3$	$n = 2$	XY	0.346	1.32	4.81	0.669
$d = 3$	$n = 3$	Heisenberg	0.362	1.39	4.82	0.705
$d = 3$	$n = \infty$	spherical	1/2	2	5	1
$d \geq 4$	all n	n-vector	1/2	1	3	1/2

Table 5.2 Exact and approximate Curie temperatures for the Ising model (in units of zJ/k_B).

lattice	d	z	mean-field	Oguchi	exact
linear chain	1	2	1	0.782	0.000
square	2	4	1	0.944	0.567
simple cubic	3	6	1	0.974	0.752
bcc	3	8	1	0.985	0.794
fcc	3	12	1	0.993	0.816

Second, the critical exponents depend on the *dimensionality* of the magnet, and mean-field exponents become exact in four or more dimensions. This is a consequence of the Ginzburg criterion (Section 5.4.3), which shows that critical fluctuations become relatively unimportant for $d \geq 4$. The dimensionality four is known as an upper critical dimension, as contrasted to the lower critical dimension that describes the onset of ferromagnetism ($d = 2$ for the Ising model). In Section 7.2.3 we will present a very simple and illustrative geometrical interpretation of the critical dimension $d = 4$. Taking into account both d and n, the models can be divided into universality classes. For example, n-vector models in $d > 4$ dimensions, mean-field models and infinite-range interactions models ($z = \infty$) all belong to the same universality class. For $d < 4$, there are different universality classes for different spin dimensionalities n (Table 5.1).

5.4.5 Renormalization group

There are various methods of calculating Curie temperatures and critical exponents. We have seen that there exist a few exact solutions, but in most cases it is necessary to use approximations or numerical simulations. Examples are series expansions and Monte-Carlo simulations (Yeomans 1992). Here we focus on the renormalization-group approach, as introduced by Kadanoff *et al.* (1967). The idea is to determine the partition function by iteratively mapping the magnet onto a simpler system. The simplest real-space renormalization approach considers the partition function of a magnet containing $N \to \infty$ spins. Summation over each second spin yields, after resizing, a magnet with $N/2$ spins. Figure 5.19 illustrates this block-spin renormalization for one- and two-dimensional magnets. The iterative character of the procedure means that the renormalization operations form a group, the renormalization group (RG). In strict sense, the renormalization group is a *semigroup* with associativity and neutral element but without inverse element, because each renormalization step leads to a loss of information.

The physics behind the block-spin renormalization is seen from the renormalization behavior of a typical paramagnetic $\uparrow$ cluster of size ξ, as in Fig. 5.19(c). The summation over each second spin *reduces* the size of the cluster, by factors $b = 2$ and $b = \sqrt{2}$ for Figs 5.19(a) and (b), respectively. Since the correlation length is largest near T_c, the renormalization means that the system moves *away* from the critical point. In terms of Fig. 5.19(c), the left part of the diagram corresponds to a temperature slightly above T_c, whereas the right part reflects a somewhat higher temperature.

Each renormalization step reduces ξ by a factor b, $\xi' = \xi/b$, so that the correlation length rapidly decreases. An exception is the critical point, where both ξ and $\xi/2$ are infinite. This can be used to determine the Curie temperature. If the summation over each second spin leaves the partition function unchanged, then $\xi' = \xi = \infty$ and $T = T_c$. In renormalization-group (RG) theory, such a point is referred to as a *fixed* point. On the temperature axis, the critical point is an unstable fixed point, because the renormalization enhances small deviations from the critical point.

The renormalization behavior is often visualized in the form of flow diagrams. Figure 5.20 deals with the zero-field Ising model and shows the RG flow on the temperature axis. Unstable fixed points are flow sources, whereas stable fixed points act as sinks. In Fig. 5.20, both systems have a relatively uninteresting stable fixed point at $T = \infty$. The one-dimensional Ising model has an unstable fixed point at $T = 0$, in

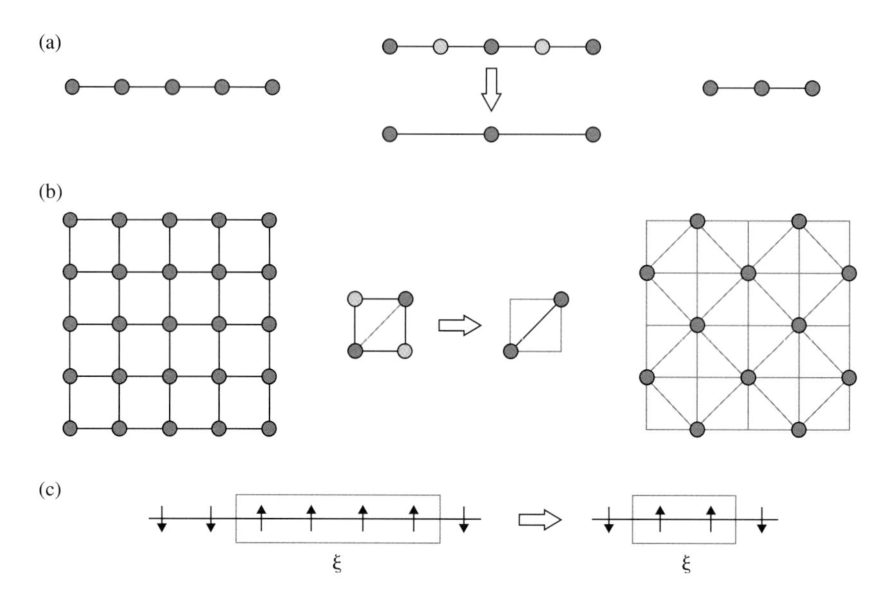

Fig. 5.19 Real-space renormalization: (a) one-dimensional magnet, (b) square lattice, and (c) behavior of the correlation length ξ. The hollow arrow represents the block-spin summation over each second spin.

$d = 1$: $T = 0$ $T \to \infty$

$d \geq 2$: $T = 0$ T_c $T \to \infty$

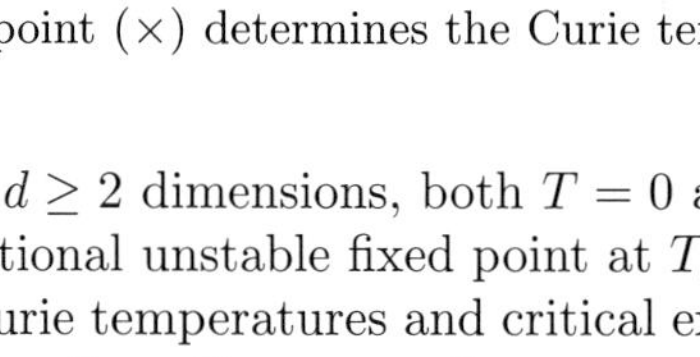

Fig. 5.20 Renormalization-group flow for Ising ferromagnets. The plot is on the $H = 0$ temperature axis. The fixed point ($\times$) determines the Curie temperature $I_c > 0$.

agreement with $T_c = 0$. In $d \geq 2$ dimensions, both $T = 0$ and $T = \infty$ are stable fixed point, and there is an additional unstable fixed point at $T_c > 0$.

To actually calculate Curie temperatures and critical exponents, we must consider the partition function and perform the summation over each second spin. It is convenient to introduce the dimensionless coupling constant $\kappa = J/k_B T$. In one dimension,

$$\sum_{s=\pm 1} \exp(\kappa\, s_1 s + \kappa\, s\, s_2) = c\ \exp(\kappa' s_1 s_2) \tag{5.71a}$$

and

$$\exp(\kappa s_1 + \kappa s_2) + \exp(-\kappa s_1 - \kappa s_2) = c\ \exp(\kappa' s_1 s_2) \tag{5.71b}$$

This equation contains two unknowns, κ' and c. Considering the two cases $s_1 = \pm s_2$ yields $c = 2$ and $\kappa' = \frac{1}{2}\ln(\cosh(2\kappa))/2$. The only fixed points of this equation are $\kappa = 0\,(T = \infty)$ and $\kappa = \infty\,(T = 0)$, in agreement with the upper part of Fig. 5.20.

While the Curie temperature of one-dimensional magnets is zero, the transformation (5.71) can be used to used to determine T_c for magnets that contain chains as building blocks, such as hypothetical cross-linked magnetic wires. The Curie temperature then exhibits a logarithmic dependence on the chain length between the links.

On a *square lattice*, an approximate block-spin summation includes the bright atoms in Fig. 5.19(b). Denoting their spins by s and s', we obtain

$$\sum_{s,s'=\pm 1} \exp(\kappa\, s_1\, s + \kappa\, s\, s_2 + \kappa\, s_1\, s' + \kappa\, s'\, s_2) = c\, \exp(\kappa'\, s_1\, s_2) \tag{5.71c}$$

and $4\cosh^2(\kappa s_1 + \kappa s_2) = c\, \exp(\kappa' s_1 s_2)$. As above, we determine κ' and c by considering $s_1 = \pm s_2$. This yields $c = 4\exp(\kappa')$ and the RG equation

$$\kappa' = \ln\cosh(2\kappa) \tag{5.72}$$

Putting $\kappa' = \kappa$ leads to one unstable fixed point, $\kappa^* = 0.60938$. The corresponding Curie temperature is given by $\kappa^* = J/k_B T_c$ and equal to $1.64\, J/k_B$. By comparison, the respective exact and mean-field results are $2.27\ J/k_B$ and $4\ J/k_B$.

Critical exponents are obtained by analyzing the vicinity of the critical point. For example, the definition of the exponent ν implies

$$\ln\left(\frac{\xi'}{\xi}\right) = -\nu \ln\left(\frac{T' - T_c}{T - T_c}\right) \tag{5.73}$$

where $\xi' = \xi/\sqrt{2}$. To determine the temperature differences, we use $\kappa = J/k_B T$ and expand (5.72) in the vicinity of κ^*. The result of the calculation is the exponent

$$\nu = \frac{1}{2}\, \frac{\ln(2)}{\ln(2) + \ln(\tanh(2\kappa^*))} \tag{5.74}$$

Since $\tanh(x) < 1$, the exponent is larger than 1/2. In the present approximation $\nu = 0.669$, as compared to the respective mean-field and exact exponents of 1/2 and 1. We see that both T_c and ν deviate from the mean-field predictions and that our simple renormalization procedure yields the correct trend. The limited numerical agreement reflects the crude treatment of mediated next-nearest neighbor interactions in the two-dimensional model.

Real-space renormalization of the partition function, as considered in this section, is not the only RG approach. In k-space renormalization, the block-spin procedure is replaced by an integration over small wave vectors. This approach is especially useful to describe dimensionalities $d = 4 - \varepsilon$, where ε is small. For the n-vector model, the ε-expansion yields the critical exponents $\beta = 1/2 - 3\varepsilon/2(n+8)$, $\gamma = 1 + \varepsilon(n+2)/1(n+8)$, $\delta = 3 + \varepsilon$, and $\nu = 1/2 + \varepsilon(n+2)/4(n+8)$. In four dimensions, the exponents are mean-field like but typically accompanied by logarithmic corrections, as rationalized by $x^\varepsilon \sim \ln(x)$ for $\varepsilon \to 0$.

Renormalization-group techniques are also used in other areas of physics, such as quantum electrodynamics. Mass renormalization prevents point-like charged particles such as electrons from possessing infinite masses $m = E/c^2$, as suggested by the divergence of the electric field energy $\int (1/r^2)^2\, dV$. The procedure adds two unobservable masses, the particle's "bare" mass and that originating from the electromagnetic self-interaction, to yield a physically meaningful renormalized mass. A similar procedure is used in many-body physics, as in the treatment of the Kondo effect (Mattuck 1976). Finally, RG approaches related to those in magnetism are used to describe soft matter, such as polymers and gels (Section 7.2).

5.5 Temperature dependence of anisotropy

Summary Room-temperature anisotropy energies per atom are much smaller than k_BT, indicating that the finite-temperature magnetic anisotropy relies on the support by interatomic exchange. The exchange suppresses the excitation of atomic spins onto states with reduced anisotropy. Simple ferromagnets, such as Fe and Co, obey the $m = n(n+1)/2$ power laws predicted by the Callen and Callen model. For example, second- and fourth-order anisotropies are characterized by $m = 3$ and $m = 10$. However, the applicability of the model is an exception rather than the rule. In the single-ion model of rare-earth anisotropy, which describes high-performance permanent magnets such as $SmCo_5$ and $Nd_2Fe_{14}B$, the temperature dependence is determined by the intersublattice exchange and qualitatively very different from the Callen and Callen predictions. Pictorially, intramultiplet excitations destroy the net asphercity of the 4f charge clouds by thermally randomizing the directions of the rare-earth moment. Other compounds have exponents that deviate from the Callen and Callen predictions, for example $m = 2$ and $m = 1$ for the 2nd-order anisotropies of $L1_0$ and actinide magnets, respectively.

By everyday standards, modern permanent magnets are very strong. For example, a compact cylindrical $Nd_2Fe_{14}B$ magnet of radius 1.5 cm is able to hold the weight of an adult, and trying to remove such a magnet from a fridge door usually results in nasty scratches across the surface. However, the strength of the anisotropy looks much less impressive if one considers the anisotropy energy per atom. Taking $K_1 = 5\,\mathrm{MJ/m^3}$ and a volume of $V_o = 12\,\mathrm{Å}^3$ per atom, corresponding to dense packing with an average atomic radius of 1.28 Å, we obtain a temperature equivalent $K_1 V_o / k_B$ of only 4.3 K. How can we explain the existence of a significant anisotropy at and above room temperature?

In reality, the atomic moments are stabilized by *interatomic exchange*. For example, the anisotropy of typical rare-earth transition-metal (RE-TM) magnets largely originates from the rare-earth sublattice. In temperature units, the intersublattice exchange J^* is of order 100 K, indicating that a substantial fraction of the anisotropy survives at and above room temperature. Figure 5.21 shows the temperature dependence of the anisotropy for typical RE-TM magnets. Since J^* increases with the number of TM neighbors per RE atom, the decay of the anisotropy is less pronounced for transition-metal-rich intermetallics. For example, the anisotropy of $SmCo_5$ decreases from about $25\,\mathrm{MJ/m^3}$ at low temperatures to about $15\,\mathrm{MJ/m^3}$ at room-temperature.

The temperature dependence of the magnetic anisotropy is of importance in many areas of magnetism and often much more pronounced than the temperature dependence $M_s(T)$ of the magnetization. For example, in simple one-sublattice ferromagnets, second- and fourth-order anisotropy contributions scale as M_s^3 and M_s^{10}, whereas the

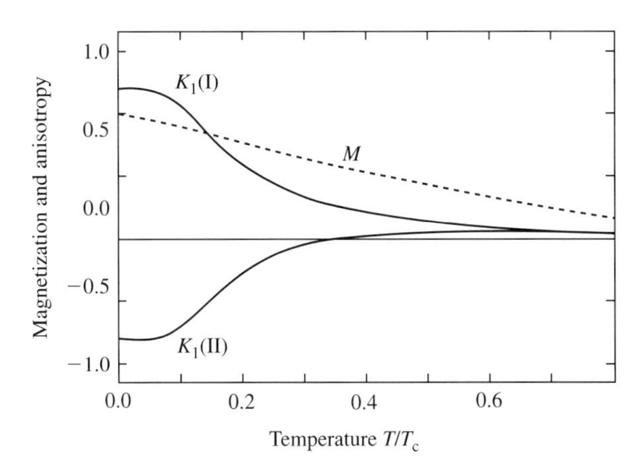

Fig. 5.21 Temperature dependence of K_1 for typical RE-TM magnets. T_c is the Curie temperature, and the curves I and II refer to materials that exhibit easy-axis and easy-plane magnetism at low temperatures, respectively.

exchange stiffness A is approximately quadratic in M_s. The anistropy reduction is even more drastic in typical rare-earth intermetallics. Many permanent magnets are used at temperatures above 100 °C, and the relatively strong temperature dependence of the Nd^{3+} anisotropy in $Nd_2Fe_{14}B$ is a major shortcoming of this otherwise excellent material. Similar requirements apply to soft magnets, sensors and, especially, ultrahigh-density recording media, where thermal stability is a major consideration. Note that the weight of higher-order anisotropy contributions decreases with increasing temperature, so that K_2 and K_3 are often negligible at and above room temperature.

Section 5.5.1 is devoted to the Callen and Callen model, which illustrates how the temperature dependence of the anisotropy is determined from the zero-temperature anisotropy energy. In Section 5.5.2, we consider the temperature dependence of the rare-earth anisotropy, whereas Section 5.5.3 deals with the finite-temperature anisotropy of $L1_0$ alloys.

5.5.1 Callen and Callen model

Simple one-sublattice ferromagnets, such as Fe and Co, are reasonably well described by the Callen and Callen model (1963). The model goes back to Akulov (1936) and predicts power laws of the type

$$K(T) = K(0) \left(\frac{M_s(T)}{M_s(0)} \right)^m \tag{5.75}$$

where n-th order anisotropy constants obey $m = n(n+1)/2$. In other words, the temperature dependence of the respective second-, fourth-, and sixth-order anisotropy contributions is proportional to the third, tenth, and twenty-first powers of the magnetization. For example, K_1 is characterized by power-law exponents $m = 3$ (uniaxial) and $m = 10$ (cubic).

To derive the Callen and Callen exponents, we consider the classical uniaxial mean-field Heisenberg model, a normalized classical spin $\mathbf{s}$ interacts with a self-consistent mean field $\mathbf{H}_{\rm MF} = J\,m\,\mathbf{e}_{\rm z}$. The magnetization is equal to $m = \langle\cos\theta\rangle$, whereas the anisotropy scales as $\langle\cos^2\theta\rangle$. The averages $\langle\cos^{\rm n}\theta\rangle$ are obtained by evaluating

$$\langle\cos^{\rm n}\theta\rangle = \frac{\int \exp(Jm/k_{\rm B}T)\cos^{\rm n}\theta\,\sin\theta\,{\rm d}\theta}{\int \exp(Jm/k_{\rm B}T)\sin\theta\,{\rm d}\theta} \tag{5.76}$$

where the integrals extend from $\theta = 0$ to π. As in the zero-temperature case (3.33), it is necessary to consider Legendre polynomials of $\cos\theta$, such as $m = \langle\cos\theta\rangle$,

$$K_1(T) = \frac{1}{2}\,K_1(0)\,(3\langle\cos^2\theta\rangle - 1) \tag{5.77a}$$

and

$$K_2(T) = \frac{1}{8}\,K_2(0)\,(35\langle\cos^4\theta\rangle - 30\,\langle\cos^2\theta\rangle + 5) \tag{5.77b}$$

At low temperatures, (5.77b) is solved by series expansion, leading to $\langle\cos^{\rm n}\theta\rangle = 1 - nk_{\rm B}T/J$. To determine the temperature dependence of K_1, we consider $m = 1 - k_{\rm B}T/J$ and $K_1(T) = K_1(0)(1 - 3k_{\rm B}T/J)$. Since $(1-x)^3 \approx 1 - 3x$ for small x, we can write $K_1(T)/K_1(0) = m^3$. This is the sought-for Callen and Callen law for uniaxial anisotropy ($n = 2$ and $m = 3$).

To check the applicability of the Callen and Callen model, we consider the classical mean-field Heisenberg model with uniaxial anisotropy. The model can be solved exactly and yields a straight $K_1(T)$ line. Figure 5.22 compares the exact solution (Skomski *et al.* 2006a) with the corresponding Callen and Callen prediction $K_1 \sim M^3$. We see that the Callen and Callen law works surprisingly well, except for the immediate vicinity of $T_{\rm c}$. For K_2, the agreement is somewhat poorer (not shown in Fig. 5.22). More generally, the anisotropies of simple metals, such as Fe, Co, and Ni, are quite well

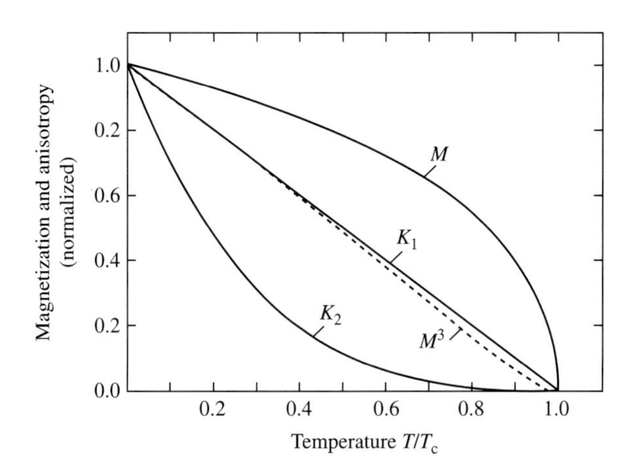

Fig. 5.22 Temperature dependence of the magnetization and anisotropy of simple magnets with uniaxial anisotropy.

described by the Callen and Callen theory, irrespective of the classical or quantum-mechanical nature of the underlying Hamiltonian and of the itinerant nature of the magnetic ground state.

However, comparison with Fig. 5.14 shows that the Callen and Callen model cannot be used to describe intermetallics alloys. For example, it is unable to reproduce the K_1(II) zero in Fig. 5.14. The main reason is that the Callen and Callen model links the anisotropy to the magnetization. Both magnetization and anisotropy are affected by this mechanism, albeit somewhat differently. In RE-TM intermetallics, the atomic origin of the magnetization (TM-sublattice) is different from the origin of the anisotropy (RE-sublattice), and the two sublattices are characterized by different exchange constants and different temperature dependences (Section 5.5.2). Similar arguments apply to other substances. For example, some magnets with very strong spin-orbit coupling, such as uranium sulfide (US), have huge low-temperature anisotropies of $K_1 = 100 \cdots 1000\,\mathrm{MJ/m^3}$ (Brooks and Johannson 1993) and are described by Callen and Callen exponent $m = 1$ (Skomski *et al.* 2006a).

Note that the failure of the Callen and Callen theory is unrelated to the single-ion character of the anisotropy. It has become popular to equate the Callen and Callen model with single-ion anisotropy, but the rare-earth anisotropy illustrated in Fig. 5.23 is a single-ion anisotropy, too. In fact, single-ion anisotropy is a good approximation for most materials, because the spin-orbit coupling is centered in the atomic cores and because the rotation of an atomic spin has very little effect on the crystal field. One exception is the dipolar anisotropy of gadolinium, where a large dipole interaction (large Gd moment) coincides with zero lowest-order anisotropy ($\alpha_J = 0$).

5.5.2 Rare-earth anisotropy

In Section 3.4 we have seen that the magnetic anisotropy of rare-earth intermetallics is largely due to the electrostatic interactions of the aspherical 4f charge clouds with the anisotropic crystalline environment. In most rare-earth ions, the spin-orbit coupling is sufficiently strong to ensure that Hund's rule is well satisfied, $J = |L \pm S|$. As mentioned in Section 2.2.3, partial exceptions are Sm^{3+} and especially Eu^{3+}, where the respective level intermultiplet splittings of 1200 K and 300 K yield some J mixing. However, the leading contribution to the temperature dependence of the anisotropy

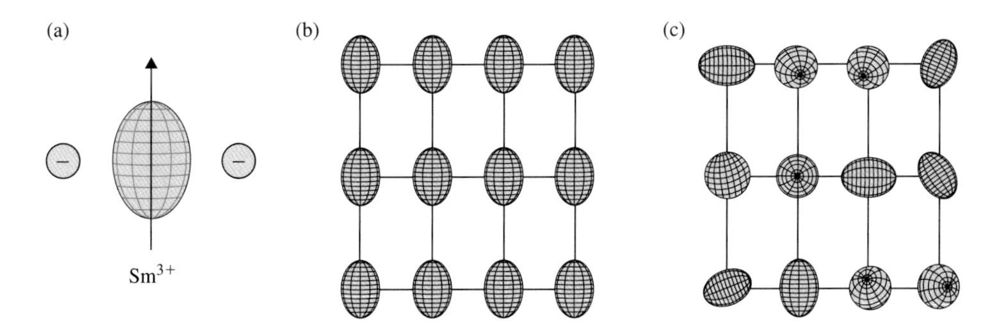

Fig. 5.23 Mechanism of finite-temperature rare-earth anisotropy: (a) origin of anisotropy, (b) zero-temperature 4f charge distribution, and (c) finite-temperature charge distribution.

comes from *intramultiplet* excitations, where $J_z < J$ (Herbst 1991). In a simple classical picture, $J_z \sim \cos\theta$ and $K_1 \sim 3\cos^2\theta - 1$, so that the temperature dependence of K_1 reflects the thermal randomization of the 4f moments. Figure 5.23 illustrates this mechanism. The low-temperature charge distribution is aspherical (a), but thermal excitations randomize the moments (b) and thermal averaging leads to spherical 4f charge clouds with zero net anisotropy.

The randomization of the rare-earth moment directions is suppressed by the intersublattice exchange J^*. Physically, the exchange field created by the transition-metal sublattice favors maximum spin alignment, $J_z = J$. In a quantum-mechanical treatment, $3\cos^2\theta - 1$ must be replaced by $3J_z^2 - J(J+1)$. More generally, it may be convenient to start from *equivalent operators* such as $\mathrm{O}_2^0 = 3\mathbf{J}_z^2 - \mathbf{J}^2$, especially when considering higher-order and nonuniaxial anisotropies (Hutchings 1964). The anisotropy is obtained by putting the magnetic ion in an exchange field and calculating $\langle 3J_z^2 - J(J+1)\rangle$. In analogy to Section 3.4, the result is

$$K_1 = -\frac{3}{2V_\mathrm{R}}\langle Q_2\rangle A_2^0 \tag{5.78}$$

where

$$\langle Q_2\rangle = \alpha_\mathrm{J}\langle r^2\rangle_{4\mathrm{f}}\langle 3J_z^2 - J(J+1)\rangle \tag{5.79}$$

is the thermally averaged quadrupole moment of the 4f charge cloud. At zero temperature, $J_z = J$, and (5.79) reduces to (3.27). At infinite temperature, all intramultiplet levels J_z are populated with equal probability and $\langle Q_2\rangle = 0$. This confirms our argument that averaging over all spin orientations in Fig. 5.23 yields a spherical charge cloud with zero net anisotropy.

The calculation of the average $\langle 3J_z^2 - J(J+1)\rangle = 3\langle J_z^2\rangle - J(J+1)$ is similar to the derivation of the Brillouin function for the magnetization:

$$\langle \mathbf{J}_z^2\rangle = \frac{1}{Z}\sum_{\mathrm{J_z=-J}}^{\mathrm{J}} J_z^2 \exp\left(-\frac{g\mu_\mathrm{o}\mu_\mathrm{B}H_\mathrm{RT}J_z}{k_\mathrm{B}T}\right) \tag{5.80}$$

where the RE-TM exchange field H_RT is proportional to J^*. The rare-earth magnetization and anisotropy both decrease, but the anisotropy decrease is much more pronounced. This explains the strong temperature dependence of the rare-earth anisotropy indicated in Fig. 5.21. There is a characteristic temperature $T_\mathrm{RT} \sim J^*/k_\mathrm{B}$ above which rare-earth (RE) magnetization anisotropy decrease rapidly. For typical transition-metal-rich rare-earth intermetallics, T_RT temperature is somewhat below room temperature. This is an important consideration for permanent magnets, where typical operating temperatures are at and above room temperature (Herbst 1991, Skomski and Coey 1999). Above T_RT, the magnetization decreases approximately as $1/T^2$. Below T_RT, the RE magnetization and anisotropy reach plateaus with an exponential approach to their zero-temperature values. The plateau is a quantum effect, caused by the finite splitting between the $J_z = J$ and $J_z = J - 1$ energy levels.

Equations (5.79–80) ascribe the rare-earth anisotropy to the paramagnetic behavior of tripositive 4f ions in the exchange field H_RT created by the transition-metal

(a) 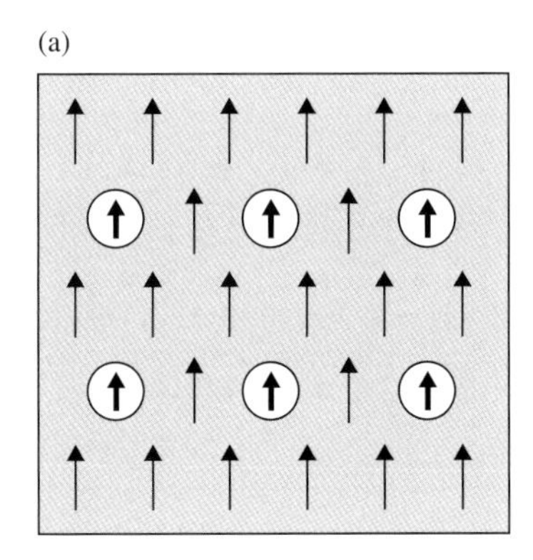

(b) 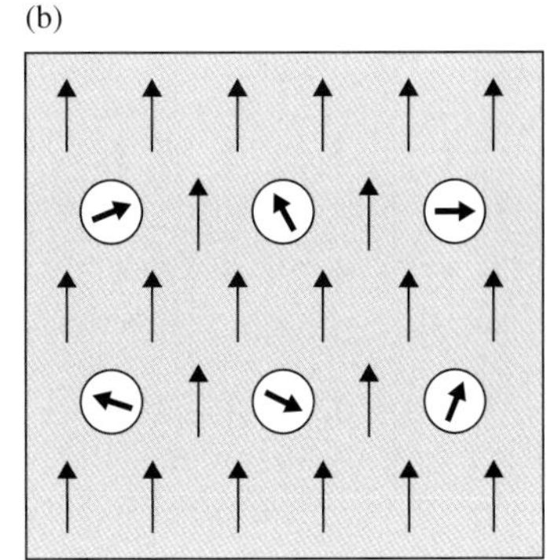

Fig. 5.24 Exchange and anisotropy in transition-metal rich RE-TM intermetallics: (a) zero temperature and (b) room temperature. The rare-earth atoms (spheres) are embedded in a transition-metal environment. The anisotropy, indicated by the thickness of the arrows, is dominated by the rare-earth atoms. Due to the relatively weak interatomic RE-TM exchange, the anisotropy breaks down well below T_c.

sublattice. Figure 5.24(a) illustrates the corresponding spin structure. Above T_{RT}, the rare-earth anisotropy contribution is small, and the less temperature-dependent anisotropy of the TM sublattice becomes the leading contribution. In addition, there may be temperature-dependent spin-reorientation transitions (Section 4.1.3) due to competing TM and second- or higher-order RE anisotropy contributions (Herbst 1991, Coey 1996). One example is the K_1(II) zero in Fig. 5.21, observed, for example, in $NdCo_5$ and caused by competing second-order easy-plane RE and easy-axis TM anisotropies.

5.5.3 Sublattice modeling

The modeling of rare-earth transition-metal alloys is relatively simple, because rare-earth 4f ions with well-defined moments ($J_z = -J, \ldots, J-1, J$) are subjected to the exchange field of the 3d sublattice. In *itinerant* magnets, the magnetic moment of the anisotropy-producing atoms becomes a separate consideration. This is not a big problem in simple ferromagnetic metals, because the Callen and Callen model relates the anisotropy directly to the magnetization, irrespective of the origin of the magnetization. However, in some alloys, such as $L1_0$ intermetallics (Section A5.1), the moment of the 4d or 5d atoms is due to spin-polarization by 3d atoms and must be determined self-consistently. This has far-reaching consequences for the temperature dependence of the anisotropy (Skomski, Kashyap, and Sellmyer 2003, Mryasov *et al.* 2005).

Heavy transition-metal elements such as Pd and Pt are exchange-enhanced Pauli paramagnets, indicating that the respective 4d and 5d electrons are close to satisfying the Stoner criterion. In ferromagnetic environments, 4d/5d electrons are easily spin-polarized by neighboring magnetic atoms and yield some contribution to the magnetization and Curie temperature. Above all, their strong spin-orbit coupling leads to a strong 4d/5d anisotropy contribution in $L1_0$ magnets such as FePd and CoPt. The magnets have anisotropies of the order of $5\,\mathrm{MJ/m^3}$ and are suitable for applications

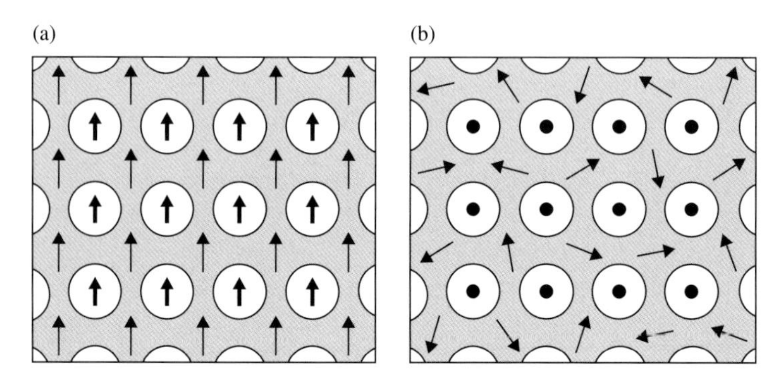

Fig. 5.25 Origin of the temperature dependence of the anisotropy of $L1_0$ magnets: (a) ferromagnetism and (b) paramagnetism. The Pt or Pd atoms (white) are embedded in an Fe or Co sublattice (gray).

such as specialty permanent magnets and magnetic recording (Klemmer *et al.* 1995, McCurrie 1994)

The 3d sublattice exerts an exchange field $J^*\langle S\rangle$ on the 4d/5d sublattice, creates a 4d/5d moment, and realizes, via spin-orbit coupling, the magnetic anisotropy. Figure 5.25 compares the zero-temperature magnetism (a) of $L1_0$ compounds with the spin structure at temperatures approaching T_c (b). At low temperatures, the 3d sublattice (gray) spin-polarizes the 4d/5d atoms (white) and yields a substantial 4d/5d magnetization M_o (thick arrows). The situation is very similar to that shown in Fig. 5.25(a) and results in a strong anisotropy. At *high temperatures*, the magnetization $M \sim \langle S\rangle$ of the 3d sublattice approaches zero, and the 4d/5d moment and anisotropy both collapse. This single-ion mechanism leads to significant deviations from the Callen and Callen behavior, roughly corresponding to a power-law exponent $m = 2$ (Skomski, Kashyap, and Sellmyer 2003, Skomski: *et al.* 2006a). Simplifying somewhat, the 4d/5d spin polarization M_o is proportional to M, so that the quadratic dependence of the anisotropy energy on M_o translates into $K_1 \sim M^2$.

Exercises

1. ***Thermal equilibrium.*** Determine and interpret the Boltzmann distribution for the two limits $T = 0$ and $T = \infty$.
2. ***Critical point and magnetic field.*** Estimate the smoothing of the $M(H)$ curve at T_c for an iron magnet in a field of 0.5 T.
3. ***Square spin cluster.*** Calculate the free energy for a square cluster of four Ising spins.
4. ***Sublattice magnetizations.*** Estimate the sublattice magnetization ratios from Fig. 5.14.

   Answer: For the three materials, the ratios M_2/M_1 are about 0 (one-sublattice ferromagnetism), 0.45, 0.9, and 1.3

5. ***Two-spin correlations in a pair model.*** Calculate the correlation finctions $\langle s_1 s_1 \rangle - \langle s_1 \rangle^2$ and $\langle s_1 s_2 \rangle - \langle s_1 \rangle \langle s_2 \rangle$ for a dumbbell of two Ising spins.
6. ***Mean-field Curie temperature of small clusters.*** Calculate the mean-field Curie temperatures for a dumbbell ($N=2$), a square ($N=4$), and a cube ($N=8$). Compare the mean-field predictions with the exact result.
7. ***High-temperature limit of magnetic anisotropy.*** Convince yourself that K_1 and K_2 approach zero at very high temperatures.

   Hint: Assume that $k_B T$ is much larger than the interatomic exchange.
8. ***Coupled spins of arbitrary length S.*** Consider two spins $\mathbf{S}_1$ and $\mathbf{S}_2$ of equal quantum number S and calculate the energy levels $-J\mathbf{S}_1 \cdot \mathbf{S}_2$ for $\mathbf{S} = \mathbf{S}_1 + \mathbf{S}_2$ (ferromagnetic coupling) and $\mathbf{S} = \mathbf{S}_1 - \mathbf{S}_2$ (antiferromagnetic coupling).

   Hint: Exploit the fact that $\mathbf{S}^2 = S(S+1)$ for any spin S.

   Answer: The respective FM and AFM energies are $-JS^2$ and $+JS(S+1)$.
9. ***Critical point.*** Why is it important to distinguish between the critical point and the critical temperature?

   Answer: The critical point requires both $T = T_c$ and $H = 0$, because finite fields destroy the critical singularity by smoothing the magnetization.
10. ***Logarithmic corrections.*** In some cases, critical exponents contain logarithmic corrections. Show that the scaling law $x^4 \ln(x)$ amounts to an exponent $4+\varepsilon$, where $\varepsilon \ll 1$.
11. ***Matrix operators.*** The quantum-mechanical partition function $Z = \mathrm{Tr}(\exp(-\mathbf{H}/k_B T))$ contains the Hamilton operator in the exponent. What is the meaning of the function $\exp(-H/k_B T)$?

    Answer: Operator functions are defined by series expansions, such as $\exp(\mathbf{x}) = \mathrm{I} + \mathbf{x} + \frac{1}{2}\mathbf{x}2 + \mathrm{O}(\mathbf{x}3)$. The expansion is straightforward but, due to the size of the energy matrices, often very cumbersome.
12. ***Paramagnetic gas.*** Calculate the thermally averaged room-temperature moment of a paramagnetic gas atom of spin 3/2 in a field of 1 T.

    Answer: 0.0115 μ_B
13. ***Entropy of a paramagnetic gas.*** Calculate the entropy for a spin-$\frac{1}{2}$ particle in a magnetic field.
14. ***Ising chain and Ising rings.*** On a closed path, the reversal of a spin creates two broken bonds. Can we use the argumentation of 5.4.1, where one bond is broken, to describe an infinite Ising ring?
15. ***Renormalization of spin fluctuations.*** Can the self-consistent renormalization of spin fluctuations be considered as a renormalization-group approach?
16. ***Universality of Curie temperature.*** Is the Curie temperature universal or nonuniversal?
17. ***Paramagnetic anisotropy.*** Crystal-field interaction and spin-orbit coupling survive above T_c. Why is the paramagnetic anisotropy of most magnets negligibly small?
18. ****Susceptibility and correlation length.*** Use the Ising model to show that the divergence of the susceptibility at $T = T_c$ is accompanied by a divergence of the correlation length (critical fluctuations).

Hint: Write the Hamiltonian as $H = H_o - \Sigma_i h_i s_i$, where $H_o(s_1, \ldots, s_N)$ is an unspecified field-independent exchange interaction, $s_i = \pm 1$, and $h_i = \mu_o \mu_B H_i$ is a local field. The basic idea is to use derivatives $\partial Z/\partial h_i$ and $\partial Z^2/\partial h_i \partial h_j$ and to "puzzle" them together to determine the so-called nonlocal susceptibility $\chi_{ij} = \partial \langle s_i \rangle / \partial h_j$ and the correlations $C_{ij} = \langle s_i\, s_j \rangle - \langle s_i \rangle \langle s_j \rangle$.

6 Magnetization dynamics

The nonequilibrium character of magnetization processes means that magnetization processes are time-dependent, even if the external magnetic field is kept constant. For example, freshly magnetized permanent magnets lose some remanence, about 0.1% per decade, soft magnets exhibit harmful high-frequency hysteresis losses, and the long-term thermal stability of stored information is important in magnetic recording. The oscillation and relaxation times involved vary from one nanosecond or less in materials for high-frequency applications to millions of years in magnetic rocks and meteorites.

Intrinsic magnetic properties, such as magnetization and anisotropy, are realized on very small length scales, typically less than 1 nm, and correspond to very fast processes of quantum-mechanical origin. They can be considered as equilibrium properties, described by the Boltzmann distribution (5.2). By contrast, extrinsic phenomena, such as hysteresis, are realized on length scales of several nanometers or more. By atomic standards, they require considerable equilibration times, and many methods familiar from equilibrium statistical mechanics become inapplicable. For example, time averages $\langle A \rangle = \int A(t)\,\mathrm{d}t/\Delta \mathrm{t}$ and ensemble averages $\langle A \rangle = \sum_\mathrm{n} A_\mathrm{n} / \sum_\mathrm{n}$ are no longer equivalent. A well-known example is glass, where ensemble averaging gives the impression of a liquid, whereas time averaging (evolution of atomic positions) yields a solid-like picture. Rather than using (5.1–3) for both time and ensemble averages, nonequilibrium calculations require the explicit solution of equations of motions.

There are two basic phenomena involved: *quantum dynamics* and *thermally activated dynamics*. Quantum dynamics follows from the time-dependent Schrödinger equation (A.3.1) and is illustrated by the single-spin model of Section 1.3. For typical atomic energy differences of order 1 eV, the characteristic times $\hbar/E$ are in the femtosecond region (10^{-15} s). Magnetic fields correspond to much lower energies, and the typical precession times are in the nanosecond region, corresponding to GHz resonance frequencies (Section 6.1). Another type of time dependence involves heat-bath degrees of freedom. This includes damping and relaxation phenomena (Section 6.2) and thermal activation over energy barriers (Section 6.3).

6.1 Quantum dynamics and resonance

Summary The dynamics of paramagnetic ions has the character of spin precession, and essential features of this picture carry over to ferromagnets. This is exploited in the modeling of magnetic resonance and spin waves. Spin waves are wave-vector dependent excited states that contribute to both finite-temperature equilibrium and nonequilibrium magnetic properties. For example, the low-temperature spontaneous magnetization of three-dimensional Heisenberg ferromagnets is significantly smaller than the mean-field prediction, and Bloch's spin-wave arguments indicate that there is no long-range isotropic ferromagnetism in two or less dimensions. The magnetization of spin waves is generally delocalized but always corresponds to an integer number of switched spins. In perfect ferromagnets, the spin waves form a continuum of delocalized states, but real-structure imperfections and nanoscale structural features lead to spin-wave localization and discrete spin-wave levels.

In principle, it is possible to predict the evolution of any physical system from the time-dependent Schrödinger equation

$$\mathrm{i}\,\hbar\frac{\partial|\Psi\rangle}{\partial t} = \mathsf{H}|\Psi\rangle \tag{6.1}$$

where H is the full many-body Hamiltonian and $|\Psi\rangle$ is the many-body wave function. However, this is a complicated and often unnecessary procedure, because most degrees of freedom described by (6.1) are irrelevant. A typical example is the reversal of the magnetization of a small particle, where we are interested in the average magnetization $\mathbf{M}(t)$ but not in the spin orientation of individual atoms. The same is true for irrelevant electronic and mechanical (phononic) degrees of freedom. Figure 6.1 illustrates the distinction between relevant degrees of freedom (M) and irrelevant degrees of freedom (x_i).

The simplest approach to quantum dynamics is to consider an isolated quantum system, that is, to ignore the gray area in Fig. 6.1(b). An example is the precession of spins in an external magnetic field. In the next subsections we will see that this approach applies to both paramagnets and ferromagnets but is unable

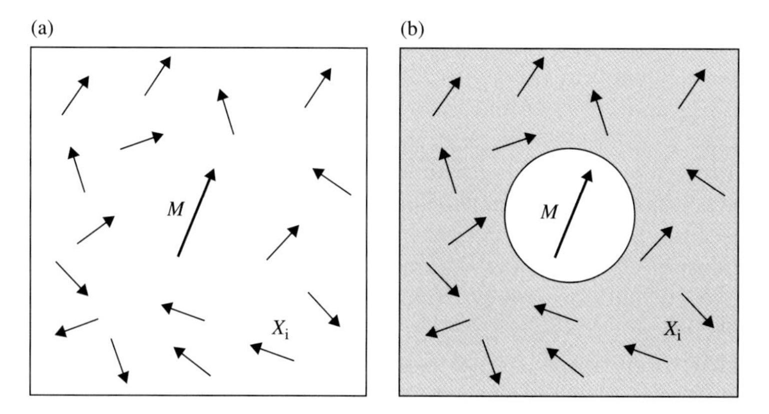

Fig. 6.1 Relevant and irrelevant degrees of freedom: (a) full description and (b) embedding of a relevant degree of freedom (M) in a heat bath describing the irrelevant degrees of freedom (x_i).

to describe damping and thermal effects. The reason is the deterministic character of the Schrödinger equation (6.1). As in the equilibrium limit, the description of finite-temperature dynamics requires the consideration of a heat bath formed by the irrelevant degrees of freedom. The corresponding coarse-graining transition from Fig. 6.1(a) to (b) will be discussed in Section 6.3.

6.1.1 Spin precession

The simplest model of magnetization dynamics is a single spin in a magnetic field. It is described by the Zeeman Hamiltonian $\mathsf{H} = -\mu_o\mu_B\mathbf{H}\cdot\boldsymbol{\sigma}$, where $\mathbf{H}$ is the external magnetic field and $\boldsymbol{\sigma}$ is the vector formed by the 2×2 Pauli matrices (2.32). The solutions of the corresponding Schrödinger equation

$$\mathrm{i}\,\hbar\frac{\partial|\psi\rangle}{\partial t} = -g\mu_o\mu_B\mathbf{s}\cdot\mathbf{H}\,|\psi\rangle \tag{6.2}$$

are two-component wave functions (ψ_1, ψ_2) where $(1,0)$ and $(0,1)$ correspond to $\uparrow$ and $\downarrow$ states. For electron spins, which are unquenched, $g = 2$ and $g\mathbf{s} = \boldsymbol{\sigma}$. Let us, for moment, fix the coordinate frame so that $\mathbf{H} = H\mathbf{e}_z$ and H is diagonal. The Schrödinger equation then assumes the form

$$\mathrm{i}\,\hbar\frac{\partial\psi_1}{\partial t} = -\mu_o\,\mu_B\,H\psi_1 \tag{6.3a}$$

$$\mathrm{i}\,\hbar\frac{\partial\psi_2}{\partial t} = +\mu_o\,\mu_B\,H\psi_2 \tag{6.3b}$$

The solution of these equations is $\psi_{1/2}(t) = \psi_{1/2}(0)\exp(\mp\mathrm{i}\,\omega t/2)$, where $\omega = 2\mu_o\mu_B H/\hbar$. The dynamics has the character of a spin precession, and the resonance frequency $\nu = \omega/2\pi$ is given by the conversion factor 28.0 GHz/T. For *nuclear* magnetic resonance (NMR), the electron spin moment must be replaced by the much smaller nuclear moment. An example is the conversion factor of 42.6 MHz/T for protons (H^+). A well-known medical application of NMR is magnetic resonance imaging (MRI).

A straightforward way of determining $\langle\mathbf{s}\rangle = \langle\psi(t)|\mathbf{s}|\psi(t)\rangle$ is to use $\langle\psi|\mathbf{s}|\psi\rangle = \sum_{ij}\psi_i^*(t)\mathbf{s}_{ij}\psi_j(t)$, where $\psi_i = (\psi_1, \psi_2)$. A more elegant and more general way is to construct an equation of motion directly from the Schrödinger equation. Using $\mathrm{i}\,\hbar\partial\langle\mathsf{A}\rangle/\partial t = \langle\mathsf{AH} - \mathsf{HA}\rangle$ (A.3.1) and exploiting the commutation rules for spin operators yields

$$\frac{\partial\langle\mathbf{s}\rangle}{\partial t} = \frac{g\,\mu_o\,\mu_B}{\hbar}\langle\mathbf{s}\rangle\times\mathbf{H} \tag{6.4}$$

In a slightly different context, this equation is known as the Landau-Lifshitz equation (Section 6.2.1). It shows that the motion of the spin is always perpendicular to the field, in agreement with experiment. A striking prediction is that fields $\mathbf{H}$ *antiparallel* to $\langle\mathbf{s}\rangle$ do not change magnetization, that is, $\mathrm{d}\langle\mathbf{s}\rangle/\mathrm{d}t = 0$. In reality, sufficiently strong reverse fields lead to magnetization reversal. The reason is the neglect of interactions such as the spin-orbit coupling, which yield nondiagonal matrix elements between $\uparrow$ and $\downarrow$ states (see below).

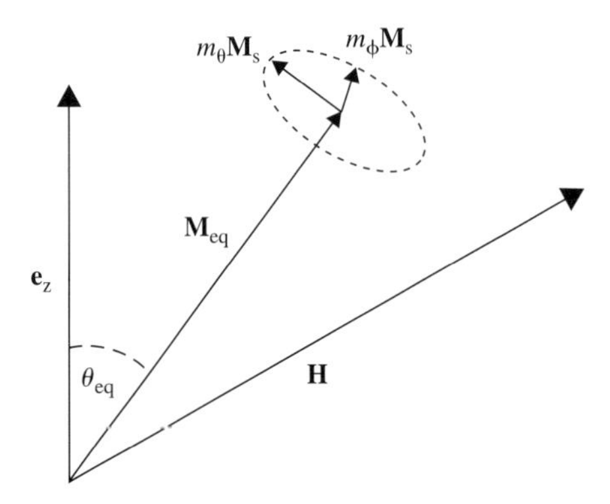

Fig. 6.2 Precession in an anisotropic ferromagnet. $\mathbf{M}_{eq}(\mathbf{H})$ is the equilibrium magnetization direction, and magnetic resonance corresponds to small perturbations m_ϕ and m_θ. In strong fields, $\mathbf{M}_{eq} \parallel \mathbf{H}$.

6.1.2 Uniform magnetic resonance

Equation (6.4) has been derived for paramagnetic spins but can also be used to describe *ferromagnets*. This is because the large exchange fields $\mathbf{H_J}$ responsible for ferromagnetism are parallel to the magnetization, so that their contribution to the cross product in (6.4) vanishes. A condition is that the local magnetization remains uniform (coherent) throughout the magnet.

In anisotropic ferromagnets, one must add the anisotropy field to the external field, and the spin precession is ellipsoidal rather than circular. Figure 6.2 illustrates that the precessing magnetization vector remains close to the equilibrium magnetization $\mathbf{M}_{eq}$. In the coordinate frame of Fig. 6.2, the equations of motion for arbitrary anisotropy are

$$\frac{\partial m_A}{\partial t} = \frac{g\mu_o\mu_B}{\hbar}\left(\frac{H + 2K_B}{\mu_o M_s}\right) m_B \tag{6.5a}$$

$$\frac{\partial m_B}{\partial t} = -\frac{g\mu_o\mu_B}{\hbar}\left(\frac{H + 2K_A}{\mu_o M_s}\right) m_A \tag{6.5b}$$

Here the indices A and B refer to the principal axes of the dashed ellipsoid in Fig. 6.2, corresponding to the anisotropy energy $E_a = -K_A m_A^2 - K_B m_B^2$. The resonance obeys

$$\omega = \frac{g\mu_o\mu_B}{\hbar}\sqrt{(H + 2K_A/\mu_o M_s)(H + 2K_B/\mu_o M_s)} \tag{6.6}$$

In general, K_A and K_B depend, via $\mathbf{M}_{eq}$, on the magnetic field.

Let us consider three examples. First, the trivial isotropic case $K_A = K_B = 0$ is described by $\omega = 2\mu_o\mu_B H/\hbar$, in agreement with Section 6.1.1. Second, uniaxial configurations are characterized by $K_A = K_B = K$ and obey

$$\omega = \frac{g\mu_o\mu_B}{\hbar}\left(\frac{H + 2K}{\mu_o M_s}\right) \tag{6.7}$$

They differ from isotropic magnetism by the addition of the anisotropy field $2K/\mu_o M_s$ to the external field. Third, the general case $K_A \neq K_B$ is exemplified by a soft-magnetic thin film in the x–y plane. The perpendicular configuration ($\theta_{eq} = 0$) is uniaxial, because the x and y directions are equivalent, and $\omega = g\mu_o\mu_B(H - DM_s)/\hbar$. To analyze the in-plane case ($\theta_{eq} = \pi/2$), we assume that the magnetization is parallel to $\mathbf{e}_x$. In free space, the y and z directions would then be equivalent. However, in the film, the demagnetizing field partially suppresses the perpendicular magnetization component to the film, so that

$$\omega = \frac{g\mu_o\mu_B}{\hbar}\sqrt{H(H - DM_s)} \tag{6.8}$$

We see that the in-plane frequency is higher by a factor $(1 - DM_s/H)^{-1/2}$ than the perpendicular frequency.

The resonance modes described by (6.5–8) occur in *perfect* ferromagnets, including thin films (Kittel 1986, Farle 1998). Real-structure features, such as surfaces and imperfections, tend to make the precession nonunifom. This includes inhomogeneous magnetic fields, because ω is a function of H. Examples are surfaces (Eshbach and Damon 1960), confined structures (Hillebrands and Ounadjela 2002), and inhomogeneous nanomagnets (Chipara, Skomski, and Sellmyer 2002, Skomski 2003).

6.1.3 Spin waves

So far, we have restricted ourselves to the uniform mode, $\mathbf{M}(\mathbf{r}, t) = \mathbf{M}(t)$. Nonuniform or incoherent magnetization modes are encountered in imperfect solids and as excited modes in homogeneous ferromagnets. In this subsection we consider excited states known as *spin waves*. Ferromagnetic exchange J_{ij}, as parameterized by the exchange stiffness A, favors parallel spin alignment between neighboring atoms, so that spin waves cost exchange energy. In the simplest case, the energy is of thermal origin, and the spin waves contribute to the temperature dependence of the magnetization.

A simple quantum-mechanical model is the ferromagnetic Heisenberg chain

$$\mathsf{H} = -2\sum_i J\mathbf{s}_i \cdot \mathbf{s}_{i+1} - g\mu_B\mu_o \sum_i \mathbf{s}_i \cdot \mathbf{H} \tag{6.9}$$

where $\mathbf{H} = H\mathbf{e}_z$ defines a quantization axis. The ground state is ferromagnetic and can be visualized as $|\uparrow\uparrow\uparrow\uparrow \ldots \uparrow\uparrow\rangle$. For noninteracting spins ($J = 0$), the lowest-lying excitations are of the type $|\uparrow\downarrow\uparrow\uparrow \ldots \uparrow\uparrow\rangle$, but in the presence of exchange, these states are no longer eigenstates of (6.9). To see this, we use the notation $\mathbf{s} = (s^x, s^y, s^z)$ and express $\mathbf{s} = \frac{1}{2}\boldsymbol{\sigma}$ in terms of the Pauli matrices (2.32), that is,

$$\sigma^x = \begin{pmatrix} 0 & 1 \\ 1 & 0 \end{pmatrix}, \quad \sigma^y = \begin{pmatrix} 0 & -i \\ i & 0 \end{pmatrix}, \quad \text{and} \quad \sigma^z = \begin{pmatrix} 1 & 0 \\ 0 & -1 \end{pmatrix} \tag{6.10}$$

The operators σ^x and σ^y mix $\uparrow$ and $\downarrow$ states, and the same is true for the operator products $s_i^x s_{i+1}^x$ and $s_i^y s_{i+1}^y$ appearing in $\mathbf{s}_i \cdot \mathbf{s}_{i+1}$. This means that the exchange adds complexity by forcing the spins to rotate.

To account for the involvement of s^x and s^y, we introduce creation and annihilation operators $\mathsf{s}^\pm = \mathsf{s}^x \pm i\mathsf{s}^y$ similar to those of Section 2.1.4. Putting $\mathsf{s}^x = (\mathsf{s}^+ + \mathsf{s}^-)/2$ and $\mathsf{s}^y = (\mathsf{s}^+ - \mathsf{s}^-)/2i$ into (6.9) yields

$$\mathsf{H} = -2\sum_i J\mathsf{s}_i^z\mathsf{s}_{i+1}^z - \sum_i J\mathsf{s}_i^+\mathsf{s}_{i+1}^- - \sum_i J\mathsf{s}_i^-\mathsf{s}_{i+1}^+ - g\mu_B\mu_o\sum_i \mathsf{s}_i^z H \tag{6.11}$$

This equation looks more complicated than (6.9) but has a transparent physical meaning. Expressing s^+ and s^- in terms of the Pauli matrices yields

$$\mathsf{s}^+ = \begin{pmatrix} 0 & 1 \\ 0 & 0 \end{pmatrix} \quad \text{and} \quad \mathsf{s}^- = \begin{pmatrix} 0 & 0 \\ 1 & 0 \end{pmatrix} \tag{6.12}$$

The operators s^+ and s^- increase and reduce the spin by one unit, respectively. This is seen very easily by using the vector notation $|\uparrow\rangle = (1, 0)$ and $|\downarrow\rangle = (0, 1)$, so that $\mathsf{s}^+|\downarrow\rangle = |\uparrow\rangle$ and $\mathsf{s}^-|\uparrow\rangle = |\downarrow\rangle$. Operator products such as $\mathsf{s}_i^+\mathsf{s}_{i+1}^-$ move spin states along the chain. An example is

$$\mathsf{s}_2^+\,\mathsf{s}_3^-\,|\uparrow\downarrow\uparrow\uparrow\uparrow\uparrow\rangle = |\uparrow\uparrow\downarrow\uparrow\uparrow\uparrow\rangle \tag{6.13}$$

where the propagation of a $\downarrow$ spin is realized by reducing the z-component of the third spin (s_3^-) and enhancing the z-component of the second spin (s_2^+). Equation (6.13) shows that spin excitations move in an insulator like electrons in a metal, except that they do not carry any charge. This is an example of the spin-charge separation introduced in Section 2.1.8.

The treatment of the wave functions depends on the density of reversed spins. When the density is small, for example at low temperatures, we can model the system as a superposition of states with one reversed spin, as in (6.13), and label the wave function by the position of the reversed spin. For example, (6.13) becomes $\mathsf{s}_2^+\,\mathsf{s}_3^-\,|2\rangle = |3\rangle$. More generally, the exchange J in (6.11) connects neighboring spins so that

$$\mathsf{H}|i\rangle = 2J|i\rangle - J|i-1\rangle - J|i+1\rangle + g\mu_B\mu_o H|i\rangle \tag{6.14}$$

Here we have ignored the physically unimportant zero-point energy $JN/2$, where N is the number of spins in the chain.

Equation (6.14) is very similar to tight-binding description of interatomic hopping in an atomic chain (Panel 6) and easily solved by putting $|k\rangle = \exp(ikR_i)|i\rangle$. These wave-vector-dependent excitations are known as *spin waves* or magnons. Each spin wave corresponds to a single reversed spin, but the spin reversal extends throughout the magnet, as contrasted to the localized reversal shown in (6.13). The energy levels are

$$E(k) = 2J(1 - \cos(ka)) + g\mu_B\mu_o H \tag{6.15}$$

where a is the interatomic distance. For small wave vectors, or long wavelengths $\lambda = 2\pi/k$, the dispersion relation simplifies to $E(k) = Jk^2a^2 + g\mu_B\mu_o H$. This equation is often written as $E(k) = Dk^2 + g\mu_B\mu_o H$, where the *spin-wave stiffness* D is closely related to exchange stiffness A (Section 4.2.3). For the ferromagnetic transition-metal elements, D is of the order of 400 meVÅ^2. In two and three dimensions, the dispersion relation has the form $E(\mathbf{k})$ and is often treated on a quasi-classical phenomenological level.

Spin waves exist not only in ferromagnets but also in ferri- and antiferromagnets. In (6.14), this requires the replacement of J by sublattice-specific exchange constants $J_{\rm ij}$, very similar to the treatment of sublattice effects in magnetic alloys (Section 5.3.6). The result is several branches of spin waves, one for each eigenmode of the system. Furthermore, in antiferromagnets, the dispersion relation is essentially linear, $E(k) \sim k$. In a magnetic field, long-wavelength spin-waves have an energy $E(0) = g\mu_{\rm B}\mu_{\rm o}H$, and a similar excitation energy, known as *anisotropy gap*, is encountered in anisotropic magnets. The gap can be used for the experimental determination of anisotropy constants.

In the absence of a magnetic field, long-wavelength magnons have very low energies $E \sim k^2$. Since each spin wave carries a moment of $\mu_{\rm B}$, thermally excited spin waves yield a strong contribution to the temperature dependence of the spontaneous magnetization (Bloch 1930). This helps to explains the failure of the mean-field model at low temperatures (Section 5.3.7) and the absence of ferromagnetism in low-dimensional isotropic magnets. The idea is to write the spontaneous magnetization as

$$M_{\rm s}(T) = M_{\rm s}(0)\left(1 - \frac{2}{N}\sum_{\mathbf{k}} n_{\mathbf{k}}(T)\right) \tag{6.16}$$

where $n_{\mathbf{k}}$ is the number of magnons of wave vector $\mathbf{k}$ and N is the total number of spins. For fermions, such as electrons, the number of particles per quantum state is zero or one. However, magnons are bosons, and any k-state can be occupied by an arbitrarily large number $n_{\mathbf{k}}$ of magnons. It is possible to artificially accumulate a large number of magnons in well-defined k-states (Demokritov *et al.* 2006), but this condensation is physically different from the Bose-Einstein condensation of light atoms such as He (Section A.3.4). For example, the Bose-Einstein condensation temperature $T_{\rm o}$ scales as $1/m^{2/3}$ (Wannier 1966), but magnons are massless, corresponding to $T_{\rm o} = \infty$. Trivial examples are magnets in uniform and periodic magnetic fields, where magnon numbers with $k = 0$ (coherent rotation) and $k > 0$ (periodic field) can be very large.

In the formalism of second quantization (Section 2.1.6), one can consider n particle-number operators with eigenvalues $n = 0, 1, \ldots, \infty$, but the total number of magnons cannot be larger than N, corresponding to complete magnetization reversal. This indicates that the assumption of noninteracting one-magnon states breaks down for large deviations from $M_{\rm s}(0)$. The corresponding nonlinear corrections go beyond the scope of simple models of magnetism and can be treated by methods such as Holstein-Primakoff transformations (Dyson 1956, Jones and March 1973). Here we stick to the approximation of noninteracting magnons, where the occupancy probabilities are multiplicative. For example, if the equilibrium probability p_1 of finding one magnon of energy $E_{\mathbf{k}}$ is $\exp(-E_{\mathbf{k}}/k_{\rm B}T)$, then that of finding two magnons in the state p_1 is $p_2 = \exp(-2E_{\mathbf{k}}/k_{\rm B}T)$. Averaging over all occupation numbers, $n_{\mathbf{k}} = \sum_{\rm n} n\, p_{\rm n} / \sum_{\rm n} p_{\rm n}$, yields the Bose-Einstein distribution $n_{\mathbf{k}} = 1/(\exp(E_{\mathbf{k}}/k_{\rm B}T) - 1)$. In infinite solids, the summation over $\mathbf{k}$ can be replaced by an integral

$$\sum_{\mathbf{k}} n_{\mathbf{k}} \sim \int \frac{1}{\exp(E_{\mathbf{k}}/k_{\mathrm{B}}T) - 1} \, \mathrm{d}\mathbf{k} \tag{6.17}$$

where k extends from zero to some cutoff of order $1/a$. Putting $E_{\mathbf{k}} = Dk^2$, substituting $x = Dk^2/k_{\mathrm{B}}T$, and exploiting $\mathrm{d}\mathbf{k} \sim k^{\mathrm{d}-1}\,\mathrm{d}k$ yields

$$\sum_{\mathbf{k}} n_{\mathbf{k}} \sim \mathrm{T}^{\mathrm{d}/2} \int \frac{x^{\mathrm{d}/2-1}}{\exp(x) - 1} \, \mathrm{d}x \tag{6.18}$$

In the long-wavelength limit, x is small, $\exp(x) \approx 1 + x$, and the integral simplifies to $\int x^{\mathrm{d}/2-2}\,\mathrm{d}x$. This expression diverges in one and two dimensions. In one dimension, the divergence scales as $1/k$, meaning that long-wavelength magnons destroy ferromagnetism at any finite temperature. This behavior is similar to that of one-dimensional Ising spins (Section 5.4.1). In two dimensions, the divergence is weak (logarithmic), indicating the onset of ferromagnetism. More generally, a theorem due to Mermin and Wagner (1966) states that there is no ferromagnetic long-range order in two-dimensional isotropic Heisenberg magnets. However, there is always some anisotropy in two-dimensional magnets, so that the Mermin-Wagner theorem is of limited practical importance. The anisotropy creates a spin-wave gap, meaning that the excitation energy of the coherent mode ($k = 0$) is finite and (6.17) no longer diverges.

In three dimensions, the integral (6.18) is finite and $\sum_{\mathbf{k}} n_{\mathbf{k}}$ scales as $\mathrm{T}^{3/2}$. This is Bloch's famous law determining the spontaneous magnetization at low temperatures. Interactions between spin waves yield a T^4 contribution in addition to Bloch's $T^{3/2}$ term. The Bloch argument can also be used to discuss the transition from one-dimensional to three-dimensional magnetism in nanowires. In an atomic wire, there is only one wave-vector direction, namely parallel to the wire axis. In nanowires, there are additional perpendicular components. The finite wire radius R puts a lower limit to the corresponding wave vectors and energies, so that these modes are frozen at low temperatures. However, when $k_{\mathrm{B}}T$ exceeds about D/R^2, the distribution of the k-vectors and the behavior of the wire become bulk-like. In a strict sense, nanowires are nonferromagnetic, but with increasing cross-sectional area, the one-dimensional features become less obvious and more difficult to measure. In the simplest case, the equilibration times (Section 6.4.7) become very long when the radius exceeds a very few nanometers.

6.1.4 Spin dynamics in inhomogeneous magnets*

Macroscopic inhomogeneities lead to a trivial superposition of resonance lines, because each region has a different resonance frequency. This adds to the broadening of the resonance lines and may even split the lines. However, the inhomogeneities compete against interactions. On macroscopic scale, inhomogeneous local fields and long-range magnetostatic interactions enhance and reduce the broadening, respectively. Macroscopic interactions of this type have been investigated for a long time, starting with Walker (1957).

Exchange becomes important on small length scales, in close analogy to micromagnetic problems. A simple example is magnetization modes in nanowires, as

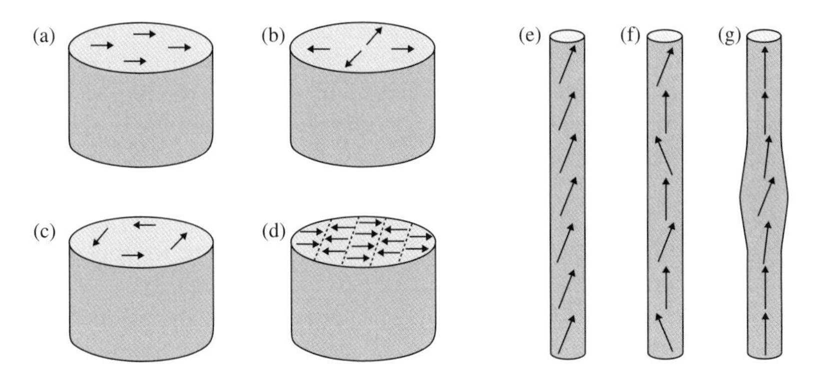

Fig. 6.3 Magnetization modes in nanowires in a magnetic field parallel to the axis of revolution: (a–d) cross-sectional view of the magnetization component perpendicular to the wire axis and (e–f) magnetization along the wire. Modes (e) and (f) differ by their respective wave vectors k = 0 (uniform) and $k > 0$ (spin wave), whereas (g) is the lowest-lying solution of (6.17).

shown in Fig. 6.3. There are many modes with wave vectors $\mathbf{k}$ parallel and perpendicular to the wire axis $\mathbf{e}_z$, or forming some angle with $0 < \theta < \pi/2$ with $\mathbf{e}_z$. The boundary conditions imposed by the surface of the wires gives rise to *spin-wave quantization* for k vectors perpendicular to the wire axis. Compared to (a), the modes (b–d) cost exchange energy, as do (f) and (g) compared to (e). This is because $J \sim A\nabla^2$ punishes magnetization inhomogeneities. However, inhomogeneous modes may be excited by thermal activation or admixed due to wire imperfections. The mode shown in Fig. 6.3(g) costs some exchange energy but yields a gain in magnetostatic energy, because the demagnetizing field of an ellipsoid of revolution with finite aspect ratio is smaller than that of an infinite wire.

A simple solution is obtained for very thin wires with uniaxial symmetry, as in Fig. 6.3(e–g). Adding the exchange energy to the anisotropy energy transforms (6.5) into

$$2A\frac{\mathrm{d}^2 m}{\mathrm{d}z^2} + \left(2K_{\mathrm{eff}}(z) + \mu_{\mathrm{o}}M_{\mathrm{s}}H - \frac{\hbar\omega}{g\mu_{\mathrm{B}}}M_{\mathrm{s}}\right) m = 0 \tag{6.19}$$

where $m(z)$ is the small magnetization component perpendicular to the wire axis and K_{eff} is the effective anisotropy (Chipara, Skomski, and Sellmyer 2002). In the example of Fig. 6.3(g), the imperfection is geometrical rather than chemical, and K_{eff} reflects the inhomogeneity of the local shape anisotropy. Wires homogeneous along the z axis are characterized by a constant function $K_{\mathrm{eff}}(z)$ and the magnetic resonance reduces to the uniform mode (6.5). For arbitrary anisotropy, (6.19) is a differential equation, and the resonance mode is determined by the competition between exchange, A, and disorder, $K_{\mathrm{eff}}(z)$. The outcome of the competition depends on the spatial extent Δz of the inhomogeneity. When Δz is much smaller than about 10 nm, the exchange ensures a nearly uniform mode. When Δz is much larger than about 10 nm, then $A\,\mathrm{d}^2m/\mathrm{d}z^2 \approx 0$ and the modes are a trivial macroscopic superposition.

As emphasized by Aharoni (1996), nucleation and resonance modes are closely related to each other, and the latter can be considered as a special case with $\omega = 0$. This helps us to understand real-structure effects in magnetic resonance. For example, (6.19) implies that the modes are superpositions of waves of the type $\mathbf{k} = (0,\ 0, k_z)$. The energy density of the lowest-lying modes *perpendicular* to the wire axis is of order A/R^2, so that modes in thin wires are confined to the z-direction, as in Fig. 6.3(e–g). In thick wires, this is a poor approximation, and one must consider wave vectors $\mathbf{k} = (k_x, k_y, k_z)$ and inhomogeneities $K_{\text{eff}}(x, y, z)$. Examples are magnetization modes in polycrystalline nanowires (Skomski *et al.* 2000) and in nanotubes.

Aside from the lowest lying nucleation ($\omega = 0$) or resonance ($\omega > 0$) modes, eq. (6.19) predicts excited modes. These modes are long-wavelength spin waves very similar to those of (6.15). More generally, ferromagnetic resonance (FMR) and spin-wave resonance (SWR) can be treated on a common footing and are then referred to as electron spin resonance (ESR). An upper limit to the exchange energy is obtained as $A\nabla^2 \sim A/a^2$, where a is the interatomic distance. This limit is known as *antiferromagnetic resonance* and described by appropriate atomic Hamiltonians of the type (6.9), with generally two or more sublattices. The corresponding modes are similar to the spontaneous magnetization modes of Section 5.3.6, but the physics is somewhat different, and special care is required to analyze what is actually observed in a given experiment. In particular, the field must be able to interact with the magnetization and have a nonzero projection onto the mode, $\int H(z)\, m(z)\, \mathrm{d}z \neq 0$ in (6.19). A well-known example is spin-wave resonance in thin films (Kittel 1986), where the magnetic field is homogeneous but surface anisotropy acts as an imperfection that yields some admixture of spin-wave character.

6.2 Relaxation

Summary The precession of magnetization vectors is damped by the interaction between relevant macroscopic (magnetic) and irrelevant or heat-bath degrees of freedom. The latter include, for example, lattice vibrations (phonons). The damping or relaxation time depends on the interactions between different subsystems, as described by Fermi's golden rule. However, the deterministic character of the Schrödinger equation forbids irreversible processes, and the same is true for the closely related Liouville-von Neumann equation. An example is Zermelo's recurrence objection, which states that any system eventually returns to its original state. The reason for the reversibility is the consideration of both relevant and irrelevant degrees of freedom. In reality, there is a separation of microscopic (reversible) and macroscopic (irreversible) time scales, and relaxation is obtained by integration over all microsopic or irrelevenat degrees of freedom. A simple mechanical analog is a system of masses coupled by harmonic springs. Relaxation proceeds towards local energy minima, as opposed to thermally activated magnetization processes, although the two phenomena have similar physical origins.

An important feature of magnetization dynamics is *relaxation*. Equations such as (6.3) predict an undamped precession, with a infinite lifetime τ. In reality, the magnetization state relaxes towards its equilibrium position. In magnetic resonance, this is an essential contribution to the line width, and in magnetic materials, it yields a time dependence of the magnetic properties. There are different relaxation mechanisms. First, nondiagonal and generally random matrix elements give rise to transitions between different magnetization directions. Second, the magnetization may change due to quantum tunneling through a magnetic energy barrier. This mechanism is related to the first mechanism but usually observed in perfect structures, such as magnetic molecules (Wernsdorfer 2006), and limited to low temperatures. Third, on a macroscopic scale, one encounters dynamic phenomena such as eddy-current losses in metallic magnets. Fourth, an important contribution is thermally activated magnetization reversal over micromagnetic energy barriers. The focus of this section is on quantum-mechanical mechanisms, especially the first one, whereas the thermally activated magnetization reversal will be treated in Section 6.4.

6.2.1 Damped precession

Phenomenological relaxation models are obtained by adding damping terms to (6.4). One example is the Landau-Lifshitz equation, which can be written as

$$\frac{\partial \mathbf{s}}{\partial t} = \frac{\omega_o}{H} \mathbf{s} \times \mathbf{H} - \frac{1}{\tau_o H} \mathbf{s} \times (\mathbf{s} \times \mathbf{H}) \tag{6.20}$$

where $\mathbf{s} = \mathbf{M}/M_s$, ω_o describes the undamped precession and τ_o is a relaxation time of order 10^{-9} s. Physically, τ_o is determined by interactions that couple spin states with different projections onto $\mathbf{H}$ (Section 6.2.2), most notably spin-orbit coupling. Equation (6.20) is algebraically equivalent to the Gilbert equation, where the term $\mathbf{s} \times (\mathbf{s} \times \mathbf{H})$ is replaced by $\mathbf{s} \times \mathrm{d}\mathbf{s}/\mathrm{d}t$ but the parameters ω_o and τ_o have a slightly different meaning (Aharoni 1996).

To study the behavior of (6.20), we consider a spin precessing around $\mathbf{H} = H\mathbf{e}_z$. Restricting ourselves to small precession angles, that is, to terms linear in s_x and s_y, we obtain

$$\frac{\partial s_x}{\partial t} = \omega_o s_y - \frac{1}{\tau_o} s_x \tag{6.21a}$$

$$\frac{\partial s_y}{\partial t} = -\omega_o s_x - \frac{1}{\tau_o} s_y \tag{6.21b}$$

The perpendicular magnetization component $m = \sqrt{s_x^2 + s_y^2}$ obeys $\partial m/\partial t = -m/\tau_o$, corresponding to an exponential decay $m(t) = m(0)\exp(-t/\tau_o)$. This shows that τ is indeed a relaxation time. Substitution of s_y into (6.22b) yields

$$\frac{\partial^2 s_x}{\partial t^2} + \frac{2}{\tau_o}\frac{\partial s_x}{\partial t} + \left(\omega_o^2 + \frac{1}{\tau_o^2}\right) s_x = 0 \tag{6.22}$$

Equations of this type are well known from mechanics and describe damped oscillations. In magnetic resonance, the damping is seen as a line-width broadening ΔH. For weak damping, $\tau_o \gg 1/\omega_o$, the corresponding linewith $\Delta H \sim 1/\tau_o$. In practice, one often uses damping parameters $\alpha \sim 1/\tau_o\omega_o$, whose inverse $1/\alpha$ is the number of oscillations before the amplitude decays to some fraction of its original value.

6.2.2 *Physical origin of relaxation

Equations (6.20–23) provide a phenomenological description of damping but do not explain the underlying physics. To understand damping, it is necessary to identify the quantum-mechanical processes that add to the unperturbed spin precession (6.2–4) and yield the damping term in the Landau-Lifshitz equation. In a nutshell, the operator $\mathsf{H} = \mathsf{H}_o$ in (6.1) must be replaced by $\mathsf{H}_o + \mathsf{V}$, where V describes the interactions of the system. Typically, V contains several contributions, such as spin-orbit, magnetostatic, electron-magnon, electron-phonon, and/or magnon-magnon interactions. The individual mechanisms depend on the investigated material, and both intrinsic factors, such as the presence of conduction electrons, and extrinsic factors (imperfections) play a role. A very simple mechanism involves magnetostatic fields created by misaligned atomic spins. In general, the fields are neither parallel nor antiparallel to H_z and therefore able to change the spin component parallel to the field.

A quantum mechanical treatment of V is based on second-order perturbation theory of the Schrödinger equation $\mathrm{i}\hbar\partial|\psi\rangle/\partial t = \mathsf{H}_o|\psi\rangle + \mathsf{V}|\psi\rangle$. It starts from the eigenfunctions $|\psi_j\rangle$ of H_o and yields transition rates $W_{ij} = W(j \rightarrow i)$ between the eigenstates of the unperturbed Hamiltonian. The result of the calculation is *Fermi's golden rule*

$$W_{ij} = \frac{2\pi}{\hbar}|\langle\psi_i|\mathsf{V}|\psi_j\rangle|^2\ \delta(E_i - E_j) \tag{6.23}$$

The delta function $\delta(E_i - E_j)$ means that the scattering is limited to states of equal energy. This is because perturbation theory involves denominators $1/(E_i - E_j)$, which are largest for $E_i - E_j$ (see appendix A.3.3). The rates scales as $W_{ij} \sim 1/\tau$ and provide a link between relaxation behavior and the quantum-mechanical matrix elements.

Models described by Fermi's golden rule exhibit a time-dependent decay of the original quantum state. The decay includes energy redistributions between different subsystems but conserves the total energy of the system. This is striking, because experience tells us that energy *changes* during relaxation. A closely related feature is that the scattering described by (6.23) mixes the wave functions but does not create the thermal disorder (entropy) expected for relaxation processes.

To study the entropy production, we start from the *Liouville-von Neumann equation* $\mathrm{i}\hbar\partial\rho/\partial t = [\mathsf{H}, \rho]$ for the density matrix, which is equivalent to the Schrödinger equation and can be derived in analogy to (A.31), and determine the time dependence of the entropy operator $\eta = -k_B \ln \rho$. Here we have used the symbol η to avoid confusion with the spin. An equation of motion is obtained by using $\langle A\rangle = \mathrm{Tr}(\mathsf{A}\rho)$:

$$\frac{\partial\langle\eta\rangle}{\partial t} = \frac{1}{\mathrm{i}\hbar}\mathrm{Tr}(\eta\mathsf{H}\rho - \eta\rho\mathsf{H}) \tag{6.24}$$

Using the identity $\mathrm{Tr}(\mathsf{ABC}) = \mathrm{Tr}(\mathsf{CAB})$ and exploiting that $[\rho, \eta] = [\rho\eta - \eta\rho] = 0$ we find that $\partial\langle\eta\rangle/\partial t = 0$. Even if we were able to solve the exact Schrödinger (or Liouville-von Neumann) equation, we wouldn't be able to reproduce the observed dissipation of energy! The ultimate reason is the reversible character of the Schrödinger equation. The Schrödinger equation does not forbid the spontaneous creation of a house from a pile of rubble, but experience tells us that this is not the case. A classical analog is Zermelo's *Wiederkehreinwand* or "recurrence objection", which was based on a Poincaré cycle after which any system returns to it original state. In fact, as argued by Boltzmann (1896), such a recurrence is possible but extremely unlikely and can safely be ignored.

How can we transform (6.24) into an irreversible equation? Consider a quantum-mechanical system with energy-level spacings $E_{\mu+1} - E_\mu \approx \Delta E$. The system has a recurrence time of order $1/\Delta\omega \sim \hbar/\Delta E$. For isolated systems, the recurrence time is finite, as exemplified by the resonance frequency of the paramagnetic gas (Section 6.1.1). However, interactions with the environment yield additional levels, and with increasing system size the number of levels and the recurrence time increase, too. Physically, the environment acts as a heat bath that absorbs the energy produced during relaxation.

In the Liouville-von Neumann equation, the embedding in a heat bath amounts to separating the degrees of freedom into two classes: relevant degrees of freedom, such as the magnetization $\mathbf{M}$ of a particle, and irrelevant degrees of freedom, such as the velocities and positions of the gas atoms in the heat bath. The heat bath exerts a random force, and this force is the reason for the irreversibility. Integration over all heat-bath degrees of freedom leaves a modified Liouville-von Neumann equation differing from the equation for the isolated system by additional random-force and relaxation terms. In practice, this procedure is complicated, particularly since the heat-bath degrees of freedom may actually be located inside the magnets (magnons and phonons). However, the next subsection presents a simple classical model that illustrates how the random-force and relaxation terms arise.

In general, the relaxation term involves a memory function $g(t-t')$ so that

$$\frac{\partial s}{\partial t} + \int_{-\infty}^{t} g(t-t')\, s(t')\, \mathrm{d}t' = 0 \tag{6.25}$$

where s is the considered relevant degree of freedom. On a macroscopic scale, one may replace the memory function by a δ-function and then obtains an ordinary relaxation equation, $\partial s/\partial t + \Gamma_\mathrm{o} s = 0$. This equation has the solution $s(t) = s(0)\exp(-t/\tau_\mathrm{o})$, where $\tau_\mathrm{o} = 1/\Gamma_\mathrm{o}$ is the relaxation time. The opposite limit of a constant memory function describes undamped oscillation, $\partial^2 s/\partial t^2 + \omega_\mathrm{o}^2 s = 0$, whereas exponential memory functions reproduce (6.22), interpolating between the two limits. As pointed out by Mori (1965), memory functions are a very fundamental aspect of nonequilibrium statistical mechanics, containing much of the physics involved.

6.2.3 *A mechanical model

On an atomic level, relaxation is due to interactions with a heat bath. For simplicity, we consider a classical model consisting of coupled harmonic oscillators. One particle has a mass M much larger than the mass m of the other particles. We assume that the particle of mass M carries a magnetic moment, so that the position s of this particle serves as relevant magnetization variable. The model Hamiltonian is

$$\mathsf{H} = \frac{M}{2}\left(\frac{\mathrm{d}s}{\mathrm{d}t}\right)^2 + \frac{q}{2}s^2 + \sum_\mathrm{i}\frac{m}{2}\left(\frac{\mathrm{d}x_\mathrm{i}}{\mathrm{d}t}\right)^2 + \sum_\mathrm{i}\frac{k_\mathrm{i}}{2}x_\mathrm{i}^2 - g\sum_\mathrm{i} s\,x_\mathrm{i} \tag{6.26}$$

Here g describes the coupling between the relevant magnetization variable s, whose relaxation is investigated, and the heat bath $x_{\rm i}$. The equations of motion are

$$\left(\frac{\mathrm{d}x_{\rm i}}{\mathrm{d}t}\right)^2 + \omega_{\rm i}^2 x_{\rm i} = \frac{g}{m} s \tag{6.27}$$

$$\left(\frac{\mathrm{d}s}{\mathrm{d}t}\right)^2 + \Omega^2 s = \frac{g}{M} \sum_{\rm i} x_{\rm i} \tag{6.28}$$

The equation for $x_{\rm i}$ has the structure $\mathrm{d}^2x/\mathrm{d}t^2 + \omega^2 x = b(t)$ and the solution

$$x(t) = x(0)\cos(\omega t) + \frac{\frac{\mathrm{d}x}{\mathrm{d}t}(0)}{\omega}\sin(\omega t) + \exp(-\mathrm{i}\omega t)\int_0^t\int_0^t \exp(2\mathrm{i}\omega t' - \mathrm{i}\omega t'')\, b(t'')\,\mathrm{d}t''\,\mathrm{d}t' \tag{6.29}$$

Putting this result and $b = sg/m$ into the equation of motion for s we obtain

$$\left(\frac{\mathrm{d}s}{\mathrm{d}t}\right)^2 + \Omega^2 s = f(t) + \frac{g^2}{mM}\sum_{\rm i} \exp(-\mathrm{i}\omega_{\rm i} t) \int_0^t\int_0^{t'} \exp(2\mathrm{i}\omega_1 t' - \mathrm{i}\omega_1 t'') s(t'')\,\mathrm{d}t''\,\mathrm{d}t' \tag{6.30}$$

Here $f(t) = g\sum_{\rm i}[x_{\rm i}(0)\cos(\omega_{\rm i}t) + \mathrm{d}x_{\rm i}/\mathrm{d}t(0)\sin(\omega_{\rm i}t)/\omega_{\rm i}]/M$ can be considered as a random force which depends on the unknown initial conditions $x_{\rm i}(0)$ and $\mathrm{d}x_{\rm i}/\mathrm{d}t(0)$. Note that $\langle f\rangle = 0$, whereas the strength of the force is given by the bath's equilibrium condition $\langle[\mathrm{d}x_{\rm i}/\mathrm{d}t(0)]^2\rangle + \omega_{\rm i}^2\langle x_{\rm i}^2(0)\rangle = k_{\rm B}T/m$.

The motion of the coupled oscillators s and $x_{\rm i}$ gives rise to a *recurrence* time scaling as $1/\Delta\omega$, where $\Delta\omega$ is a typical frequency spacing. However, the thermodynamic limit the number of heat-bath oscillators goes to infinity and $\Delta\omega = 0$. We can therefore replace the summation $\sum_{\rm i}\ldots$ by an integral $\int D(\omega)\ldots\mathrm{d}\omega$. Furthermore, we assume that g is small, so that $s(t'')$ can be replaced by the unperturbed expression $s(0)\cos(\Omega t'')$. Then the t'' integration yields delta-functions $\delta(\omega \pm \Omega)$ and we can exploit $\mathrm{d}\sin(\Omega t'')/\mathrm{d}t = \Omega\cos(\Omega t'')$ to obtain

$$\left(\frac{\mathrm{d}s}{\mathrm{d}t}\right)^2 + \Gamma\left(\frac{\mathrm{d}s}{\mathrm{d}t}\right) + \Omega^2 s = f(t) \tag{6.31}$$

where

$$\Gamma = \frac{\pi \mathrm{g}^2 D(\Omega)}{mM\Omega^2} \tag{6.32}$$

This equation shows how relaxation reflects the coupling to a heat bath. Note that this treatment and the result are classical but similar to the more complicated quantum-mechanical treatment (Zwanzig 1961, Mori 1965).

6.3 Coarse-grained models

Summary The time-dependent many-body Schrödinger equation can, in principle, be used to predict the evolution of any magnetic system. However, this is neither practical nor necessary, because individual heat-bath degrees of freedom do not contain any specific information about the magnetic behavior. Coarse-grained models abstract from quantum-mechanical features operative on atomic length and time scales. Simplifying somewhat, there are descriptions based on three different types of equations: (i) master or rate equations, (ii) Fokker-Planck equations, and (iii) Langevin equations. These approaches are physically largely equivalent, although the master equation is able to describe macroscopic magnetization jumps, whereas the Fokker-Planck and Langevin equations interpret macroscopic magnetization changes as a chain of microscopic events. Like the master equation, the Fokker-Planck equation deals with the probability of magnetization configurations, whereas the Langevin equation governs the local magnetization vector as a function of random thermal forces.

The large number of heat-bath degrees of freedom makes the solution of the full Schrödinger equation practically impossible and essentially useless. To make meaningful predictions about the relevant magnetic degrees of freedom, such as the position of a domain wall, one must treat the irrelevant degrees of freedom as a heat bath. This procedure has been outlined in the previous section. The corresponding procedure is also known as *coarse graining*, because it maps the full phase space onto a simplified or "coarse-grained" phase space. The coarse graining is accompanied by the introduction of random forces and relaxation times which the interaction with the heat bath. The corresponding models ignore quantum-mechanical features such as the evolution of individual atomic wave functions. However, coarse-grained parameters, such as the relaxation time, have a sound quantum-mechanical basis and can, in principle, be calculated by quantum-mechanical methods.

Figure 6.4 illustrates the modeling for a mechanical system and a magnetic particle. The forces f_{i} and fields h_{i} are of atomic origin but usually modeled as a white noise,

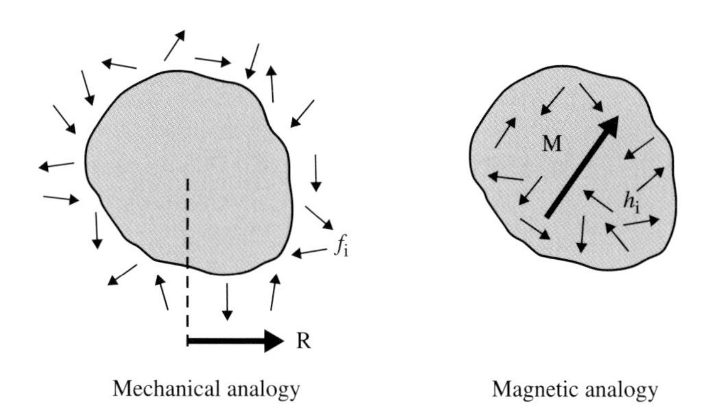

Fig. 6.4 Random forces f_{i} and random fields h_{i} in coarse-grained models.

that is, the investigated relaxation phenomenon is assumed to be much slower than the underlying quantum-mechanical interactions. For example, very fast processes require the replacement of the parameter τ by the memory function $g(t - t')$.

6.3.1 Master equation

A conceptually very simple but powerful approach to dynamics is based on transition rates $W(s, s') = W(s' \to s)$ between states s' to s. The bookkeeping of such transitions leads to the rate or *master* equation

$$\frac{\partial P(s)}{\partial t} = \int [W(s, s')\, P(s') - W(s', s)\, P(s)]\, \mathrm{d}s' \tag{6.33}$$

for the probability $P(s)$. For example, s may correspond to a domain-wall position, and the $W(s, s')$ then describes domain-wall jumps from s to s'. In discrete systems, the integration must be replaced by a summation of discrete states s_i and s_j (A.2.4). Figure 6.5 illustrates the meaning of this equation.

The basic assumption behind the master equation is the Markov character of the evolution, that is, $P(s,t)$ is a function of $P(s, t - \mathrm{d}t)$ but independent of earlier states, such as $P(s, t - 2\mathrm{d}t)$. Equation (6.25) indicates that memory functions $g(t - t')$ are, in general, non-Markovian. For example, (6.33) excludes oscillations, which are described by a constant g. Fortunately, the assumption of a needle-shaped memory function is often a good approximation, and (6.33) provides a adequate description of many systems.

Equilibrium, $\partial \mathrm{P}(s)/\partial t = 0$, is realized by the *detailed-balance* condition

$$W(s, s')\, P(s') = W(s', s)\, P(s) \tag{6.34}$$

This equation must be satisfied for each pair of states s and s', or s_i and s_j. The corresponding probability flux is shown Fig. 6.5(b). Detailed balance, as the limiting case of the models in this section, must not be confused with the physically different realization of $\partial P(s)/\partial t = 0$ by steady-state processes, as in Fig. 6.5(c). Since the equilibrium probability $P(s)$ is proportional to $\exp(-E(s)/k_\mathrm{B}T)$, the detailed-balance condition can also be written as

$$W(s, s') = W(s', s)\, \exp\left(\frac{(E(s') - E(s))}{k_\mathrm{B}T}\right) \tag{6.34a}$$

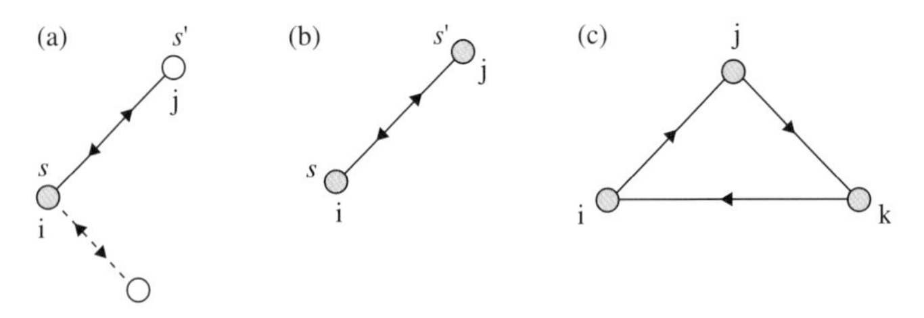

Fig. 6.5 Master equation: (a) summation over all final and initial (b) detailed balance, and (c) steady state.

For example, Fermi's golden rule (6.23) describes transitions without energy change, $E(s') = E(s)$, so that $W(s, s') = W(s', s)$.

A very simple master-equation model assumes two states of energy $E_{1/2}$ separated by a saddle point of energy $E_o > E_{1/2}$. The model has many applications in magnetism and beyond. For example, by identifying the states 1 and 2 with atomic positions it provides an approximate description of viscoelasticity. The master equation is

$$\frac{\mathrm{d}P_1}{\mathrm{d}t} = W_{12}P_2 - W_{21}P_1 \tag{6.35a}$$

$$\frac{\mathrm{d}P_2}{\mathrm{d}t} = W_{21}P_1 - W_{12}P_2 \tag{6.35b}$$

where

$$W_{12} = \Gamma_o \exp\left(\frac{E_2 - E_o}{k_B T}\right) \tag{6.36a}$$

$$W_{21} = \Gamma_o \exp\left(\frac{E_1 - E_o}{k_B T}\right) \tag{6.36b}$$

It is an easy exercise to show that W_{21} and W_{12} obey the detailed-balance principle. The energy differences in (6.36) have a very simple interpretation as activation energies. For example, when going from state 1 to state 2, as described by (6.36b), the activation energy (or energy barrier) is equal to $E_o - E_1$.

Let us consider the magnetization variable $s = P_2 - P_1$, which obeys $-1 \leq s \leq +1$. Subtracting (6.35a) and (6.35b) and taking into account that $P_1 + P_2 = 1$ yields the equation of motion

$$\frac{\mathrm{d}s}{\mathrm{d}t} = \Gamma(s_{eq} - s) \tag{6.37}$$

Here the equilibrium magnetization has the familiar form $s_{eq} = \tanh((E_1 - E_2)/2k_B T)$, and the relaxation rate

$$\Gamma = \Gamma_o \exp\left(\frac{E_1 - E_o}{k_B T}\right) + \Gamma_o \exp\left(\frac{E_2 - E_o}{k_B T}\right) \tag{6.38}$$

Equations (6.37) and (6.38) are used, for example, to describe *superparamagnetic* particles (Section 6.4.6), where $s = \cos\theta$ and the magnetization fluctuates over an energy barrier. For aligned Stoner-Wohlfarth particles, $E_{1/2} - E_o = KV(1 \pm H/H_a)^2$ and

$$\Gamma = 2\Gamma_o \exp\left(-\frac{KV}{k_B T}\left(1 + \frac{H^2}{H_a^2}\right)\right) \cosh\left(\frac{\mu_o M_s H V}{k_B T}\right) \tag{6.39}$$

In small fields $\Gamma \ll \Gamma_o$, but when the field approaches the anisotropy field H_a, then equilibrium is established with a rate approaching Γ_o.

Often $E_1 \gg E_2$, that is, the final state has an energy much lower than the initial state. Then $s = 1$ in equilibrium and (6.38) reduces to the Arhhenius law $\Gamma_o \exp(E_a/k_B T)$. Here $E_a = E_1 - E_o$ is the energy barrier that must be overcome to reach equilibrium. In the discussion of slow magnetization processes (Section 6.4), we will make extensive use of this limit.

The Ising model, characterized by $s_i = \pm 1$, has no intermediate states and, therefore, no inherent dynamics. However, it is possible to introduce Ising dynamics by assuming specific transition rates W_{ij}. One example is the *Glauber* model (1963), defined by $W(s_i \to -s_i) = \frac{1}{2}\Gamma_o(1 - s_i \tanh(s_i h_i + s_i \sum_j J_{ij}\, s_j))$. It yields the equation of motion $\Gamma_o \mathrm{d}\langle s_i \rangle / \mathrm{d}t = -\langle s_i \rangle + s_i^o$, where $s_i^o = \langle \tanh(h_i + \sum_j J_{ij}\, s_j) \rangle$ is the equilibrium magnetization. On a mean-field level, $s_i^o = \tanh(h_i + \sum_j J_{ij} \langle s_j \rangle)$ and, in the vicinity of the critical point,

$$\Gamma_o \frac{\mathrm{d}\langle s_i \rangle}{\mathrm{d}t} = -\langle s_i \rangle + h_i + \sum_j J_{ij} \langle s_j \rangle \tag{6.40}$$

This dynamics is of the Ornstein-Zernike type (Section 5.3.3)

6.3.2 Fokker-Planck equations

The transition rates $W(s', s)$ of the master equation contain both deterministic (drift) and random (diffusion) terms. In the simplest case, the diffusion is described by rates $W(s', s) = \frac{1}{2}\Gamma_o \delta(s' - s - \Delta s) + \Gamma_o \delta(s' - s + \Delta s)$, corresponding to small jumps $\pm\Delta s$ with equal probability. The δ-function makes the integration in (6.33) trivial and yields the probability distribution shown in Fig. 6.6(a). In the discrete case, the diffusion is described by Pascal's triangle, whereas the continuous distribution is Gaussian.

To describe drift, as created by an external force or field, we must consider different rates $\Gamma_+ \neq \Gamma_-$ for jumps $\pm\Delta s$. These jumps corresponds to a biased random walk. The rates obey $\Gamma_+ + \Gamma_- = \Gamma_o$ and generally depend on the starting point s. Putting $W(s', s) = \Gamma_+\, \delta(s' - s - \Delta s) + \Gamma_-\, \delta(s' - s + \Delta s)$ into the master equation and assuming small jumps Δs yields the *Fokker-Planck equation*

$$\frac{\partial P}{\partial t} = -\frac{\partial}{\partial s}(K_D P) + D \frac{\partial P^2}{\partial s^2} \tag{6.41}$$

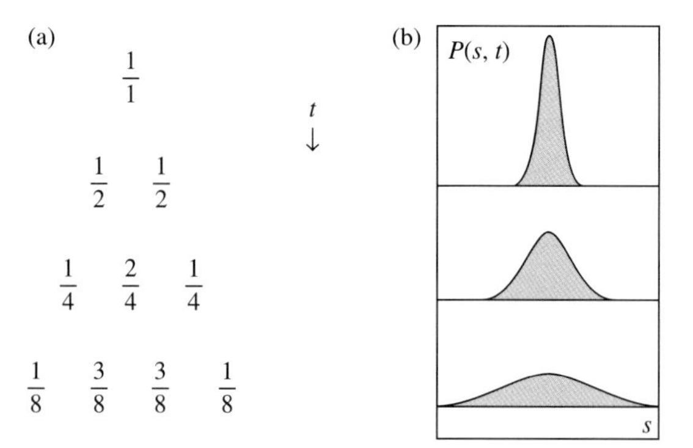

Fig. 6.6 Probability distribution and diffusion: (a) Pascal's triangle in the discrete case and (b) Gaussian distribution in the continuum limit. In magnetism, the diffusion usually corresponds to a fluctuation of the magnetization direction.

Here $K_D = (\Gamma_+ - \Gamma_-)\Delta s$ is the drift coefficient and $D = \Gamma_o^2 \Delta s/2$ is the diffusion coefficient. Note that master and Fokker-Planck equations were first derived to describe fluids but are now used in many areas of science (Risken 1989).

The drift term in (6.41) is due to an external potential $E(s)$ which competes against the random thermal forces. It is often convenient to use the Fokker-Planck equation in the form

$$\frac{\partial P}{\partial t} = \frac{\Gamma_o}{k_B T} \frac{\partial}{\partial s} \left(\frac{\partial E}{\partial s} P \right) + \Gamma_o \frac{\partial P^2}{\partial s^2} \tag{6.42}$$

The energy E competes against the thermal disorder, as described by the diffusion term. By putting $\partial P/\partial t = 0$ and integrating over s it is easy to show that the Fokker-Planck equation reproduces the correct equilibrium limit $P \sim \exp(-E(s)/k_B T)$. In the opposite limit of nonequilibrium states captured in deep potential valleys, with activation energies E_a much larger than $k_B T$, the dynamics approaches the Arrhenius limit with relaxation rates $\Gamma_o \exp(-E_a/k_B T)$. The description of this regime is also known as Kramers' escape-rate theory (1940) and, in magnetism, as the Arrhenius-Brown-Néel theory.

In the absence of an effective field, the Fokker-Planck equation reduces to an ordinary diffusion equation. A simple example is a plate-like thin-film particle with in-plane magnetization and zero in-plane anisotropy (Skomski, Zhou, and Sellmyer 2005). In other words, the magnetization is confined in the film plane but free to rotate. Let us start from the initial state $\phi(0) = 0$, and consider the magnetization projection $M_x(t) = M_s \cos\phi$. The probability $P(\phi)$ is obtained from $\partial P/\partial t = \Gamma_o \, \partial^2 P/\partial \phi^2$:

$$P(\phi, t) = \frac{1}{\sqrt{4\pi\Gamma_o t}} \exp\left(-\frac{\phi^2}{4\Gamma_o t} \right) \tag{6.43}$$

Evaluating the integral $M_x(t) = M_s \int P(\phi, t) \cos\phi \, d\phi$, where ϕ extends from $-\infty$ to $+\infty$, yields $M_x(t) = M_s \exp(-\Gamma_o t)$. This means that the average magnetization projection decays exponentially with a relaxation time $\tau_o = 1/\Gamma_o$, in spite of the Gaussian character of $P(\phi, t)$.

In most cases of practical interest, the magnetic phase space is multidimensional, and the derivatives in (6.42) must be replaced by vector expressions (Section A.2.4). For example, $\partial E/\partial s$ becomes $\nabla_s E = \partial E/\partial s_i$. Figure 6.7 compares a two-dimensional problem, namely the diffusion of a magnetization vector on the unit sphere, with damped spin precession. In the absence of an external potential E (zero field and zero anisotropy), the probability distribution is essentially Gaussian, and $\langle m^2 \rangle$ increases linearly with time.

To conclude this subsection, we mention that a formal derivation of the Fokker-Planck equation is based on the Kramers-Moyal expansion of the master equation. Writing $W(s, s') = w(s; \xi)$, where $\xi = s' - s$, transforms (6.23) into

$$\frac{\partial P(s)}{\partial t} = \int w(s - \xi; \xi) \, P(s - \xi) \, d\xi - P(s) \int w(s; -\xi) \, d\xi \tag{6.44}$$

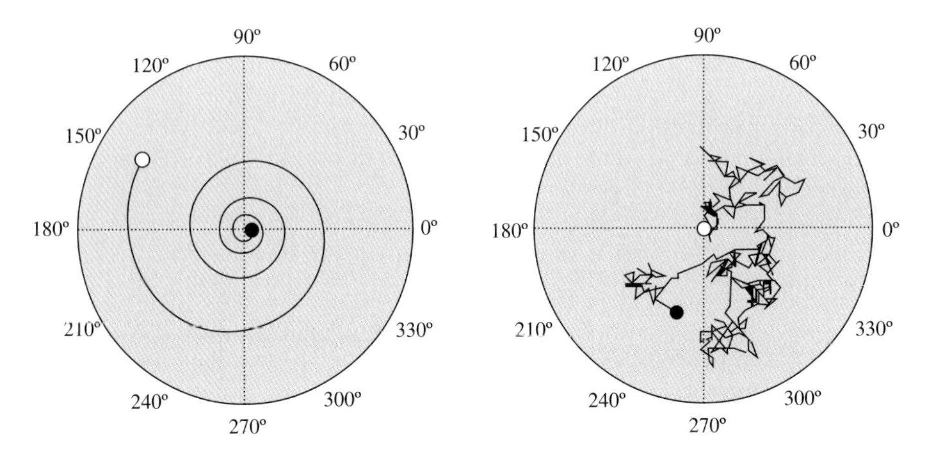

Fig. 6.7 Magnetization dynamics: damped precession (left) and diffusion in spin space (right). The curves are simulations for a nanoparticle, covering a time of order 0.1 ns. In both polar plots, the direction of the motion is from the white circles to the black circles.

Expanding the first integral into small deviations ξ from s yields

$$\frac{\partial P(s)}{\partial t} = -\frac{\partial P}{\partial s} \int \xi w(s;\xi)\, P(s)\, \mathrm{d}\xi + \frac{1}{2}\frac{\partial^2}{\partial s^2} \int \xi^2 w(s;\xi)\, P(s)\, \mathrm{d}\xi \tag{6.45}$$

The integrations no longer involve $P(s)$, and introducing the jump moments

$$\alpha_\mathrm{n}(s) = \int \xi^\mathrm{n} w(s;\xi)\, \mathrm{d}\xi \tag{6.46}$$

we obtain the Fokker-Planck equation

$$\frac{\partial P(s)}{\partial t} = -\frac{\partial}{\partial s}[\alpha_1(s)\, P(s)] + \frac{1}{2}\frac{\partial^2}{\partial s^2}[\alpha_2(s)\, P(s)] \tag{6.47}$$

By comparison with (6.41–42) we see that α_1 and α_2 describe the respective drift and diffusion contributions to the dynamics.

Diffusion equations—including the diffusive part of the Fokker-Planck equation—can be derived by considering concentrations (or probabilities), as fluids. The flux $\mathbf{j} = \rho\nu$ is subject to the continuity requirement $\mathrm{d}\rho/\mathrm{d}t + \nabla\cdot\mathbf{j} = 0$ and, in the absence of external forces $\mathbf{f}(\mathbf{r})$, proportional to the concentration gradient, $\mathbf{j} = -D\nabla\rho$ (Fick's equation). We will use this approach in the discussion of diffusion and other transport processes (Section 7.2). To obtain the drift part, one must add a local force $\mathbf{f}(\mathbf{r})$.

6.3.3 Langevin models

In our initial derivation of the Fokker-Planck equation we have assumed that thermal excitation yields small but random magnetization jumps $\pm\Delta s$. Solving the Fokker-Planck equation yields the probability $P(s,t)$, from which averages such as $\langle s(t)\rangle$ and $\langle s(t)s(t')\rangle$ are obtained by integration. This leads us to an intriguing question: Can we avoid the calculation of $P(s)$ and determine $s(t)$ directly from the jumps $\pm\Delta s$? The answer is yes, and the corresponding equation is the *Langevin equation*

$$\frac{\partial s}{\partial t} = -\frac{\Gamma_o}{k_B T}\frac{\partial E}{\partial s} + \sqrt{2\Gamma_o}\,\xi(t) \tag{6.48}$$

Here the random thermal forces $\xi(t)$ have the character of a delta-correlated white noise and obey $\langle \xi(t) \rangle = 0$ and $\langle \xi(t)\,\xi(t') \rangle = \delta(t - t')$.

As an example, let us consider a small Stoner-Wohlfarth particle near the nucleation field (Section 4.1.1), where $E = (K_1 - \mu_o M_s H/2) V s^2$ and

$$\frac{\partial s}{\partial t} = -\frac{\Gamma_o}{k_B T}(2K_1 - \mu_o M_s H) V s + \sqrt{2\Gamma_o}\,\xi(t) \tag{6.49}$$

For any realization $\xi(t')$, the solution of (6.49) is

$$s(t) = \sqrt{2\Gamma_o}\int_{-\infty}^{t} \exp(-\Gamma(t - t'))\,\xi(t')\,\mathrm{d}t' \tag{6.50}$$

where $\Gamma = \Gamma_o(2K_1 - \mu_o M_s H)V/k_B T$. Since $\langle \xi(t) \rangle = 0$, the average $\langle s(t) \rangle$ vanishes. However, the correlation function is

$$\langle s(t) s(t') \rangle = \frac{k_B T}{(2K_1 - \mu_o M_s H)V}\exp(-\Gamma|t - t'|) \tag{6.51}$$

In equilibrium, the average energy $\langle E \rangle = (K_1 - \mu_o M_s H/2)\,V \langle s^2 \rangle$ is equal to $k_B T/2$, as expected for a single quasiclassical degree of freedom.

When the reverse field H approaches the nucleation field $H_a = 2K_1/\mu_o M_s$, both the relaxation time $\tau = 1/\Gamma$ and the magnitude $\langle s^2(t) \rangle$ of the fluctuations diverge. However, the magnitude of the effect is usually small, because typical particles contain thousands of atoms and V is large by atomic standards. In fact, a more prominent effect is caused by the tails of the Gaussian distribution, in combination with sufficiently long waiting times (Section 6.4).

In multidimensional systems, eqs. (6.50) and (6.51) must be replaced by the eigenmodes of the systems. In particular, the single relaxation rate Γ must be replaced by the eigenvalues of energy matrix $\partial^2 E/\partial s_i \partial s_j \sim \Gamma_{ij}$ (Section A.2.2) and the long-time behavior is determined by the smallest eigenvalues of Γ_{ij}. Figure 6.8 illustrates the real-space meaning of the corresponding fluctuations.

Equation (6.42) indicates the relaxation processes may become very slow near transition points, accompanied by a divergence of correlations, as in (6.51). A similar scenario is realized near continuous phase transitions, where critical fluctuations diverge at the critical point (Section 5.4). An interesting critical phenomenon is the *critical slowing down* of the order parameter. On a mean-field level one starts from the Ornstein-Zernike theory (Section 6.4), based on a linearized mean-field equation reminiscent of (6.49). The long-wavelength limit $k = 0$ is characterized by the relaxation time

$$\tau_o \sim \frac{1}{|T - T_c|} \tag{6.52}$$

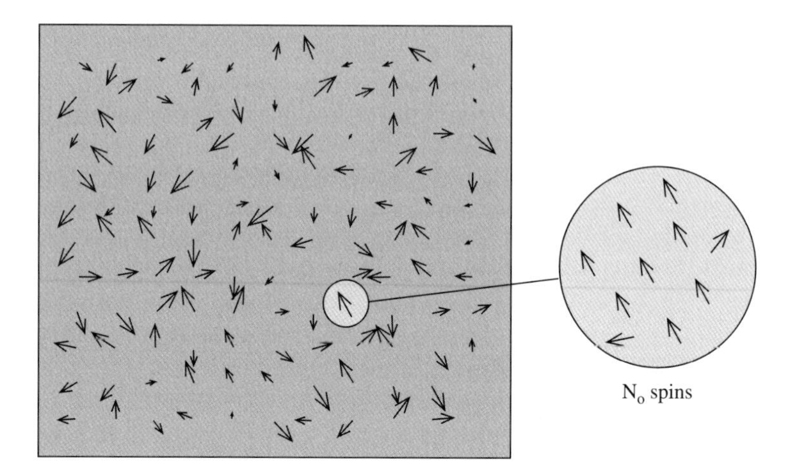

Fig. 6.8 Cooperative spin blocks. When the size N of the particle becomes too big, then thermal activation leads to the formation of cooperative units. The system then exhibits a number of eigenmodes with different relaxation times.

whereas finite wave vectors yield, at T_c,

$$\tau_{\mathbf{k}} \sim \frac{1}{1/\tau_o + Ak^2} \tag{6.53}$$

The same result is obtained by analyzing (6.40). As their equilibrium counterparts, the mean-field equations (6.40) and (6.52–53) are subject to renormalization corrections.

6.4 Slow magnetization dynamics

Summary The nonequilibrium character of magnetic hysteresis leads to a time dependence of the extrinsic properties known as magnetic viscosity. The corresponding activated magnetization reversal reflects the cooperative thermal excitation of nanoscale volumes and is a small correction to the leading static magnetization processes. For example, freshly magnetized permanent magnets lose a few mT of their magnetization within the first few hours, and the coercivity decreases with decreasing sweep rate $\mathrm{d}H/\mathrm{d}t$. As a crude rule, magnetic viscosity is described by a logarithmic time dependence. A simple derivation of the logarithmic law assumes an ensemble of independent relaxation processes, but it can also be considered as an inverted Arrhenius law. The involved energy barriers are smaller than or comparable to 25 $k_\mathrm{B}T$, so that the reversal usually requires the support by a magnetic field. The field dependence of the energy barriers is described by a power-law exponent 3/2, although highly symmetric energy landscapes yield an exponent 2. These two exponents cover a wide range of coherent and incoherent magnetization processes, including various types of pinning and nucleation. The slow dynamics of nanoparticles is a consideration in several areas of magnetism, such as magnetic recording and ferrofluids. On length scales of a very few nanometers, the behavior blends into equilibrium thermodynamics and acquires the character of giant thermodynamic fluctuations.

The nonequilibrium character of magnetic hysteresis means that thermal excitations drive the magnetization towards equilibrium. For example, permanent magnets lose some percentage of their magnetization after saturation, and the coercivity depends on the sweep rate $\mathrm{d}H/\mathrm{d}t$ of the applied magnetic field. This affects the long-term stability of permanent magnets (Skomski and Coey 1999) and recording media (Comstock 1999, Weller *et al.* 2000). They are also important in small particles, where they dominate the hysteresis (superparamagnetism), and in magnetic rocks, where equilibration times reach millions of years. The Landau-Lifshitz equation describes the precession of the magnetization as well as its relaxation towards the local or global energy minima. However, they are not able to describe thermally activated jumps over energy barriers.

The time and temperature dependence of extrinsic properties reflects a variety of mechanisms. First, the atomic-scale intrinsic temperature dependence of K_1, A, and M_s translates into an intrinsic temperature dependence of hysteretic properties. This determines the so-called "static" coercivity, at which a local free-energy minimum vanishes. For example, in Section 5.5 we have seen that the magnetic anisotropy tends to exhibit a pronounced decrease with increasing temperature, accompanied by a reduction in coercivity. However, this coercivity reduction is time-independent and easily incorporated into micromagnetic calculations, by taking appropriate equilibrium values of $K_1(T)$, $A(T)$, and $M_s(T)$. Second, thermally activated jumps over metastable free-energy barriers yield dynamic or "extrinsic" corrections to the static hysteretic behavior. Thermally activated jumps yield only small corrections, because typical energy barriers in ferromagnets are much larger than $k_B T$. Figure 6.9 compares the two mechanisms.

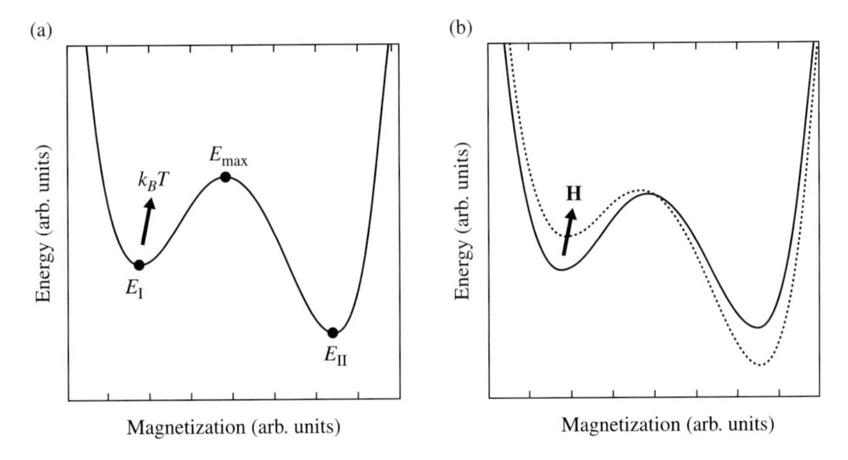

Fig. 6.9 Magnetization processes: (a) thermally activated and (b) field-induced. In most systems, thermal activation is a small correction to the leading field-dependent or "static" mechanism.

In some cases, time-dependent magnetization changes are only indirectly related to magnetism. One example is eddy currents and the skin effect in high-frequency magnetic materials, which follow from Maxwell's equations and mean that one must use insulating oxides rather than metals in microwave and other applications. Another example is the diffusion aftereffect in steel. The coercivity of steels is largely determined by domain-wall pinning involving interstitial carbon atoms. Since the domain-wall energy depends on the local carbon concentration, the application of an external magnetic field promotes a diffusive rearrangement of the carbon atoms and of the wall position until the total energy is minimized. This adjustment, also known as the Snoek effect, influences both magnetic and mechanical relaxations. However, in the following, our emphasis is on magnetic excitations.

In good approximation, the thermal excitation over energy barriers is described by the Arrhenius or Néel-Brown law

$$\tau = \tau_o \exp\left(\frac{E_a}{k_B T}\right) \tag{6.54}$$

where E_a is the activation energy associated with the energy barrier and $\tau_o = 1/\Gamma_o$ is an inverse attempt frequency of order $10^{-10} \ldots 10^{-11}$ s (Section 6.2). Figure 6.9 illustrates the meaning of the energy barrier $E_a = E_{max} - E_I$. Equation (6.54) is a nonequilibrium analog of the Boltzmann factor $\exp(-E/k_B T)$. Kramers' escape-rate theory (1940), originally developed to describe chemical reactions, shows that (6.54) is an exact low-temperature solution of the Fokker-Planck equation ($k_B T \ll E_a$). Aside from the activation energy E_a, there is also an activation *entropy* S_a proportional to the logarithm of the number of paths over the energy barrier (exercise on activation entropy). In a strict sense, (6.54) must be replaced by $\tau = \tau_o \exp((E_a - TS_a)/k_B T)$, but the activation entropy is usually incorporated into τ_o. By analyzing Kramers' theory it can be shown that S_a reflects the curvature of the saddle point defining the activation energy.

Inverting (6.54) yields the energy barrier

$$E_a = k_B T \ln\left(\frac{\tau}{\tau_o}\right) \tag{6.55a}$$

accessible after some waiting time τ. Local minima with energy barriers larger than (6.55a) can be considered as frozen. For typical laboratory-scale experiments, the time scale τ is about 100 s, as compared to $\tau_o \sim 10^{-9}$ s. This means that magnetization reversal occurs for energy barriers up to about $\ln(\tau/\tau_o) \approx 25\, k_B T$.

$$E_a = 25\, k_B T \tag{6.55b}$$

At room temperature, the corresponding energy barrier $E_a/k_B = 7{,}500$ K. This is significantly smaller than energy barriers encountered in most materials, which are of order 100,000 K. Exceptions are, for example, superparamagnetic particles, where $E_a \approx K_1 V$ is small due to the small particle size (Section 6.4.6). Magnetic recording, which corresponds to waiting times of order 100 years, requires energy barriers in excess of $40\, k_B T$ (Weller and McDaniel 2006), whereas geological time scales of about one million years correspond to $52\, k_B T$.

6.4.1 Magnetic viscosity and sweep-rate dependence

The time dependence of the remanent magnetization is known as magnetic viscosity, magnetic aftereffect, or ageing. Experiment shows that the time dependence of the magnetization is well approximated by the logarithmic magnetic-viscosity law

$$M(H,t) = M(H,t_o) - S\ln\left(\frac{t}{t_o}\right) \tag{6.56}$$

where S is the magnetic-viscosity constant (Becker and Döring 1939, Givord and Rossignol 1996). Equation (6.56) determines the stability of the information stored in magnetic and magneto-optical recording media (Sellmyer *et al.* 1998, Weller and Moser 1999) and yields time-dependent remanence losses in permanent magnets (Skomski and Coey 1999). As mentioned above, typical permanent magnets lose a small fraction of their magnetization, typically a few tenths of a percent, within the first few hours after production. Magnetization changes similar to (6.56) are also encountered in soft magnets and in systems such as disordered magnets and spin glasses (Section 7.1.4).

The magnetic after effect was, in fact, discovered as early as 1889 (Ewing). Among the initially discussed magnetic-viscosity mechanisms were eddy-current losses, but in the early twentieth century it became clear that the main mechanism was thermal activation over energy barriers, as described by (6.55a). A striking feature is the logarithmic character of the decay. It means that the magnetization decay is initially very fast but then slows down considerably, corresponding to the decay of some percentage of the magnetization per decade. For example, a remanence loss of 0.5% between one hour and ten hours after the production of a permanent magnet leads to the prediction of a matching additional loss of 0.5% between the twentieth and two-hundredth days. In the next subsection we will see that the logarithmic law reflects the broad distribution of activation energies (energy-barrier distribution) encountered in most magnets.

Thermal activation over energy barriers affects not only the magnetization but also the coercivity: the higher the sweep rate $\eta = \mathrm{d}H/\mathrm{d}t$ of the external magnetic field, the higher the coercivity (Sellmyer *et al.* 1998). The dependence is approximately logarithmic

$$H_c(\eta) = H_c(\eta_o) + \Delta H_c \ln\left(\frac{\eta}{\eta_o}\right) \tag{6.57}$$

where $\Delta H_c = k_B T/\mu_o M_s V^*$ has the character of a fluctuation field. The quantity V^* is referred to as activation volume and used to evaluate experimental data. Its physical meaning will be discussed below.

Both the sweep-rate dependence (6.57) and the logarithmic magnetic-viscosity law (6.56) have the same physical origin, namely (6.54), and similar orders of magnitude. However, (6.56) relies on an energy-barrier distribution, whereas (6.57) is obtained for both wide and narrow distributions of the activation energy.

6.4.2 Superposition model of magnetic viscosity

A simple magnetic-viscosity model consists of $\mathrm{i} = 1 \ldots N$ independent magnetization or switching processes described by individual activation energies $E_a(\mathrm{i})$ and relaxation times $\tau_\mathrm{i} = \tau_\mathrm{o} \exp(E_\mathrm{a}(\mathrm{i})/k_\mathrm{B}T)$. The physical nature of the magnetization reversal is of secondary importance but may be visualized as coherent rotation in an ensemble of magnetic grains or as domain-wall jumps in a bulk magnet. Based on our master equation analysis of Section 6.3, the equation of motion $\mathrm{d}M_\mathrm{i}/\mathrm{d}t = -(M_\mathrm{i} - M_\mathrm{io})/t$, where M_i is the magnetization contribution of the i-th process. For $M_\mathrm{i} = M_\mathrm{io}$ at $t = 0$, the solution is

$$M_\mathrm{i}(t) = -M_\mathrm{io} + 2M_\mathrm{io} \exp\left(\frac{-t}{\tau_\mathrm{i}}\right) \tag{6.58}$$

Next we assume a continuous distribution of energy barriers ($N = \infty$) and average over all processes. Writing $\tau_\mathrm{o} = 1/\Gamma_\mathrm{o}$ we obtain

$$M(t) = -M_\mathrm{s} + 2M_\mathrm{s} \int_{-\infty}^{\infty} P(E)\, \mathrm{e}^{-\Gamma_\mathrm{o} t \exp(-E/k_\mathrm{B}T)}\, \mathrm{d}E \tag{6.59}$$

where $P(E_\mathrm{a})$ is the energy-barrier distribution. A straightforward but lengthy approach is to choose a model distribution, such as a rectangular distribution $P(E) = 1/E_\mathrm{o}$ between $-E_\mathrm{o}/2$ and $+E_\mathrm{o}/2$, and to analyze the asymptotics of the time dependence of the magnetization (Becker and Döring 1939).

A more elegant approach is to exploit that the energy-barrier distribution is much broader than $k_\mathrm{B}T$. We start from the identity $\Gamma_\mathrm{o} t \exp(-E/k_\mathrm{B}T) = \exp[-(E - k_\mathrm{B}T \ln(\Gamma_\mathrm{o}t))/k_\mathrm{B}T]$ and use $\exp(-\exp(-x/\varepsilon)) \approx \Theta(\mathrm{x})$, where $\Theta(\mathrm{x})$ is the step function, $\Theta(\mathrm{x} < 0) = 0$ and $\Theta(\mathrm{x} > 0) = 1$. The step function converts the exponential term into a finite limit of integration

$$M(t) = -M_\mathrm{s} + 2M_\mathrm{s} \int_{k_\mathrm{B}T \ln(\Gamma_\mathrm{o}t)}^{\infty} P(E)\, \mathrm{d}E \tag{6.60}$$

It is convenient to subtract the magnetization $M(t_\mathrm{o})$ at some reference time t_o, so that

$$M(t) = M(t_\mathrm{o}) - 2M_\mathrm{s} \int_{k_\mathrm{B}T \ln(\Gamma_\mathrm{o}t_\mathrm{o})}^{k_\mathrm{B}T \ln(\Gamma_\mathrm{o}t)} P(E)\, \mathrm{d}E \tag{6.61}$$

Since $\ln(\Gamma_\mathrm{o}t) \approx \ln(\Gamma_\mathrm{o}t_\mathrm{o})$, we can replace $P(E)$ by $P(E_\mathrm{o})$ and obtain the logarithmic law

$$M(t) = M(t_\mathrm{o}) - 2M_\mathrm{s}k_\mathrm{B}TP(E_\mathrm{o}) \ln\left(\frac{t}{t_\mathrm{o}}\right) \tag{6.62}$$

or $S = 2M_\mathrm{s}k_\mathrm{B}TP(E_\mathrm{o})$. Figure 6.10 illustrates the meaning of (6.60–62).

The energy barriers are field-dependent and the energy-barrier distribution is related to the switching-field distribution, $P(E)\,\mathrm{d}E = P_\mathrm{SF}(H)\,\mathrm{d}H$. As mentioned in Section 4.3.3, $P_\mathrm{SF}(H)$ is closely related to the irreversible part of the susceptibility, $\chi_\mathrm{irr}(H) = 2M_\mathrm{s}P_\mathrm{SF}(H)$. With $\mathrm{d}E = \partial E/\partial H\, \mathrm{d}H$, this yields the Street and Woolley version of (6.62), epitomized by the magnetic-viscosity constant (Givord and Rossignol

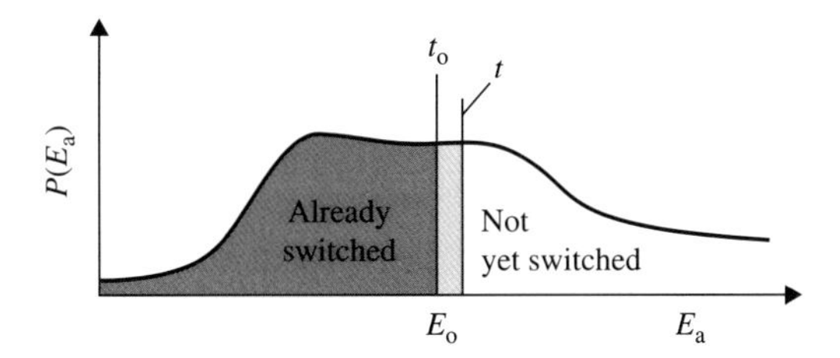

Fig. 6.10 Energy-barrier distribution and magnetic viscosity. Processes with low energy barriers easily switch (dark gray), whereas processes with high activation energies cannot be activated thermally (with). Magnetic viscosity is due to a small boundary region (bright gray).

1996, Lyberatos and Chantrell 1997)

$$S = \frac{k_B T \chi_{irr}}{\partial E_a / \partial H} \tag{6.63}$$

Both χ_{irr} and S reach a pronounced maximum near the coercivity. To abstract from this maximum, one writes $S = \chi_{irr} S_v$, where the magnetic-viscosity coefficient S_v is only weakly field-dependent. S_v has the dimension of a magnetic field and is also known as the *fluctuation field* (Néel 1951). S_v varies between less than 1 μT in soft magnets and about 10 mT in hard magnets.

6.4.3 Asymptotic behavior*

This subsection deals with the field dependence of magnetic-viscosity and sweep-rate contributions and their long-time asymptotics. (Since some explicit and transparent examples will be presented below, the reader may prefer to skip these paragraphs and use them for reference at a later state.) Let us start with taking into account that the field dependence of $P(E)$ has two contributions: the static switching-field distribution PSF(H_o) and, for each H_o, an energy barrier depending on $H - H_o$. We express this dependence in terms of a function f, so that $E = f(H - H_o)$. The energy-barrier distribution $P(E)$ is obtained by averaging over all values of H_o:

$$P(E) = \int P_{SF}(H_o)\, \delta(E - f(H - H_o))\, dH_o \tag{6.64}$$

Here the delta function (Section A.2.4) ensures that the integration is limited to processes whose activation energy is equal to E. Putting (6.64) into (6.61)

$$M(t) = M(t_o) + \chi_{irr}[g(k_B T \ln(\Gamma_o t_o)) - g(k_B T \ln(\Gamma_o t))] \tag{6.65}$$

In this equation, g is the inverse function of f, that is, $f(g(E)) = E$. In general, f and g are complicated nonlinear functions, and $M(t)$ goes beyond the logarithmic law (Skomski, Kirby, and Sellmyer 1999). However, linearizing (6.65) with respect to $\ln(\Gamma_o t) - \ln(\Gamma_o t_o) = \ln(t/t_o)$ reproduces the logarithmic law. Some explicit energy-barrier models will be discussed in Section 6.4.4.

A similar equation is obtained for the sweep-rate dependence of the coercivity. Exploiting $\eta/\eta_o \sim t_o/t$, using $E_a = f(H - H_o)$, and equating H with H_c we obtain

$$H_c = H_o + g\left(k_B T \ln\left(\frac{\eta_o}{\eta}\right)\right) \tag{6.66}$$

In the next subsection, we will specify the functions f and g, and discuss a few examples.

A side from nonlinear contributions, the logarithmic law breaks down for both $t = 0$ and $t = \infty$, where it predicts $M(t) = \pm\infty$. A better approximation is (Skomski and Christoph 1989)

$$M(t) = 2M_s \left(\frac{t}{\tau_o}\right)^{-k_B T/E_o} - M_s \tag{6.67}$$

Since $x^\varepsilon - 1 = \varepsilon \ln x$ for small exponents, the intermediate regimes of (6.62) and (6.67) are barely distinguishable.

6.4.4 Energy-barrier models

To investigate the field dependence of slow magnetization dynamics, we must specify the function $E_a = f(H - H_o)$. A frequently used and, as we will see, physically meaningful energy expression is

$$E_a = K_o V_o \left(1 - \frac{H}{H_o}\right)^m \tag{6.68}$$

Here K_o, V_o, H_o and m are physical parameters that derive from the magnet's real structure. (The exponent m used in this section must not be confused with the magnetic moment.) Zero-temperature switching occurs when the energy barrier E_a vanishes, that is, for a reverse field H_o. Substituting E_a into (6.54) and equating the field H with H_c yields (Kneller 1966)

$$H_c = H_o \left(1 - \left(\frac{K_B T}{K_o V_o} \ln\left(\frac{\tau}{\tau_o}\right)\right)^{1/m}\right) \tag{6.69}$$

In magnetic recording, this relation is also known as the Sharrock equation. The power-law exponents m and $1/m$ correspond to the functions f and g in the previous subsection and are a simple example of the formalism (6.64–66). Linearizing this equation with respect to $\ln(\tau/\tau_o)$ reproduces the experimentally observed linear dependence of H_c on $1/\tau \sim dH/dt$. The linearization works fairly well for experiments covering a few orders of magnitude of τ but breaks down for very long time scales (Skomski, Kirby, and Sellmyer 1999). For example, a magnetic-viscosity experiment covering 0.01 mT/s to 10 mT/s amounts to a change of $\ln(10/0.01) = 6.91$. The ratio $\ln(10^4)/\ln(10^{11}) = 0.27$ amounts to corrections of the order of 27% and indicates that the linearized logarithmic law is probably applicable. However, comparing a time scale of 100 s (laboratory-scale experiments) with 50 years (magnetic data storage) yields corrections of order 63%, which turns the V^* description into a very crude estimate.

To derive the power law (6.68) and the exponent m, one must start from the micromagnetic energy landscape, as in Fig. 6.9. The idea is to expand the energy landscape in the vicinity of the saddle point that defines E_a. The expansion is meaningful, because $k_B T \ln(\tau/\tau_o) \approx 25\, k_B T$ (or roughly 7,500 K) is much smaller than typical energy barriers. Taking, for example, an anisotropy constant of 1 MJ/m^3 and an activation volume of $10 \times 10 \times 10\,\mathrm{nm}^3$ yields a temperature equivalent of about 100,000 K. This means that thermally activated reversal needs the support by an external field (Fig. 6.9). The field reduces the energy barrier until the static switching condition $H = H_o$ is nearly satisfied and thermal activation becomes effective.

Let us, for simplicity, consider a single degree of freedom x, such as a Stoner-Wohlfarth rotation angle or a domain-wall position. Including linear, quadratic, and cubic terms, the energy is

$$E(x) = a_o + a_1 x + \frac{a_2}{2} x^2 + \frac{a_3}{3} x^3 - b_o H x \tag{6.70}$$

where the phenomenological parameters a_o, a_1, a_2, a_3, and b_o describe the real-structure of the magnet. They depend on $K_1(\mathbf{r}, T)$, $A(\mathbf{r}, T)$, and $M_s(\mathbf{r}, T)$ and must be determined separately.

The function (6.70) has no extremum ($H > H_o$), or one minimum E_{min} and one maximum $E_{max}(H < H_o)$. The field-dependent energy barrier $E_a = E_{max} - E_{min}$ is calculated by putting $\partial E/\partial x = 0$. The result is (6.68) with $m = 3/2$ and parameters $K_o V_o$ and H_o that depend on a_o, a_1, a_2, a_3, and b_o. For symmetric energy barriers, $E(-x) = E(x)$, the coefficient $a_3 = 0$, and one must include the $a_4 x^4$ term in (6.70). Straightforward calculation shows that this changes the exponent to $m = 2$.

The exponents $m = 3/2$ and $m = 2$ are well established for a variety of systems, including Stoner-Wohlfarth particles and strong domain-wall pinning. The exponent $m = 3/2$, which was first derived by Néel in 1950, is quite common and describes a variety of coherent and incoherent magnetization processes. Examples are strong domain-wall pinning and the reversal of misaligned Stoner-Wohlfarth particles (Gaunt 1983, Victora 1989, Skomski 2003). The exponent $m = 2$ describes, for example, aligned Stoner-Wohlfarth particles. In this case, V_o and K_o are equal to the particle's volume and anisotropy, respectively, and H_o is equal to the anisotropy field $2K_1/\mu_o M_s$ (see exercise).

6.4.5 *Linear and other laws

Linear energy barrier laws, characterized by $m = 1$ in (6.68), are occasionally assumed in semiphenomenological approaches, but their derivation from physically meaningful energy landscapes has remained a challenge. Nonanalytic energy landscapes $E(x)$, for example at grain boundaries, are candidates for linear laws, but they are convoluted with the smooth domain-wall profile. This is seen by putting a "needle-shaped" anisotropy inhomogeneity $\delta(x)$ into (4.38). Other approaches start from unrealistic or ill-defined energy landscapes and yield pathological predictions such as infinite zero-temperature coercivities. For example, series expansion in the vicinity of the static switching field reduces

$$E_a = \mu_o M_s V_o H_o \left(\frac{1}{H} - \frac{1}{H_o} \right) \tag{6.71}$$

to an $m=1$ law, but for $H=0$ it amounts to the unphysical prediction of an infinite energy barrier.

It is also worthwhile emphasizing that the linear law $E_a \sim H_o - H$ looks like a Zeeman energy, but in contrast to widespread belief, the Zeeman interaction does *not* yield a linear field dependence. The Zeeman energy is an energy level rather than an energy barrier. To obtain E_a, one must add the magnet's internal energy to the Zeeman energy, find the minima and maxima of the total energy, and finally calculate the energy difference. This is the procedure leading to (6.70), rather than just looking at the Zeeman energy. In fact, restricting the consideration to the Zeeman energy yields $E_a=0$, because there is no energy barrier for free spins in a magnetic field.

The above-mentioned postulation of micromagnetic models with $m=1$ must not be confused with the reasonable use of the linear law to evaluate experimental data. The starting point is

$$E_a(H) = \mu_o M_s V^*(H - H_o) \tag{6.72a}$$

where V^* is the above-introduced effective or activation volume. V^* is frequently used as a phenomenological fitting parameter. For example, (6.72a) leads to (Givord and Rossignol 1996)

$$S = \frac{K_B T \chi_{irr}}{\mu_o M_s V^*} \tag{6.72b}$$

However, experimental values of V^* depend not only on the physical switching volume V_0 but also on temperature. An example is aligned Stoner-Wohlfarth particles, where $V^* = (25 k_B T V/K_1)^{1/2}$. More generally, (6.68) yields $V^* \sim V_o^{1/m} T^{1-1/m}$, meaning that $V \neq V_o$ unless $m=1$. To further complicate matters, V_0 is not necessarily the volume of a single particle or grain—due to cooperative and localization effects it may be smaller or larger than the particle volume (Section 7.4.4). An alternative method of deriving V^* is to exploit the relation $\mu_o M_s V^* = -\partial E_a/\partial H$ (Street and Wooley 1949), where the derivative is taken at coercivity. Typical orders of magnitude of room-temperature activation volumes are $500\,\mathrm{nm}^3$ in permanent magnets (Givord and Rossignol 1996) and $2000\,\mathrm{nm}^3$ in transition-metal nanowires.

6.4.6 Superparamagnetism

Very small magnetic particles exhibit a gradual transition to paramagnetism. For sufficiently small numbers N of atoms, the interatomic exchange ensures that the spins are all parallel, so that the particle behaves like a *superparamagnetic* single spin or "macrospin" of length N. As outlined by Bean and Livingston (1959), superparamagnetism involves two phenomena: the Langevin-type alignment of macrospins in an external field (Section 5.2.4 and Section 5.4.2) and the superparamagnetic blocking or freezing of the magnetization reversal. In a magnetic measurement, both mechanisms lead to an error δH in the hysteresis loop. The Langevin smoothing is of order $\delta H_{th} = 3 k_B T / N \mu_o M_s V_{at}$, where V_{at} is the volume per atom. The blocking effect is described by $\delta H_b = H_a (25 k_B T / K_1 N V_{at})^{1/m}$, where the exponent m is equal to 3/2 or 2 (Section 6.4.4). The Langevin and blocking corrections both decrease with increasing particle size, but the blocking effect dominates in large particles,

$K_1V > k_BT/5\,(m=3/2)$ and $K_1V > k_BT/25\,(m=2)$. In the absence of a magnetic field, the Langevin character of the room-temperature magnetization becomes important below particle sizes of about 2 nm, whereas the superparamagnetic blocking occurs between about 3 nm (very hard magnets) and 30 nm (very soft magnets). From a micromagnetic point of view, superparamagnetic particles are Stoner-Wohlfarth particles, and their properties are essentially described by (6.37–39).

Superparamagnetic blocking is of interest in many areas of magnetism, from magnetic recording and biomedicine to magnetic rocks. Note that ensembles of small magnetic particles, or fine-particle magnets, are also known as "elongated single-domain particles" (ESD). This term is unfortunate, because it gives the false impression that coherent rotation and single-domain magnetism are just two names for the same phenomenon. In fact, as discussed in Section 4.2.6, hard-magnetic powder particles having radii slightly smaller than R_{sd} are single-domain but exhibit incoherent nucleation.

Small magnetic particles in stable colloidal suspensions are known as *ferrofluids* (Charles 1992). They are used, for example, in bearings and loudspeakers, and to monitor magnetic fields and domain configurations. A variety of materials can be used, including transition-metal elements and oxides, and a typical particle size is 10 nm. Most ferrofluids are based on hydrocarbons or other organic liquids, whereas water-based ferrofluids are more difficult to produce. A characteristic feature of the magnetization dynamics of ferrofluid particles is the distinction between Brownian relaxation and Néel relaxation. Néel relaxation involves jumps over magnetic energy barriers, as discussed in this section, whereas Brownian relaxation reflects the mechanical rotation of the particles due to Zeeman interaction. The Brownian relaxation time is $\tau_B = 3V\eta/k_BT$, where η is the mechanical viscosity of the embedding liquid.

6.4.7 *Fluctuations

At zero temperature, the magnetization reversal is realized by the path with the lowest saddle-point energy. Figure 6.11 illustrates this point for an arbitrary energy landscape. In (a), the external field is somewhat smaller than the switching field, although thermal activation may be effective in overcoming the energy barrier determined by the lowest-lying mode (solid line). In both the static and thermally activated regimes, the reversal proceeds in the direction of the lowest-lying mode, as indicated by the arrow in (b), and excited modes (dashed lines) can almost always be ignored.

Let us consider the example of a domain wall simultaneously pinned at two sites I and II. The wall has two propagation options: depinning from site I (followed by depinning from site II) and depinning from site II (followed by depinning from site I). Since pinning centers are real-structure features, the respective activation energies E_{I} and E_{II} are somewhat different. Without loss of generality we can assume that $E_{\text{I}} < E_{\text{II}}$, so that the static wall propagation starts at site I. At low but nonzero temperatures, propagation starting from site II is not impossible but very unlikely, because the corresponding probability $\exp((E_{\text{I}} - E_{\text{II}})/k_BT)$ is very small. What is a "low temperature" by micromagnetic standards? Typical micromagnetic energy barriers are of order $K_1\delta_B^3$, so let us assume that $E_{\text{I}} = 0.9\,K_1\delta_B^3$ and $E_{\text{II}} = 1.1K_1\delta_B^3$. Taking $A = 10$ pJ/m and $K_1 = 0.1$ MJ/m^3 we find that $E_{\text{II}} - E_{\text{I}}$ corresponds to a temperature

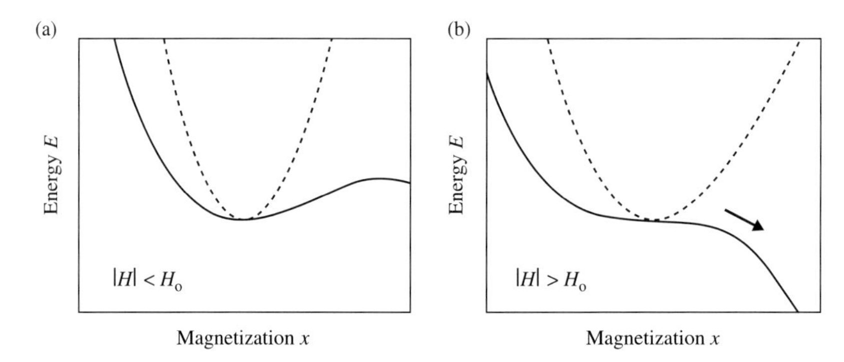

Fig. 6.11 Magnetization reversal in a multidimensional energy landscape: (a) metastable energy minimum and (b) vanishing of the metastable minimum at the static coercivity (b). The reversal (arrow) is realized by the lowest-lying mode (solid lines), as contrasted to excited modes (dashed). The modes may have different physical meanings, such as the angles ϕ and θ for a particle of irregular shape, the domain-wall position and curvature in a weak-pinning magnet, and the domain-wall positions at different pinning sites for strong-pinning magnets.

equivalent of 45,000 K. By microscopic standards, this is a high energy, indicating the involvement of thousands of spins and meaning that only the lowest-lying mode I is important at low temperatures.

The involvement of excited micromagnetic modes has the character of "giant thermodynamic fluctuations". Figure 6.12(a) shows an everyday example, namely the breaking of a cup, whose spontaneous repair due to thermal forces is possible but extremely unlikely. In the magnetic analogy, the giant fluctuations correspond to nuclei containing thousands of atoms (b). This must be contrasted to the thermally activated switching of individual spins, characterized by a large probability of order $\exp(-zJ/k_BT)$ but not resulting in magnetization reversal.

As emphasized by Aharoni (1996), modes such as those shown in Fig. 6.12(b) correspond to unreasonably high coercivities, and the same is true at low but nonzero temperatures (Skomski 2003). Reversal modes are solutions of a micromagnetic problem and cannot be postulated on intuitive grounds. Other modes, exemplified by Fig. 6.12(b) and proposed, for example, by Braun (1993), Braun and Bertram (1994), and Hinzke and Nowak (1998), are energetically unfavorable and correspond to nucleation fields higher than exact micromagnetic solutions. Simplifying somewhat, these modes cost a significant amount of exchange energy but enhance rather than reduce the nucleation field. This means that static magnetization reversal involving exact nucleation modes occurs *before* the thermally assisted dynamic reversal involving excited modes. A popular "counterargument" is infinite activation volume for both coherent rotation and curling. Experimental activation volumes are always finite, but this is primarily a zero-temperature effect, caused by structural imperfections and unrelated to thermal excitations. If one could fabricate an ideal ellipsoid of revolution with infinite aspect ratio, then the lowest-lying mode would be delocalized and the activation volume infinite.

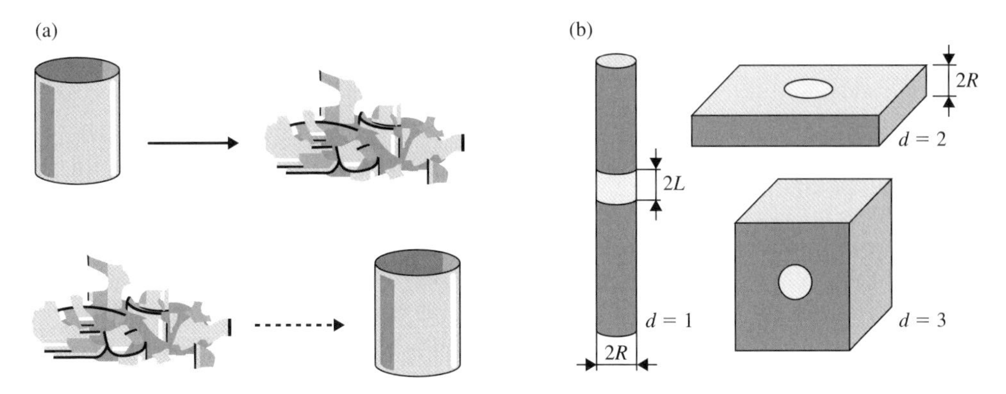

Fig. 6.12 Giant thermodynamic fluctuations: (a) broken cup and (b) nuclei in perfect nanostructures.

On the other hand, in Section 5.4.1 we have seen that one-dimensional magnets, including nanowires, are nonferromagnetic. How can we reconcile the infinite extension of the curling and coherent-rotation modes with the absence of long-range ferromagnetic order in one-dimensional magnets? The explanation is the disproportionately strong effect of fluctuations in one dimension. For example, in a long monatomic spin chain, the reversal of a single spin yields macroscopic magnetization changes. A simple but essentially correct argument is to consider a nanowire of radius R and to assume that thermal activation leads to the reversal of a segment of length $2L$, as shown in Fig. 6.12(b). This is paid by the creation of two domain walls of combined energy $2\pi R^2\gamma$, where $\gamma = 4(AK_1)^{1/2}$ is the domain-wall energy. Equating this energy to $25\,k_BT$ yields the transition temperature

$$T_o = \frac{8\pi R^2\sqrt{AK_1}}{25\,k_B} \tag{6.73}$$

above which giant fluctuations destroy the magnetization. For a typical ferromagnetic materials with $A = 10\,\mathrm{pJ/m}$ and $K_1 = 1\,\mathrm{MJ/m^3}$ and radii of 1 nm and 10 nm, we obtain $T_o = 57.6\,\mathrm{K}$ and $T_o = 5,760\,\mathrm{K}$, respectively. Alternatively, anisotropies of 0.1 and $10\,\mathrm{MJ/m^3}$ lead to the onset of room-temperature stability at radii of 4.0 and 1.3 nm, respectively.

Below T_o, thermal excitations lead to magnetization fluctuations whose range L is larger than the domain-wall thickness parameter $\delta_o = (A/K_1)^{1/2}$. A calculation similar to that by Skomski *et al.* (2000) yields

$$L = \frac{\delta_o}{\sqrt{1 - H/H_o}} \tag{6.74}$$

In the limit of static magnetization reversal, where $H = H_o$, this equation reproduces the coherent-rotation mode ($L = \infty$). This reconciles the dynamic behavior with the exact nucleation mode. At small temperatures, the fluctuations obey $L \sim 1/T$ and yield small coercivity corrections of order T^2.

In *two and three dimensions*, the propagation of the domain wall becomes a major consideration. For example, adding the Zeeman and domain-wall energies for a spherical nucleus yields

$$E(L) = 4\pi\gamma L^2 - 2\mu_o M_s H \left(\frac{4\pi R^3}{3}\right) \tag{6.75}$$

This expression differs from the energies considered in Section 4.3.2 by describing the expansion of a *free* domain wall, as contrasted to the pinning of domain walls by structural defects. Using $\partial E/\partial L = 0$ we obtain the critical radius $L_p = \gamma/\mu_o M_s H$ above which domain-wall propagation is favorable. For example, the thermally activated switching of single atomic spins—a quite frequent event—does not translate into domain-wall expansion. In two dimensions (circular nuclei in thin films), the critical radius is similar, $L_p = \gamma/2\mu_o M_s H$.

Small fields correspond to large L_p and mean that huge energies are necessary to create stable domain walls. This limit can safely be excluded from the present consideration. Fields approaching H_o facilitate domain expansion, so that that thermally activated domain-formation must be considered separately. In two dimensions, the corresponding activation energy $E_a \approx 2\pi At$, where $t = 2R$ is the film thickness. This energy is field-independent and corresponds to a cylindrical domain of length t and radius $L \approx 2\delta_o$. Using $E_a = 25\,k_B T$ and $A = 10\,\text{pJ/m}$ we obtain temperature estimates of 46 and 1,820 K for film thicknesses of 0.25 nm and 10 nm, respectively. In $d > 2$ dimensions, the activation energy increases with the applied reverse field, and thermal excitations are *de facto* negligible.

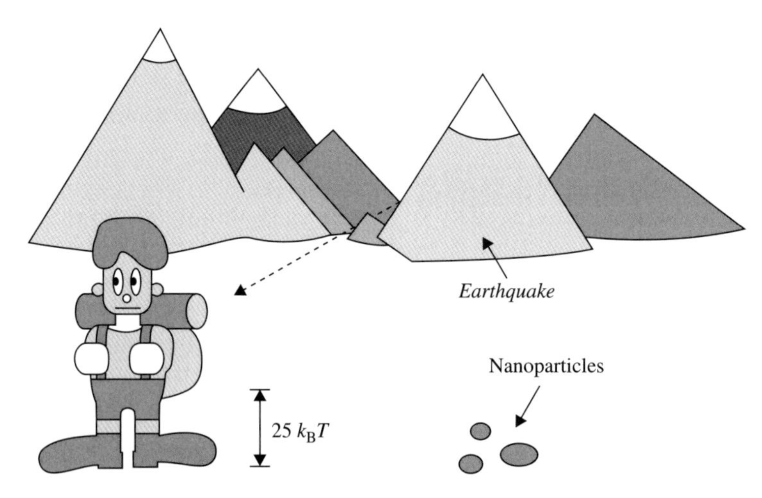

Fig. 6.13 Role of fluctuations. In lowest order, magnetization processes are realized by wandering over the saddle point of lowest energy (dashed line), helped by small thermal excitations (25 $k_B T$) and taking into account that the energy landscape exhibits an intrinsic temperature dependence. The thermally assisted paths over other saddle points (excited modes) and time-dependent fluctuations or "earthquakes" yield corrections to the leading contributions.

Figure 6.13 illustrates the role of temperature in magnetization processes. Thermal fluctuations involving paths others than that with the lowest saddle-point energy affect magnetization processes in ultrathin films and elongated nanoparticles. The latter is important in magnetic recording, because it reduces the energy barrier from the Stoner-Wohlfarth expression $K_1 V$ to $\pi R^2 \gamma$. If the length of the particle exceeds $4\delta_o = 4\delta_B/\pi$, then domain-wall creation is more favorable than Stoner-Wohlfarth rotation. The corresponding energy barrier $E_a = 4\pi R^2 (AK_1)^{1/2}$ translates into maximum recording density scaling as γ/T. Real systems may also be treated by numerical simulations, especially at temperatures approaching T_c and in the case of very fast magnetization processes (Nowak *et al.* 2005). However, the numerical modeling has remained nontrivial unless one focuses on very small particles or adjusts the exchange to lower values, which reduces the number of atoms to be considered. Otherwise, one must consider thousands of atoms, and fluctuations involving excited modes are often small corrections to the leading contributions due to intrinsic temperature dependences and low-lying modes.

Finally, it is in order to emphasize that sublattice effects yield a disproportionately strong contribution to the thermal instability. As we have seen in Section 3.3, the anisotropy of alloys such as $Nd_2Fe_{14}B$ and PtCo involves the spin-orbit coupling of electrons in the partially filled shells of heavy transition-metal elements. The heavy transition-metal elements are only loosely coupled to the iron-series transition-metal ions, so that thermal excitation have a strong impact on anisotropy and coercivity (Section 5.5). This finding is not limited to equilibrium. In nonequilibrium, it means that atomic fluctuations of the type shown in Fig. 5.23 yield local energy-barrier changes, similar to $K_1(\mathbf{r})$ profiles created by structural disorder. Figure 6.13 illustrates that this effect can be compared to earthquakes that change altitudes of the mountains, saddle points, and valleys.

Exercises

1. ***Creation operator acting on spin-up state.*** What happens on applying the creation operator s^+ to a spin-up state $|\uparrow\rangle$?
2. ***Equation of motion for a single spin.*** Derive the equation of motion for a single spin ($S = 1/2$) in a magnetic field.
   Hint: Use the angular-momentum commutation rules to determine $\mathrm{d}\langle S\rangle/\mathrm{d}t$ from the Schrödinger equation, as outlined in Section 6.1.1.
3. ***Landau-Lifshitz relaxation.*** Show that the Landau-Lifshitz equation predicts any small perpendicular magnetization component $m = (s_\mathrm{x}^2 + s_\mathrm{y}^2)^{1/2}$ to decay with a relaxation time τ.
   Hint: Use $m^2 = s_\mathrm{x}^2 + s_\mathrm{y}^2$ and $\mathrm{d}(m^2)/\mathrm{d}t = 2m\,\mathrm{d}m/\mathrm{d}t$.
4. ***Spin waves in square nanodots.*** Calculate the lowest-lying spin-wave modes in a square nanodot of area $a \times a$ and thickness t. Assume that the nanodot is very hard, with an easy axis perpendicular to the square.
5. ***Spin precession.*** Use the equation of motion (6.3) to calculate $\langle\sigma_\mathrm{x}(t)\rangle$ and $\langle\sigma_\mathrm{z}(t)\rangle$ for the initial state $\psi(0) = (1, 1)/\sqrt{2}$.
   Answer: $\langle\sigma_\mathrm{x}(t)\rangle = \cos(2\omega t)$ and $\langle\sigma_\mathrm{z}(t)\rangle = 0$.
6. ***Magnetic resonance.*** Use the Gilbert equation $\mathrm{d}\mathbf{M}/\mathrm{d}t = \gamma(\mathbf{M} \times \mathbf{H}) - \alpha(\mathbf{M} \times \mathrm{d}\mathbf{M}/\mathrm{d}t)/M_\mathrm{s}$ to calculate the magnetization components M_x, M_y, M_z for a typical

ferromagnetic resonance experiment.What happens if an anisotropy field is added to H?
Hint: Choose a coordinate frame where the **H** is in the z-direction and consider the small magnetization components $M_x = M_s m_x$ and $M_y = M_s m_y$.
Answer: Linearization yields $dm_x = \gamma m_y H + \alpha dm_y/dt$ and $dm_y/dt = -\gamma m_x H - a\, dm_y/dt$ and a damped circular precession, $m_x = m_o \sin \omega t$ and $m_y = m_o \cos \omega t$, where $\omega = \gamma H/(1 + a^2)$ and $m_o(t) = m_o(0) \exp(-\alpha \omega t)$. Anisotropy may cause nonequivalence of the x and y directions and lead to elliptical precession with two different frequencies (Section 6.1.2).

7. ***Spin-wave gap.*** Estimate the spin-wave gap for a magnetic material with a uniaxial anisotropy of $1\,\mathrm{MJ/m^3}$.
8. ***Eddy-current losses.*** Identify a few classes of materials where the eddy-current contributions to the magnetization dynamics are important or unimportant.
   Answer: Eddy-current phenomena such as the skin effect are a serious consideration in soft-magnetic materials for high-frequency applications. This is because the energy of the currents induced in each cycle must be dissipated, yielding a hysteresis similar to Section 1.5. Eddy-current contributions are negligible in many permanent magnets, in low-frequency soft magnets, and in insulating oxides.
9. ***Glauber model.*** The Glauber model of Ising-type magnetization dynamics is defined by the master-equation transition rates $W(-s, s) \sim (1 - s \tanh(h/k_B T))$. Show that these rates obey the detailed-balance principle.
10. ***Explicit equation of motion for two-level master equation.*** Show that $ds/dt = \Gamma(s_{eq} - s)$ can also be written as $ds/dt = \Gamma' \sinh((E_1 - E_2)/2k_B T) - \Gamma' s \cosh((E_1 - E_2)/2k_B T)$ where $\Gamma' = 2\Gamma_o \exp((E_1 + E_2 - 2E_o)/2k_B T)$.
11. ***Superparamagnetic blocking.*** Calculate the room-temperature blocking radii for Fe, Co, and Ni.
12. ***Energy barrier in small elongated particles.*** In the last part of Section 6.4.7, the energy barrier is estimated as $\pi R^2 \gamma$, half the value introduced for an inifinte wire. Why?
13. ***Activation entropy.*** Two chambers are separated by a thin wall having a hole of radius $R = 0.1\,\mathrm{mm}$. One chamber contains gas atoms that diffuse into the second chamber. Calculate the change in activation entropy if the hole is enlarged to $R = 2\,\mathrm{mm}$.
14. ***Energy-barrier laws.*** Calculate the power-law exponent m for the following energy barriers E_a: $1/H$, $\exp(-(H - H_o)/\Delta H)$, $\ln(H - H_o)$, $1/H$, and $1/H - 1/H_o$. Discuss the results.
15. ***Energy barrier of Stoner-Wohlfarth particle.*** Show that the energy barrier for an aligned Stoner-Wohlfarth particle is $E_B = K_1 V(1 - H/H_a)^2$.
    Hint: The problem is easily solved by tracking the energy minima and maxima of the Stoner-Wohlfarth (free) energy.
16. ***Long-time stability of magnetization.*** A small single-crystalline spherical particle is characterized by a uniaxial anisotropy of $K_1 = 5\ \mathrm{MJ/m^3}$. Calculate the diameter and volume above which the magnetization remains stable for (a) 2 hours, (b) 20 years, and (c) 20 million years.
    Hint: Use $\tau = \tau_o \exp(-E_a/k_B T)$ where $E_a = K_1 V$, $T = 300\,\mathrm{K}$, and $\tau_o = 1\,\mathrm{ns}$.

17. ***Orbital moment of spin pairs.*** Compensated $\uparrow\downarrow$ electron pairs have a zero spin moment. Can they have a nonzero orbital moment?
18. ***Fermions and spin waves.*** Refute or confirm the following argumentation: Since electrons are fermions and spin waves are electronic excitations, spin waves must be fermions.

7 Special topics and interdisciplinary models

7.1 Disordered magnets and spin glasses

Summary Atomic disorder has far-reaching consequences for the behavior of magnetic materials. It modifies the electronic structure but does not necessarily destroy ferromagnetism, as exemplified by amorphous ferromagnets. Spin glasses combine disorder with competing exchange, and their ground state is neither ferromagnetic nor antiferromagnetic. The equilibrium and nonequilibrium properties of spin glasses have remained a complex problem, and several models have been developed, such as the Edwards-Anderson (EA) and Sherrington-Kirkpatrick (SK) models. On a mean-field level, the determination of ordering and spin-glass temperatures involves the diagonalization of large random matrices.

Disorder is a key aspect of condensed-matter physics. In Chapter 3 we have seen that small imperfections may have a drastic effect on the coercivity, and many materials are disordered solids rather than perfect or nearly perfect crystals. Depending on the material, there are chemical disorder, structural disorder, or combinations of the two types. Examples in magnetism are substituted transition-metal and rare-earth magnets (chemical disorder) and magnetic glasses such as amorphous Fe-B (structural disorder).

In many systems, the disorder adds new physics. First, the ground state may change due to phenomena such as electron localization. Second, the finite-temperature behavior of disordered magnets is often different from that of ordered magnets, and it may not be possible to define a Curie temperature in disordered magnets. Third, the dynamics of disordered magnets is generally very different from that of ordered magnets, as epitomized by glass-like behavior in many systems. As mentioned in Chapter 6, this indicates that time and ensemble averages are different and adds considerable complexity.

There is an extensive review literature in the fields of disordered solids and spin glasses, such as Ziman (1979), Moorjani and Coey (1984), Fischer and Hertz (1991), and Economou (1990). Much information is now obtained from computer experiments (simulations), but the numerical calculations are demanding, as are many models. This section focuses on a number of very transparent and instructive models, mentioning more complicated models in due course. The emphasis of this section is on exchange effects; nanoscale disorder is treated in Section 7.4.

7.1.1 Atomic disorder and electronic structure

The lattice periodicity of perfect solids means that the electron-wave functions have the character of Bloch states $\psi_{\mathbf{k}}$. Physically, the electron states are extended waves that are reflected at lattice planes. In disordered solids, this picture is no longer valid, because the electrons are reflected at imperfections, too, and the resulting eigenstates mix different k vectors. For very weak disorder, one can replace the crystal potential $V(\mathbf{r})$ by its spatial average $\langle V(\mathbf{r})\rangle$, so that the lattice periodicity is conserved and the system remains diagonal in k-space. This is known as the *virtual-crystal* approximation. However, the potential $V(\mathbf{r}) = V_i(\mathbf{r} - \mathbf{R}_i)$ associated with atomic disorder is usually strong, and the determination of the one-electron eigenstates is a key problem in the description of disordered solids.

For simplicity, we consider the extreme tight-binding model for a solid containing N atoms and one orbital per atom (panel 6). Writing the wave functions in terms of atomic orbitals yields $\psi(\mathbf{r}) = \sum_i c_i \phi(\mathbf{r} - \mathbf{R}_i)$, where the c_i are expansion coefficients. In the bra-ket or vector notation, this is $|\psi\rangle = \sum_i c_i |i\rangle$, meaning that the N atomic orbitals $|i\rangle$ define a vector space (Hilbert space), and that the coefficients c_i are the components of the wave function $|\psi\rangle$. The Hamiltonian can be expanded in terms of the atomic orbitals $|\mathrm{i}\rangle$, so that

$$\mathsf{H} = \sum_{ij} |i\rangle E_{ij} \langle j| \tag{7.1}$$

where the E_{ij} remain to be specified. By exploiting the orthogonality relation $\langle i|j\rangle = \delta_{ij}$, we obtain the Hamiltonian in form of an $N \times N$ matrix, $E_{ij} = \langle i|\mathsf{H}|j\rangle$. Explicitly, the time-independent Schrödinger equation

$$E|i\rangle = \sum_j t_{ij}|j\rangle + V_i|i\rangle \tag{7.2}$$

Here t_{ij} is the hopping integral between the i-th and j-th atom and V_i is the crystal potential of the i-th atom. In explicit matrix form, $E\,c_i = \sum_j t_{ij}c_i + V_i c_i$.

If all $V_i = V_o$ for all atoms, then the crystal potential V_o yields a physically unimportant shift of the zero-point energy but leaves the eigenfunctions unchanged. Here we focus on chemical disorder (V_i) and ignore the structural disorder shown in Panel 6, so that N atoms form a periodic lattice. To derive the eigenstates, we take into account that $t_{ij} = t(\mathbf{R}_i - \mathbf{R}_j)$ and evaluate $\sum_j t_{ij}\, c_i$. Using the *ansatz* $c_i(\mathrm{k}) = \exp(\mathrm{i}\mathbf{k} \cdot \mathbf{R}_i)$ and substituting $\mathbf{R}_{ij} = \mathbf{R}_i - \mathbf{R}_j$ we convince ourselves that the eigenfunctions are indeed diagonal in k-space, and that the eigenvalues are $E(\mathbf{k}) = \sum_j t(\mathbf{R}_{ij}) \exp(\mathrm{i}\mathbf{k} \cdot \mathbf{R}_{ij})$.

Next, we consider chemical disorder described by $V_i = V_o \pm \delta V$. This case is frequently encountered in binary alloys, $A_x B_{1-x}$. For strong disorder, $\delta V \gg t_{ij}$, the energy levels approach $V_o \pm \delta V$. For disorder of intermediate strength, the eigenstates and eigenfunctions exhibit a complicated dependence on V_o and x. They also depend on the dimensionality of the lattice and on the range of the hopping integral t_{ij}. As a rule, the degree of electron localization increases with δV, especially in low-dimensional systems, because B-neighbors of A-atoms (or *vice versa*) act as potential barriers. However, even for $\delta V = \infty$ the localization remains imperfect. This is because an A-atom may have one or more A-neighbors, so that hopping onto these sites remains easy and leads to the formation of so-called cluster-localized eigenstates.

The are several models and approximations dealing with the electronic structure of disordered magnets and nonmagnets. One question is how impurity atoms affect energy gaps and band edges. For example, at band edges, disorder leads to exponentially decaying tails in the density of states known as Urbach tails. In the limit of low impurity concentrations, $x \ll 1$, one can consider isolated A atoms interacting with the host lattice. An example will be presented in the next subsection. A powerful approach to arbitrary concentrations is the coherent-potential approximation (CPA). It can be considered as a self-consistent effective-medium approach similar to the mean-field approximation, but due to its single-site character, it fails to reproduce cluster localization and Urbach tails in the alloys model. A very simple example of an inhomogeneous *magnetic* model is the antiferromagnet of Fig. 2.1. However, the antiferromagnetic model reduces to a relatively simple 2×2 matrix, whereas disorder corresponds to $N = \infty$.

The Anderson model describes 3d atoms embedded in and interacting with a broad conduction band of approximately constant density of states (DOS). The magnetic moment depends on the hybridization strength (s–d interaction) and on the position of the atomic 3d levels relative to the Fermi energy. Levels far above and far below E_F correspond to empty and occupied 3d orbitals, respectively, so that the magnetic moment is zero. For the 3d levels close to E_F, the 3d atom carries a nonzero moment if the intra-atomic Stoner exchange dominates the s–d coupling. For strong hybridization, the impurity states become delocalized and the moment vanishes. Another model exploring the relation between impurity and band states will be discussed in the next subsection.

There is complex interplay between disorder and correlation effects. A well-known example is the metal-insulator transition (Mott 1974). It means that electrons reduce their Coulomb energy by sticking to individual atoms, rather than hopping to neighboring sites (Section 2.1.7). The corresponding Mott localization adds to the Anderson localization described by (7.2). However, as other correlation effects, the metal-insulator or Mott transition goes beyond the independent-electron approximation.

7.1.2 *Green Functions

A powerful approach to deal with eigenfunctions and eigenvalues is the use of *Green functions*. They can be defined for a broad variety of matrix and differential equations and describe, for example, the scattering of waves and the energy levels of electrons in imperfect crystals. They are also known as propagators, because they have the character of wave-like perturbations. Here we focus on one special type of Green functions, namely quantum-mechanical Green functions for tightly bound electrons.

The mathematical idea behind the approach is to simplify calculations by inverting rather than diagonalizing an operator. The Green function belonging to the time-independent Schrödinger equation is defined as

$$\mathsf{G}(E) = \frac{1}{E - \mathsf{H}} \tag{7.3a}$$

Here H is the Hamiltonian of the investigated system and E is an energy parameter that may or may not be equal to an eigenenergy. Once the Green function is exactly or approximately known, it can be used to calculate physical quantities such as energies, densities of states, and correlation functions in a straightforward way.

For any given system, there are various representations of the Green function. For example, it may convenient to use the matrix $\mathsf{G}_{ab} = \langle a|\mathsf{G}|b\rangle = \int \psi_a^*(\mathbf{r})\,\mathsf{G}\,\psi_b(\mathbf{r})\,dV$, where the $\psi_a(\mathbf{r})$ is an orthogonal set of functions (A.2.2). One choice is the use of *eigen functions* of H, $|\mu\rangle$ or $\psi_\mu(\mathrm{r})$, so that

$$\mathsf{G}(E) = \sum_\mu |\mu\rangle \frac{1}{E - E_\mu} \langle\mu| \tag{7.3b}$$

In addition to the choice of wave functions, it is often convenient to use derived Green functions, such a $G^\pm(E) = G(E \pm \mathrm{i}\varepsilon)$, where ε is a small parameter that helps to deal with the poles of the function. Figure 7.1 shows $G^\pm(E)$ for the Hubbard Green function $G(E) = 2/(E + (E^2 - W^2/4)^{1/2})$.

Free-electron gases have Green functions of the type $G(\mathbf{r}, \mathbf{r}', E) = G_o(|\mathbf{r} - \mathbf{r}'|, E)$. They can be calculated exactly, by using plane waves in (7.4b) and integrating over all k-states. In three dimensions, $G_o(\mathbf{R}, E) \sim \exp(\mathrm{i}\kappa(E)/R)/R$, where $\kappa = (2m_e E/\hbar^2)^{1/2}$, and similar expressions exist for one and two dimensions. In the tight-binding approximation, the Green function is a matrix, $\mathsf{G}(E) = G_{ij}(E)$.

$$G_{ij}(E) = \frac{1}{E\delta_{ij} - E_{ij}} \tag{7.4}$$

This equation can also be written as $\mathsf{G}(\mathrm{E}) = \Sigma_{ij}|i\rangle 1/(E\delta_{ij} - E_{ij})\langle j|$.

A property of particular importance in magnetism is the density of states (Section 2.4.2). It is a useful tool for considering the *local* density of states, $\mathsf{D}_i(E)$, because it contains more information than the total DOS, $\mathsf{D}(E) \sim \sum_i \mathsf{D}_i(E)$. For example, the local electron density is $2\int \mathsf{D}_i(E)\,dE$, where $E \leq E_F$. In a spin-polarized

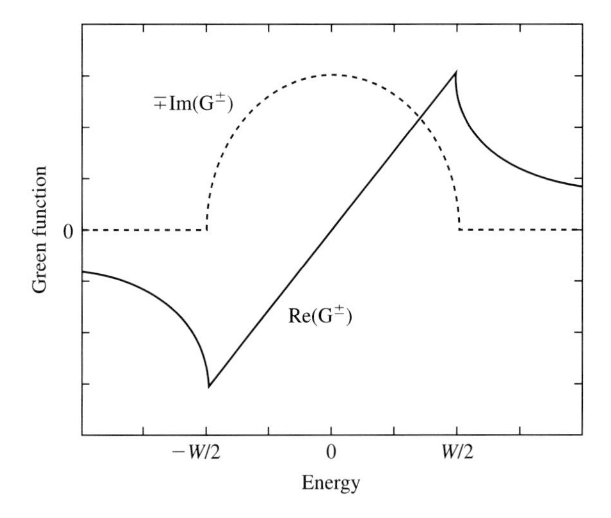

Fig. 7.1 Real and imaginary parts of the diagonal elements of the Hubbard Green function. The imaginary part is essentially the local density of states, and W is the bandwidth.

picture, $\mathsf{D_i}(E)$ determines whether an atom develops a magnetic moment, in analogy to the Stoner criterion. The local DOS is defined as the projection of the total DOS onto the i-th atom and obtained from $\mathsf{D_i}(E) = \sum_\mu \langle i|\mu\rangle\, \delta(E - E_\mu)\, \langle\mu|i\rangle$. For each eigenstate $|\mu\rangle$, the delta-function checks whether the state contributes to the DOS, and $\langle i|\mu\rangle\langle\mu|i\rangle$ is the probability that an electron in the state μ is located on the i-th atom. The involvement of $|\mu\rangle$ makes an explicit calculation of the local DOS cumbersome, but comparison with (7.3b) shows that the local DOS and the Green function involve $|\mu\rangle$ in a similar way.

By using the definition of $\mathsf{D_i}(E)$ and taking into account that $G_{\mathrm{ii}} = \langle i|\mathsf{G}|i\rangle$, one can show easily that the local DOS and the Green function are linked to each other by

$$G_{\mathrm{ii}}(E) = \int \frac{\mathsf{D_i}(E)}{E - \lambda}\, \mathrm{d}\lambda \tag{7.5}$$

Inverting this equation by considering $\pm\mathrm{i}\varepsilon$ yields, after short calculation

$$\mathsf{D_i}(E) = \frac{-1}{\pi}\, \mathrm{Im}\, G_{\mathrm{ii}}^{+}(E) \tag{7.6}$$

In other words, the local density of states is determined by the imaginary part of the diagonal elements of the Green function. An example is the Hubbard Green function in Fig. 7.1, which corresponds to a semicircular DOS.

To describe crystal imperfections, (7.3) must be evaluated for $\mathsf{H} = \mathsf{H_o} + \mathsf{V}$, where V describes the imperfection. Introducing $\mathsf{G_o} = 1/(E - \mathsf{H_o})$ and using the series expansion $1(1 - x) = 1 + x + x^2 + \cdots$ we obtain

$$\mathsf{G} = \mathsf{G_o} + \mathsf{G_o V G_o} + \mathsf{G_o V G_o V G_o} + \cdots \tag{7.7}$$

Let us embed a single impurity of energy $V_{\mathrm{ij}} = V_{\mathrm{o}}\delta_{\mathrm{ij}}$ in a host of energy $E = 0$. Exploiting $\mathsf{G_o} V_{\mathrm{o}} \mathsf{G_o} = V_{\mathrm{o}} \mathsf{G_o^2}$ we obtain the closed expression

$$\mathsf{G} = \frac{\mathsf{G_o}}{1 - V_{\mathrm{o}}\mathsf{G_o}} \tag{7.8}$$

Note that Green functions are often written as $\mathsf{G} = \mathsf{G_o} + \mathsf{G_o T G_o}$, where T is the t-matrix. In the present case, $\mathsf{T} = V_{\mathrm{o}}/(1 - V_{\mathrm{o}}\mathsf{G_o})$. The use of the t-matrix is particularly useful for two or more impurities, where (7.7) contains mixed terms and (7.8) no longer applies.

For the Green function of (7.8), the local DOS at the impurity site simplifies to $\mathsf{D_i}(E) = \mathsf{D_o}(E)/|1 - V_{\mathrm{o}}\mathsf{G_o}(E)|^2$, where $\mathsf{D_o}(E)$ and $\mathsf{G_o}(E) = \langle i|G_{\mathrm{o}}|i\rangle$ refer to the unperturbed host lattice (Economou 1990). Using the Hubbard Green function (Fig. 7.1) yields

$$\mathsf{D_i}(E) = \frac{4}{\pi}\, \frac{\sqrt{W^2 - 4E^2}}{W^2 + 16V_{\mathrm{o}}(V_{\mathrm{o}} - E)} \tag{7.9}$$

for the impurity site. Figure 7.2 shows this function, which is centered around the energy $E = 0$ of the host. In addition, for attractive potentials $V_{\mathrm{o}} < -W/4$, there is a sharp peak at $E = V + W^2/16V$.

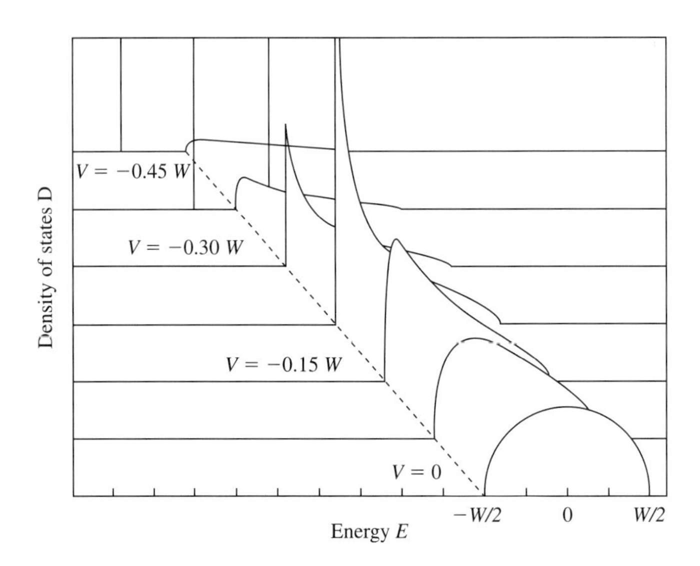

Fig. 7.2 Local density of states for a single impurity of energy $V = V_o$ in a band of width W.

Physically, $V = 0$ means that impurity and host atoms are the same and $\mathsf{D}_i(E) = \mathsf{D}_o(E)$. With increasing attractive potential, there is a skewing of the DOS, accompanied by the formation of a relatively sharp peak. This peak has the character of a resonance and indicates that the impurity electron remains integrated in the band. At a critical potential $V_o = -W/4$, a discrete impurity level splits off the continuum. Below this value, the impurity electron continues to hop onto neighboring sites and partially occupies the band, but the corresponding probability decreases as $W^2/16V_o^2$.

7.1.3 Ferromagnetic order in inhomogeneous magnets

On a mean-field level, it is straightforward to determine the critical temperature T_c from the interactions J_{ij}. This includes not only ferro- and antiferromagnets but also disordered magnets. Let us consider the mean-field equation for the zero-field Ising model,

$$m_i = \tanh\left(\frac{\sum_i J_{ij} m_i}{k_B T}\right) \tag{7.10}$$

where $m_i = \langle s_i \rangle$. By definition, $m_i = 0$ for $T \geq T_c$, so that we can linearize (7.9):

$$k_B T m_i = \sum_j J_{ij} m_j \tag{7.11}$$

This equation establishes an eigenmode problem (A.2.2), solved by diagonalization of the $N \times N$ matrix J_{ij}. There are N eigenvalues $J_\mu = k_B T_\mu$, one per spin, and the critical temperature is given by the largest eigenvalue, $T_c = \max(J_\mu/k_B T)$. The other eigenvalues have no transparent physical meaning, because $m_i \neq 0$ below T_c and (7.10) is no longer a reasonable approximation. The eigenmode $m_i(T_c)$ describes the spin structure that develops below T_c.

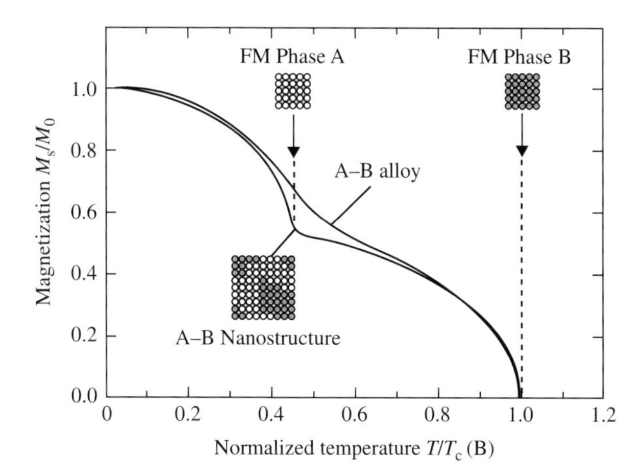

Fig. 7.3 Phase transitions and nanostructuring.

A simple example is the Curie temperature $T_c = zJ/k_B$ of ordinary one-sublattice ferromagnets, which corresponds to the homogeneous mode $m_i = m_o$. In antiferromagnets, $T_c = T_N$ is the Néel temperature (Section 5.3.6), and $m_i = \pm m_o$, depending on the sublattice. Similar relations exist for other structures, such as ferrimagnets and two-sublattice ferromagnets (Smart 1966). More generally, it is often possible to characterize magnetization states by wave vectors k, including ferromagnetism ($k = 0$), antiferromagnetism ($k = \pi/a$), and different noncollinear and incommensurate spin structures ($0 < k < \pi/2$). At surfaces, $m_i = m(z)$, where z is the distance from the surface, and one may encounter phenomena such as a separate surface Curie temperature (Section 7.4.3). A similar z dependence is encountered in multilayers.

In disordered ferromagnets, the quantitative character of the phase transition depends on the length scale of the disorder, as compared to the correlation length ξ. Figure 7.3 illustrates this point by comparing nanostructures with alloys. Macroscopic mixtures of two ferromagnets A and B are two-phase like, with two well-defined Curie temperatures. In this regime, a inhomogeneous ferromagnet is difficult to distinguish from a mixture of macroscopic phases. In the opposite limit of atomic inhomogeneities (A-B alloys), there is a smooth $M(T)$ dependences with a single Curie temperature.

In a strict sense, the transition from single-phase to two-phase behavior occurs at infinite length, because $\xi = \infty$ at T_c. However, in practice, structuring on a length scale L of a few nanometers makes the $M(T)$ curves two-phase like, because $\xi(T_A) \sim 1/|T_B - T_A|^\nu$ is very small (Skomski 1999, 2003). However, there is only one Curie temperature $T_c \approx T_B$. Note that Curie-temperature changes in nanostructures ($T_c \neq T_B$) are a small effect, in contrast to the substantial nanoscale energy-product enhancement discussed below. It is therefore not possible to improve the Curie temperature of a phase by nanostructuring.

The lower single-phase phase T_A does *not* correspond to a critical temperature, because neighboring B regions kill the Curie transition by exerting a small exchange field, similar to the effect of an external magnetic field. The field strongly decreases with increasing L but is always finite and smoothes the phase-transition singularity.

Since no assumptions have been made concerning the exponent ν, this argumentation is not limited to mean-field models.

This example shows that intrinsic properties are realized on fairly small length scales, even if the range of critical fluctuations goes to infinity. It also explains why many two-phase nanostructures, such as the multilayers, mimic the coexistence of two independent magnetizations. An alternative but equivalent explanation is that exchange-energy differences associated with long-range fluctuations are quite small and cannot compete against nanoscale local features.

7.1.4 Spin glasses

Spin-glass behavior is observed in a variety of disordered magnetic materials and involves both competing exchange and disorder. This adds a considerable degree of complexity, comparable to the intriguing equilibrium and nonequilibrium properties that distinguish mechanical glasses from crystalline solids. Examples of spin glasses are iron-series transition-metal atoms in noble-metal hosts, such as Fe_xAu_{1-x} and Mn_xCu_{1-x}, metallic glasses, such as amorphous Fe-Sn, and chemically disordered oxides and sulfides, for example $Eu_{1-x}Sr_xS$ (Moorjani and Coey 1984, Fischer and Hertz 1991).

Competing exchange *without* disorder, as encountered in rare-earth elements, gives rise to spin structures characterized by some wave vector k, whose absence is a characteristic feature of spin glasses. In the opposite extreme, randomness without competing exchange tends to destroys the periodicity of the system but may continue to support ferromagnetism, as observed in amorphous transition metals. The canonical or exchange spin glasses discussed in this section are loosely related to random-field and random-anisotropy magnets, which will be treated in Section 7.4.3. There are also links to superparamagnetic clusters and particles, with or without magnetostatic dipole interactions. Figure 7.4 shows some spin structures encountered in two-dimensional Ising models: (a) ferromagnets, (b–c) dipolar magnets, and (d–f) spin glasses. In ferromagnets, there are two ground states ($\uparrow$ and $\downarrow$), in dipolar magnets, the number of degenerate ground states is small, and in spin-glasses, the number is large.

Due to the randomness of the interactions J_{ij}, we expect the average magnetization to vanish. However, averages involving squares of the magnetization are generally nonzero. As in mechanical glasses, it is necessary to distinguish between *ensemble and time averages* (introduction to Chapter 6). Ensemble averages of glasses are liquid-like, meaning that atomic-scale snapshots of glasses and liquids look equally disordered, but time averages are solid-like, because the atomic positions are frozen. Spin glasses are described by the Edwards-Anderson order parameter $q_{EA} = [\langle s_i(t) \cdot s_i(0)\rangle]_{av}$, where the average $[\ldots]_{av}$ involves the limits $N \to \infty$ and $t \to \infty$. In the spin-glass phase, $q_{EA} > q$, where q is the equilibrium order parameter $[\langle s_i\rangle^2]_{av}$. By comparison, the high-temperature phase (paramagnet or "liquid") exhibits $q_{EA} = q$.

The parameters q_{EA} and q correspond to time and ensemble averages, respectively, and $q_{EA} > q$ phase indicates *broken ergodicity* in the spin-glass phase. We have encountered nonergodicity in the context of the Curie transition (Section 5.4), where the magnetization remains captured in one of the two ferromagnetic states ($\pm M_s$). However, spin-glass ergodicity is much less transparent, because the random J_{ij} create a very complicated energy landscape with many valleys and a large number of nearly

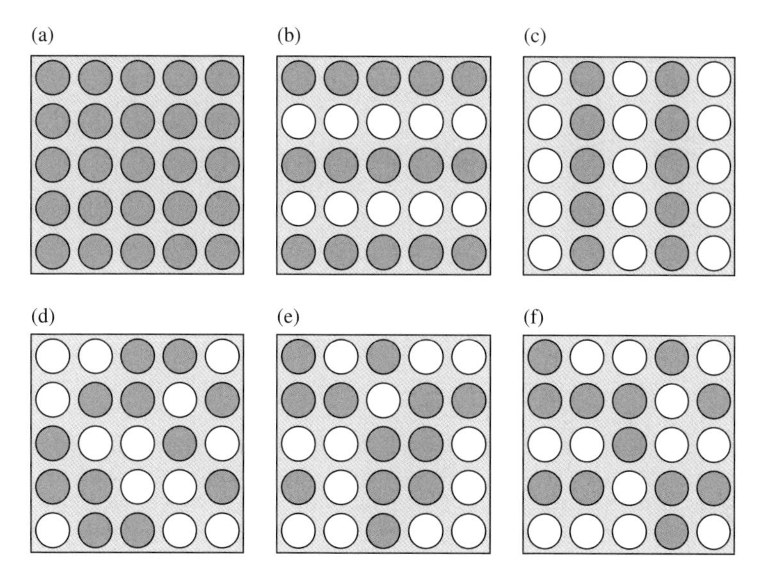

Fig. 7.4 Schematic ground states of Ising magnets: (a) ferromagnetic coupling, (b–c) dipolar interaction and (d–f) spin glasses. Dark and bright circles indicate ↑ and ↓ spins.

degenerate low-lying states (ground states). Broken ergodicity means that the spin configuration remains captured in a valley even for $t \to \infty$.

The fluctuation-response theorem (5.1.2) indicates that averages over squared magnetization variables are closely related to the susceptibility, and the question arises whether spin-glass freezing is similar to superparamagnetic freezing or constitutes a phase transition with a well-defined glass transition temperature T_f and with a susceptibility singularity at T_f. A similar question is encountered in the field of polymers and inorganic glasses: Is the low-temperature viscosity η of glasses exponentially large (supercooled liquid) or infinite (glass phase)? The latter case is often approximated by the Vogel-Fulcher-Tamman relation $\eta \sim \exp(E_\mathrm{a}/(T - T_\mathrm{o}))$, where T_o is roughly equivalent to T_f.

There are many spin-glass models, classified not only by spin dimensionality n and real-space dimensionality d (Section 5.2) but also by the range and character of the interactions. A "trivial" spin-glass model is the Mattis model, where $J_\mathrm{ij} = J_\mathrm{o}\tau_\mathrm{i}\tau_\mathrm{j}$ and $\tau_\mathrm{i} = \pm 1$. Using the gauge transformation $s'_\mathrm{i} = \tau_\mathrm{i} s_\mathrm{i}$ it is straightforward to make all bonds ferromagnetic (Mattis 1976). The phase-transition behavior of the model is that of an ordinary ferromagnet, except that the ferromagnetic mode $s_\mathrm{i} \sim const.$ is replaced by the gauge-transformed mode $s_\mathrm{i} \sim \pm\tau_\mathrm{i}$. However, the Mattis model ignores one essential feature of random exchange, namely frustration. Consider a square of Ising spins where three nearest-neighbor bonds are ferromagnetic and the fourth bond is antiferromagnetic. It is not possible to simultaneously satisfy all bonds or, equivalently, to find a gauge transformation mapping the frustrated square onto a ferromagnetic square.

Nontrivial spin-glass behavior is obtained for bonds with Gaussian distributions $P(J_\mathrm{ij}) \sim \exp(-(J_\mathrm{ij} - J')^2/2\Delta J'^2)$, where $J' = J_\mathrm{o}/z$ and $\Delta J' = \Delta J/\sqrt{z}$. Examples are

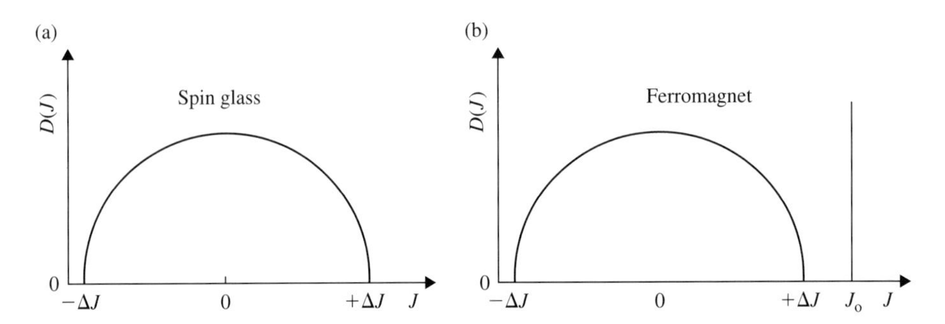

Fig. 7.5 Density of states of the interaction matrix J_{ij} for the SK model: (a) zero ferromagnetic exchange and (b) nonzero ferromagnetic exchange $J_o > 0$. The freezing temperature is given by the largest eigenvalue.

the short-range Edwards-Anderson (EA) model (1975), which assumes Heisenberg spins, and the Sherrington-Kirkpatrick (SK) model (1975), where Ising spins are coupled by long-range interactions, $z = N$. A simple approach to the $J_o = 0$ SK model is to assume that the mean field $h_i = \sum_j J_{ij} m_j$ obeys a Gaussian distribution where $[h_i^2]_{av}$ is approximately equal to $[m_i^2]_{av} \sum_j J_{ij}^2 = q\Delta J$. Together with $m_i = \tanh(h_i/k_B T)$, this yields the self-consistent Sherrington-Kirkpatrick equation

$$q = \frac{1}{\sqrt{2\pi}} \int \exp\left(-\frac{\xi^2}{2}\right) \tanh^2\left(\frac{\Delta J \sqrt{q}\xi}{k_B T}\right) d\xi \tag{7.12}$$

Close to T_f, we expect q to be small, so that $q \approx \Delta J^2 q / k_B^2 T^2$ and $T_f = \Delta J / k_B$.

Figure 7.5 shows the eigenvalues of the SK interaction matrix J_{ij} for zero J_o (complete randomness) and nonzero J_o (partially ferromagnetic bonds). As in other mean-field interaction models, the critical temperature is given by the largest eigenvalue of J_{ij}. For complete randomness, the density of states $D(J)$ obeys Wigner's semicircle law and $T_f = \Delta J / k_B$, in agreement with (7.12). The corresponding eigenstate (ground state) is spin-glass like, with zero average magnetization, $[\langle s_i \rangle]_{av} = 0$. For nonzero ferromagnetic or "coherent" exchange J_o, the ground state exhibits $[\langle s_i \rangle]_{av} > 0$, and for $J_o > \Delta J$, a single ferromagnetic splits off the density of states and $T_f = T_c = J_o / k_B T$. This corresponds to the transition from spin-glass behavior to disordered ferromagnetism.

Due to the infinite interaction range of SK model, the above mean-field transition temperature is exact. However, in the EA model, the interactions are generally short range, and the mean-field critical temperature may overestimate the trend towards the formation of a separate spin-glass phase. For example, there are small regions with predominantly ferromagnetic exchange and a mean-field critical temperature of order zJ/k_B. These regions with local magnetic order respond to an external magnetic field and yield what is known as *Griffiths* singularities (1969). Figure 7.6 illustrates how a gauge transformation transforms a spin-glass mode into a Griffiths-like ferromagnetic droplet at relatively high temperatures of order $\Delta J / k_B$ (Skomski 2003). Note that the droplet formation does not amount to ferromagnetism, because the interaction between different droplets is nonferromagnetic. At low temperatures, a spin glass can

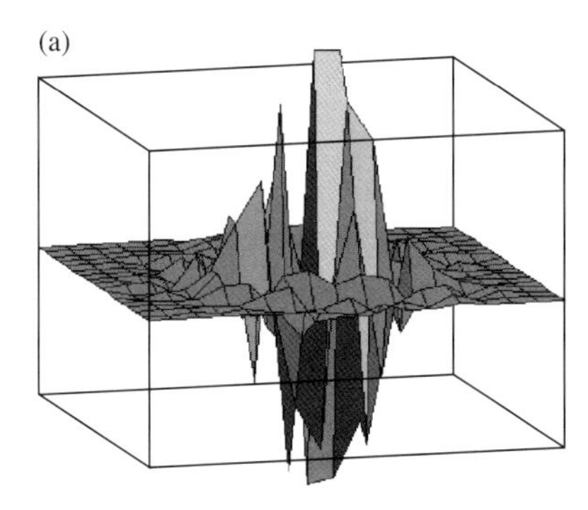

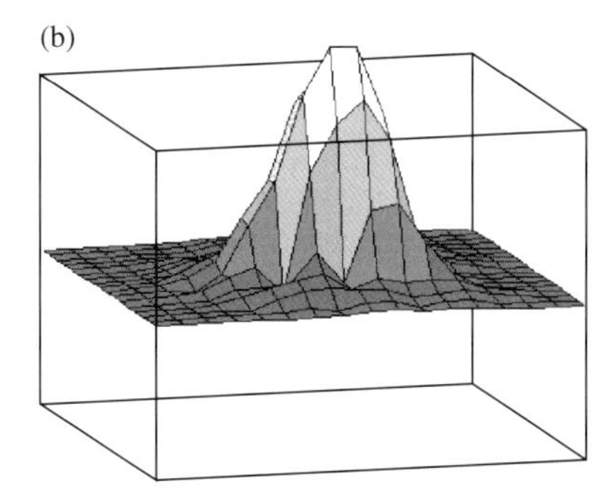

Fig. 7.6 Droplet formation in a two-dimensional Ising spin glass with short-range interaction: (a) original droplet and (b) after gauge transformation. The figure shows a droplet mode $s_\mathrm{i}(x, y)$ at relatively high temperatures.

be considered as an ensemble of droplets with temperature-dependent effective interactions (Fisher and Huse 1986).

The prediction of spin-glass transitions for individual models and dimensionalities is nontrivial, but in some cases there exist transparent solutions (Fischer and Hertz 1991). One-dimensional spin glasses do not exhibit spin-glass transitions, as one may guess from the very general argument of Section 5.4.1. For short-range interactions, mean-field theory becomes exact in six or more dimensions, indicating that the Ginzburg criterion cannot be used in its original form. Phase transitions in three-dimensional spin glasses with short-range interactions have remained a matter of controversy (Mézard *et al.* 1984, Fisher and Huse 1986), but renormalization-group and numerical calculations indicate that at least some three-dimensional models undergo an nontrivial spin-glass transition (Marinari, Parisi, and Ruiz-Lorenzo 1998).

7.2 Soft matter, transport, and magnetism

Summary There are many links between magnetism and transport properties. Finite-temperature magnetism exhibits a randomness reminiscent of diffusion processes and polymer chains, and there has been cross-fertilization in both directions. For example, self-avoiding polymer chains can be mapped onto an n-vector model with $n = 0$, and the self-interaction behavior of polymer chains is a real-space analog to the onset of mean-field behavior in four-dimensional ferromagnets. Other phase-transition analogies are percolation and gases in metals, where the links are both mathematical (lattice-gas description) and physical (interstitial permanent magnets). Magnetoresistance based on various mechanisms (AMR, GMR, CMR, TMR) is of practical importance, as is superconductivity.

There are many links between models of magnetism, soft condensed matter, and transport, inspite of the physical differences between these systems. We have already mentioned that the $n = 0$ vector-spin model corresponds to a polymer. To rationalize

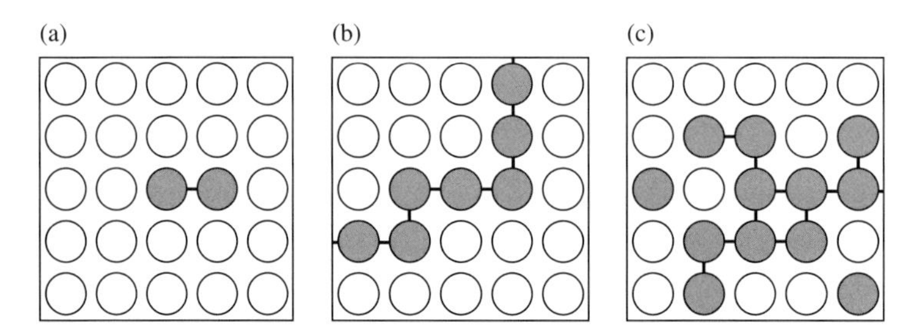

Fig. 7.7 Lattice animals: (a) dimer, (b) chain, and (c) cluster that looks like a small dog. In polymer science, the square lattice is known as the (two-dimensional) Flory-Huggins lattice.

this connection, we consider the high-temperature expansion of the Ising model. Writing

$$\exp\left(\frac{\sum_{\mathrm{i}>\mathrm{j}} J_{\mathrm{ij}} s_{\mathrm{i}} s_{\mathrm{j}}}{k_{\mathrm{B}} T}\right) = \prod_{\mathrm{i}>\mathrm{j}} \exp\left(\frac{J_{\mathrm{ij}} s_{\mathrm{i}} s_{\mathrm{j}}}{k_{\mathrm{B}} T}\right) \tag{7.13}$$

and exploiting $\exp(J_{\mathrm{ij}} s_{\mathrm{i}} s_{\mathrm{j}}/k_{\mathrm{B}}T) = \cosh(J_{\mathrm{ij}}/k_{\mathrm{B}}T) + s_{\mathrm{i}} s_{\mathrm{j}} \sinh(J_{\mathrm{ij}}/k_{\mathrm{B}}T)$ for $s_{\mathrm{i}} s_{\mathrm{j}} = \pm 1$, we obtain the partition function as a series. The series contains products such as $J_{\mathrm{ij}}\, J_{\mathrm{mn}}$, which can be visualized as graphs. Figure 7.7 shows a few examples of these graphs, which are also known as lattice animals. The contribution of these clusters to partition function and free energy depends on a variety of factors, such as the presence of an external magnetic field. However, we note that the cluster (b) looks like a chain. The chains may be investigated as separate entities, and their statistics describes, for example, polymers and diffusion processes. As we will see, there is a crucial distinction between random-flight or *random-walk* (RW) *chains*, with self-intersections, and *self-avoiding walk* (SAW) chains, without self-intersections.

7.2.1 Random walks, polymers, and diffusion

Polymer chains and diffusion processes have much in common. The diffusion of a particle can be described by the position $\mathbf{R}_{\mathrm{i}} = \mathbf{R}(t_{\mathrm{i}})$, where t_{i} is the time after the i-th diffusion step. On a square or simple-cubic lattice, $t_{\mathrm{i}} = i\tau$, where τ is the time between two collisions, and $\mathbf{R}_{\mathrm{i}} = \mathbf{R}_{\mathrm{i}-1} + \mathbf{l}_{\mathrm{i}}$, where $|\mathbf{l}_{\mathrm{i}}| = a$ is the lattice constant. In the polymer analogy, a has the character of a statistical segment length. By definition, the average $\langle \mathbf{l}_{\mathrm{i}} \rangle = 0$ for random processes, so that $\langle \mathbf{R}_{\mathrm{N}} \rangle = \sum_{\mathrm{i}=1\ldots N} \langle \mathbf{l}_{\mathrm{i}} \rangle$ vanishes. Successive diffusion events are independent of each other, so that $\langle \mathbf{l}_{\mathrm{i}} \cdot \mathbf{l}_{\mathrm{j}} \rangle = 0$ for $i \neq j$ (Markov character, Section 6.3.1). However, for $i = j$ the averages are nonzero, $\langle \mathbf{l}_{\mathrm{i}} \cdot \mathbf{l}_{\mathrm{i}} \rangle = a^2$, and $\langle \mathbf{R}_{\mathrm{N}}^2 \rangle = \sum_{\mathrm{i,j}=1\ldots N} \langle \mathbf{l}_{\mathrm{i}} \cdot \mathbf{l}_{\mathrm{j}} \rangle$ reduces to

$$\langle \mathbf{R}_{\mathrm{N}}^2 \rangle = N a^2 \tag{7.14}$$

Since $N = t/\tau$, this equation can also be written as $\langle \mathbf{R}^2 \rangle = t\,a^2/\tau$. The probability distribution $p(\mathbf{R}, t)$ obeys the diffusion equation $\partial p/\partial t = D\,\nabla^2 p$, where $D = a^2/2d\tau$ is the *diffusion coefficient* of the model. For a one-dimensional example, $p(x, t)$, see Fig. 6.6. For noninteracting particles, $p(\mathbf{r}, t)$ is essentially the concentration $c(\mathbf{r}, t)$.

Equation (7.14) describes polymers if one interprets N as the chain's total or contour length $L = Na$, in analogy to the diffusion time $t = N\tau$. In the polymer analogy, a is known as the Kuhn or statistical segment length. The total length $L \sim N$ must be distinguished from the end-to-end distance R. In three dimensions, the end-to-end distance obeys the probability distribution

$$p(\mathbf{R}, N) = \frac{1}{(2\pi N a^2/3)^{\mathrm{d}/2}} \exp\left(-\frac{3\mathbf{R}^2}{2Na^2}\right) \tag{7.15}$$

This equation is a solution of the above-mentioned diffusion equation and can also be considered as a continuum version of Pascal's triangle (Fig. 6.6).

This probability distribution is closely related to the configurational entropy $S = k_\mathrm{B}\ln(p)$ and to the mechanical properties of polymers. For example, rubbers consist of crosslinked polymers chains whose contour length L is fixed chemically, by creating sulfur bridges. The simplest model, the phantom network, assumes an ensemble of noninteracting chains. The mechanical force obeys $\mathbf{f} = \partial F/\partial \mathbf{R}$, where $F = E - TS$ is the free energy. In steel and many other materials, the leading contribution is from the energy E, but rubber is an exception, because the links between neighboring segments $\mathbf{l}_\mathrm{i}$ and $\mathbf{l}_{\mathrm{i}+1}$ do not store energy (freely jointed or Kuhn segments). As a consequence, energy changes are negligible, and the mechanical force $\mathbf{f}$ is of entropic origin (Treloar 1973). The $\mathbf{f} = -k_\mathrm{B}T\,\partial\ln(p)/\partial\mathbf{R}$ is equal to $3k_\mathrm{B}T\mathbf{R}/Na^2$, and since the N is inversely proportional to the density ν_c of chains, Young's modulus is essentially equal to $3\nu_\mathrm{c}k_\mathrm{B}T$. This is exploited in the production of rubbers, where high sulfur contents yield high crosslink densities and hard rubbers. A similar mechanism is responsible for the mechanical force created by muscle tissue (see below). The equations of this paragraphs describe rubbers on a mean-field level, neglecting possible corrections from self-avoiding-walk correlation (Section 7.2.3) and from chain entanglement.

The entropy elasticity of rubber is closely related to the pressure of the ideal gas and the susceptibility of the paramagnetic case. In all three cases, the external force or field leaves the internal energy unchanged but reduces the entropy by forcing the system in a less random state. Stretching a piece of rubber means work against thermal collisions, as contrasted to work against chemical interactions. However, in contrast to molecular and paramagnetic gases, the constraint $|\mathbf{l}_\mathrm{i}| = a$ introduces mean-field-type correlation effects (Section 7.2.3).

7.2.2 *The $n = 0$ vector-spin model

Before returning to polymers and diffusion, it is in order to discuss the relation between magnetism and polymer statistics. In Section 5.2.2, we have seen that Ising and Heisenberg models can be interpreted as $n = 1$ and $n = 3$ vector-spin models, respectively, and mentioned that the $n = 0$ vector-spin model describes polymer chains. In terms of Fig. 7.7, this means that the lattice animals are "snakes" (b), as opposed to the "dog" shown in (c). This is a considerable simplification, because it limits the number on configurations that contribute to the partition function.

Let us write the partition function as $\mathrm{Z} = 2^\mathrm{N}\langle\exp(J_\mathrm{ij}\mathbf{s}_\mathrm{i}\mathbf{s}_\mathrm{j}/k_\mathrm{B}T)\rangle_\mathrm{o}$, where $\langle\ldots\rangle_\mathrm{o}$ denotes an equally weighted average over all spin configurations. Series expansion of the exponential function yields averages $\langle\ldots\mathbf{s}_\mathrm{i}\ldots\mathbf{s}_\mathrm{j}\ldots\mathbf{s}_\mathrm{m}\ldots\rangle_\mathrm{o}$, which reduce to powers of individual spins, $\langle\mathbf{s}^\alpha\rangle = \langle\mathbf{s}_\mathrm{i}^\alpha\rangle$. It is convenient to use the normalization $\mathbf{s}^2 = n$

and to start from the characteristic function or generating functional

$$f(\mathbf{k}) = \langle \exp(\mathrm{i}\mathbf{k} \cdot \mathbf{s}_\mathrm{i}) \rangle_\mathrm{o} \tag{7.16}$$

Spin averages are easily obtained from f by taking derivatives at $k=0$. For instance, $\partial^4 f/\partial k_\mathrm{x}^4 = -\langle s_\mathrm{ix}^4 \exp(\mathrm{i}\mathbf{k}\cdot\mathbf{s}_\mathrm{i})\rangle_\mathrm{o}$ yields $\langle s_\mathrm{ix}^4\rangle_\mathrm{o}$. Next, we consider $\nabla^2 f(\mathbf{k}) = -\langle \mathbf{s}^2 \exp(\mathrm{i}\mathbf{k}\cdot\mathbf{s}_\mathrm{i})\rangle_\mathrm{o}$, where $\nabla^2 = \nabla_\mathrm{k}^2$ is the Laplace operator in k-space. Exploiting $\mathbf{s}_\mathrm{i}^2 = n$, we obtain the differential equation $\nabla^2 f = -n\, f$. A simple example is the Ising model, where $\partial^2 f/\partial k^2 = -f$ and $f(k) = \cos(k)$. For even and odd integer exponents α, the respective averages $\langle s_\mathrm{i}^\alpha\rangle_\mathrm{o}$ are equal to 1 and 0, as expected for spins $s_\mathrm{i} = \pm 1$.

The polymer limit $(n=0)$ is described by the solution $f(k) = 1 - k^2/2$ of $\nabla^2 f = 0$. The only nonvanishing derivative at $k=0$ is the second derivative, so that $\langle \mathbf{s}_\mathrm{i}^\alpha\rangle = 0$ for $\alpha > 2$. The corresponding graphs are chains without self-intersections (de Gennes 1979). More generally, correlation functions $\langle \mathbf{s}_\mathrm{i}\cdot\mathbf{s}_\mathrm{j}\rangle$ correspond to self-avoiding chains of end-to-end distance $|\mathbf{R}_\mathrm{i} - \mathbf{R}_\mathrm{j}|$.

7.2.3 Polymers and critical dimensionality

The critical exponent ν describes spin correlations $\langle \mathbf{s}_\mathrm{i}\cdot\mathbf{s}_\mathrm{j}\rangle$ or, in the polymer chains, the end-to-end distance $R(N) = (\langle \mathbf{R}_\mathrm{N}^2\rangle)^{1/2}$. Equation (7.14) implies $R \sim N^{1/2}$ and corresponds to the mean-field exponent $\nu = 1/2$. This is not surprising, because the segments interact with nearest neighbors only. Distant neighbors are ignored, in spite of the possibility of long-range self-intersections, $\mathbf{R}_\mathrm{i} = \mathbf{R}_\mathrm{j}$ for $|i-j| \gg 1$. In polymer science, these repulsive interactions are known as *excluded volume*. As in the magnetic analogy, long-range correlations must be considered if one is attempting to obtain realistic exponents. The effect of long-range correlations is seen most clearly by comparing random walks (Fig. 7.8 left) with self-avoiding walks (Fig. 7.8 right). In one dimension, the picture is very clear. Random walks are diffusion-like, with $R^2 = Na^2$ in any dimension, including $d=1$. The one-dimensional self-avoiding walk couldn't be more different. The first step is random, to $+a$ or $-a$, but then the direction is fixed, because the chain is not allowed to go back, and $R = Na$ and $\nu = 1$.

Simulations and experiments confirm that the deviations from $\nu = 1/2$ mean-field value occur in less than four dimensions, as in the magnetic n-vector model. In

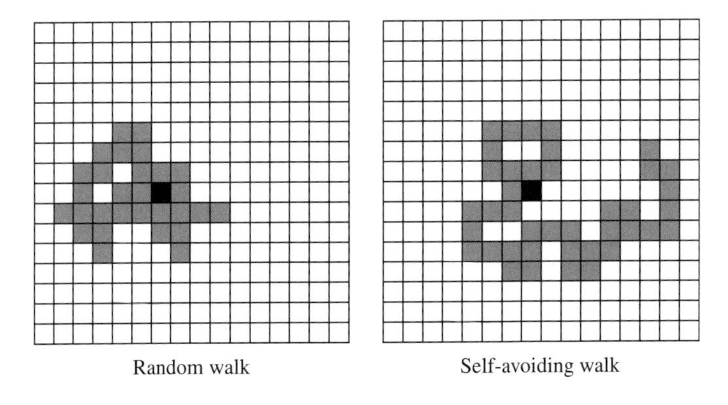

Fig. 7.8 Random and self-avoiding walks in two dimensions.

three-dimensions, the exponent $\nu = 0.588$ has been verified for polymers in solution, indicating a certain degree of "swelling" due to the excluded volume. Surprisingly, polymers in the melt keep the mean-field exponent. This has been observed by deuterium-labeled neutron diffraction and means that the swelling is compensated by the interaction with other chains.

An fascinating point is the relation between short-range and long-range interactions in self-avoiding polymer chains. As in magnets, the critical exponent ν is independent of short-range interactions and of geometrical details such as the square or triangular character of the lattice. In the polymer analogy, (7.14–15) are based on the assumption of freely jointed chain segments, with $\langle \mathbf{l}_{i+1} \cdot \mathbf{l}_i \rangle = 0$. This is nontrivial, because chemical interactions give rise to correlations, such as $\langle \mathbf{l}_{i+1} \cdot \mathbf{l}_i \rangle = l_o^2 \cos\theta$ between neighboring monomers. An example is a valence-angle chain formed from C-C bonds, where $\theta = 109.5°$ and $\cos\theta = 1/3$. However, the corresponding correlations decay exponentially, $\langle \mathbf{l}_{i+m} \cdot \mathbf{l}_i \rangle = l_o^2 \cos^m\theta$, and for large N the end-to-end distance approaches $\langle \mathbf{R}_N^2 \rangle = N\, l_o^2 (1 + \cos\theta)/(1 - \cos\theta)$. This indicates that chemical interactions change a but leave $\nu = 1/2$ unaffected. By contrast, long-range correlations mean that $\langle \mathbf{l}_{i+m} \cdot \mathbf{l}_i \rangle \neq 0$ for $m \sim N$ and $\nu > 1/2$.

The entropic force introduced in Section 7.2.1 is roughly proportional to $1/Na^2 = 1/La$, where the chain length L is fixed and the effective or Kuhn segment length a depends on chemical interactions. In a nutshell, this is the mechanism responsible for muscle contractions (Wöhlisch 1940). Extended muscles are characterized by large segment lengths a, which correspond to large end-to-end distances, $Na^2 = La$. The energy necessary to achieve the uncontracted state is provided chemically, by an interaction of adenosine triphosphate (ATP) with the muscle proteins (Huxley 1974). On releasing the energy stored in the segments, a becomes small and thermal forces lead to an entropic contraction of the muscle. Both the basic mechanism and the realized tensile stresses, of order 1 MPa, are very similar to rubber elasticity.

It is interesting to note that a relatively simple theoretical approach by Flory yields $v = 3/(2 + d)$ and quite accurately reproduces the critical exponents for $d \leq 4$. The idea is to compare a repulsive excluded-volume term of order $R^d (N/R^d)^2$ with an attractive entropy-elastic term of order R^2/N. However, as remarked by de Gennes (1979), the accuracy of the Flory exponents is accidental, because the repulsive and attractive terms are both overestimated by a large amount. The two errors cancel each other, and refined calculations of the repulsive and attractive terms have actually reduced the quality of the predictions. The accidental canceling of errors in model calculations is a problem in magnetism, too, and ill-controlled model assumptions and approximations may lead to misinterpretation of the physics of the investigated system. An example is the relation $H_c = 2K_1/\mu_o M_s$, where poor modeling may over- or underestimate both the coercivity and the anisotropy.

To discuss the critical dimensionality of self-avoiding walks it is convenient to start from the fractal dimension d^*. By definition, the number N of particles or segments depends on the linear dimension as $N \sim R^{d*}$. In solid materials, d^* is equal to the real-space dimensionality d, but for random-walk chains it follows from (7.14) that $d^* = 2$. This means that random walks are "space-filling" in two dimensions but sparse in three dimensions, where $d^* < d$. Figure 7.9 illustrates the two-dimensional character of random walks.

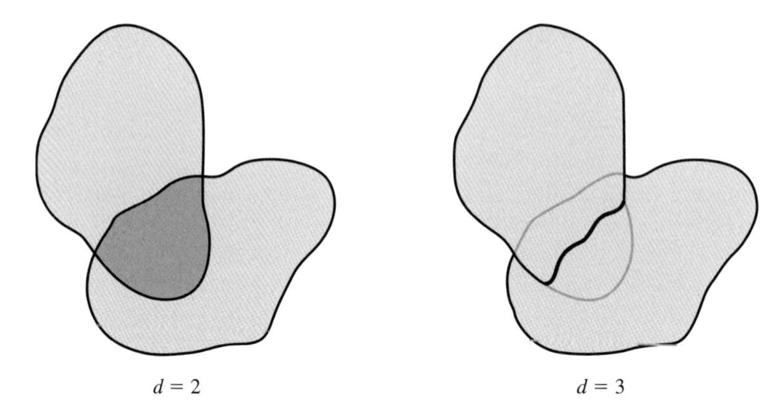

Fig. 7.9 Intersections of random walks. The fractal dimensionality of random walks is two, and in four dimensions, their (self-)intersections become zero-dimensional. This means that ferromagnetic mean-field critical exponents are exact for $d \geq 4$, aside from logarithmic corrections in four dimensions.

How can we understand that mean-field critical exponents are exact in $d \geq 4$ dimensions? In the polymer analogy, mean-field behavior and critical fluctuations correspond to random and self-avoiding walks, respectively, and critical fluctuations have the character of self-intersections between distant parts of a chain. Let us consider two well-separated subchains with $N' \sim N$ segments each. Since $N' \gg 1$, the subchains are essentially two-dimensional objects, and their intersection determines whether the polymer chain can be considered as a random walk. In two dimensions, the intersection is two-dimensional, as shown in Fig. 7.9(a). The substantial overlap (gray area) indicates that self-interactions are very important in two dimensions. In three dimensions, the intersection is one-dimensional, as illustrated by the dark line in Fig. 7.9(b), and therefore less predominant than in two dimensions. More generally, for $2 \leq d \leq 4$, the dimensionality of the intersection is equal to $4 - d$. In four dimensions, the excluded-volume overlap is zero-dimensional, that is, confined to isolated segments. Consequently, $d = 4$ is the critical dimension, with mean-field like exponents and small (logarithmic) corrections due to self-interactions. In more than four dimensions, the chains are unlikely to self-interact, and the excluded volume is unimportant.

The excluded volume can also be discussed in terms of a *burglar's walk*, Fig. 7.10. In one dimension, for example on a long island, the burglar may select the initial direction of his raid, but as people become aware of him, he cannot return to previously beleaguered places. In two dimensions, the burglar has some freedom but must remain careful to avoid places visited earlier. In three dimensions, for example in a big inner-city department store, his life is even easier, and in four dimensions, a burglar's accidental return to a place raided in the past is unlikely.

7.2.4 Percolation

An interesting phenomenon is the behavior of "swiss-cheese" systems once the volume fraction p of the holes reaches the percolation threshold. Examples are the transport of liquids (oil) and gases (carbon monoxide) through porous rocks, the sol-gel transition

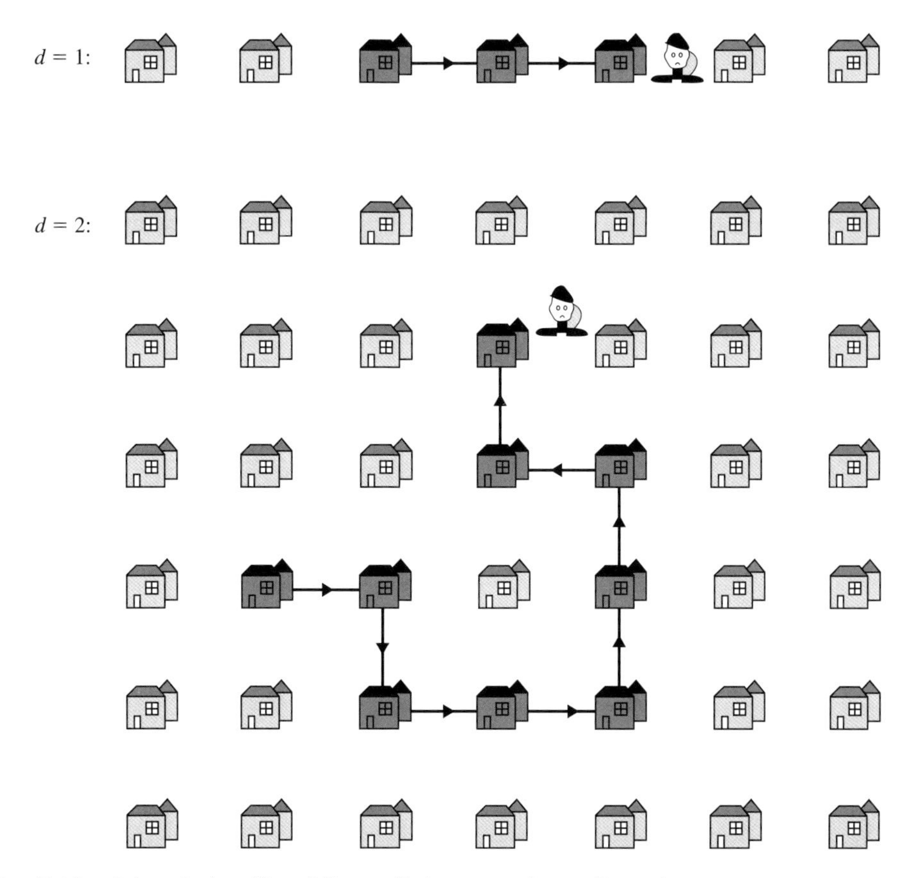

Fig. 7.10 A burglar's self-avoiding walk in one and two dimensions.

from a highly viscous liquid to an elastic solid, the onset of normal conductivity in metal-insulator composites, and the onset of superconductivity in a composite containing normal-metal and superconducting phases. Percolation exhibits some similarities with magnetic phase transitions. For instance, the volume fraction p and the size ξ of the largest "conducting" cluster, white area in Fig. 7.11(a), are analogous to the temperature and the correlation length, respectively. Percolation occurs at percolation threshold p_c, defined by $\xi \to \infty$ and analogous to the Curie temperature.

There are two frequently considered types of percolation, site percolation, as shown in Fig. 7.11, and bond percolation, where neighboring sites are disconnected or connected by a line. Another type is continuum percolation. The percolation threshold depends on the type of percolation (Kirkpatrick 1973, Stauffer and Aharony 1992). A mean-field argument predicts percolation if a given site is connected to two or more neighboring sites, $zp_c \geq 2$, so that the percolating cluster extends to infinity. Another mean-field estimation predicts $p_c = 1/(z-1)$ for high-dimensional hypercubic lattices ($z = 2d$). This value is obtained from the Bethe lattice, where each site has z neighbors without self-intersections (Cayley tree).

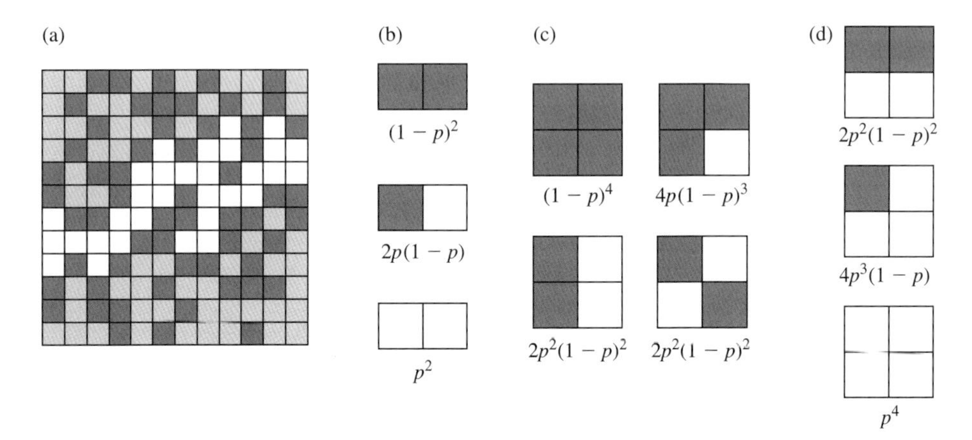

Fig. 7.11 Percolation: (a) percolating cluster, (b) renormalization in one dimension, (c) insulating blocks in two dimensions, and (d) conducting blocks in two dimensions. In (a), the white and bright gray sites are conducting, whereas the dark sites are insulating.

Table 7.1 Percolation thresholds for various lattices. The last two columns are mean-field type estimates (see main text).

	d	z	p_c (site)	p_c (bond)	$2/z$	$1/(z-1)$
chain	1	2	1	1	1	1
honeycomb	2	3	0.6962	0.6527	0.667	1
square	2	4	0.5927	1/2	0.5	0.5
triangular	2	6	1/2	0.7473	0.333	0.333
diamond	3	4	0.43	0.388	0.5	0.2
simple cubic	3	6	0.3116	0.2488	0.333	0.333
bcc	3	8	0.246	0.1803	0.25	0.2
fcc	3	12	0.198	0.191	0.167	0.143
(hypercubic)	4	8	0.197	0.1601	0.25	0.091
(hypercubic)	5	10	0.141	0.1182	0.2	0.143
(hypercubic)	6	12	0.107	0.0942	0.167	0.111
(hypercubic)	7	14	0.089	0.0787	0.143	0.091

Table 7.1 shows that p_c depends not only on the dimensionality of the lattice but also on the lattice type. Mean-field critical exponents become exact for $d \geq 6$. For example, the size ξ of the percolating cluster is described by

$$\xi \sim (p_c - p)^{\nu} \tag{7.17}$$

where the mean-field exponents $\nu = 1/2$. For bond percolation, ν is equal to 1 ($d=1$), 4/3 ($d=2$), and 0.89 ($d=3$).

Like the Curie transition, the critical behavior near the percolation transition can be described by renormalization-group analysis. Figures 7.11(b) and (c–d) illustrate this for one- and two-dimensional site percolation, respectively. During the procedure, blocks containing 2^d sites are mapped onto a new lattice characterized by $\xi' = \xi/2$ and $p' = p'(p)$. The renormalization-group equation $p' = p'(p)$ is obtained by deciding which block sites are "conducting". For $d = 1$, there is only one conducting configuration, bottom of Fig. 7.11(b), so that $p' = p^2$. This equation has two fixed points, a stable fixed point with $p = 0$ and $\xi = 0$ and an unstable fixed point with $p = 1$ and $\xi = \infty$. As in the ferromagnetic analogy, we are interested in the unstable fixed point, corresponding to the percolation threshold $p_c = 1$. This result is not surprising, because a single insulating site kills the conductivity of the chain. In two dimensions, p' is obtained by adding the probabilities of Fig. 7.11(d):

$$p' = 2p^2 - p^4 \tag{7.18}$$

Putting $p = p'$ yields two stable fixed points, $p = 0$ and $p = 1$, and one unstable fixed point, $p = 0.61803$. This means that $p_c = 0.61803$, close to the exact value of 0.5927. The critical exponent ν is obtained from (7.18) by linearization (see exercise on critical exponents for percolation). In two dimensions, $\nu = 1.6353$, as compared to the exact value 4/3. The deviation indicates that the small-cell real-space renormalization leading to (7.18) is only approximate, similar to the two-dimensional block-spin renormalization of Section 5.4.5.

The main purpose of this subsection was to elaborate the close relationship between magnetic phase transitions, polymer chains, and percolation. A unifying feature is the self-similarity near the critical point, epitomized by $\xi' = \xi/2$ and $\xi = \infty$. Paramagnetic spin fluctuations (Fig. 5.17), polymer chains, and percolating clusters are statistically self-similar (fractal), that is, rescaling leaves the object essentially unchanged. This is not the case for ferromagnets at low temperatures. Consider, for example, a cubic ferromagnetic particle having a size of $10 \times 10 \times 10\,\mathrm{nm}^3$. The first renormalization step yields a particle of $5 \times 5 \times 5\,\mathrm{nm}^3$, and four additional renormalization steps reduce the particle to a single atom.

7.2.5 Diffusive transport

In the context of the Fokker-Planck equation (Section 6.3.2), we have seen how thermal excitations change the probability $P(s, t)$ of realizing a magnetization state s. The same approach can be used to determine the concentration $c(\mathbf{r}, t)$ of diffusing particles and the transport of electrical charges and heat. The corresponding diffusion equation derives in close analogy to Section 6.3.2, but it is easier to combine the particle flux $\boldsymbol{j} = -D\nabla c$ (Fick's law) with the continuity equation $\partial c/\partial t + \nabla \cdot \boldsymbol{j} = 0$, so that

$$\frac{\partial c}{\partial t} = \nabla \cdot (D\nabla c) \tag{7.19}$$

The diffusion constant $D \approx l^2/2d\tau$, where l is the mean free path, d is the dimensionality of the system, and τ denotes the relaxation time. During diffusion, individual particles perform a random walk, and for simple diffusion on a lattice, $l \sim a$. For one-dimensional diffusion, a particle with $x(0) = 0$ obeys $\langle x(t)\rangle = 0$ and $\langle x^2(t)\rangle = 2Dt$.

Electrical conductivity is described by $\mathbf{j} = \sigma_e \mathbf{E}$, where $\mathbf{E}$ is the electric field and σ_e is the conductivity. This definition is equivalent to Ohm's law $U = RI$, where $R = l_w/A_w\sigma_e$ is the resistance of a resistor of cross section A_w and length l_w.

Electrical conduction means that electrons traveling through the lattice are scattered by magnetic or nonmagnetic imperfections, and the mean free path l describes the average distance between the defects. Since the velocity l/τ of the electrons is roughly equal to the Fermi velocity $v_F = \hbar k_F/m_e$ (Section 2.4.1), electrical conduction is a diffusion process with $D \sim l\hbar k_F/m_e$. The spin-dependent diffusion of electrons is important in spin electronics (Section 7.2.7).

An important aspect of diffusion is energy dissipation. In the resistor analogy, the dissipated energy is equal to $\int UI\mathrm{d}t = R\int I^2\mathrm{d}t$. Unless energy is provided externally, for example by a battery, the system moves towards equilibrium and the current dies rapidly. This distinguishes diffusive transport from other scattering mechanisms.

There are many techniques for dealing with diffusion problems, especially in disordered systems. We note that (7.19) is very similar to the Schrödinger equation if one replaces t by it. It is therefore possible to use diagrammatic approaches (Feynman diagrams) and quantum-mechanical Green-function methods. This includes the path- or functional-integral formulation, where propagators (Green functions) are written as integrals over trajectories or diffusion paths $x(t)$. Let us consider the one-dimensional transition probabilities $p(x, x_o; t_o)$, where x and x_o are the respective final and initial positions of the particle. When t is small, $t = t_o$, then

$$p(x, x_o; t_o) = \frac{1}{\sqrt{2\pi D t_o}} \exp\left(-\frac{(x-x_o)^2}{2Dt_o}\right) \tag{7.20}$$

in close analogy to (7.15). The limitation to small time differences t_o accounts for spatial variations of D. However, we can exploit the Markov character of diffusion, $p(x, x_o; 2t_o) = \int p(x, x'; t_o)\, p(x', x_o; t_o)\, \mathrm{d}x'$, and compute any transition probability $p(x, x_o; t_o)$ by repeated integration. For example,

$$p(x, x_o; 3t_o) = \frac{1}{(2\pi D t_o)^{3/2}} \int \exp\left(-\frac{(x-x')^2}{2D(x)t_o} - \frac{(x'-x'')^2}{2D(x')t_o} - \frac{(x''-x_o)^2}{2D(x'')t_o}\right) \mathrm{d}x'\, \mathrm{d}x'' \tag{7.21}$$

Eventually, one obtains the *functional integral* $p(x, x_o; t) = \int \exp(-O/2) Dx$, where $Dx = \mathrm{d}x'\, \mathrm{d}x'' \ldots \mathrm{d}x^{\mathrm{n}}$ and the Onsager-Machlup function $O = O_o + \int D^{-1}(\mathrm{d}x/\mathrm{d}t')^2 \mathrm{d}t'$ involves integration from $t = 0$ to $t' = t$ (Onsager and Machlup 1953).

7.2.6 Gases in magnetic metals

Some transition metals and many intermetallic alloys are able to accommodate atoms such as H, C, and N on interstitial crystal sites. A well-known example is steel, where the C atoms occupy the small octahedral interstices at the center of the faces of the bcc unit cell. However, since the interstices are relatively small, the carbon concentration is low and the occupancy restricted to one pair of adjacent faces, for example $z = 0$ and $z = a$. This leads to a tetragonal lattice distortion known as *martensitic* phase transition and is responsible for the mechanical and magnetic hardness of steel. Hydrogen goes into many metals and alloys, and sometimes modifies the magnetic properties of the alloys (Fast 1976).

One way of introducing gases into metals is gas-phase interstitial modification, by heating the metal in gases such as N_2, H_2, NH_3, and CH_4. For example, heating in ammonia amounts to the reaction $2\,NH_3$ (gas) $\rightarrow 2\,N$ (interstitial) $+ 3\,H_2$ (gas). The decomposition of ammonia costs energy but is entropically favorable, because three hydrogen atoms have a higher gas-phase entropy than the two ammonia atoms (see exercise on ammonia formation). For this reason, ammonia easily decomposes at metallic surfaces, especially if the surface is able to absorb nitrogen. This method was originally used to produce nitrogen martensite (Jack 1951) but later extended to intermetallic alloys for permanent magnets (Coey and Sun 1990). It is also possible to use N_2 (Coey and Skomski 1993) and to introduce elements such as carbon from the melt or by solid-state reaction (Skomski 1996b).

From the surface, the gas diffuses into the bulk of the material, with a diffusion constant $D = D_o \exp(-E_a/k_B T)$, where $D_o \sim 1\,\mathrm{mm}^2/\mathrm{s}$. The final concentration x increases with gas pressure, but the main consideration is the reaction energy. Figure 7.12 illustrates several cases. An interesting feature is that elastic interactions between interstitial atoms give rise to a phase segregation (spinodal decomposition) into phases with high and low interstitial concentrations, as illustrated in the bottom part of Fig. 7.12(c). The transition occurs below some critical temperature, about 300 °C for PdH_x (Fast 1976), and is reminiscent of a ferromagnetic phase transition. It can be described by an Ising-type lattice-gas model (Alefeld 1969, Skomski 1996b) and exhibits features such as negative diffusion coefficients, meaning that a concentration gradient in the vicinity of a phase boundary *increases.* (By contrast, ordinary diffusion reduces existing concentration gradients.) The critical temperature is of order $E\Delta v^2/V$, where E is Young's modulus, Δv is the volume expansion per interstitial atom, and V is the crystal volume per interstitial site. Due to the long-range character of the elastic interactions, the transition exhibits *mean-field* critical exponents.

The behavior of hydrogen atoms in metals is essentially metallic. Let us compare two isolectronic metals, Pd and Ni. The formation of PdH_x is exothermic, so that palladium absorbs huge amounts of hydrogen ($x \approx 0.6$). The 0.6 hydrogen electrons per Pd atom fill the holes in the 4d band (Mott and Jones 1936), shift the Fermi level, and reduce the Pauli susceptibility. The formation of NiH_x is endothermic, $x \ll 1$,

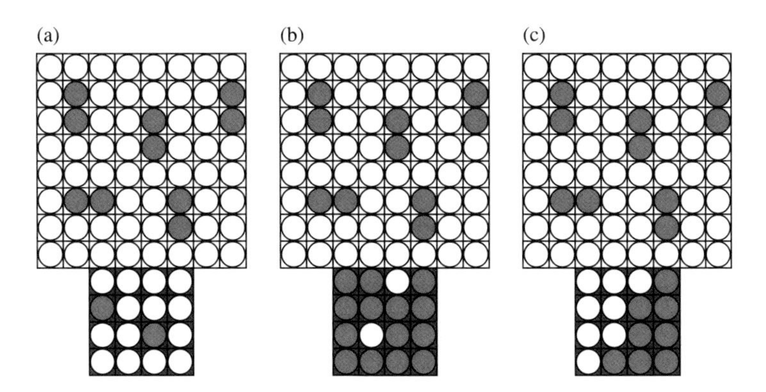

Fig. 7.12 Gases in metals: (a) endothermic reaction, (b) exothermic reaction, and (c) phase segregation (spinodal decomposition). The big white and smaller dark boxes symbolize the gas and solid phases, respectively, and the gray spheres are H or N atoms. The shown configurations refer to equilibrium, after diffusion in the solid phase.

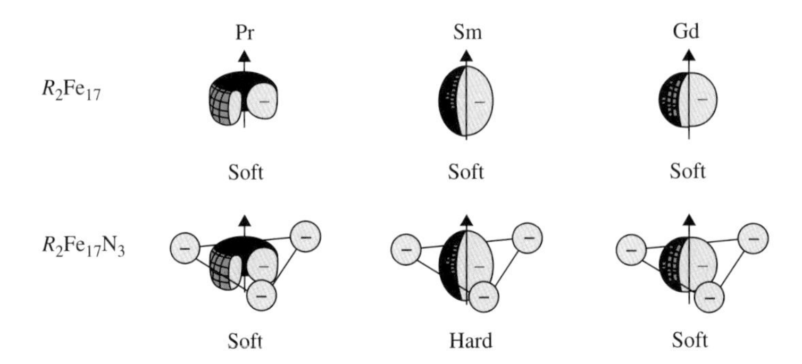

Fig. 7.13 Nitrogen coordination in $R_2Fe_{17}N_x$ magnets (R = Pr, Sm, Gd). The hard ($K_1 > 0$) or soft ($K_1 \sim 0$) character of the intermetallics is determined by the electrostatic repulsion between the nitrogen atoms and the rare-earth 4f shells. Interstitial nitrogen changes the preferential magnetization direction of Sm_2Fe_{17} from in-plane to easy-axis.

but the concentration increases with hydrogen pressure. The hydrogen electrons fill the 0.6 ↓ holes of the Ni 3d shell and reduce the Ni moment by $x\,\mu_B$. This is very similar to the reduction of the moment per atom $Ni_{1-x}Cu_x$. Nitrogen is much more electronegative and may actually *enhance* the moment. For example, iron in thin-film $Fe_{16}N_2$ has a moment of order $2.8\,\mu_B$ (Kim and Takahashi 1972, Coey *et al.* 1994).

Gases in metals can be used to substantially improve magnetic properties. Hydrogen easily goes into many rare-earth transition-metal intermetallics, and some of these materials, for example $LaNi_5$, are of interest in the field of hydrogen storage. However, the corresponding changes in the magnetic properties are usually small. Bigger effects can be obtained by using nitrogen. For example, interstitial modification of Sm_2Fe_{17} yields $Sm_2Fe_{17}N_3$, with significantly improved Curie temperature, magnetization, and anisotropy (Coey and Sun 1990, Skomski 1996b). Particularly impressive is the improvement of the anisotropy constant K_1, from $-0.8\,\mathrm{MJ/m^3}$ to $+8.6\,\mathrm{MJ/m^3}$.

Rare-earth transition-metal intermetallics with the nominal composition R_2Fe_{17} exist in two closely related structures, rhombohedral Th_2Zn_{17} (light rare earth) and hexagonal Th_2Ni_{17} structure (heavy rare earth). Both structures consist of alternating Fe and R-Fe layers, similar to the $Nd_2Fe_{14}B$ structure (Fig. 5.15), but the stacking of the layered building blocks is different, ABC (rhombohedral) and AB (hexagonal). In both structures, the nitrogen occupies interstitial sites in the R-Fe layers, forming triangles that coordinate the rare-earth atoms (Fig. 7.13). As discussed in Section 3.4, the leading contribution to the rare-earth transition-metal anisotropy reflects the crystal-field interaction of the aspherical rare-earth 4f shells. The electronegative nitrogen atoms act as strong crystal-field sources, and examination of Fig. 7.13 shows that the only compound with stable c-axis anisotropy is $Sm_2Fe_{17}N_3$.

7.2.7 Magnetoresistance

In Section 7.2.5, we have emphasized the diffusive character of Ohm's law, $I = U/R$ or $j = E/\rho_e$, where E is the electric field. Electrons are scattered by impurities, lattice distortions (phonons), and other inhomogeneities, and this random scattering translates into the dissipation of energy. The motion of the electrons, and therefore the resistance, depends on the external magnetic field and on the magnetization of the material. This phenomenon, known as *magnetoresistance*, is of practical importance in fields of magnetic recording, sensors, and spin electronics. Let us start with ordinary resistance and then consider spin-dependent mechanisms.

A qualitative explanation of metallic conductivity is provided by the quasiclassical Drude theory. The motion of the electrons obeys $m\mathrm{d}^2x/\mathrm{d}t^2 = \mathrm{e}E$, so that the average velocity of the electrons $\langle v \rangle = \mathrm{e}Et/m$. According to this equation, the electron velocity and the corresponding conductivity $\sigma_\mathrm{e} = 1/\rho_\mathrm{e}$ would diverge with increasing time. In reality, scattering yields a finite relaxation time τ after which the electrons' acceleration is interrupted, and the average velocity is limited to $\langle v \rangle = \mathrm{e}E\tau/m$. The scalar current density is defined as $j = ne\langle v \rangle$, so that $\sigma_\mathrm{e} = ne\langle v \rangle/E$ and

$$\rho_\mathrm{e} = \frac{m_\mathrm{e}}{n\,\mathrm{e}^2\tau} \tag{7.22}$$

In spite of its classical origin, this equation provides a correct explanation of the resistance in terms of the relaxation time τ. Basically, it expresses the conductivity in terms of the mean free path time $l = v_\mathrm{F}\tau$, where $v_\mathrm{F} = \hbar k_\mathrm{F}/m_\mathrm{e}$ is the Fermi velocity. In good metals, $n = 10^{28}\,\mathrm{m}^{-3}$, $v_\mathrm{F} \approx 10^6\,\mathrm{m/s}$, and $\rho = 1/\sigma = 1\mu\Omega\,\mathrm{cm}$. The corresponding mean free path, $l \approx 100\,\mathrm{nm}$, is much larger than the interatomic distance. In poor conductors, l is reduced, and in insulators, it approaches the interatomic distance (metal-insulator transition).

One aspect of quantum-mechanical theories of conduction is to determine t from the Schrödinger equation

$$\mathrm{i}\hbar\frac{\mathrm{d}\psi}{\mathrm{d}t} = \mathsf{H}\psi + V(\mathbf{r},t)\psi \tag{7.23}$$

where H is the unperturbed one-electron Hamiltonian. To calculate the relaxation time we start from the scattering rate $W_{\mathbf{k}'\mathbf{k}}$ of an electron of wave vector k into a state of wave vector k′. For sufficiently weak time dependence $V(t)$ we can use Fermi's golden rule

$$W_{\mathbf{k}'\mathbf{k}} = \frac{2\pi}{\hbar}\,\delta(E_\mathbf{k} - E_{\mathbf{k}'})\,|\langle\psi_\mathbf{k}\rangle V(\mathbf{r})|\psi_{\mathbf{k}'}\rangle|^2 \tag{7.24}$$

where the $\langle\psi_\mathbf{k}|V(\mathbf{r})|\psi_{\mathbf{k}'}\rangle = V_{\mathbf{k}\mathbf{k}'}$ are nondiagonal matrix elements. The next step is to evaluate the master or rate equation

$$\frac{\mathrm{d}p(\mathbf{k})}{\mathrm{d}t} = \sum_{\mathbf{k}'} W_{\mathbf{k}\mathbf{k}'}(p(\mathbf{k}') - p(\mathbf{k})) \tag{7.25}$$

This equation has form of a one-electron equation, but it can be shown to satisfy the Pauli principle (Mott and Jones 1936). In other words, the resistivity of dense electron gases is not affected by the Fermi statistics of the electrons. A similar phenomenon governs the interstitial diffusion in the limit of large concentrations, Fig. 7.12(b). Intuitively, one might expect that the diffusion slows down on approaching full occupancy, because double occupancy of interstitial sites is forbidden. However, for noninteracting interstitial atoms, the corresponding on-site repulsion terms cancel, and the diffusivity is independent of concentration.

In the case of free electrons, the spherical symmetry of the Fermi surface simplifies the calculation and yields, with, $\mathrm{d}p/\mathrm{d}t = -p/\tau$, the inverse relaxation time

$$\frac{1}{\tau} = \sum_{\mathrm{s}\mathbf{k}'} \left(\frac{1 - k'_{\mathrm{x}}}{k_{\mathrm{x}}} \right) W_{\mathbf{k}\mathbf{k}'} \tag{7.26}$$

where the index s means that the integration is restricted to the Fermi surface. The factor $1 - k'_{\mathrm{x}}/k_{\mathrm{x}}$ indicates that scattering is most effective if k'_{x} is negative. This equation can be evaluated for different types of scattering, such as impurity scattering ($\rho \sim T$) and electron-phonon scattering ($\rho = const.$).

The magnetoresistance of simple metals reflects the curvature of electron trajectories in a magnetic field and is often very small. In ferromagnets, magnetoresistance is largely due to the spin dependence of the matrix elements $\langle \psi_{\mathrm{k}} | V(\mathbf{r}) | \psi_{\mathrm{k}'} \rangle$, which enter (7.24). A widespread and important mechanism is *anisotropic magnetoresistance* (AMR). It is caused by spin-orbit interaction (Campbell, Fert, and Jaoul 1970) and means that aspherical electron orbitals (Figs 3.8, 5.23, and 7.13) lead to an anisotropic electron scattering. For example, a prolate charge distribution yields a scattering and resistance minimum if the magnetization is parallel to the current. AMR is typically of moderate strength, with room temperature anisotropies of order 1%.

Giant magnetoresistance (GMR) exploits the fact that ↑ and ↓ electrons are scattered differently in regions with opposite magnetization (Baibich *et al.* 1988). Consider, for example, the two-current model for multilayers with (i) ferromagnetic coupling and (ii) antiferromagnetic coupling. The model is defined by the assumption of two separate spin channels, ↑ and ↓. In the ferromagnetic case, ↑ (or ↓) electrons can easily travel through the lattice, depending on which spin channel exhibits the lowest resistance. In the antiferromagnetic case, there is no privileged spin channel, because both ↑ and ↓ electrons must pass through regions of parallel and antiparallel spins. This amounts to a mixing of channels with low and high resistance and means that the resistance of the AFM configuration is higher than that of the FM configuration. GMR exists in several variants, including multilayer GMR, spin-valve GMR (without RKKY coupling), and granular GMR. Related mechanisms are tunnel magnetoresistance (TMR), where ferromagnetic regions are separated by an insulator (Moodera and Mathon 1999), and colossal magnetoresistance, which involves magnetic phase transitions (Coey, Viret, and von Molnár 1999).

A potential source of magnetoresistance is spin-dependent electron scattering scattering at domain walls and other micromagnetic features. In a crude approximation, the scattering ability of a micromagnetic inhomogeneity scales as $(\nabla\mathbf{M})^2$, but the smooth character of conventional domain walls makes this contribution very small. Increasing the anisotropy is no option, because it enhances $\nabla\mathbf{M}$ at the expense of higher operating fields (Skomski 2001). A more promising way to enhance $\nabla\mathbf{M}$ is to use junctions and interfaces with reduced exchange stiffness A', as discussed in Section 4.4.

A somewhat different phenomenon is ballistic magnetoresistance. It is a specific regime common to several magnetoresistance mechanisms and means that electrons are scattered (reflected) without energy dissipation. The regime is realized on length scales smaller than the mean free path $l = v_{\mathrm{F}}\tau$. It is related to the Landauer formula for

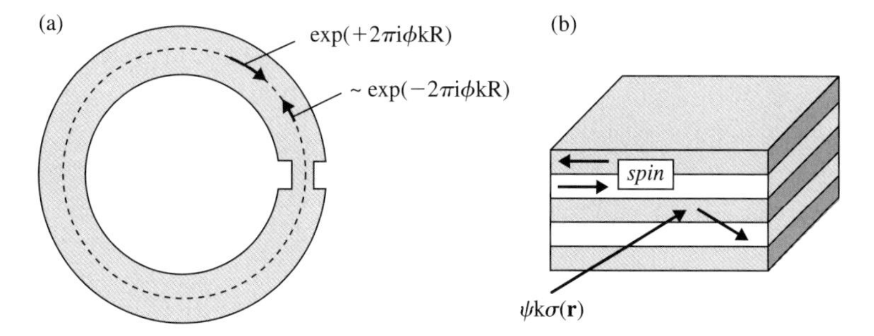

Fig. 7.14 Ballistic electron scattering: (a) ring with nanojunction and (b) strictly periodic multilayers. In both cases, the resistance is zero, and additional imperfections are necessary to dissipate energy.

the conductance, $G = e^2 T/h$, where T is the transmittance (Landauer 1970, Tsymbal, Mryasov, and LeClair 2003). Ballistic electron scattering may or may not yield resistance. Figure 7.14(a) shows a metallic ring where electrons are scattered at a nanojunction but do not dissipate energy. A current existing at $t = 0$ persists to infinity, and $R = 0$. To realize nonzero resistance, the electrons must be scattered elsewhere in the ring, and the function of the junction is then to reflect the electrons into regions where their energy can be dissipated. A similar situation is encountered in ideal multilayers whose resistance is zero because the wave functions of the conduction electrons are eigenfunctions of the system. This is analogous to the absence of resistance in strictly periodic solids, where $V_{\mathbf{kk}'} = 0$.

7.2.8 Other transport phenomena involving magnetism

Let us, finally, mention a few transport phenomena that involve magnetism but whose treatment goes beyond this book. One aspect is the wide range of materials of interest in magnetism. Our focus has been on metallic conductivity, but there are other interesting classes of materials, such as magnetic semiconductors, halfmetallic oxides, and superconductors.

Magnetic semiconductors have briefly been mentioned in the context of RKKY exchange (Section 2.3.2). Examples are substances such as GaN, ZnO, and SnO_2 doped with 3d atoms such as V, Cr, Mn, Fe, and Co (Dietl *et al.* 2000). For some magnetic semiconductors, T_c exceeds room temperature, which makes them potentially useful for spin electronics. The investigation of these materials is demanding, both structurally and from the point of view of electronic structure, magnetism, and transport. One question is how the magnetic atoms arrange in the host lattice and whether there is a phase transition similar to Fig. 7.12. In *halfmetallic oxides*, where one spin channel is conducting and the other spin channel is (almost) insulating (de Groot and Buschow 1986). Examples are CrO_2 (Coey and Venkatesan 2002), $(La_{0.7}Sr_{0.3})MnO_3$, and the semi-Heusler alloy NiMnSb. Ideally, this amounts to an infinite magnetoresistance ratio, but in reality this ratio is greatly reduced by interface effects (Dowben and Skomski 2003) and mechanisms such as finite-temperature spin mixing (Skomski and Dowben 2002).

Transport effects of importance in heterostructures for electronic applications are spin injection and spin torque. The question is whether and how spin-polarized electrons are injected from one phase (usually a ferromagnet) into another phase. The spin injection is, essentially, a quantum effect, because it involves the wave-vector dependent reflection and transmission of electrons (Ziese and Thornton 2001, Tsymbal, Mryasov and LeClair 2003). After injection, the electrons undergo relaxation, characterized by Drude-type mean free paths (Section 7.2.7) However, most scattering events do not flip the spin, and the corresponding room-temperature spin-diffusion lengths can be very large. In simple metals, where the spin-orbit coupling is small, the spin-diffusion length may exceed 1 μm, but in transition metals, the spin-orbit coupling yields a more effective spin flipping, and spin-diffusion length is smaller. A rough estimate for late iron-series trandition metals is somewhat less than 10 nm, but this value is real-structure dependent and usually quite different for $\uparrow$ and $\downarrow$ electrons.

Spin-polarized currents in ferromagnets reflect the different $\uparrow$ and $\downarrow$ densities of states and exert a torque in inhomogeneous structures, especially in the vicinity of interfaces. Examples are multilayers (Slonczewski 1996) and nanowires (Wegrowe *et al.* 1999). As outlined in Section 2.1.6, charges in metals attract or repel conduction electrons to ensure charge neutrality, so that electron-density fluctuations $\delta n_\uparrow(\mathbf{r}) + \delta n_\downarrow(\mathbf{r})$ are screened on an atomic scale. However, the spin accumulation $\delta n_\uparrow(\mathbf{r}) - \delta n_\downarrow(\mathbf{r})$ remains unchanged until spin-flip scattering changes the distribution of $\uparrow$ and $\downarrow$ electrons.

Superconductivity is a well-developed research field involving pairing interactions between electrons of opposite spin spins (see. e.g. Ashcroft and Mermin 1976). The Bardeen-Cooper-Schrieffer (BCS) theory explains the formation of electron pairs (Cooper pairs) by electron-phonon coupling. In a nutshell, the electrons' electrostatic field modifies the positions of the atomic cores, and the lattice vibrations (phonons) yield an attractive interelectronic interaction V_{eff}. The coupled electrons have opposite spins, so that the Cooper pairs are zero-spin bosons which can form a superconducting Bose-type condensate below some critical temperature T_{o}. Since the attractive interaction is mediated by phonons, the critical temperature scales as the Debye temperature θ_D rather than the Fermi temperature T_{F}, and the BCS theory yields $T_{\mathrm{c}} = 1.14\,\theta_{\mathrm{D}} \exp(-1/V_{\mathrm{eff}}\mathsf{D}(E_{\mathrm{F}}))$, where $V_{\mathrm{eff}}\mathsf{D}(E_{\mathrm{F}}) \ll 1$. Here the involvement of the density of states indicates that the pairing interaction is limited to electrons near the Fermi level. Typical BCS superconductors are Al, Pb, and Nb_3Ge, where the respective critical temperatures are 1.2 K, 7.2 K, and 23 K. High-temperature superconductors, ceramic cuprates such as $YBa_2Cu_3O_{7-\delta}$ ($T_{\mathrm{c}} = 93$ K), go beyond the BCS theory. Some aspects of high-temperature superconductivity are qualitatively understood in terms of the so-called $t-J$ model (Fulde 1991). The model considers deviations from half-filling in strongly correlated Heisenberg antiferromagnets (Section 2.1), where the additional carriers compete with the antiferromagnetic order.

In Section 2.1.8, we have discussed that interactions between highly correlated and independent electrons give rise to Kondo behavior, and this behavior is accompanied by a resistance minimum at the Kondo temperature. Correlations are also responsible for metal-insulator or Mott transition (Section 2.1.7). Loosely related to the magnetoresistance of simple metals is the Hall effect, where a magnetic field creates a Lorentz field, curves the trajectories of the electrons, and yields a voltage.

7.3 Bruggeman model

Summary The Bruggeman model describes linear mechanical, magnetic, electrical, and transport properties for a broad variety of composites. The idea is to start from exact solutions for small volume fractions of a second phase in a main or matrix phase. Arbitrary volume fractions are then treated by self-consistently embedding the phases in an effective medium. The theory yields materials parameters as functions of the volume fractions and geometries of the phases. Each system is described by a single interaction parameter g, which is equal to the percolation threshold of the composite. The predictions of the Bruggeman model, especially the behavior near the percolation threshold, are mean-field like.

A frequently occurring problem is the determination of effective materials parameters for composites, ranging from naturally occurring biological structures and traditional materials to artificial materials used in transport, space, microelectronic, and other applications. For example, the composite structure of bones and wood ensures stiffness without brittleness, and the same principle is exploited in artificial mechanical materials, such as concrete and reinforced polymers. In mechanical materials, one often considers the elastic moduli, such as Young's modulus, and the corresponding inverse moduli (compliances). In magnetism, one encounters, for example, the problem of finding the effective permeability or susceptibility of a composite material. A simple volume averaging over the volume fractions of the phases is often a poor approximation. Consider, for example, the resistivity ρ_e of a metal-insulator composite. Since $\rho_e = \infty$ in the insulating phase, the volume average $\langle \rho_e \rangle$ is infinite, in striking contrast to the finite resistivity above the percolation threshold. Another example is the well-known distinction between parallel and serial connections of resistors, which cannot be reduced to a volume average.

The *Bruggeman model* (1935) makes it possible to treat two-phase composites on a mean-field level. The idea is to start with small volume fractions ϕ of one phase. For example, in three-dimensional composites, one may consider a single spherical particle (phase II) in an infinite matrix (phase I), Fig. 7.15. This geometry is relatively easy to treat, because the phase-II inclusion is isolated and does not interact with other inclusions. The transition from small to arbitrary volume fractions is realized by embedding phase-II (and phase-I) particles in an effective matrix with properties intermediate between those of phases I and II, Fig. 7.16(e–f). Similar to Section 5.3, the embedding must be self-consistent, that is, the calculated materials constant must be equal to the materials constant assumed for the effective medium.

7.3.1 Static and dynamic properties

The Bruggeman approach describes a variety of static and dynamic electromagnetic, mechanical, and diffusive phenomena. Examples are magnetism (effective susceptibility), electrical conduction (conducting composites, insulating inclusions in metallic matrix, metal-superconductor composites), thermal conduction (heat insulation using composite construction materials), diffusion (hydrogen and nitrogen transport in intermetallic composites), electrodynamics (static and dynamic dielectric response of inhomogeneous media), rheology (viscosity of colloidal suspensions, such as blood,

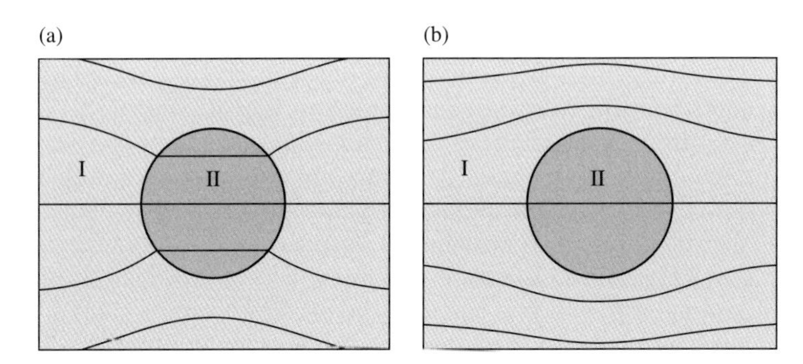

Fig. 7.15 Flux-line modification due to inhomogeneities: (a) flux-line attraction in regions with large compliance and (b) flux-line repulsion in regions with small compliance.

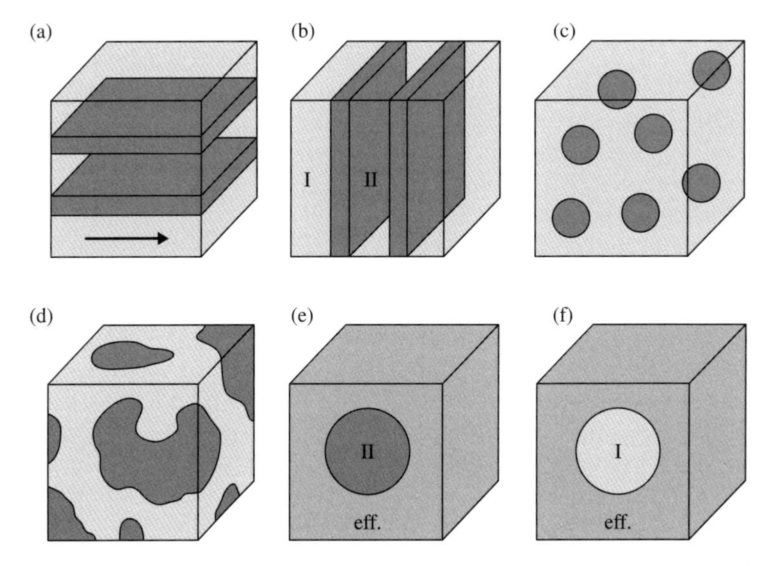

Fig. 7.16 Bruggeman models: (a) parallel geometry, (b) series geometry, (c–d) three-dimensional composites, and (e–f) self-consistent embedding. The arrow shows the direction of the applied field or force.

food, gels), and mechanical composites (elasticity of reinforced construction materials, filled polymers, such as car tires).

The treatment of these phenomena is simplified by two features: conjugate relations, also known as dual or inverse relations, and complex materials constants. Examples of *conjugate quantities* are the resistivity ρ_e and the conductivity $\sigma_e = 1/\rho_e$ of metal-insulator composites, and the shear modulus G and the shear compliance $J = 1/G$ of mechanical composites. Materials constants such as σ_e and J are known as compliances, whereas conjugate constants such as ρ_e and G are referred to as stiffnesses. By definition, compliances describe the magnitude of the system's reaction to an external field or force, so that the magnetic susceptibility is a compliance, not a stiffness. Conjugate relations are very useful in the discussion of composites. For example, parallel geometries, Fig. 7.16(a), are characterized by the additivity of compliances, whereas serial geometries, Fig. 7.16(b), lead to the additivity of stiffnesses. In mechanics, effective materials constants for series and parallel geometries are known as Reuss and Voigt (or Kelvin) averages.

Complex materials constants relate static and dynamic properties. For example, a simple model of viscoelasticity is defined by the materials equation

$$\sigma_{\mathrm{m}} = G\varepsilon + \eta_{\mathrm{m}} \frac{\mathrm{d}\varepsilon}{\mathrm{d}t} \tag{7.27}$$

where σ_{m} is the shear stress, G denotes the shear modulus, and η_{m} is the mechanical viscosity. Fourier transformation reduces this equation to the complex materials equation $\sigma^* = G^* \varepsilon^*$, where $G^* = G + \mathrm{i}\omega\eta_{\mathrm{m}}$ is the complex shear modulus and $\tau = \eta_{\mathrm{m}}/G$ is the relaxation time of the system (see e.g. Ward and Hadley 1993). Other examples are the linear model of hysteresis (Section 1.5) and the conductance quantity $\sigma_{\mathrm{e}}/\omega$, which has the character of an imaginary permittivity. Fortunately, complex materials constants are easy to deal with if one properly accounts for the imaginary parts. For example, $J^* = 1/G^*$, where $G^* = G' + \mathrm{i}G''$ and $J^* = J' + \mathrm{i}J''$.

7.3.2 *Parameterization

The Bruggeman model assumes two phases of materials constants $Q_{\mathrm{I/II}}$ and volume fractions $\phi_{\mathrm{I/II}}$. For convenience, we write $\phi_{\mathrm{II}} = \phi$ and $\phi_{\mathrm{I}} = 1 - \phi$. Many problems are characterized by stationary or static equations of the type $\nabla(Q(\mathbf{r})\nabla\Phi) = 0$, where Q is a materials constant and Φ is a potential. For example, diffusion processes are characterized by $Q = D$ (diffusion constant) and $\Phi = c$, whereas linear magnetic media are characterized by $Q = \chi$ (or $Q = \mu$) and $\Phi = \Phi_{\mathrm{M}}$. Similar equations exist for other transport processes and for elastic media (Christensen 1979). For simple geometries, such as Fig. 7.16(a) and (b), the problem is solved very easily by considering the serial and parallel behavior of the considered materials parameters, but complicated geometries, such as Fig. 7.16(d), require complex calculation and averaging procedures.

For small volume fractions ϕ, it is relatively easy to calculate the effective materials constant by explicitly solving the underlying differential equation. For example, the response of isolated magnetic spheres is essentially given by the demagnetizing factor $D_{\mathrm{mag}} = 1/3$, which is intermediate between the demagnetizing factors $D_{\mathrm{mag}} = 0$ and $D_{\mathrm{mag}} = 1$ for the parallel and serial geometries, respectively. In the most general case, we can write

$$Q_{\mathrm{eff}} = Q_{\mathrm{I}} \left(1 + \frac{a_1 Q_{\mathrm{II}} + a_2 Q_{\mathrm{I}}}{a_3 Q_{\mathrm{II}} + a_4 Q_{\mathrm{I}}} \phi \right) \tag{7.28}$$

For example, isotropic dielectric composites are characterized by $a_1 = 3$, $a_2 = -3$, $a_3 = 1$, and $a_4 = 2$.

The constants a_1, a_2, a_3, and a_4 are not universal but depend on the physical phenomenon and on the geometry and dimensionality of the composite. Surprisingly, there is only one independent parameter. For nearly homogeneous composites, the effective materials constants are equal to the volume average, $Q_{\mathrm{eff}} = (1-\phi)Q_{\mathrm{I}} + \phi Q_{\mathrm{II}}$ (Landau and Lifshitz 1984). Realizing the limit $Q_{\mathrm{II}} \approx Q_{\mathrm{I}}$ in (7.28) yields $a_1 = a_3 + a_4$,

$a_2 = -(a_3 + a_4)$, and

$$Q_{\text{eff}} = Q_{\text{I}} \left(1 + \phi \, \frac{Q_{\text{II}} - Q_{\text{I}}}{g Q_{\text{II}} + (1-g) Q_{\text{I}}} \right) \tag{7.29}$$

where $g = a_3/(a_3 + a_4)$. For the above-mentioned dielectric composite, $g = 1/3$, whereas the parallel and series configurations of Fig. 7.16(a) and (b) are characterized by $g = 0$ and $g = 1$, respectively.

7.3.3 *Self-consistent materials equations

Figure 7.16(e–f) illustrates the self-consistent embedding procedure. The embedding is done for *both* phases, because any region of the composite is surrounded by a mixed or "effective" environment. Explicitly,

$$Q_{\text{eff,I}} = Q_{\text{eff}} \left(1 + \phi_{\text{o}} \, \frac{Q_{\text{I}} - Q_{\text{eff}}}{g Q_{\text{I}} + (1-g) Q_{\text{eff}}} \right) \tag{7.30a}$$

and

$$Q_{\text{eff,II}} = Q_{\text{eff}} \left(1 + \phi_{\text{o}} \, \frac{Q_{\text{II}} - Q_{\text{eff}}}{g Q_{\text{II}} + (1-g) Q_{\text{eff}}} \right) \tag{7.30b}$$

where ϕ_{o} is the fictitious volume fraction of the phase-I and phase-II regions in the effective matrix. From this equation, Q_{eff} is obtained by averaging over the two configurations of Fig. 7.16(e) and (f), that is, $Q_{\text{eff}} = (1-\phi) Q_{\text{eff,I}} + \phi Q_{\text{eff,II}}$. This yields

$$\begin{aligned} Q_{\text{eff}} = {} & \phi \, Q_{\text{eff}} \left(1 + \frac{Q_{\text{II}} - Q_{\text{eff}}}{g Q_{\text{II}} + (1-g) Q_{\text{eff}}} \, \phi_{\text{o}} \right) \\ & + (1-\phi) Q_{\text{eff}} \left(1 + \frac{Q_{\text{I}} - Q_{\text{eff}}}{g Q_{\text{I}} + (1-g) Q_{\text{eff}}} \, \phi_{\text{o}} \right) \end{aligned} \tag{7.31}$$

and

$$\phi \, \frac{Q_{\text{II}} - Q_{\text{eff}}}{g Q_{\text{II}} + (1-g) Q_{\text{eff}}} + (1-\phi) \frac{Q_{\text{I}} - Q_{\text{eff}}}{g Q_{\text{I}} + (1-g) Q_{\text{eff}}} = 0 \tag{7.32}$$

We are pleased to see that this equation no longer contains the fictitious volume fraction ϕ_{o}.

Mathematically, (7.32) is a quadratic equation for Q_{eff}. Aside from the sign of the physically relevant solution, which must be determined separately,

$$Q_{\text{eff}} = \frac{Q_{\text{II}}(\phi - g) + Q_{\text{I}}(1 - g - \phi)}{2(1-g)} \left(1 \pm \sqrt{1 + \frac{4 Q_{\text{II}} Q_{\text{I}} g (1-g)}{(Q_{\text{II}}(\phi - g) + Q_{\text{I}}(1 - g - \phi))^2}} \right) \tag{7.33}$$

This equation interpolates between $Q_{\text{I}}(\phi = 0)$ and $Q_{\text{II}}(\phi = 1)$. For $Q_{\text{I}} \approx Q_{\text{II}}$, this equation reduces to $Q_{\text{eff}} = \langle Q \rangle$, but in general the dependence on ϕ is strongly nonlinear.

7.3.4 *The response parameter g

Equation (7.33) yields the effective materials constant $Q_{\rm eff}$ as a function of $Q_{\rm I}$, $Q_{\rm II}$, ϕ, and g. The parameter g depends on the geometry and dimensionality of the considered composite. For compliances described by $\nabla(Q(\mathbf{r})\nabla\Phi) = 0$, it is equal to the *demagnetizing factor* of the corresponding magnetic system. This includes magnetic and dielectric susceptibilities, diffusion, and conductivity. For example, the geometries of Fig. 7.16(c–f) all correspond to *spheres* embedded in an effective matrix, so that $g = 1/3$. Similarly, Fig. 7.16(a) and (b) correspond to the above-mentioned respective parameters $g = 0$ and $g = 1$. For macroscopically isotropic composites in d dimensions (inhomogeneous wires, inhomogeneous films, and bulk composites), $g = 1/d$.

Response parameters g for various elastic, viscous, and viscoelastic properties are obtained by comparing the results of continuum calculations (Christensen 1979) with (7.30). For spherical inclusions, the shear modulus G and the bulk modulus K are described by

$$g = \frac{2(4 - 5\,\nu_{\rm o})}{15(1 - \nu_{\rm o})} \quad \text{and} \quad g = \frac{1 + \nu_{\rm o}}{3(1 - \nu_{\rm o})} \tag{7.34}$$

respectively. In both equations, $\nu_{\rm o}$ is the Poisson ratio of the matrix. Incompressible materials obey $\nu_{\rm o} = \frac{1}{2}$ and $K = \infty$, and the shear modulus is described by $g = 2/5$. When the second phase is very hard, $G = \infty$, then (7.33) predicts $G_{\rm eff} = G_{\rm I}(1 + 2.5\phi)$. This equation has become the starting point for the description of the elastic behavior of carbon-reinforced rubber, as used in car tires (Erman and Mark 1997). It was first obtained by Einstein (1911), in the context of the effective viscosity of suspensions of spherical particles in a liquid.

Conjugate materials constants, such as J and G, are linked by the simple transformation $g \to 1 - g$. For example, the conductivity of a granular composite is described by $g = 1/3$, because it has the character of a compliance. The corresponding stiffness, namely the resistivity, is characterized by $g = 2/3$. Real and imaginary parts of complex materials constants have the same g.

7.3.5 *Percolation in the Bruggeman model

The Bruggeman model describes percolation (Section 7.2.4) on a mean-field level. In (7.33), this is achieved by putting $Q_{\rm II} = 0$ or $Q_{\rm II} = \infty$, depending on the physical context and on whether one considers a compliance or a stiffness. For $Q_{\rm II} = 0$, the percolation threshold $\phi_{\rm c} = 1 - g$ and $Q_{\rm eff} = Q_{\rm I}(1 - \phi/\phi_{\rm c})$, whereas $Q_{\rm II} = \infty$ yields $\phi_{\rm c} = g$ and $Q_{\rm eff} = Q_{\rm I}/(1 - \phi/\phi_{\rm c})$. An example is the resistivity of a metal-insulator composite, which diverges when the volume fraction of the insulating phase approaches $g = 2/3$. For reinforced rubbers, the theory predicts a divergence of G when the volume fraction of the hard phase reaches 40%. This is close to the experimental value of slightly more than 45% (Christensen 1979).

The Bruggeman model describes the percolation transition on a mean-field level. Each region of the composite interacts with an average local environment only, whereas the long-range correlations (percolating cluster) are ignored. It is also important to keep in mind that the Bruggeman theory refers to a linear equation of state. Hysteretic effects are approximated in a very crude way, as outlined in Section 1.5. Furthermore, the model is characterized by the *absence* of characteristic length scales, such as domain-wall widths in magnetic systems and end-to-end distances in polymers. In the simplest case, equations of the type $\nabla(Q(\mathbf{r})\nabla\Phi) = 0$ must be replaced by $\nabla(Q(\mathbf{r})\nabla\Phi) - \kappa^2\Phi = 0$ (Skomski 2004). We will return to this question in the next section.

7.4 Nanostructures, thin films, and surfaces

Summary The magnetism of nano- and thin-film structures is intermediate between atomic and macroscopic magnetism but cannot be reduced to a superposition of the two limits. Characteristic length scales are of order $\alpha/a_o = 7.52\,\text{nm}$, that is, comparable to magnetic domain-wall widths. Phenomena of atomic origin are often important on somewhat smaller length scales of 1 to 2 nm. For example, magnetic nanoparticles embedded in a nonmagnetic metallic matrix experience a significant RKKY coupling, in spite of the rapidly oscillating character of the RKKY interaction. Competing nanoscale exchange and anisotropy are described by random-anisotropy models, whereas other micro- and nanomagnetic models describe composite nanostructures and thin-film nanostructures, such as exchange-coupled multilayers. One example is hard-soft two-phase permanent magnets, whose performance goes beyond what is expected from the volume fractions of the involved hard and soft phases.

Magnetism is, to a large extent, a *nanoscale* phenomenon. This refers not only to naturally occurring and artificial nanostructures but also to traditional hard and soft magnets. Figure 7.17 shows some schematic nanostructures, ranging from particles and ferrofluids (Charles 1992, Dormann, Fiorani, and Tronc 1999, Williams *et al.* 2003) to multilayers (Astalos and Camley 1998, Fullerton, Jiang, and Bader 1999), nanowires (Sellmyer, Zheng, and Skomski 2001), nanotubes and nanorings (Sui *et al.* 2004, Sorge *et al.* 2005), patterned thin films (Ross *et al.* 2002, Sellmyer 2002, Lodder 2006) and networks (Hirohata *et al.* 2000). Some systems combine both thin-film and particulate features, especially composite thin films (Al-Omari and Sellmyer 1995, Sellmyer *et al.* 2001, Sellmyer *et al.* 2002).

Natural nanomagnetism is encountered, for example, in magnetostatic bacteria, which live in dark environments and contain chains of magnetite particles. The chains have sizes of the order of 40 to 100 nm and are used for vertical orientation. Similar particles have been found in the brains of other animals, such as bees and pigeons, and it is being investigated whether they contribute to flight orientation. Some nanostructures, such as ferrofluids and sintered permanent magnets, do not occur in nature but are relatively easy to produce. Others require sophisticated processing methods.

Magnetic nanostructures offer rich physics, as reviewed by Solzi, Ghidini, and Asti (2002) and Skomski (2003). An intriguing feature is that nanomagnetism cannot be reduced to a superposition of atomic and macroscopic effects. This is seen most clearly by analyzing nanomagnetic lengths, which roughly scale as $a_o/\alpha \approx 7.52\,\text{nm}$. Here a_o is the Bohr's hydrogen radius and α is Sommerfeld's fine-structure constant (Skomski 2003). This length comprises many interatomic distances without being macroscopic and leads to new questions. By definition, the correlation length ξ approaches infinity at the Curie point. How can T_c be defined in nanostructures, and how does ξ interfere

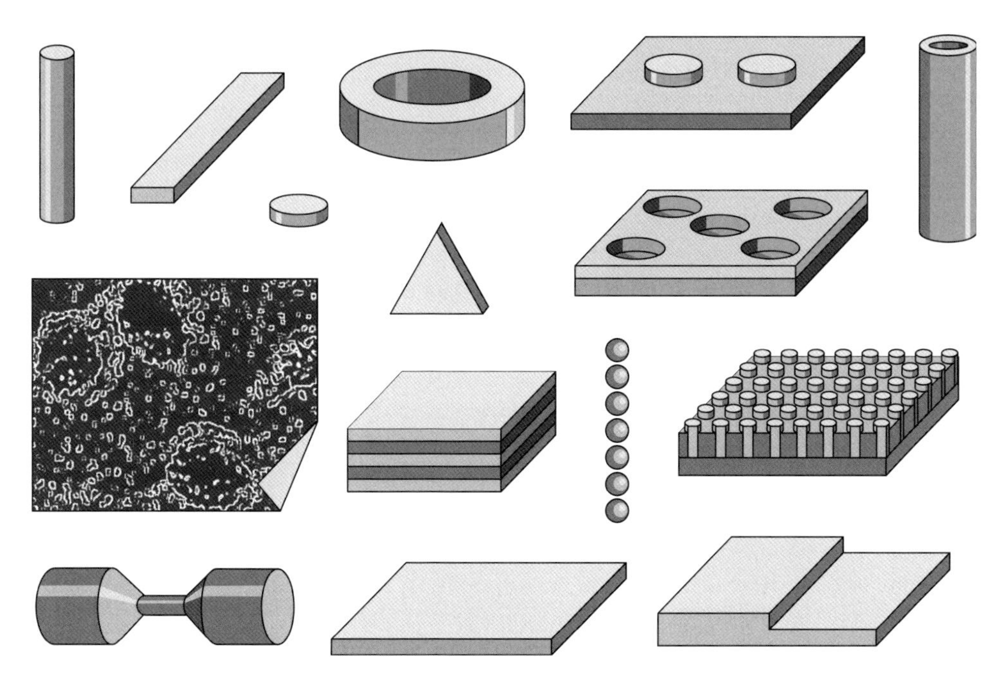

Fig. 7.17 Some magnetic nanostructures (schematic). The structures exist on various length scales and in different chemical compositions.

with a_o/α? More generally, how many atoms are necessary to make a nanostructure indistinguishable from a bulk magnet? Similarly, Bloch wave functions extend to infinity, and the question arises how the electronic structure is modified in nanostructures.

Another fascinating aspect of magnetic nanostructures is the wide range of geometries (Fig. 7.17), chemical compositions, and fabrication techniques (Skomski 2003, Nalwa 2002, Sellmyer *et al.* 2002). In a broader sense, nanostructures include molecular magnets (Wernsdorfer 2006) and thin-film structures, such as ultrathin films (Oepen and Kirschner 1989, Gradmann 1993, Bland and Heinrich 1994, Krams *et al.* 1994, Málek and Kamberský 1958, Bander and Mills 1988), multilayers (Grünberg, *et al.* 1986), and stepped and rough surfaces (Chuang, Ballantine, and O'Handley 1994, Sander *et al.* 1996). For each basic geometry, there exist countless homogeneous and inhomogeneous chemical compositions, and the fabrication techniques range from comparatively easy-to-produce bulk nanocomposites to demanding artificial nanostructures.

Magnetic nanostructures have many applications, and nanostructuring opens the door for completely new materials and technologies. This materials-by-design concept includes both top-down approaches (artificial nanostructuring) and bottom-up strategies, such as self-assembly. Present or potential applications include but are not limited to permanent magnets (Coehoorn *et al.* 1988, Skomski and Coey 1993, Manaf, Buckley, and Davies 1993, Liu *et al.* 1998), soft magnets (Yoshizawa, Oguma, and Yamauchi 1988, Herzer 1992, Suzuki and Cadogan 1999), recording media (Sellmyer *et al.* 1998, Terris *et al.* 1999, Weller and McDaniel 2006), and functional structures and materials for micromechanical, spin electronics, and other applications (Cowburn and Welland 2000, Tsymbal, Mryasov, and LeClair 2003, Ziese and Thornton 2001, Liou and Yao 1998, Coey *et al.* 1998). For example, magnetoresistive random-access memories (MRAM) for non-volatile information storage use cells where fixed and soft magnetic layers are coupled by a magnetic tunnel junction (MTJ, Parkin *et al.* 1999). More generally, magnetic nanodots are a potential alternative to current-based electronics, by exploiting magnetic degrees of freedom to store and process information classically (Cowburn and Welland 2000, Sorge *et al.* 2004) or quantum-mechanically (Skomski *et al.* 2004c). Nanoparticle ferrofluids are being considered for cancer treatment, guided by a magnet and delivering high local doses of drugs or radiation (Panyam *et al.* 2004).

7.4.1 Length scales in nanomagnetism

In Section 4.2.5, we analyzed typical micromagnetic length scales, such as the domain-wall width and the exchange length. These lengths determine the range of magnetic interactions and decide whether a spin structure is atomic, nanoscale, or macroscopic. As a rule, phenomena on length scales smaller than 1 to 2 nm are governed by exchange and other atomic interactions. This corresponds to roughly 5 interatomic distances, depending on the considered material, and ensures that atomic characteristics, such as the density of states, approach their bulk values. At the upper end, nanomagnetism blends into submicron magnetism (less than 1,000 nm), whereas lengths larger than 1,000 nm (1 μm) are usually considered macroscopic. There are, however, many borderline cases and exceptions, for example atomic phenomena on length scales somewhat larger than 2 nm. Some phenomena involve several length scales and are tackled by multiscale modeling (Garcia-Sanchez *et al.* 2005).

To derive a typical nanomagnetic length scale, we take into account that magnetic phenomena are often described by differential equations of the type

$$\nabla^2\phi - \kappa^2\phi = f(\mathbf{r}) \tag{7.35}$$

where $1/\kappa$ is an interaction length. This equation must be compared with the inhomogeneous Laplace or Poisson equation $\nabla^2\phi = f(\mathbf{r})$, obtained from (7.35) by putting $\kappa = 0$. The Poisson equation implies macroscopic length scales $1/\kappa = \infty$ and describes, for example, magnetostatic phenomena.

As we can see from nucleation-field equations such as (4.29), the κ^2 term in (7.35) is the ratio of a local energy (anisotropy) and an interaction energy (exchange). In many systems, local and interaction energies are comparable, and $1/\kappa$ is then roughly equal to the interatomic distance a. In metals, the Fermi wave vector $k_F \sim 1/a$, so that $\kappa \sim k_F$. Examples of this atomic limit are the magnetic moments in narrow bands (Section 2.4) and correlations far below the Curie temperature. Nanoscale effects mean that the κ^2 local term is unable to compete against the interaction term, because the local energy is too small by physical origin or by cancellation (Skomski 2004). Examples of the latter mechanism are the Curie transition and exchange-enhanced Pauli paramagnets (2.4.3), where the cancellations involve exchange and thermal energies (Curie transition), and Coulomb interactions and hopping (paramagnons). In nanomagnets, κ^2 is small by physical origin, $\kappa^2 \sim K_1/A \ll 1$. In A.3.5 it is outlined that K_1, A a, and

a scale as $m_e\alpha^4c^2$, $m_e\alpha^2c^2$, and a_o, respectively, where $\alpha = 4\pi\varepsilon_o e^2/\hbar c \approx 1/137$ is Sommerfeld's fine-structure constant. This yields the characteristic length $a_o/\alpha \approx 7.52$ nm. Of course, this derivation is rather crude, but a_o/α correctly epitomizes the magnetism of typical nanostructures. For example, in flattened ellipsoids of revolution (thin films) with perpendicular anisotropy, curling is more favorable than coherent rotation when the cross section of the films exceeds some value scaling as $(a_o/\alpha)^2$ (Skomski, Oepen, and Kirschner 1998).

7.4.2 Nanomagnetic effects of atomic origin

In the previous subsection, we have argued that atomic magnetism is limited to length scales up to about 1 or 2 nm. There are, however, some exceptions. We have already discussed long-range magnetization correlations near T_c, which become nanoscale and eventually macroscopic on approaching the critical point. However, this phenomenon is of secondary practical interest, because it is supported by very small (free) energy differences and limited to the immediate vicinity of T_c. A quantum-mechanical example is the magnetic anisotropy of metallic magnets, which sensitively depends on the position of the Fermi level between nearly degenerate levels (Section 3.4.4). Nanostructuring may lead to small changes in the level positions, accompanied by disproportionately strong anisotropy changes.

Most *quantum effects* are operative on an atomic scale. They affect the macroscopic behavior of solids in a quasiclassical way, without specific reference to quantum states. For example, interatomic exchange is a quantum effect, but macroscopic ferromagnets can be treated classically, by considering an continuum magnetization $\mathbf{M}(\mathbf{r})$. Similarly, the mechanical hardness of solids is due to quantum-mechanical interactions, but there is no need to analyze the trajectory of a falling stone quantum-mechanically. In fact, with the exception of small-scale molecular magnets (Wernsdorfer 2006) and clusters (Reddy, Khanna, and Jena 1999), it is usually very difficult to see quantum effects in magnetic nanostructures. This includes effects such as quantum tunneling in nanoparticles (Chudnovsky and Gunther 1988, Zhang *et al.* 1992, Wernsdorfer *et al.* 1997) and wave-function entanglement between nanodots (Skomski *et al.* 2004c). These effects have present or potential applications in areas such as quantum computing (Nielsen and Chuang 2000, Engel and Loss 2005).

Surfaces and interfaces are atomic perturbations with far-reaching consequences. This refers not only to surface (and interface) anisotropy (Section 3.5.3) but also to the magnetic moment. In Section 2.4.2 we have seen that the local density of states (DOS) depends on the local atomic environment, and surfaces are no exception (panel 6). Based on the moments theorem, we have found that the width W of the local DOS scales as $\sqrt{z}$, where z is the number of nearest neighbors. More generally, the moment's theorem yields the correct bandwidth but ignores details of the band structure, such as peaks in the density of states. Increasing the number of neighbors improves the resolution of the density of states and makes it possible to distinguish between bulk sites and sites close to surfaces.

Since narrow bands (small W) favor ferromagnetism (Section 2.4.3), the reduction of z at surfaces yields enhanced magnetic moments for the surface atoms of some materials. Another phenomenon, loosely related to the narrowing of the local DOS, is the formation of surface states, which are localized in the z direction normal to the surface but extended in the $x-y$ plane (Desjonquères and Spanjaard 1993, Himpsel *et al.* 1998) As a rule, surface states and surface moments do not extend very far into the bulk, seldom more than a few interatomic distances. For an example, see the modification of the moment and of the effective interatomic exchange in random exchange Fe-Pt magnets, as obtained from first-principle electronic-structure calculations (Sabiryanov and Jaswal 1998).

A thermodynamic rather than quantum-mechanical effect is the enhanced *surface Curie temperature* encountered in some magnets, such as Gd (Mills 1971, Binder and Hohenberg 1972). The existence of a separate surface Curie temperature is caused by enhanced exchange at the surface, and the magnetization modes (7.10) responsible for surface transitions are analogous to quantum-mechanical surface states. In terms of Panel 6, the matrices $J_{\rm ij}$ and $t_{\rm ij}$ are physically different but mathematically very similar, and the corresponding magnetization modes decay exponentially into the bulk (Skomski, Waldfried, and Dowben 1998).

Surface transitions are closely related to the crossover from three- to two-dimensional behavior in magnetic thin films, which involves the ratio of correlation length ξ and film thickness t. Close to $T_{\rm c}$, ξ is comparable to or larger than t, and the behavior of the film becomes two-dimensional. The Curie transition also depends on the film's magnetic anisotropy—the two-dimensional Heisenberg model is non-ferromagnetic, but an arbitrarily small anisotropy takes over near the critical point and yields a single anisotropy-dependent but nonzero Curie temperature (Bander and Mills 1988).

Another effect involving the matrix $J_{\rm ij}$ is the spin-wave quantization in systems such as nanowires and multilayers. For example, in long magnetic nanotubes (Sui *et al.* 2004), magnons with wave vectors parallel to the tube axis form a continuum. By contrast, magnons with tangential wave vectors are quantized by the $\mathbf{m}(\phi_{\rm a} + 2\pi) = \mathbf{m}(\phi_{\rm a})$, where $\mathbf{m}$ describes the magnetization of the spin wave and $\phi_{\rm a}$ is the azimuthal angle of the tube. The resulting level spacing scales as a^2/R^2 and is very small for nanotubes with radii exceeding 10 nm.

Figure 7.18 illustrates interaction effects in nanoparticles (a) and at surfaces (b). In both cases, an infinitely thin perturbation is assumed, $f(\mathbf{r}) \sim \delta(z)$, where the z-axis is normal to the surface. The solution of (7.35) is then, approximately, $\phi(z) \sim \exp(-\kappa z)$, and the only difference between atomic and nanoscale effects is the Ornstein-Zernike interaction range $1/\kappa$. If the mode reaches far into a sphere, then one must use spherical Bessel functions (Zhou *et al.* 2005b). Typical atomic phenomena are characterized by $1/\kappa \sim 0.5\,\text{nm}$, and the perturbation decays after a few interatomic distances. Nanomagnetic phenomena are characterized by much larger decay lengths $1/\kappa \sim 10\,\text{nm}$. Consider, for example, a magnetic nanoparticle of radius 10 nm. Only a thin layer contributes to the surface anisotropy, which is an atomic effect, but the corresponding nucleation mode covers the whole particle. The net effect on the hysteresis can be quite strong, even in big nanoparticles, because the anisotropy energy per surface atom is often much larger than the anisotropy energy per bulk atom.

An interesting phenomenon is *nanoscale RKKY* exchange. In Section 2.3.2 we have seen that the exchange $J(\mathbf{R})$ is long-range ($\sim 1/R^3$) rather than exponentially decaying. Depending on the considered metal, this may or may not have a big effect on the Curie temperature, but it affects the asymptotics of the $J(R)$. For magnetic

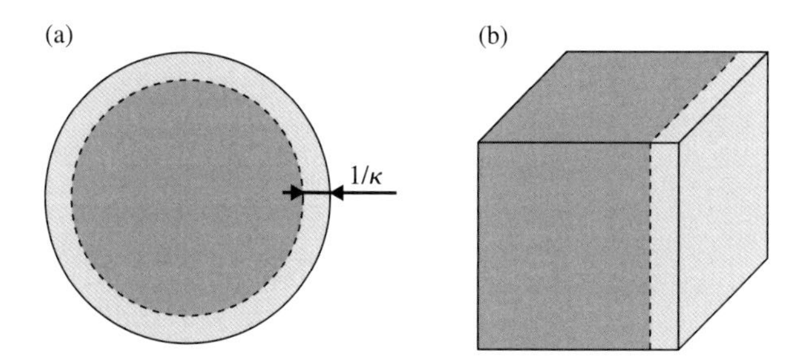

Fig. 7.18 Range of interaction effects: (a) nanoparticle and (b) surface. For typical atomic and nanomagnetic phenomena, $1/\kappa \approx 0.5\,\text{nm}$ and $1/\kappa \approx 10\,\text{nm}$.

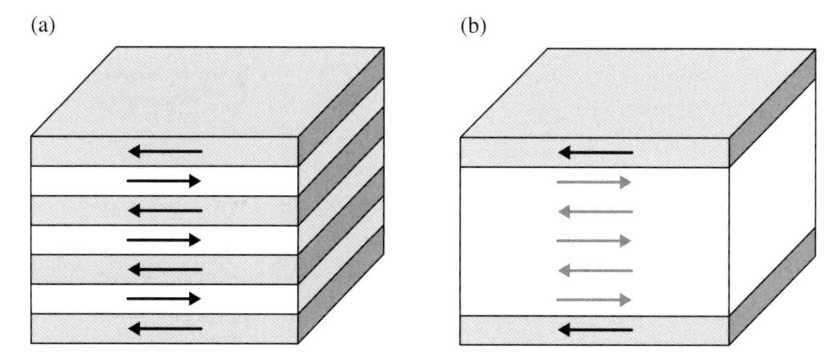

Fig. 7.19 Interlayer exchange in thin-film nanostructures: (a) "trivial" Ornstein-Zernike exchange and (b) RKKY exchange through a nonmagnetic spacer layer of thickness L (gray arrows).

nanoparticles embedded in nonmagnetic metallic matrix, the oscillatory character of the RKKY exchange reduces the interparticle interaction, and it may be tempting to assume that the net interaction between nanoparticles is close to zero. Surprisingly, the net interaction *increases* with particle size, albeit less rapidly than the magnetostatic interaction (Skomski 1999, de Toro *et al.* 2001). The net RKKY exchange scales as R^{μ}, where $\mu < 6$ depends on the geometry of the particles. As a rule, for particles larger than about 1 nm the RKKY interaction is less pronounced than the magnetostatic interaction.

Let us next consider RKKY interactions between ferromagnetic layers separated by a nonmagnetic metallic spacer layer of thickness L. Figure 7.19 compares the RKKY mechanism (b) with the net coupling due to the antiferromagnetic exchange between adjacent layers. In (a), the net interaction depends on whether the number of layers is odd or even, and the periodicity of the oscillation is equal to the bilayer thickness. Thin-film RKKY interactions (b) are oscillatory, too, but the distance dependence differs from that of bulk magnets. For perfect layers, the net interaction is

$$J(L) \sim \frac{1}{L^2}\sin(2k_{\mathrm{F}}L) \tag{7.36}$$

A straightforward way to derive this result is to start from (2.38) and to perform an integration over the film plane, that is over all x and y values while keeping $z = L$. The coupling can also be interpreted as a quantum-well problem, with electron wave functions confined to the spacer layer (gray arrows in Fig. 7.19) but otherwise similar to Fig. 2.16.

For rough surfaces, one encounters a *biquadratic* exchange favoring a relative magnetization angle of approximately 90°, as compared to 0° (FM) and 180° (AFM). The basic mechanism (Slonczewski 1991) is similar to the rare-earth noncollinearity of Section 2.3.3, except that the competing exchange is not between nearest and next-nearest layers but between lateral thin-film regions (patches) with interlayer exchange $\pm\delta J$. In the simplest case, one must consider four sublattices. There are two magnetic layers, labeled t (for top) and b (for bottom), and each layer contains two types of lateral regions. The magnetization is slightly canted in each patch and described by the in-plane angles $\theta_{t+} = \delta\theta$, $\theta_{t-} = -\delta\theta$, $\theta_{b+} = 90° - \delta\theta$, and $\theta_{b-} = 90° + \delta\theta$. The canting by the fluctuation angle $\pm\delta\theta$ allows the patches to benefit from the interlayer exchange without being punished by a huge increase in intralayer exchange J_o. Minimizing the energy

$$E = -2J_o \cos(2\delta\theta) - \delta J \cos\left(\frac{\pi}{2} - 2\delta\theta\right) + \delta J \cos\left(\frac{\pi}{2} + 2\delta\theta\right) \tag{7.37}$$

yields the fluctuation angle $\delta\theta = \delta J/2J_o$ and the biquadratic coupling energy $\delta J^2/J_o$.

Exchange-coupled thin-film nanostructures are important in magnetic recording and spin electronics (Grünberg *et al.* 1986, Baibich *et al.* 1988). An example is spin-valve sensors, where a switching soft layer changes the electrical resistance (Section 7.2.7). Some films exhibit exchange bias similar to that encountered in the Co:CoO system (Section 3.5.2). Exchange-bias phenomena may actually be very strong, giving rise to inverted hysteresis loop (proteresis).

7.4.3 Random anisotropy

Many magnetic nanostructures are characterized by completely or partially random easy axes $\mathbf{n}(\mathbf{r})$, with far-reaching effects on the magnetism. A less stringent type of random anisotropy is encountered in crystallographically oriented or *aligned* composites where $\mathbf{n}(\mathbf{r}) = \mathbf{e}_z$ but $K_1 = K_1(\mathbf{r})$. Random-anisotropy magnets are relatively easy to produce and are used in various applications, including hard and soft magnets. In hard magnets, exchange coupling between randomly oriented nanograins gives rise to remanence enhancement (Coehoorn *et al.* 1988), and in soft magnets, one exploits the fact that the average anisotropy of isotropic random-anisotropy magnets is zero (Herzer 1992, 1995).

Atomic-scale random anisotropy was first discussed in the context of random-field magnetism (Imry and Ma 1975), in spin glasses (Harris, Plischke, and Zuckermann 1973), and amorphous ferromagnets on a mean-field level (Callen, Liu, and Cullen 1977) and beyond (Alben, Chi, and Becker 1978). However, random-field and random-anisotropy magnets cannot be considered as canonical spin glasses, because the exchange is basically ferromagnetic. Later, the atomic random-anisotropy models were generalized to magnetic nanostructures (Chudnovsky, Saslow, and Serota 1986, Richter and Skomski 1989, Herzer 1992, Suzuki and Cadogan 1999), and there exists a rich

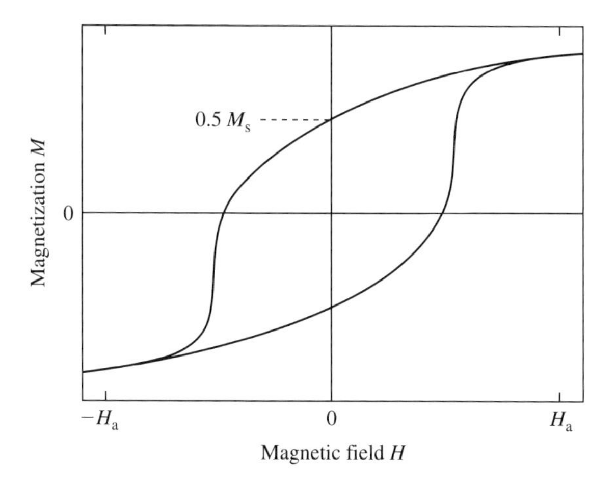

Fig. 7.20 Hysteresis loop of an ensemble of randomly oriented uniaxial Stoner-Wohlfarth particles ($K_1 > 0$, $K_2 = 0$)

review literature on these systems (Moorjani and Coey 1984, Fischer and Hertz 1991, Sellmyer and O'Shea 1992, Herzer 1995).

Let us first consider a noninteracting ensemble of randomly oriented Stoner-Wohlfarth particles. The corresponding hysteresis loop, Fig. 7.20, is a superposition of Stoner-Wohlfarth loops of the type of Fig. 4.5. The remanence, $M_r = M_s/2$, is obtained by evaluating the integral

$$M_r = M_s \int_0^{\pi/2} \cos\theta \, \sin\theta \, d\theta \tag{7.38}$$

Since the energy product scales as M_r^2, the remanence reduction by a factor 2 amounts to an energy-product decrease by a factor 4. The coercivity, $H_c = 0.479\, H_a$, is smaller than the coercivity $H_c = H_a$ of aligned Stoner-Wohlfarth particles, but in very hard magnets it remains sufficient to support a large energy product. The strategy in permanent magnetism has therefore been to sacrifice some coercivity in order to enhance the remanent magnetization. For cubic magnets with iron-type ($K_1 > 0$) and nickel-type ($K_1 < 0$) anisotropy, the respective remanence ratios M_r/M_s are 0.832 and 0.866, respectively. However, the low anisotropy of cubic magnets makes it difficult to exploit this remanence in permanent magnets.

Ferromagnetic *interactions* compete against the random easy axes and enhance the magnetization. In the remanent state, the intergranular exchange adds to the external field and aligns the magnetization of the grains or particles. The initial or virgin state, which corresponds to thermodynamic equilibrium, is more complicated, because $\langle \mathbf{n} \rangle = 0$ in isotropic magnets and the assumed mean field (exchange field) may overestimate the magnetization. Figure 7.21 compares the spin structure of noninteracting ensembles in the virgin and remanent states (a–b) with that of interacting random-anisotropy particles (c). In the context of magnetic glasses, the spin structure

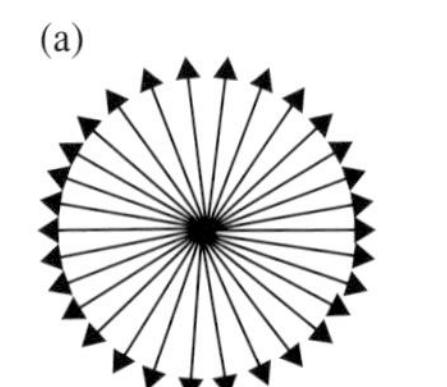

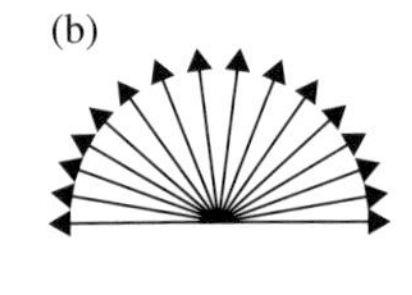

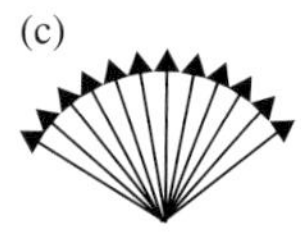

Fig. 7.21 Ensembles of random-anisotropy grains or particles: (a) virgin state of a noninteracting ensemble, (b) remanent state of a noninteracting ensemble, and (c) remanent state in the presence of ferromagnetic interactions. Each arrow represents one grain or particle.

Fig. (7.21a) is known as a speromagnet, whereas those of (b) and (c) are asperomagnetic (Moorjani and Coey 1984).

It is convenient to describe the intergranular interactions in terms of an effective exchange constant J_{eff}. The effective exchange is obtained by integration over the free-energy density of adjacent grains (Skomski 2003). For fine-grained single-phase materials, one can use the estimate $J_{\text{eff}} \approx AR$, where R is the average grain radius and A is the exchange stiffness. Grain boundaries with reduced exchange stiffness yield a reduction of J_{eff}, and it is necessary to use more sophisticated approximations (Section 4.4). A popular estimate, $J_{\text{eff}} \approx JR^2/a^2$, is obtained by considering the atomic exchange J between pairs of adjacent atoms of interatomic distance a. However, this approximation overestimates the exchange, because it ignores the energy stored in the magnetization tails of Fig. 4.24.

The *weak-coupling* limit is defined by $J_{\text{eff}} \ll K_1 V$, where V is the grain volume. The intergranular exchange has then the character of a small interaction field, and the magnet can be treated on a mean-field level. The remanence ratio $M_{\text{r}}/M_{\text{s}} = \frac{1}{2} + J_{\text{eff}}/6K_1 V$, and the coercivity does not change very much as a function of J_{eff}. Weak coupling is rarely realized on an atomic scale, because $R \sim a$ and $J_{\text{eff}} \approx J \gg K_1 a^3$ for most materials. In nanostructures, where $R \gg a$, the weak-coupling regime is encountered more frequently. A rough criterion for weak coupling, $R > \delta_{\text{B}}$, is obtained by taking into account that $J \approx Aa$. The criterion is often satisfied in isotropic permanent magnets, where $\delta_{\text{B}} \approx 5\,\text{nm}$ and $R > 5\,\text{nm}$. The energy product of these magnets benefits from the remanence enhancement, which occurs without much loss in coercivity (Coehoorn *et al.* 1988, Manaf, Buckley, and Davies 1993). In fact, some of these magnets exhibit huge coercivities of more than 4 T (Kuhrt *et al.* 1992), in spite of the absence of long-range ferromagnetic order. This is another indication that ferromagnetic order and hysteresis are only loosely related to each other.

To investigate the *strong-coupling* limit, we start with a brief discussion of the random-field analogy (Imry and Ma 1975). In the random-field model, individual atoms or particles (index i) are subjected to fields H_{i} of magnitude $\pm H_{\text{o}}$. If the field were the only consideration, then the magnetization would follow the local field. However, spatial variations of the magnetization are opposed by the interatomic exchange, and the competition between the random-field and exchange energies leads to magnetically correlated regions containing N atoms. The correlated regions are characterized by the averages $\langle H \rangle = 0$ and $\langle H^2 \rangle = H_{\text{o}}^2/N$. In other words, there remains a net contribution of magnitude $\pm H_{\text{o}}/\sqrt{N}$, reminiscent of the standard deviation in statistical data analysis. The stronger the exchange, the larger N and the smaller the net random field (or net random anisotropy).

To determine N, we must minimize the average energy density E/V. Let us return to the picture of nanoscale random-anisotropy magnets, where

$$\frac{E}{V} = \frac{A}{\xi^2} - K_1 \frac{1}{\sqrt{N}} \tag{7.39}$$

and ξ is the micromagnetic correlation length. In d dimensions, the volume of the correlated regions is of the order of $L^{\rm d} \approx N\,R^{\rm d}$. Minimizing the resulting expression

$$\frac{E}{V} = \frac{A}{\xi^2} - K_1 \frac{R^{\rm d/2}}{\xi^{\rm d/2}} \tag{7.40}$$

with respect to ξ yields the scaling relation

$$\xi \sim R \left(\frac{\delta_{\rm B}}{R}\right)^{4/(4-{\rm d})} \tag{7.41}$$

The strong-coupling regime is characterized by $R \ll \delta_{\rm B}$. Correlations $L \gg R$ are predicted by (7.41) in less than four dimensions.

In analogy to random fields, $K_1/\sqrt{N}$ can be considered as an effective anisotropy. Since the coercivity scales as $H_{\rm c} \sim 2K_1/M_{\rm s}$, the reduced effective anisotropy is accompanied by a coercivity reduction:

$$H_{\rm c} \sim \frac{2K_1}{\mu_{\rm o} M_{\rm s}} \left(\frac{R}{\delta_{\rm B}}\right)^{2{\rm d}/(4-{\rm d})} \tag{7.42}$$

In three dimensions, this means that the coercivity is proportional to R^6 and, for fixed R, to K_1^4. For grains with reduced grain-boundary exchange (Section 4.4), the scaling laws are $\xi \sim R^{(2-{\rm d})/(4-{\rm d})}$ and $H_{\rm c} \sim R^{{\rm d}/(4-{\rm d})}$.

As pointed out by Herzer (1992, 1995), random-anisotropy nanostructuring is a powerful tool to improve the coercivity of soft magnetic materials. Once the grain size is sufficiently small to realize the strong-coupling limit ($R \ll \delta_{\rm B}$), the coercivity is strongly reduced. In permanent magnetism, the coercivity reduction due to random anisotropy is undesired, and one attempts to create nanostructures with $R \sim \delta_{\rm B}$.

7.4.4 *Cooperative magnetization processes

Magnetic nanostructures often exhibit a competition between interparticle interactions and disorder. From an atomic point of view, all magnetization processes are cooperative, because the interatomic exchange suppresses the disorder-induced reversal of individual spins. On a mesoscopic scale, the outcome of the competition depends on the ratio of the competing energy contributions. For example, remanence enhancement (Section 7.4.5) is caused by intergranular exchange competing with easy-axis disorder and accompanied by an increase of the micromagnetic correlation length ξ (Section 7.43). This corresponds to a breakdown of the picture of individual grains, and the magnetization processes become cooperative. Similar micromagnetic localization effects exist in other magnetic systems, such as hard-soft composites (Skomski,

Liu, and Sellmyer 1999). The increase of ξ with exchange is a very general feature, mathematically akin to the delocalization of electrons in a solid. In other words, the Anderson localization of electrons in a disordered lattice (Anderson 1958) and the "micromagnetic" localization of magnetization modes (Section 4.3.1) have much in common.

In the hysteresis loop, disorder corresponds to the switching field distribution ΔH of individual particles, whereas intergranular exchange can be described by an interaction field $H_{\rm ex}$. The ratio of the two fields decides whether the behavior of the magnet is noncooperative ($\Delta H > H_{\rm ex}$) or cooperative ($\Delta H < H_{\rm ex}$). In a crude approximation, the micromagnetic susceptibility $\mathrm{d}M/\mathrm{d}H \approx M_{\rm s}/\Delta H$, so that cooperativity is most likely for large susceptibilities. As mentioned, the cooperative character refers to *nanoscale* magnetization processes. On an atomic scale, ferromagnetic reversal is cooperative, because field energies of order $\mu_{\rm o}\mu_{\rm B}H$ are unable to compete against the interatomic exchange.

In the noncooperative limit, the exchange is of the mean-field type and can be added to the external magnetic field. An example is microcrystallites, which can often be described as particles subject to an interaction field. This is the idea behind the Preisach model (Section 4.3.3). Unfortunately, the addition of the exchange field is not possible for *cooperative* systems, where one may encounter huge interaction fields without any effect on the hysteresis loop. Physically, the exchange yields a rigid coupling between neighboring particles or grains, and the exchange-coupled grains behave as a single units. The same is true for atomic exchange fields, where the addition of the mean field $H_{\rm eff} \sim zJ/k_{\rm B}$ to the external field would overestimate the coercivity by several orders of magnitude.

Figure 7.22 illustrates the onset of cooperative behavior for an ensemble of aligned particles with common c-axis. The anisotropy $K_1(\mathbf{r})$ is random and the particles exhibit some exchange interaction. The "coherent" part of the anisotropy, $\langle K_1(\mathbf{r})\rangle = K_{\rm o}$, corresponds to an anisotropy field $H_{\rm a} = 2K_{\rm o}/\mu_{\rm o}M_{\rm s}$, whereas the random part, $\Delta K(\mathbf{r}) = K_1(\mathbf{r}) - K_{\rm o}$, competes against the interparticle exchange. For small exchange

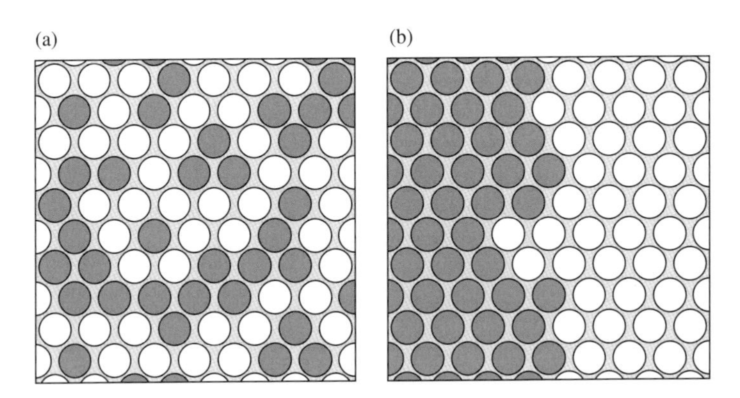

Fig. 7.22 Magnetization reversal in an ensemble of aligned and interacting particles: (a) Stoner-Wohlfarth-like reversal in the limit of small intergranular exchange and (b) discrete pinning for relatively large exchange. White and black particles have $\uparrow$ and $\downarrow$ magnetizations, respectively.

(a), the particles switch according to their individual anisotropy, starting with the softest particles. When the exchange is larger than ΔK (b), then the switching of individual particles acquires the character of a "discrete pinning" (Zhou *et al.* 2005a), even if the reversal of individual particles remains Stoner-Wohlfarth-like. In a slightly different context, the dark and bright quasi-domains shown in (b) are known as *interaction domains* (Craik and Isaak 1960, Cullity 1972).

The distinction between Fig. 7.22(a) and (b) is only loosely related to that between strong and weak pinning (Section 4.3.2), and to the strong- and weak-coupling limits of the previous subsection. There are also "mixed" systems with more than one kind of disorder. When the interparticle exchange is realized by a ferromagnetic matrix with reduced exchange stiffness $A' < A$, then the spin structure in the quasi-domain walls is essentially given by Fig. 4.24. In the limit of infinite exchange, the system behaves like a homogeneous magnet with an effective anisotropy constant K_o (Skomski and Coey 1993). We will return to this limit in Section 7.4.3.

As electron localization, micromagnetic localization depends on the dimensionality of the system. In one and, marginally, two dimensions, all eigenmodes are localized due to arbitrarily small disorder. One example is *nanowires* electrodeposited into porous alumina, which have lengths of order $z_{\max} = 1{,}000$ nm and diameters down to 10 nm or less (Sellmyer, Zheng, and Skomski 2001). The large aspect ratio makes the wires quasi-one-dimensional, and the nucleation mode $m(\mathbf{r}) = m(z)$ is described by (4.29). For a single imperfection located at $z = z_o$, the mode decays as $m(z) = m_o \exp(-|z - z_o|/\xi)$, and it is straightforward to show that the decay length increases with decreasing strength and extension of the imperfection (Skomski *et al.* 2000, Zheng *et al.* 2000). A similar localization determines the energy barrier of the mode, and the localization is accompanied by a coercivity reduction very similar to the coercivity mechanism discussed in Section 4.3.1.

For infinitesimally small imperfections, ξ approaches infinity, in agreement with the coherent-rotation reversal in needle-shaped ellipsoids of revolution (Aharoni 1996). Other modes do *not* correspond to a physically meaningful nucleation mode but may be excited thermally (Section 6.4.7). In the opposite limit, strong imperfections yield localization lengths ξ comparable to the spatial extent of the imperfection. In practice, the localization tends to occur at the wire ends, which are a strong perturbation. Fixing the magnetization at the wire ends, for example by hard-magnetic caps, would move the mode away from the wire ends, without making it delocalized. Experimentally, both magnetization and magnetic viscosity experiments indicate that the ξ is comparable to a few wire diameters, corresponding to geometrical features such as thickness fluctuations in the middle of the wires or at the wire ends. Furthermore, wires are often polycrystalline, with coercivity corrections due to random anisotropy.

7.4.5 Two-phase nanostructures

There are many types of magnetic two-phase nanostructures, with a variety of applications in permanent magnetism, soft magnetism, sensors, spin electronics, and, more recently, magnetic recording. They occur in many variations and geometries, including but not limited to multilayers and granular composites. Here we focus on aligned *hard-soft* two-phase nanostructures, characterized by a local anisotropy $K_1(\mathbf{r})$ in combination with a common easy axis ($\mathbf{n} = \mathbf{e}_z$). The idea is to create two-phase

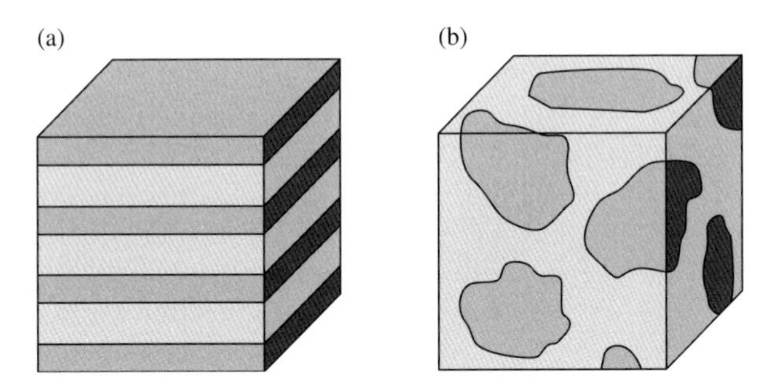

Fig. 7.23 Hard-soft nanostructures: (a) multilayer and (b) granular composite. The dark and bright regions are hard and soft phases with the respective volume fractions f and $1-f$. In both cases, the hard regions act as a skeleton to stiffen the magnetization direction of the soft phase.

magnets whose energy product (Fig. 4.2) goes beyond what is expected from the volume fractions of the involved phases. In other word, adding a soft-magnetic phase to a hard-magnetic phase *improves* the performance of the hard phase. This is possible because the energy product $(BH)_{\max} \leq \mu_o M_s^2/4$, even for infinite coercivity. In fact, some rare-earth transition-metal magnets have huge anisotropies and coercivities, but the alloying with rare earths reduces the magnetization M_s and the energy product. It is therefore advantageous to sacrifice some of the surplus anisotropy to improve M_s. However, the addition of the soft phase must be realized carefully, because coercivities $H_c < M_s/2$ are harmful to the energy product.

Figure 7.23 shows two schematic realizations of hard-soft nanostructures. The total magnetization is equal to the volume average of the magnetizations of the two phases, but this is not necessarily true for other quantities, such as the exchange stiffness and the coercivity. For multilayers, the underlying nucleation problem was first considered by Nieber and Kronmüller (1989), but the slightly different idea of coupling a soft or "exchange-spring" layer to a hard phase goes back to Goto *et al.* (1965). Arbitrary geometries (Fig. 7.23) with anisotropy, magnetization, and exchange inhomogeneities were first treated by Skomski and Coey (1993). The last paper also contains explicit energy-product estimates for rectangular hysteresis loops with of coercivity $H_c = M_s/2$. Experimental energy-product enhancement has been obtained in an Fe-Pt system (Liu *et al.* 1998), but the realization of rectangular loops has remained a challenge for most systems.

The nucleation field $H_N \approx H_c$ is obtained by solving the differential equations of the type (4.19) and (4.29). Equation (4.34) predicts a decrease $H_N \sim 1/L^2$ as a function of the L of the soft inclusions (or soft-layer thickness). For large values of L, the soft regions switch easily and H_N is small. This reduces the energy product—by initiating the magnetization reversal of the magnet ($H_c \approx H_N$) and by giving rise to an unfavorable "shouldered" or "wasp-like" hysteresis-loop shape. The corresponding switching of the soft regions is also known as *exchange-spring magnetism* (Kneller 1991, Sawicki *et al.* 2000). If L is smaller than about twice the domain-wall width of the hard phase, or approximately 10 nm, then the magnetization direction is nearly constant on a local scale and we can use the "virtual-crystal" or effective-anisotropy approximation, $K_1(\mathbf{r}) \approx \langle K_1 \rangle$. This yields $H_c = 2\langle K_1 \rangle / \mu_o \langle M_s \rangle$ and a closed expression for the energy product as a function of the volume fraction f of the soft phase. Initially, $(BH)_{max}$ increases with f, but for large values of f the average anisotropy becomes too small to sustain the necessary coercivity, and $(BH)_{max}$ decreases.

Aside from the size of the soft inclusions, one must account for the "tunneling" of the nucleation mode from the soft phase into the hard phase, similar to Fig. 4.19. Small corrections to the virtual-crystal or effective-anisotropy approximation involve the correlation function $\langle K_1(\mathbf{r}) K_1(\mathbf{r}') \rangle$. Note that the nucleation problem is very different from the percolation problem. In the latter, the onset of transport occurs when the percolating backbone is fomed. In hard-soft two-phase nanostructures, (4.29) means that main criterion is the distance to the nearest hard region, irrespective of whether the geometry is granular or layered—there is no "percolation" through small-scale soft regions.

Hard-soft nanocomposites are often obtained in form of thin films, as granular materials or continuous multilayers (Al-Omari and Sellmyer 1995, Astalos and Camley 1998, Liu *et al.* 1998, Fullerton, Jiang, and Bader 1999, Davies *et al.* 2005). Aside from the potential use of these materials in permanent magnets, they are of interest as sensors and in magnetic recording (Garcia-Sanchez *et al.* 2005). Note that the improvement of magnetic properties due to nanostructuring is limited to extrinsic properties, such as the energy product, which are realized on length scales of order α/a_o (Section 4.2.5). Intrinsic properties, such as the Curie temperature, are realized on atomic length scales of about 1 nm and cannot be improved by nanostructuring (Section 7.1.3).

Panel 8 The nanomagnetic hemisphere model

Micromagnetism reflects the competition between exchange, magnetostatic interactions, and nanoscale disorder. This is true not only for nanostructures but also for traditional bulk magnets, because typical pinning and nucleation centers have sizes comparable to the domain-wall width. Let us consider a spherical particle of radius R, which we divide into two hemispheres I and II. The hemispheres are characterized by anisotropies $K_{I/II}$ and a common easy axis $\mathbf{n}_{I/II} = \mathbf{e}_z$. The moments $m_{I/II} = M_s V/2$ of the hemispheres are assumed to reside at $\mathbf{r}_{I/II} = \pm \frac{1}{2} R \mathbf{e}_x$, and the magnetizations $M_{I/II} = M_s(\cos\theta_{I/II}\, \mathbf{e}_z + \sin\theta_{I/II}\, \mathbf{e}_y)$. In the figure, the dotted line connecting the hemisphere moments is along the x axis.

For small magnetization angles $\theta_{I/II}$, the magnetic energy density of the sphere is

$$\frac{E}{V} = \left(\frac{A}{R^2} - \frac{\mu_o M_s^2}{24}\right)(\theta_I - \theta_{II})^2 + \frac{1}{2} K_I \theta_I^2 + \frac{1}{2} K_{II} \theta_{II}^2 + \frac{\mu_o M_s H}{4}(\theta_I^2 + \theta_{II}^2)$$

From this equation, the nucleation modes and coercivities $H_c = H_N$ are obtained by the diagonalization of the 2×2 interaction matrix (exercise). For coherent rotation, this yields the exact result $\theta_I = +\theta_{II}$ and $H_c = 2K_1/\mu_o M_s$. The curling-type mode exhibits $\theta_I = -\theta_{II}$ and $H_c = 2K_1/\mu_o M_s + 8A/\mu_o M_s R^2 - M_s/3$. The curling nucleation field is approximate but close to the exact solution (Section 4.3.1), as is the coherence radius $R_{coh} = \sqrt{24A/\mu_o M_s^2}$.

Continued

Panel 8 *Continued*

(a) 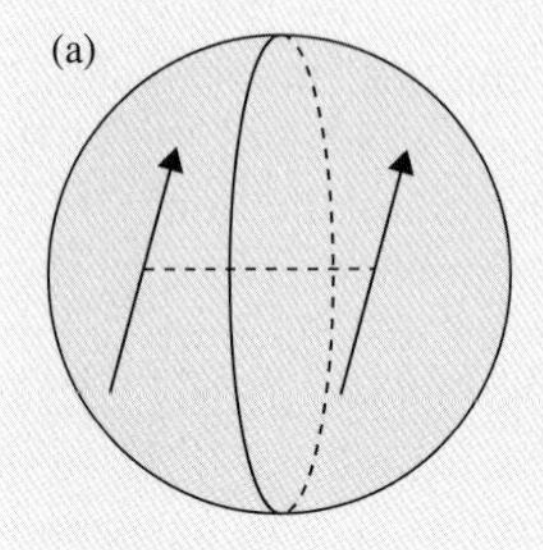 (b) 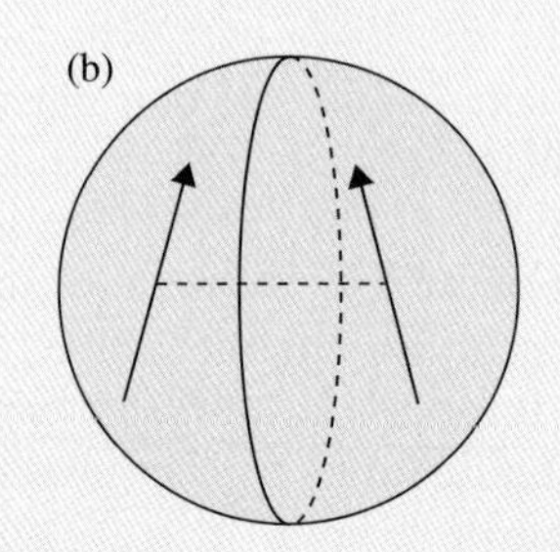 (c) 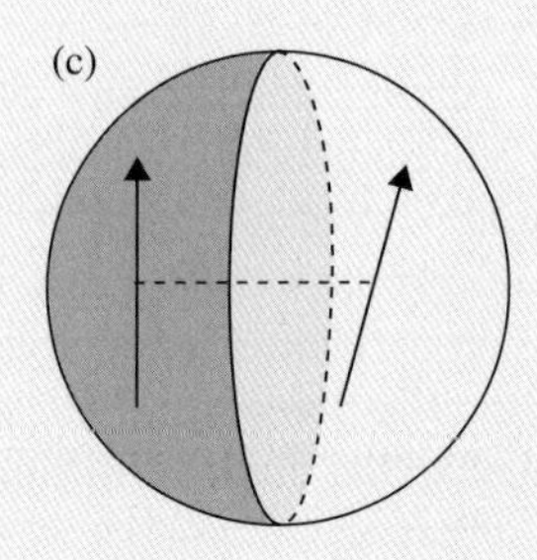

Nucleation modes in the hemisphere model: (a) coherent rotation, (b) curling, and (c) localized nucleation.

Let us next consider a two-phase particle (c), where the left hemisphere is hard (dark area, $K_{\mathrm{II}} > 0$) and the right hemisphere is soft (bright area, $K_{\mathrm{I}} = 0$). In this case, the soft phase switches first, and the nucleation field is roughly proportional to A/R^2. This mechanism is localized, as contrasted to the cooperative mechanisms (a) and (b). A similar model exists for the magnetic resonance (Section 6.1) of coupled nonequivalent layers or particles. In the cooperative limit of very thin layers, the two eigenmodes are ferro- and antiferromagnetic, whereas the resonance lines of thick layers look like a superposition of two phases.

Exercises

1. Extract and diagonalize the 2×2 interaction matrix from the energy expression for the hemisphere model. Discuss the applicability of the model to real nanostructures by finding one or two examples where the model can *not* be used.
2. Compare the model of this panel with the diatomic pair model of Section 2.1 and with the quantum-mechanical two-level model of Section A.2.3.

7.5 Beyond magnetism

Summary The transparent phase space of magnetic models has led to various applications of magnetic modeling in other areas of human knowledge. Examples are metallurgy, population dynamics, neurology, and sociology. Order-disorder transitions, gases in metals, and spinodal decomposition in alloys can essentially be described by the Ising model, whereas models of population dynamics are often of the diffusion or Fokker-Planck type. There is a close link between neurology and spin-glass models, as epitomized by the Hebb rule. As in magnetism, the quality of a model depends on the parameterization and ranges from a crude qualitative understanding to quantitative predictions. In turn, magnetic modeling has been influenced and reinvigorated by developments in other areas of science and technology.

Magnetic models have found widespread applications in other areas of physical sciences and beyond, for example, in the social sciences. As in magnetism, the choice of model depends on the phase space of the phenomenon. A widely used model is the Ising model, as parameterized by exchange constants J_{ij} and local fields h_i. For example, Ising variables $s_i = \pm 1$ can be used to describe the behavior of voters (index i) in a two-party political system. Both equilibrium and nonequilibrium phenomena are considered, the latter often in the form of Fokker-Planck dynamics. The models describe features such as interactions, nonlinearity, local bias, and randomness. Ideas from magnetism and from other areas of physics, such as fluid science, have now established the fundamental role of randomness, nonlinearity, and nonequilibrium in many areas of human knowledge. For example, hysteresis occurs not only in magnetism but also during economical cycles.

Quantitatively, the models make more or less accurate predictions, depending on the specific range of applicability. A general challenge is the actual parameterization of complex phenomena. For example, fictitious temperatures make it possible to model randomness, but the modeling of stock-market fluctuations in terms of a heat bath is a very crude approach. We know this from magnetism, where models describe some aspects of reality very well but fail to reproduce features that lie beyond the scope of the model.

This section focuses on a few specific examples. Magnetic models have also contributed to our general understanding of physics, from quantum mechanics and phase transitions to nonlinear phenomena, but this aspect goes beyond the scope of this section. We also ignore some complex systems and phenomena loosely related to magnetism, such as self-organization, chaos, ferroelectrics and multiferroics, liquid crystals, and materials with negative index of refraction. For example, photonic materials have a negative index of refraction $n = (\mu\varepsilon)^{-1/2}$ when both $\mu(\omega)$ and $\varepsilon(\omega)$ are negative (Shelby, Smith, and Shultz 2001).

Ferroelectrics are an electrostatic analog to ferromagnets, with dipole moments determined by the positions of the involved ions. Some features of ferroelectrics are therefore similar to ferromagnets, for example the involved nature of coercivity, but there are also differences (Ducharme *et al.* 2000). For example, the nanoscale character of micromagnetic phenomena requires the anisotropy to originate from spin-orbit coupling (Section 7.4.1), as epitomized by the Bloch-wall width $\pi(A/K_1)^{1/2}$. The anisotropy of ferroelectrics is unrelated to anisotropy and corresponds to abrupt ferroelectric domain walls (Padilla, Zhong, and Vanderbilt 1996), broadened by elastic and other energies of electrostatic origin. Multiferroics aim at simultaneously exploiting magnetic and other degrees of freedom, most notably ferromagnets and ferroelectrics (Baettig and Spaldin 2005). The coupling between the magnetic and electric degrees of freedom is usually realized by elastic strain, but in principle it is also possible to exploit crystal-field interactions, as in Fig. 7.13.

7.5.1 Metallurgy

The Ising model can be used to describe metallurgical phenomena such as order-disorder transitions and spinodal phase segregation in binary alloys, such as NiAu and FeCo. The phase space of the Ising model, $s_i = \pm 1$, is easily mapped on the

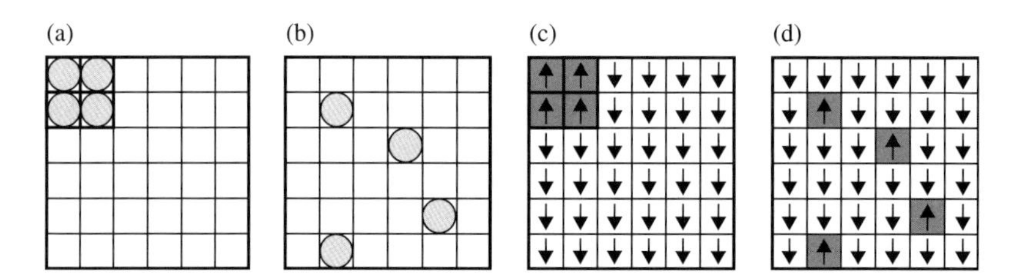

Fig. 7.24 Entropy: (a–b) gas phase or solution and (c–d) Ising ferromagnet. The entropy of (a) and (c) is lower than that of (b) and (d). Related systems are antiferromagnets (Fig. 5.12), gases in metals (Fig. 7.12) , polymers (subject to connectivity constraints) and AB alloys with order-disorder transitions and spinodal decomposition.

site occupancies of A and B atoms in a binary alloy. The interactions between neighboring pairs of $A-A$, $A-B$, and $B-B$ atoms favor segregation into A and B phases if $A-A$ and $B-B$ bonds are energetically more favorable than $A-B$ bonds. In the opposite case of strong $A-B$ bonds, the interactions yield an ordered $A-B$ alloy. Since A and B occupancies correspond to $\uparrow$ and $\downarrow$ spin states, the phase segregation into A and B phases corresponds to the onset of ferromagnetism. Similarly, the formation of ordered $A-B$ alloys is analogous to antiferromagnetic order. A similar mechanism is responsible for the behavior of gases in metals (Section 7.2.6), except that the respective spin states $\uparrow$ and $\downarrow$ now correspond to occupied and empty interstitial sites (Alefeld 1969). As indicated in Fig. 7.12, the interstitial atoms may segregate to form clusters or are randomly occupy the interstitial sites. Figure 7.24 compares interstitial alloys with magnets.

Above the critical temperature T_c, which is proportional to interaction strength, the system forms a random solid solution (A–B alloys) or a lattice gas of interstitial atoms. Below T_c, the sign of the interaction decides whether the random solid solution segregates (spinodal decomposition) or forms ordered A–B alloys. Examples are NiAu (spinodal decomposition, $T_c = 800\,°\mathrm{C}$) and FeCo (order-disorder transition, $T_c = 725\,°\mathrm{C}$). The interaction between interstitial gas atoms is of elastic origin and *attractive*, so that the gas atoms cluster below T_c (phase segregation). A well-known example is PdH_x, where $T_c = 292\,°\mathrm{C}$.

The temperature-concentration phase diagram looks like Fig. 5.9(a), except for a shift $c = (1 + s)/2$. The mean-field version of the approach is known as the Gorsky-Bragg-Williams theory. As in magnetism, the applicability of the mean-field theory depends on the range of the interactions. The elastic interactions between gas atoms in metals are long-range, as contrasted to short-range electronic interactions between atoms in alloys, and the mean-field theory yields the correct critical exponents.

The metallurgical analog to the magnetic field is the chemical potential of the gas or metal atoms. It fixes the average concentration, as the magnetic field determines the average magnetization. For gases in metals, the chemical potential is proportional to the logarithm of the pressure (Fast 1976). In a crude approximation, the lattice-gas model of Fig. 7.24 can also be used to describe the phase transition from a liquid (a) to a gas (b). The corresponding mean-field predictions are essentially equivalent to the van-der-Waals theory.

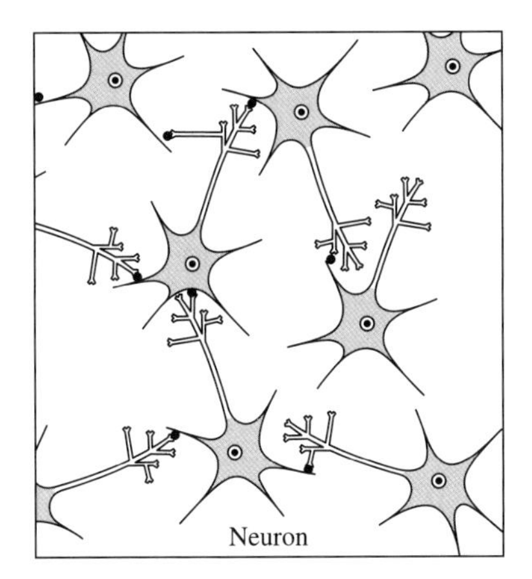

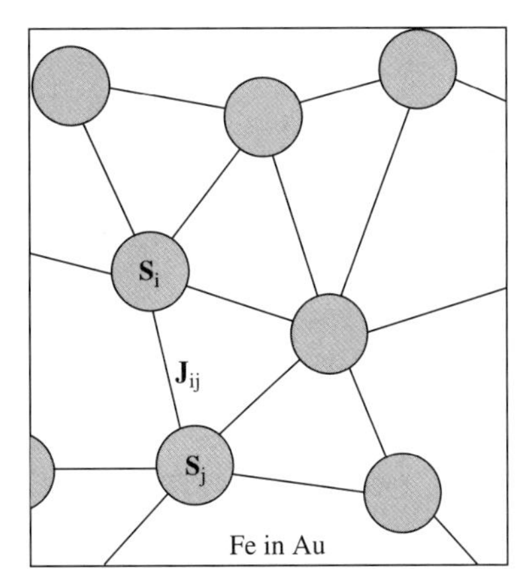

Fig. 7.25 Schematic comparison of neural networks (left) and spin glasses (right).

7.5.2 Biology and medicine

Magnetic models have various applications and generalizations in biology and neurology, and many models enjoy shared interest in magnetism, biology, and other areas of science (Haken 1983, Berg 1993). Neural networks, such as the Hopfield model, simulate the human brain, and there is a close analogy between such networks and the magnetic models. Figure 7.25 illustrates the structural and functional similarity between brain cells (neurons) and spin glasses. In fact, neural networks were one reason for the interest in spin glasses (Fischer and Hertz 1991). A specific question the associative character of brain activity. For example, unlike typical computer algorithms, the brain easily distinguishes between a number of persons, such as relatives and colleagues. In Section 7.1.4, we have mentioned that spin glasses have many low-lying states, and the identification of a person corresponds to the relaxation into the corresponding energy minimum.

The information of the brain is stored in the interactions J_{ij}, whereas the visual input has the character of an external magnetic field H_i. In the brain, the J_{ij} are realized chemically, whereas the spin states ↑ and ↓ correspond to the electrical potential of the neurons. Learning consists in the tuning of the J_{ij}. Let us consider the distinction between $m = 1 \ldots p$ persons, each characterized by a vector or "image" $\xi_i(m)$. The simplest approach is to use the Hebb rule

$$J_{ij} = \frac{1}{N} \sum_{m=1}^{p} \xi_i(m)\, \xi_j(m) \tag{7.43}$$

where N is the total number of spin or neurons. Glauber-type dynamics (Section 6.3.1) causes the model defined by (7.43) to relax into the nearest state $\xi_i(m)$, thereby associating the visual input H_i with the m-th person. The maximum number p of persons or objects that can be distinguished depends on the overlap $\xi_i(m)\, \xi_i(n)$. If all input pictures $\xi_i(m)$ were orthogonal, then $p \sim N$. However, common human features such as two eyes and mouths rather than beacons indicate a considerable degree of overlap, and $p \ll N$. In the brain, this is compensated by the large number of neurons, $N \sim 10^{11}$.

Fokker-Planck and other models have found applications in areas as population dynamics and spread of diseases (next subsection). In addition, there are many biological applications of magnetism, for example in the fields of self-organization, symbiosis, and food supply. An example of a medical application is targeted drug delivery and cancer therapy using magnetic nanoparticles (Pankhurst *et al.* 2003, Panyam *et al.* 2004)

7.5.3 Social sciences

Compared to natural sciences, social sciences are characterized by complex problems that are difficult to access by experiment. A destroyed sample is rarely a problem in physics, but extending this experimental approach to human society is not acceptable. It is, however, possible to model empirical data, such as birth rates and consumer prices. We have already mentioned Ising modeling of the voting behavior in two-party systems. In this case, H_i is a local political bias or tradition, whereas the J_ij describe the interactions between the members of the society.

This kind of modeling has been applied to a wide range of problems, from stock-market fluctuations, traffic flow, and propagation of forest fires to sociology and linguistics (Haken 1983). In economics, the approach is also known as econophysics. Such modelling cannot replace the "microscopic" search for the origin of social or economic developments, but help to understand their consequences. Two specific examples are the spread of crime (Glaeser, Sacerdote and Scheinkman 1996) and financial stability (Bornholdt and Wagner 2002). The basic idea may be illustrated by the following cinematographic analogy. A good guy (↑) interacts with a bunch of bad guys (↓), and due to an external force, introduced by the movie director, the good guy gains influence. Eventually, everybody is spin-up, corresponding to the film's happy end. However, the development towards the happy end is affected by nonlinear developments, hysteresis, and randomness.

Exercises

1. ***Phase transition in an A–B alloy.*** Construct an Ising Hamiltonian for an equiatomic A–B alloy on a simple cubic lattice and determine the mean-field order-disorder and spinodal-decomposition temperatures.
   Hint: $c_{\mathrm{iA/B}} = (1 \pm s_\mathrm{i})/2$.
2. ***Critical exponents for percolation.*** Use the real-space renormalization-group equations of Section 7.2.4 to estimate the critical exponents ν for one- and two-dimensional percolation.
   Hint: Write $p = p_\mathrm{c} - \varepsilon$ and $p' = p_\mathrm{c} - \varepsilon'$ and exploit that $\nu = -\ln(\xi'/\xi)/\ln(\varepsilon'/\varepsilon)$.
   Answer: $\varepsilon' = 2\varepsilon$ and $\nu = 1$ for $d = 1$, and $\varepsilon' = 1.52768\varepsilon$ and $\nu = 1.6353$ for $d = 2$.
3. ****Surface states.*** Find the s-band surface states for a simple-cubic (001) surface in the tight-binding approximation.
4. ****Extended polymer chain.*** Calculate the force acting on an almost extended polymer chain.
   Hint: Since $R \leq N_\mathrm{a}$, the chain is no longer Gaussian. What is the magnetic analog to the extended chain?

5. **Ammonia formation and decomposition.* The formation and decomposition of ammonia exhibits a number of interesting features. For example, ammonia synthesis is usually performed at relatively low temperatures and high pressures, and ammonia can be used to produce interstitial permanent magnets. Explain these features by analyzing the energy and entropy of ammonia. Why can ammonia be used to produce interstitial magnets such as FeN_x and $Sm_2Fe_{17}N_3$?
6. **Hubbard Green function and Wigner's semicircle law.* The real part of the Hubbard Green function has a semicircular shape. The same shape, known as Wigner's semicircle law, is obtained for random matrices with Gaussian-distributed couplings t_{ij}. Since the Green function provides a complete physical description of the system, one may conclude that both the Hubbard and Wigner models are equivalent. What is wrong with this conclusion?
7. **Micromagnetic localization in a nanowire.* Determine the nucleation mode in a thin nanowire of infinite length and anisotropy $K = 5\,\mathrm{MJ/m^3}$. The easy axis is parallel to the wire axis, and the wire contains a cylindrical imperfection of length 2 nm where the anisotropy is reduced by 20%.
   Hint: Assume that the shape anisotropy is incorporated into K, as reasonable for very thin wires.
8. **Reduced surface Curie temperature.* There are magnets that exhibit an enhanced surface Curie temperature due to enhanced exchange at the surface. However, there are no magnets with *reduced* surface Curie temperature, in spite of the realistic possibility of reduced exchange at the surface. Why?
9. **Generating function.* Calculate the generating function $f(k)$ for n-vector spin averages $\langle \mathbf{s}_i^\alpha \rangle$ as a function of n and convince yourself that higher-order contributions become unimportant as n approaches zero. For simplicity, restrict the calculation to quartic terms.
   Hint: Conversion of $\nabla_\mathbf{k}^2$ into spherical coordinates yields $\partial^2/\partial k^2 + (n-1)k^{-1}\partial/\partial k$, and the resulting differential equation is solved most easily by using the *ansatz* $f = 1 + ak^2 + bk^4$.
   Answer: $f(k) = 1 - k^2/2 + nk^4/8(n+2)$.

Appendix

A.1 Units and constants

A.1.1 Units systems and notation

Most scientists now use the international or SI system, although the cgs system has remained popular in some countries, especially in the USA and in China. The *SI system* was fully elaborated in the twentieth century, but the use of the units meter, kilogram, and second go back to the late 1790s. These mechanical units are complemented by one electromagnetic base unit, the ampere (A). All other mechanical and electromagnetic units can be written as powers of m, kg, s, and A, but some units carry separate names. Examples are the energy unit joule ($1\,\mathrm{J} = 1\,\mathrm{kgm}^2/\mathrm{s}^2$), the charge unit coulomb ($1\,\mathrm{C} = 1\,\mathrm{As}$) and the flux-density unit tesla ($1\,\mathrm{T} = 1\,\mathrm{kg/As}^2$). The SI system is logical, transparent, unambiguous, and, with the exception of the cumbersome field and magnetization unit A/m, convenient to use.

The first version of the cgs system was proposed in the 1830s and later modified to include electromagnetic units. The basic units of the cgs system are cm, g, and s, and energies are measured in erg ($1\,\mathrm{erg} = 10^{-7}\,\mathrm{J}$). The cgs system exists in several versions, such as the Gaussian system, which mixes magnetic (emu) and electric (esu) units. In the Gaussian system, Maxwell's equations (A51a–d) assume the form

$$\nabla \cdot \mathbf{D} = 4\pi\rho \tag{A.1}$$

$$\nabla \cdot \mathbf{B} = 0 \tag{A.2}$$

$$\mathrm{c}\,\nabla \times \mathbf{E} = -\frac{\partial \mathbf{B}}{\partial \mathrm{t}} \tag{A.3}$$

$$\mathrm{c}\,\nabla \times \mathbf{H} = 4\pi\mathbf{j} + \frac{\partial \mathbf{D}}{\partial \mathrm{t}} \tag{A.4}$$

The material equations are $\mathbf{D} = \mathbf{E} = 4\pi\mathbf{P}$ and, in a common notation, $\mathbf{B} = \mathbf{H} + 4\pi\mathbf{M}$. The Lorentz force changes from $q\mathbf{v} \times \mathbf{B}$ (SI) to $q\mathbf{v} \times \mathbf{B}/c$ (cgs), and there is no μ_o in the Zeeman energy.

Note that emu is not a unit but an indicator of the used unit system. As a consequence, multiplication by a dimensionless number may change the unit of a physical quantity, which is counterintuitive and not allowed in the SI system. For example, the cgs equation $\mathbf{B} = \mathbf{H} + 4\pi\mathbf{M}$ means that the flux density measured in gauss (G) is equal to the field measured in oersted (Oe) plus 4π times the magnetization measured in $\mathrm{emu/cm}^3$.

An inconvenience of the present-day SI system is that both H and M are measured in A/m. This reflects the nineteenth-century idea that magnetism is due to moving charges and interprets the spin in terms of unphysical currents. A simple solution is to consider $\mu_o H$ and $\mu_o M$, respectively, which are both measured in tesla. Some authors write these combinations as $J = \mu_o M$, and $B_o = \mu_o H$, but this raises the problem of a possible confusion with J (exchange) and B (flux density).

Notation. Throughout the book, a consistent and transparent set of symbols is used. Howover, some commonly used symbols must be identified carefully, most notably the following cases. (I) In expressions such as $\mathrm{d}V = \mathrm{d}\mathbf{r} = \mathrm{d}x\,\mathrm{d}y\,\mathrm{d}z$, V is a volume, but in some sections, especially in Chapter 2, V denotes the electron potential and then usually contains a subscript. (II) The vector $\mathbf{m}$ is the magnetic moment and, in most sections, $m = |\mathbf{m}|$. Howver, in some paragraphs of Chapter 5, m denotes the power law exponent for energy barriers. Some pairs are carefully separated in the text, so that they should not yield confusion, such as (III) t for time and hopping integral, (IV) D for diffusivity and density of states, and (V) R for radius and resistance.

The mathematical symbols $\approx$ and $\sim$ mean "approximately equal" and "proportional", respectively, and the latter is also used for order-of-magnitude estimates and scaling relations. RT means room temperature (about 298 K).

A.1.2 Unit conversions

There are various SI–cgs conversion rules for magnetic units, such as $1\,\mathrm{T} = 10\,\mathrm{kOe}$ (field), $1\,\mathrm{J/T} = 1\,\mathrm{Am}^2 = 1000\,\mathrm{emu}$ (moment), and $1\,\mathrm{kA/m} = 1\,\mathrm{emu/cm}^3$ (magnetization). For example, the room-temperature magnetization of iron, $1.707\,\mathrm{MA/m} = 1707\,\mathrm{emu/cm}^3$, corresponds to a flux density of 2.15 T or 21.5 kG. The useful relations $1\,\mathrm{Oe} \approx 80\,\mathrm{A/m}$ (field) and $1\,\mathrm{MGOe} \approx 8\,\mathrm{kJ/m}^3$ (energy product) involve the factor $1000/4\pi = 79.577$. In the cgs, demagnetizing factors and susceptibilities carry a factor 4π. For example, $D = 1/2$ is written as $D = 2\pi$ in the cgs system.

Some useful *energy conversions* are $\mathrm{eV} = 1.602 \times 10^{-19}\,\mathrm{J}$, $\mathrm{K} = 1.381 \times 10^{-23}\,\mathrm{J}$, $\mathrm{kJ/mole} = 1.66 \times 10^{-21}\,\mathrm{J}$ $(= 120.3\,\mathrm{K})$, $\mathrm{kcal} = 4184\,\mathrm{J}$, $\mathrm{Ry} = 2.180 \times 10^{-18}\,\mathrm{J}$ $(= \mathrm{e}^2/8\pi\varepsilon_o a_o = 13.6\,\mathrm{eV})$, $1/\mathrm{cm} = 1.99 \times 10^{-22}$ $(= 1.439\,\mathrm{K})$, and $\mathrm{Hz} = 6.626 \times 10^{-34}\,\mathrm{J}$.

A.1.3 Physical constants

The magnitude of the charge of the electron is $\mathrm{e} = 1.602 \times 10^{-19}$ C. The magnetic field constant $\mu_o = 4\pi \times 10^{-7}\,\mathrm{J/mA}^2$ can also be written as $\mu_o = 4\pi \times 10^{-7}$ H/m, where H (henry) is the SI unit of the inductance. The Bohr magneton, $\mu_\mathrm{B} = \mathrm{e}\hbar/2m_\mathrm{e}$, is equal to $9.274 \times 10^{-24}\,\mathrm{J/T}$, and the speed of light $c = (\varepsilon_o \mu_o)^{-1/2} \approx 3 \times 10^8\,\mathrm{m/s}$.

A.2 Mathematics

A.2.1 Linear equations

As in other areas of science, linear equations are of great importance in magnetism. First, nonlinear relations can often be approximated by linear equations. Second, quantum mechanics is essentially linear, as epitomized by the Schrödinger equation (A.29). The simplest linear equation is $C\,m = f$, where C is a force constant (or materials parameter), f is an external force, and the amplitude m describes the reaction of the system. An example is the linear susceptibility relation $M = \chi H$, where M, χ, and H correspond to m, $1/C$, and f, respectively. Other examples are mentioned in Section 7.3.

It is often necessary to consider vectors $m_{\rm i}$ rather than scalars m. For example, the local saturation magnetization may be written as $M_{\rm s,i} = M_{\rm s}(\mathbf{R}_{\rm i})$, where $\mathbf{R}_{\rm i}$ is the position of the i-th atom. In this section, we focus on the N-dimensional vector defined by $i = 1 \ldots N$ and ignore that m is often an n- or d-dimensional vector itself. An example is $\mathbf{M}(\mathbf{R}_{\rm i}) = M_{\mu {\rm i}}$, where $\mu = 1 \ldots n$ and n is the spin dimensionality (Section 5.2). The reason for focusing on scalars is practical rather than fundamental: we want to circumvent cumbersome multiple indices and avoid the ambiguous use of bold characters. A formal way of solving the vector notation problem is to use multi-indices such as $i = 1 \ldots n\,N$, corresponding to n times N degrees of freedom. However, these multi-indices are cumbersome, too, and we will stick to the simpler case of scalar relations. However, aside from these awkward but peripheral bookkeeping issues, vector spaces in physics are transparent and convenient.

Let us start with the vectors themselves. There are various equivalent ways of writing N-dimensional vectors $\mathbf{m} = |{\rm m}\rangle = m_{\rm i} = m(\mathbf{r}_{\rm i})$

$$m_{\rm i} = (m_1, m_2, \ldots, m_{\rm N}) \tag{A.5a}$$

$$\mathbf{m} = m_1\mathbf{e}_1 + m_2\mathbf{e}_2 + \cdots + m_{\rm N}\mathbf{e}_{\rm N} \tag{A.5b}$$

$$|m\rangle = m_1|1\rangle + m_2|2\rangle + \cdots + m_{\rm N}|N\rangle \tag{A.5c}$$

$$m(\mathbf{r}) = m_1\phi_1(\mathbf{r}) + m_2\phi_{\rm N}(\mathbf{r}) + \cdots + m_{\rm N}\phi_{\rm N}(\mathbf{r}) \tag{A.5d}$$

Equations (A.5a) and (A.5b) are very common in mathematics and physics, respectively, whereas (A.5c) and (A.5d) are widely used in quantum mechanics. As we will discuss below, (A.5d) is actually a special case of (A.5c), with $|i\rangle = \phi_{\rm i}(\mathbf{r})$. The cases (A.5a–d) may all be interpreted as generalizations of 'ordinary' vectors $\mathbf{r} = (x, y, z)$.

In the respective notations, the scalar product $c = a_1^* b_1 + a_2^* b_2 + \cdots + a_{\rm N}^* b_{\rm N}$ assumes the form

$$c = a_{\rm i}^+ b_{\rm i} \tag{A.6a}$$

$$c = \mathbf{a} \cdot \mathbf{b} \tag{A.6b}$$

$$c = \langle a|b\rangle \tag{A.6c}$$

$$c = \int a^*(\mathbf{r})\, b(\mathbf{r})\, {\rm dV} \tag{A.6d}$$

In particular, $c = 0$ for *orthogonal* states. The consideration of complex quantities $a = a' + {\rm i}\, a''$ and $a^* = a' - {\rm i}\, a''$ is necessary in quantum mechanics (Hilbert space), but otherwise the scalar product keeps its real-space meaning. In particular, $\mathbf{a} \cdot \mathbf{b}$ can be interpreted as a projection of $\mathbf{b}$ onto $\mathbf{a}$, or *vice versa*.

Next, we generalize the linear relation $C\, m = f$ to vectors:

$$\sum_{\rm j} C_{\rm ij}\, m_{\rm j} = f_{\rm i} \tag{A.7a}$$

$$\mathsf{\mathbf{Cm}} = \mathbf{f} \tag{A.7b}$$

$$\mathsf{C}|m\rangle = |f\rangle \tag{A.7c}$$

$$\int \mathsf{C}(\mathbf{r},\mathbf{r}')m(\mathbf{r}')\,\mathrm{d}\mathbf{r}' = f(\mathbf{r}) \tag{A.7d}$$

$$\mathsf{C}m(\mathbf{r}) = f(\mathbf{r}) \tag{A.7e}$$

Here (A.7a) and (A.7b) are known as the matrix and vector notations, respectively. The so-called bra-ket notation (A.7c) originates from quantum mechanics, whereas (A.7d–e) is most useful when dealing with continuous systems. For example, $\mathsf{C} = c^2\nabla^2 - \partial^2/\partial t^2$, where $\nabla = \mathbf{e}_x\partial/\partial x + \mathbf{e}_y\partial/\partial y + \mathbf{e}_z\partial/\partial z$, specifies (A.7e) as an inhomogeneous wave equation.

Throughout this appendix, we will use 2×2 matrices ($N = 2$) as an example. Equations (A.7a–b) then assume the form

$$C_{11}m_1 + C_{12}m_2 = f_1 \tag{A.8a}$$

$$C_{21}m_1 + C_{22}m_2 = f_2 \tag{A.8a$'$}$$

$$\begin{pmatrix} C_{11} & C_{12} \\ C_{21} & C_{22} \end{pmatrix} \begin{pmatrix} m_1 \\ m_2 \end{pmatrix} = \begin{pmatrix} f_1 \\ f_2 \end{pmatrix} \tag{A.8b}$$

One problem is to determine $\mathbf{m}$. This is achieved by solving the set of equations (A.8a) or, equivalently, inverting the matrix C: $\mathbf{m} = \mathsf{C}^{-1}\mathbf{f}$. When C is close to some multiple c of the unit matrix

$$\mathsf{I} = \begin{pmatrix} 1 & 0 \\ 0 & 1 \end{pmatrix} \tag{A.9}$$

then one can use the expansion $1/(1+\mathsf{x}) = 1 - \mathsf{x} + \mathsf{x}^2 - \mathsf{x}^3 \pm \cdots$ and obtains, in lowest order, $\mathbf{m} = 2\mathbf{f}/c - \mathsf{C}\mathbf{f}/c^2$. Another problem is to decouple the two equations (A.8a) and (A.8a$'$). This diagonalization problem will be discussed in the following subsection.

A.2.2 Eigenmode analysis

In many areas of physics and mathematics it is convenient to consider diagonal equations of the type

$$C_\mathrm{a}\, m_\mathrm{a} = f_\mathrm{a} \tag{A.10a}$$

$$C_\mathrm{b}\, m_\mathrm{b} = f_\mathrm{b} \tag{A.10a$'$}$$

rather than (A.8). This is because the two modes m_a and m_b are decoupled and because the eigenvalues $C_{\mathrm{a/b}}$ tend to have a transparent physical meaning. Examples are normal modes of coupled oscillators, the determination of quantum-mechanical eigenvalues and eigenstates (Section 2.1, Section 7.1), and magnetization modes near the Curie point (Section 5.36, Section 7.1). The following paragraphs summarize some basic features of eigenmode analysis.

The transformation from (A.8a) to (A.10a) can be considered as a *rotation* in vector space. Consider, for example, the captain of a boat who wants to predict the motion of his vessel. In general, the external force $\mathbf{f} = f_x \mathbf{e}_x + f_y \mathbf{e}_y$ forms a nonzero angle with the direction $\mathbf{v}$ of the keel. To solve the problem, one rotates the coordinate frame from $\{x, y\}$ to the frame $\{v, v_\perp\}$ of the boat, where $\mathbf{v} \cdot \mathbf{v}_\perp = 0$. In this frame, only the projection $f_v = \mathbf{f} \cdot \mathbf{v}$ contributes to the acceleration; because the projection $\mathbf{f} \cdot \mathbf{v}_\perp$ perpendicular to the keel has little influence on the motion of the vessel. We note that neither the orthogonality between the axes nor the magnitude $|\mathbf{e}_i| = 1$ of the unit vectors is changed by this procedure. This is very convenient in calculations.

After diagonalization, (A.7) becomes $C_\mu m_\mu = f_\mu$ where the index $\mu = 1 \ldots N$ labels the eigenmodes of C_{ij}. The eigenmodes can also be written as $|\mu\rangle = \sum_i \mu_i |i\rangle$,

$$|\mu\rangle = \sum_i \langle i|\mu\rangle \, |i\rangle \tag{A.11}$$

or $|\mu\rangle = \mathsf{U}|i\rangle$, where $\mathsf{U} = \sum_i \langle i|\mu\rangle$ is the unitary or "rotation" matrix that diagonalizes C. Equation (A.11) is a consequence of the completeness of eigenfunctions, that is, any vector can be represented as a sum of eigenfunctions; it can also be written as $|\mu\rangle = \sum_i |i\rangle\langle i|\mu\rangle$. Here the operator $\sum_i |i\rangle\langle i| = 1$ is also known as a "nutritious one", because it can be inserted in any expression to simplify the calculation. Physically, it projects any vector onto the orthogonal unit vectors $|i\rangle$ and then adds all components together to reproduce the original vector.

In practice, eigenmode problems are solved by considering the equation

$$\sum_j C_{ij} \, \mu_j = C_\mu \, \mu_i \tag{A.12}$$

which has solutions if the determinant $\det(C_{ij} - \lambda \, \delta_{ij}) = 0$. Here λ is any of the eigenvalues C_μ and δ_{ij} is the unit matrix, $\delta_{ij} = 1$ if $i = j$ and $\delta_{ij} = 0$ if $i \neq j$ (Kronecker symbol). For example, the diagonalization of a 4×4 matrix amounts to the solution of the secular equation $\det\,(\mathsf{C} - \lambda) = (C_1 - \lambda)(C_2 - \lambda)(C_3 - \lambda)(C_4 - \lambda)$. An explicit example will be discussed in the next subsection. The diagonalization may also be performed by exploiting the diagonal character of $\mathsf{U}^+\mathsf{C}\mathsf{U}$, where the the $\mathsf{U}^+ = \mathsf{U}^{\mathrm{T}^*}$ (conjugate transpose), but the corresponding unitary transformation U is usually unknown.

An example of a unitary transformation is *Fourier transformation*, where $\mathsf{U} \sim \exp(i\mathbf{k} \cdot \mathbf{r})$. It transforms real-space states into wave-vector states: $f_\mathbf{k} = \int \exp(i\mathbf{k} \cdot \mathbf{r}) \, f(\mathbf{r}) \, dV$. Functions localized in k-space are delocalized in real space, and vice versa. Fourier transformation makes it possible to treat periodic and, to a lesser degree, aperiodic systems. Examples are electrons in solids and spin waves.

A.2.3 Real 2 × 2 matrices

Many physical systems can be treated as two-mode systems, similar to (A.8). In some cases, the treatment is only approximate, based on the projection of a space with large N onto a space with $N = 2$. Of course, the selection of the two states depends on the involved physics. For example, at low temperatures it is often sufficient to consider the ground state and the first excited state. A real-space equivalent of this procedure is the projection of an airplane $\{x, y, z\}$ onto a radar screen $\{x, y\}$.

The eigenvectors and eigenvalues of (A.8b) are obtained by solving

$$\begin{pmatrix} C_{11}-\lambda & C_{12} \\ C_{21} & C_{22}-\lambda \end{pmatrix} = (C_{11}-\lambda)(C_{22}-\lambda) - C_{12}\,C_{21} = 0 \tag{A.13}$$

The two roots of this quadratic equation are the eigenvalues of C, and the corresponding eigenvectors are obtained by solving (A.8). It can be shown that matrices with $C_{\mathrm{ji}} = C^*_{\mathrm{ij}}$ have real eigenvalues (Hermitian matrices). For example, the Pauli matrices (2.32, 6.10) are complex but hermitian and have the real eigenvalues $\pm$ 1, corresponding to $\uparrow$ and $\downarrow$ states. However, the eigenfunctions are generally complex. For simplicity, we will restrict ourselves to real eigenfunctions, which correspond to $C_{\mathrm{ji}} = C_{\mathrm{ij}}$.

Any real 2×2 matrix C_{ij} can be written as $\mathsf{C} = c_{\mathrm{o}}\mathsf{I} + \mathsf{A}$, where $c_{\mathrm{o}} = (C_{11} + C_{22})/2$, I is the unit matrix, and

$$\mathsf{A} = \begin{pmatrix} a & b \\ b & -a \end{pmatrix} \tag{A.14}$$

Here, $b = C_{12}$ and $a = (C_{11} - C_{22})/2$. Note that the c_{o} term shifts the eigenvalues but leaves the eigenfunctions unchanged, so that we can restrict ourselves to A. Its eigenvalues

$$A_{\pm} = \pm\sqrt{a^2 + b^2} \tag{A.15}$$

For small and large off-diagonal matrix elements b, the respective eigenvalues approach $\pm a$ and $\pm b$. For small b, the splitting between the levels increases from $2a$ to $2a + b^2/a$. Since off-diagonal matrix elements correspond to interactions, we draw the conclusion that interactions lead to a *level repulsion*. For large level separations a, the repulsion is small, b^2/a, but for nearly degenerate levels ($a \approx 0$) it approaches $2b$. This finding is of far-reaching importance in physics, because it means that interactions are most important for degenerate or nearly degenerate energy levels.

The eigenvectors of A can be written as $\mu_+ = (\cos\chi,\ \sin\chi)$ and $\mu_- = (-\sin\chi,\ \cos\chi)$, where the mixing angle

$$\chi = \mathrm{a}\tan\left(\frac{\sqrt{a^2 + b^2} - a}{b}\right) \tag{A.16}$$

With increasing b, χ increases from 0 ($b = 0$) to $\pi/2$ ($b = \infty$). In the quantum-mechanical analogy, it describes the degree of hybrization. The corresponding unitary transformation

$$\mathsf{U} = \begin{pmatrix} \cos\chi & -\sin\chi \\ \sin\chi & \cos\chi \end{pmatrix} \tag{A.17}$$

describes two-dimensional rotations in the space defined by A. Note that the columns of U are the eigenvectors of A and that $\mathsf{U}^{\mathrm{T}}\mathsf{A}\mathsf{U}$ is diagonal. This feature, as well as the orthonormality and completeness of eigenfunctions, are common to all eigenmode problems. However, in the case of degenerate eigenvalues, $\mu_{\mathrm{i}} = \mu_{\mathrm{j}}$, the orthogonalization must be ensured separately, because it corresponds to a diagonal submatrix.

In this book, we have seen many applications of (A.14–17). Examples are the diatomic pair model (Section 2.1), s-d hybridization (Section 2.4), ferri- and antiferromagnetism at finite temperatures (Section 5.3.6), and the hemisphere model of panel 8.

A.2.4 Vector and functional calculus

Equations (A.5–8) indicate a close relationship between scalars, vectors, and functions. The first part of this subsection is devoted to common features, whereas the second part deals with specific operations such as the vector or cross product, which is limited to three-dimensional vector spaces.

The ordinary derivative $\mathrm{d}f/\mathrm{dx} = f'$ is defined as the $\Delta = 0$ limit of $(f(x+\Delta) - f(x))/\Delta$. It may be determined numerically or by exploiting relations such as $\mathrm{d}x^2/\mathrm{d}x = 2x$. There are two different generalizations to multidimensional systems. First, the function F(x) may be replaced by $f(\mathbf{r}) = f(x, y, z)$. The corresponding partial derivatives $\partial f/\partial x$, $\partial f/\partial y$ and $\partial f/\partial z$ are often used in form of the nabla or del operator

$$\nabla f = \frac{\partial f}{\partial x}\,\mathbf{e}_\mathrm{x} + \frac{\partial f}{\partial y}\,\mathbf{e}_\mathrm{y} + \frac{\partial f}{\partial z}\,\mathbf{e}_\mathrm{z} \tag{A.18}$$

A second generalization is obtained by considering f as a vector or function. An example is the local magnetization of a one-dimensional Ising magnet, whose respective vector and functional forms are $M_\mathrm{i} = M(x_\mathrm{i})$ and $M(x)$. This distinction is sometimes confusing. For example, the magnetization $\mathbf{M}_\mathrm{i} = \mathbf{M}(\mathbf{R}_\mathrm{i})$ involves three vectors: an n-dimensional magnetization vector ($\mathbf{M}$), a d-dimensional real-space vector ($\mathbf{R}$), and an N-dimensional vector denoted by the index i. An example is $\nabla\mathbf{M} = \nabla M_\mathrm{x}\,\mathbf{e}_\mathrm{x} + \nabla M_\mathrm{y}\,\mathbf{e}_\mathrm{y} + \nabla M_\mathrm{z}\,\mathbf{e}_\mathrm{z}$ (Chapter 4).

As a mechanical example, let us consider a system of N coupled particles of energy is $F(y_\mathrm{i})$. The force on the i-th particle, $f_\mathrm{i} = \partial F/y_\mathrm{i}$, is zero in equilibrium, that is, $\partial F/y_\mathrm{i} = 0$. For a continuous system $y(x)$, such as a string, this condition becomes $\delta F/\delta y(x) = 0$, where $\delta F/\delta y(x)$ is a *functional derivative.* As for ordinary derivatives, there exist rules to evaluate functional derivatives. For example, $F = \int \eta(f, \nabla f)\,\mathrm{d}V$ yields the Euler-Lagrange equations

$$\frac{\delta F}{\delta f(\mathbf{r})} = \frac{\partial \eta}{\partial f} - \nabla\left(\frac{\partial \eta}{\partial \nabla f}\right) \tag{A.19}$$

To prove this equation, one may replace the continuous function F by a discrete function $F(f_\mathrm{i})$ and the derivatives by differences of the type $f_{\mathrm{i}+1} - f_\mathrm{i}$. A frequently occurring energy expression is

$$F = \int \left(\frac{a}{2}(\nabla f)^2 + \frac{b}{2}f^2\right)\mathrm{d}V \tag{A.20}$$

Its functional derivative $\delta F/\delta f(\mathbf{r}) = -\nabla(a\nabla f) + bf$ is frequently used, for example in the determination of nucleation fields (Section 4.3.1) and spin waves (Section 6.1).

The functional equivalent of the unit matrix I is $\delta(x - x')$, where the *delta function* or "needle" function $\delta(x)$ is defined by $\delta(x \neq 0) = 0$ and $\int \delta(x)\,\mathrm{d}x = 1$. The delta function can also be considered as the derivative of the step or theta function,

$\delta(x) = \mathrm{d}\Theta(x)/\mathrm{d}x$, where $\Theta(x < 0) =$ and $\Theta(x > 0) = 1$. Theta functions are very useful in the evaluation of integrals, because an integral of $f(x)$ from a to b can also be written as an integral of $f(x)\,\Theta(x-a)\,\Theta(b-x)$ from $-\infty$ to $+\infty$. Partial integration based on $(fg)' = fg' + f'g$ and $g = \Theta$ can then be used to transform the theta functions into delta functions and to simplify the integrals by exploiting that $\int g(x)\,\delta(x - x_o)\,\mathrm{d}x = g(x_o)$.

In three dimensions, the $\int F(x)\,\mathrm{d}x$ corresponds to the volume integral $\int F(\mathbf{r})\,\mathrm{d}V = \int F(\mathbf{r})\,\mathrm{d}x\,\mathrm{d}y\,\mathrm{d}z$, also written as $\int F(\mathbf{r})\,\mathrm{d}\mathbf{r}$. Its generalization is the *functional integral*

$$\int F(f(\mathbf{r}))\,\mathrm{D}f = \int F(f\mathrm{i})\,\mathrm{d}f_1\,\mathrm{d}f_2 \cdots \mathrm{d}f_\mathrm{N} \tag{A.21}$$

Functional integrals are also known as path integrals, because the function $f(\mathbf{r})$ has the character of a path. Consider, for example, two villages A and B separated by a range of mountains. There are many paths joining the villages, and the total probability of going from A to B has the character of a functional integral. Many problems can be formulated as functional integrals, including the solution of the Schrödinger equation (Feynman path integrals), diffusion, and other nonequilibrium problems, as exemplified by the Onsager-Machlup function, and polymer statistics (Section 7.2). One class of exactly solvable functional integrals, where the integrand is of the type

$$F(f(\mathbf{r})) = F_o \exp\left(-\frac{1}{2}\int A(\mathbf{r},\mathbf{r}')f(\mathbf{r})f(\mathbf{r}')\,\mathrm{d}V\,\mathrm{d}V' + \int b(\mathbf{r})f(\mathbf{r})\,\mathrm{d}V\right) \tag{A.22}$$

On diagonalization of A, the exponent becomes a sum over eigenfunctions, and the functional integral reduces to a product of one-dimensional integrals

$$\int \exp\left(-\frac{1}{2}A_\mu f_\mu^2 + b_\mu f_\mu\right)\mathrm{d}f_\mu = \sqrt{\frac{2\pi}{A_\mu}}\exp\left(\frac{b_\mu^2}{2A_\mu}\right) \tag{A.23}$$

Since the quadratic (harmonic) exponent in (A.22–23) corresponds to linear equations of state, Gaussian functional integrals are frequently encountered in linearized problems. Examples are Ornstein-Zernike type free-energy calculations (Section 5.3.3) and the self-consistent renormalization of spin fluctuations (Section 5.2.5).

In three dimensions, it is possible to define cross products $(\mathbf{a} \times \mathbf{b})$ and the curl or rotation $(\nabla \times \mathbf{a})$. Explicitly,

$$\mathbf{a} \times \mathbf{b} = (a_\mathrm{x} b_\mathrm{y} - a_\mathrm{y} b_\mathrm{x})\,\mathbf{e}_\mathrm{z} + (a_\mathrm{y} b_\mathrm{z} - a_\mathrm{z} b_\mathrm{y})\,\mathbf{e}_\mathrm{x} + (a_\mathrm{z} b_\mathrm{x} - a_\mathrm{x} b_\mathrm{z})\mathbf{e}_\mathrm{y} \tag{A.24}$$

and

$$\nabla \times \mathbf{b} = (\nabla_\mathrm{x} b_\mathrm{y} - \nabla_\mathrm{y} b_\mathrm{x})\mathbf{e}_\mathrm{z} + (\nabla_\mathrm{y} b_\mathrm{z} - \nabla_\mathrm{z} b_\mathrm{y})\mathbf{e}_\mathrm{x} + (\nabla_\mathrm{z} b_\mathrm{x} - \nabla_\mathrm{x} b_\mathrm{z})\mathbf{e}_\mathrm{y} \tag{A.25}$$

where we have used the shorthand notation $\nabla_\mathrm{x} = \partial/\partial x$. The restriction of this calculus to three spatial dimensions is a consequence of the Dirac equation (A.3.5). This is also the reason for the occurrence of cross products in the Maxwell equations (A.4.1). By contrast, the divergence $\nabla \cdot \mathbf{a}$ and the gradient ∇a can be defined for any vector space. $\nabla \cdot \mathbf{a} \neq 0$ means that the vector field $\mathbf{a}(\mathbf{r})$ has a local source or drain, whereas $\nabla \times \mathbf{a} \neq 0$ means that $\mathbf{a}(\mathbf{r})$ contains closed loops, as in a vortex. Integrals over divergences and curls are related to surface and contour integrals by the Gauss's and Stokes's theorems, respectively. Gauss's theorem, $\int \nabla \cdot \mathbf{a}\, \mathrm{d}V = \oint \mathbf{a} \cdot \mathrm{d}\mathbf{S}$, means that the sum of all source and drain contributions in a volume V is equal to the integral component of the vector field normal to the surface $\mathbf{S}$ of the volume. Stokes's theorem, $\int \nabla \times \mathbf{a} \cdot \mathrm{d}\mathbf{S} = \oint \mathbf{a} \cdot \mathrm{d}\mathbf{l}$, relates the line or contour integral along the boundary of an arbitrary surface to the curl density on the surface.

A.2.5 Useful formulae

To evaluate the factorial $N! = 1 \cdot 2 \cdot 3 \ldots N$ for large numbers N one can exploit *Stirling's formula* $N! \approx \sqrt{2\pi N}\,(N/\mathrm{e})^{\mathrm{N}}$. This is useful in many areas of physics and mathematics (diffusion, polymers, entropy, probability calculations).

Translation-invariant interactions $J(\mathbf{r}-\mathbf{r}')$, such as RKKY interactions, are easily diagonalized by Fourier transformation. In the integral $J_{\mathbf{k}} = \int \exp(\mathrm{i}\mathbf{k}\cdot\mathbf{R})\, J(\mathbf{R})\, \mathrm{d}\mathbf{R}$ one introduces the angle θ between $\mathbf{k}$ and $\mathbf{R}$, so that

$$J_{\mathbf{k}} = 2\pi \int \exp(\mathrm{i}kR\cos\theta)\, J(R)\sin\theta\, \mathrm{d}\theta\, R^2\, \mathrm{d}R \tag{A.26}$$

With $\cos\theta = q$, this yields $\int \exp(\mathrm{i}krq)\,\mathrm{d}q = 2\sin(kr)/kr$ and the relatively simple integral $J_{\mathbf{k}} = (4\pi/k)\int \sin(kR)\, J(R) R\, \mathrm{d}R$. In the determination of the free-electron RKKY interaction (Section 2.3.2), one starts from $J_{\mathbf{k}}$ and then calculates $J(\mathbf{r}-\mathbf{r}')$ by inverse Fourier transformation, exploiting that $\sin(x) - x\cos(x) = -x^2(\sin(x)/x)'$. Two other useful relations are $\nabla \exp(\mathrm{i}\mathbf{k}\cdot\mathbf{r}) = \mathrm{i}\mathbf{k}\exp(\mathrm{i}\mathbf{k}\cdot\mathbf{r})$ and the integral $\int 1/(1-z^2)\,\mathrm{d}z = \mathrm{a}\tanh(z)$.

A.3 Basic quantum mechanics

Experiment shows that the motion of any particle of momentum p and energy E is wavelike, described by the wave vector $k = 2\pi/\lambda = p/\hbar$ and the circular frequency $\omega = 2\pi f = E/\hbar$. In a nutshell, this is the basic idea of *quantum mechanics*. There are many textbooks dealing with general quantum mechanics and quantum-mechanical aspects of metals (Pauling and Wilson 1935, Mott and Jones 1936, Hurd 1981), but it is in order to provide a brief introduction.

Planck's constant is very small, so that the wave character is not manifest on a macroscopic scale, but microscopic particles and phenomena are described by *wave functions* such as

$$\psi = \psi_{\mathrm{o}} \exp(\mathrm{i}\omega t - \mathrm{i}\mathbf{k}\cdot\mathbf{r}) \tag{A.28}$$

Physical quantities, or observables, are described by *operators* such as the Hamilton operator (Hamiltonian) $\mathsf{H} = \mathrm{i}\hbar\,\partial/\partial t$ and the momentum operator $\mathbf{p} = -\mathrm{i}\hbar\partial/\partial\mathbf{r} = -\mathrm{i}\hbar\nabla$. Operators act on the wave function ψ, and the expectation value or quantum-mechanical average $A = \langle \mathsf{A} \rangle$ of any operator A is equal to $\langle \psi | \mathsf{A} | \psi \rangle = \int \psi^*(\mathbf{r})\, \mathsf{A}\psi(\mathbf{r})\, \mathrm{d}V$. Similar to (A.2.2–3), one can also use the matrix notation $\langle \mathsf{A} \rangle = \sum_{\mathrm{ij}} \psi_{\mathrm{i}}^* A_{\mathrm{ij}} \psi_{\mathrm{j}}$. The quantity $\rho = \sum_{\mathrm{i}} \psi_{\mathrm{i}}^* \psi_{\mathrm{i}}$ is the probability of finding the system in the quantum state $|\psi\rangle$.

A.3.1 Time dependence

The wave function is obtained from the *Schrödinger equation*. Its time-dependent form is

$$\mathrm{i}\hbar \frac{\partial}{\partial t}|\psi\rangle = \mathsf{H}|\psi\rangle \tag{A.29a}$$

or

$$\mathrm{i}\hbar \frac{\partial}{\partial t}\psi_i = \sum_{\mathrm{j}} E_{\mathrm{ij}}\psi_{\mathrm{j}} \tag{A.29b}$$

Diagonalization yields the modes $|\psi_\mu(t)\rangle = |\psi_\mu(0)\rangle \exp(\mathrm{i}E_\mu t/\hbar)$, and the complete wave function $|\psi(t)\rangle$ is a superposition of these modes. If the system is originally in the μ-th eigenstate, $|\psi_\mu(0)\rangle \neq 0$ but $|\psi_{\mu'}(0)\rangle = 0$ for $\mu' \neq \mu$, then the system remains in the μ-th eigenstate at $t > 0$. Similarly, adding a small perturbation to a system originally in an eigenstate gives rise to transitions to other states. Let us consider the 2×2 Hamiltonian

$$\mathsf{H} = \begin{pmatrix} E_\mathrm{o} + \Delta E & V \\ V & E_\mathrm{o} - \Delta E \end{pmatrix} \tag{A.30}$$

where V is a small perturbation, such as a hopping integral. The unperturbed eigenvalues (energies) are $E_\mathrm{o} \pm \Delta E$, whereas the perturbation V leads to a oscillating hybridization of the states. When the unperturbed Hamiltonian is degenerate, $\Delta E = 0$, then the initial condition $\psi(0) = (1, 0)$ yields the wave function is $\psi = (\cos(\omega t), \sin(\omega t))$, where $\omega = V/\hbar$. Applied to electrons in solids, the larger the hopping integral, the faster the motion of the electron from atom to atom. In narrow bands (small V), the hopping frequency is much smaller than the frequency associated with the orbital motion in the atom, and the electron is practically "owned" by the atom.

An explicit equation of motion is obtained by putting the time derivative of $\langle \mathsf{A}\rangle = \int \psi * (\mathbf{r})\mathsf{A}\psi(\mathbf{r})\,\mathrm{d}V$ into (A.29):

$$\mathrm{i}\hbar \frac{\partial}{\partial t}\langle \mathsf{A}\rangle = \langle \mathsf{AH} - \mathsf{HA}\rangle \tag{A.31}$$

This equation indicates that *commutators* $[\mathsf{A}, \mathsf{B}] = \mathsf{AB} - \mathsf{BA}$ are generally nonzero in quantum mechanics. Conjugate quantities, such as the position and momentum operators, obey commutation rules of the type $[\mathbf{x}, \mathsf{p}_\mathrm{x}] = \mathrm{i}\,\hbar$. This equation is easily verified by taking into account that $\mathsf{p}_\mathrm{x}\mathsf{x}\psi \sim \partial(x\psi)/\partial x = \psi + x\partial\psi/\partial x$. Commutating observables obey $[\mathsf{A}, \mathsf{B}] = 0$ and mean that A and B have common sets of eigenfunctions (next subsection) and can be measured simultaneously. Otherwise, the measurement implies a Heisenberg uncertainty of order $\hbar$. A mechanical analog is the position x and wave vector k of a classical wave packet, where the exact determination of the wave vector ($\Delta k = 0$) requires an extended wave ($\Delta x = \infty$).

A.3.2 Eigenvalues and eigenfunctions

The replacement $\mathrm{i}\hbar\partial\psi/\partial t = E\psi$ transforms (A.29) into the time-independent Schrödinger equation

$$E|\psi\rangle = \mathsf{H}|\psi\rangle \tag{A.32}$$

The diagonalization of this equation yields the energy eigenvalues E_μ and the eigenfunctions $|\psi_\mu\rangle$. A simple example is a free particle of mass m_e, where $\mathsf{H} = \mathbf{p}^2/2m_e$. Since $\mathbf{p} = -\mathrm{i}\,\hbar\nabla$, the Schrödinger equation

$$E\psi = -\frac{\hbar^2}{2m}\nabla^2\psi \tag{A.33}$$

Using $\nabla \exp(\mathrm{i}\mathbf{k}\cdot\mathbf{r}) = \mathrm{i}\mathbf{k}\exp(\mathrm{i}\mathbf{k}\cdot\mathbf{r})$ it is an easy exercise to see that $\psi = \exp(\mathrm{i}\mathbf{k}\cdot\mathbf{r})$ is an eigenfunction of (A.2.2). The corresponding eigenvalues $E_\mathrm{k} = \hbar^2 k^2/2m_e$ are labeled by the wave vector $\mathbf{k}$, which assumes the role of the index μ in (A.2).

Aside from complex wave functions $\exp(\mathrm{i}\mathbf{k}\cdot\mathbf{r})$, (A.33) has real solutions $\cos(\mathbf{k}\cdot\mathbf{r})$ and $\sin(\mathbf{k}\cdot\mathbf{r})$, or $\exp(\mathrm{i}\mathbf{k}\cdot\mathbf{r}) \pm \mathrm{i}\exp(\mathrm{i}\mathbf{k}\cdot\mathbf{r})$. Complex wave functions are characterized by a nonzero net momentum $\langle \mathsf{p}\rangle$ (running waves), whereas real wave functions obey $\langle \mathsf{p}\rangle = 0$ (standing waves). Particle-in-a-box states are standing waves, because the particle is reflected at the boundaries. The picture carries over to the angular momentum $\mathbf{L} = \mathbf{r}\times\mathbf{p}$ or $\mathsf{L} = -\mathrm{i}\hbar\mathbf{r}\times\nabla$ (Section 2.2), where real and complex wave functions correspond to quenched and unquenched orbital moments, respectively (Section 3.3.4). Complex wave functions are also realized in conduction and tunneling scenarios, such as that of Section 7.2.

In most cases, the particle moves in a potential $V(\mathbf{r})$, and the Schrödinger equation assumes the form

$$E\,\psi = -\frac{\hbar^2}{2m}\nabla^2\psi + V(\mathbf{r})\,\psi \tag{A.34}$$

Examples are the hydrogen-like potential proportional $1/r$ (Section 2.2.1) and the periodic crystal potential of metals (Section 2.4.2). Projecting $\psi(\mathbf{r})$ on tight-binding states $|i\rangle$ and using the particle-number representation $\mathsf{n}_\mathrm{i} = \mathsf{a}_\mathrm{i}^+\mathsf{a}_\mathrm{i}$ briefly discussed in Section 2.1.7, we can write the Hamiltonian as

$$\mathsf{H} = \sum_{\mathrm{ij}} t_{\mathrm{ij}}\,\mathsf{a}_\mathrm{i}^+\,\mathsf{a}_\mathrm{j} + V_\mathrm{i}\mathsf{n}_\mathrm{i} \tag{A.35}$$

where the t_ij are the hopping integrals. Applications and solutions of this Hamiltonian are used throughout this book.

A.3.3 Perturbation theory

There are various ways of treating time-dependent and time-independent perturbations. In the latter case, one considers a Hamiltonian $\mathsf{H} = \mathsf{H}_\mathrm{o} + \mathsf{V}$ and assumes that the eigenfunctions $|\mu\rangle$ and eigenvalues $E_{\mu\mathrm{o}}$ of the unperturbed Hamiltonian H_o are known. For the eigenvalues, one obtains the expansion

$$E_\mu = E_{\mu\mathrm{o}} + \langle\mu|\mathsf{V}|\mu\rangle + \sum_{\mu'}\frac{\langle\mu'|\mathsf{V}|\mu\rangle\,\langle\mu|\mathsf{V}|\mu'\rangle}{E_{\mu\mathrm{o}} - E_{\mu'\mathrm{o}}} + \cdots \tag{A.36}$$

In lowest order, the eigenvalue corrections are therefore equal to the diagonal elements of the interaction matrix V. Similar expressions exist for the eigenfunctions.

Perturbation theory must be distinguished from variational approaches, where one uses a normalized trial wave function $|\psi(\lambda)\rangle$. Here λ denotes one (or more) freely adjustable parameters. An approximate ground-state eigenfunction, corresponding to the smallest eigenvalue, is obtained by minimizing the energy with respect to λ, $\partial\langle\psi|\mathrm{H}|\psi\rangle/\partial\lambda = 0$. Using this method to determine excited states, one must ensure that the excited trial functions are orthogonal to the ground-state wave function.

A.3.4 Quantum statistics

Quantum statistics is similar to classical statistics, as introduced in Section 5.1. Averages have the character of traces, $\langle A\rangle = \mathrm{Tr}(\mathsf{A}\rho)$, where

$$\rho = \sum_\mu |\psi_\mu\rangle \, p_\mu \, \langle\psi_\mu| \tag{A.37}$$

is the *density matrix* of the system. The role of the density matrix (or density operator) is very similar to that of the probability in classical physics. An example is the equilibrium density matrix

$$\rho = \frac{\exp(-\mathsf{H}/k_\mathrm{B}T)}{\mathrm{Tr}(\exp(-\mathsf{H}/k_\mathrm{B}T))} \tag{A.38}$$

By using the Hamiltonian in its diagonalized matrix form, that is, $E_{\mu\nu} = E_\mu$ for $\mu = \nu$ and $E_{\mu\nu} = 0$ for $\mu \neq \nu$, we can replace the trace by a sum over all eigenvalues. This reproduces the familiar expressions $p_\mu = \exp(-E_\mu/k_\mathrm{B}T)$, $\langle A\rangle = (1/Z)$ $\sum_\mu A_\mu \exp(-E_\mu/k_\mathrm{B}T)$, and $Z = \sum_\mu \exp(-E_\mu/k_\mathrm{B}T)$.

An equation of motion for the density matrix (A.38) is obtained by applying $\partial/\partial t$ and using the Schrödinger equation. The result is the *Liouville-von Neumann equation*

$$\mathrm{i}\hbar\frac{\partial\rho}{\partial t} = [\mathsf{H}(t), \rho(t)] \tag{A.39}$$

This equation is the basis for various approximations and specific models, especially in nonequilibrium statistics.

Equations (A.37–39) apply to both single- and many-particle systems. However, in many-body systems, the density matrix is generally very complicated. A free particle of momentum $p = \hbar k$ behaves as a wave of wavelength $\lambda = 2\pi/k$. Together with the energy $\hbar^2k^2/2m_\mathrm{e}$ of free particles (Section 2.4.1), the thermal energy $k_\mathrm{B}T$ defines a thermal wavelength λ of order $\hbar/(m_\mathrm{e}k_\mathrm{B}T)^{1/2}$. Macroscopic particles have large masses m and thermal wavelengths much smaller than their interparticle distances. In this limit, quantum-mechanical interaction effects are negligible. However, quantum effects are important for small particles, and atoms such as H and He may become superfluid at temperatures of order 1 K or below. The mass of electrons is even smaller than that of nuclei, and quantum effects are important at room temperature and above. In particular, the high electron densities n in metals translate into Fermi temperatures $E_\mathrm{F}/k_\mathrm{B}$ of order 37,700 K for Na and 81,600 K for Cu.

An important simplification is obtained when noninteracting particles occupy well-defined single-particle states. This is approximately the case for electrons in metals, where the zero-temperature occupancy covers all states with $E \leq E_\mathrm{F}$ (Section 2.4). The physics behind this approximation is that $E_\mathrm{F} \sim n^{2/3}$ is much larger than the Coulomb interaction $U \sim n^{1/3}$. However, electrons are fermions, characterized by half-integer spins and antisymmetric wave functions. A fermionic level may be empty or occupied, but two or more particles per quantum state are forbidden by symmetry. The corresponding wave functions have the character of *Slater determinants*, such as $|\psi\rangle|\phi\rangle - |\phi\rangle|\psi\rangle$, and determinants involving two particles in the same state, for example $|\phi\rangle = |\psi\rangle$, are zero. We have encountered Slater determinants in the context of the independent-electron approximation, Section 2.1.5,

The average occupancy of the levels is described by the quantum distribution function

$$g(E) = \frac{\sum_{\mathrm{n}} n \exp(-n\,E/k_{\mathrm{B}}T)}{\sum_{\mathrm{n}} \exp(-n\,E/k_{\mathrm{B}}T)} \tag{A.40}$$

where n includes all possible occupancies, and averages are obtained as

$$\langle \mathsf{A} \rangle = \int A(E)\, g(E)\, D(E)\, \mathrm{d}E \tag{A.41}$$

Here D is the density of states, which depends on the dispersion relation $E(k)$ and on the dimensionality. Examples are $D(E) \sim E^{\mathrm{d}/2-1}$ for free electrons, where $E \sim k^2$ and $D(E) \sim E^{\mathrm{d}-1}$ for phonons, where $E \sim k$.

The distribution function is obtained by evaluating (A.40) For fermions ($\mathrm{n} = 0, 1$) one obtains the Fermi or Fermi-Dirac distribution

$$g(E) = \frac{1}{\exp(E/k_{\mathrm{B}}T) + 1} \tag{A.42}$$

where bosons ($n = 0, 1, 2, \ldots, \infty$) obey the Bose, Bose-Einstein, or Planck distribution

$$g(E) = \frac{1}{\exp(E/k_{\mathrm{B}}T) - 1} \tag{A.43}$$

At low temperatures, both (A.42) and (A.43) exhibit pronounced deviations from the classical Boltzmann or Maxwell-Boltzmann distribution, $g(E) = \exp(-E/k_{\mathrm{B}}T)$.

If the number of particles is fixed, $\langle \mathsf{n} \rangle = \int n(E)\, g(E)\, D(E)\, \mathrm{d}E = N$, then the energies E must be adjusted by a chemical potential E_{o}, $E \to E - E_{\mathrm{o}}$. Examples of systems with fixed particle numbers are electrons in metals (fermionic), and superconducting electron pairs and superfluid helium (bosonic). For fermions, $g(E)$ is smaller than the classical value $\exp(-E/k_{\mathrm{B}}T)$, and additional levels must be provided to accommodate the particles. As a consequence, $E_{\mathrm{o}} = E_{\mathrm{F}}$ is positive and—for metallic electron densities—very large (Section 2.4.1), and metallic electrons cannot be treated classically. Bosonic systems are able to accommodate more particles than classical systems, and the corresponding mass-dependent chemical potential is negative or zero (Wannier 1966). However, even for $E_{\mathrm{o}} = 0$, the integral $\langle \mathsf{n} \rangle$ may be too small to accommodate all particles, and the remainder then forms a Bose-Einstein condensate of zero energy. Quasiparticles, such as phonons (sound waves) and magnons (spin waves), do not obey particle-number conservation and are described by $E_{\mathrm{o}} = 0$.

A.3.5 Relativistic quantum mechanics

Phenomena such as spin-orbit coupling and anisotropy are essentially relativistic. In a nutshell, relativistic physics means that the considered quantities exhibit four-vector symmetry. One example is the propagation of light, $x^2 + y^2 + z^2 = c^2t^2$, where the introduction of the four-vector $\mathbf{X} = (x, y, z; ct)$ yields the highly symmetric form $\mathbf{X}^2 = 0$. The only difference between space and time coordinates is a factor -1 in the scalar product (Minkowski metric). This four vector is sometimes written as $(x, y, z, \mathrm{i}ct)$, but i may give rise to confusion with the imaginary part of wave functions and operators. Other examples are the four-vector of momentum and kinetic energy, $\mathbf{P} = (p_\mathrm{x}, p_\mathrm{y}, p_\mathrm{z}, E/c)$, and electromagnetic fields (see below). The kinetic energy obeys $\mathbf{P}^2 = -m^2c^2$, where m is the mass of the particle (rather than the magnetic moment). The energy can also be written as

$$E = \pm\sqrt{m^2c^4 + p^2c^2} \tag{A.44a}$$

where the respective sign $\pm$ corresponds to matter and antimatter. Considering the positive sign (matter), we can expand this equation and obtain

$$E = mc^2 + \frac{1}{2}mv^2 - \frac{1}{8}\frac{v^2}{c^2}mv^2 \pm \cdots \tag{A.44b}$$

with $v = p/m$. The first three terms (b) are the rest energy, the classical kinetic energy, and the lowest-order relativistic correction to the kinetic energy.

It is easy to see that the Schrödinger equation violates relativistic invariance. For example, time and space coordinates appear as first and second derivatives, respectively, which is at odds with the four-vector symmetry of space-time. To find a relativistic generalization of the Schrödinger equation, we start from the four-vector equivalent of the nabla operator, $\nabla_4 = (\nabla_\mathrm{x}, \nabla_\mathrm{y}, \nabla_\mathrm{z};\ c^{-1}\partial/\partial t)$. A natural choice is to start from $\mathbf{P}^2 = -m^2c^2$ and to exploit that $\mathbf{P} = -\mathrm{i}\hbar\nabla_4$. The result is the *Klein-Gordon* equation

$$\nabla^2\psi - \frac{\partial^2}{c^2\partial t^2}\psi = \frac{m^2c^2}{\hbar^2}\psi \tag{A.45}$$

This relativistic wave equation describes bosons of mass m, for example mesons. The solutions of (A.45) involve an interaction length $1/\kappa = \hbar/mc$. Due to the smallness of $\hbar$, this length is usually subatomic, but photons exhibit $m = 0$ and $1/\kappa = \infty$. Note that (A.45) is reminiscent of micromagnetic equations, such as (4.29), and $1/\kappa$ is essentially a Bloch-wall width.

Fermions, such as electrons, are described by the *Dirac equation*, which is relativistically invariant but linear rather than quadratic in $\partial/\partial t$. If our real space were one-dimensional, $\nabla = \partial/\partial x$, we could use $c^2\nabla^2 - \partial^2/\partial t^2 = (c\nabla + \partial/\partial t)(c\nabla - \partial/\partial t)$ to transform the second-order differential equation (A.45) into a set of two first-order equations. The two operators $c\nabla \pm \partial/\partial t$ and wave functions $\psi_\pm$ correspond to matter and antimatter, respectively. Unfortunately, the three-dimensional nabla operator is a vector, and $c\nabla$ cannot be added to a scalar. It is therefore necessary to generalize $a^2 - b^2 = (a + b)(a - b)$ to three-dimensional vectors $\mathbf{b}$, a problem solved by Hamilton in the 1840s. He introduced algebraic units i, j, and k (quaternions) which obey angular-momentum relations such as $\mathrm{k} = \mathrm{ij}$ and $\mathrm{ij} = -\mathrm{ji}$. These units are generalizations of the imaginary unit i, created by the introduction of a unit j, so that—in our space-time formulation

$$X = ct + \mathrm{i}\,x + (y + \mathrm{i}\,z)\mathrm{j} \tag{A.46a}$$

$$\mathrm{X} = ct + \mathrm{i}\,x + \mathrm{j}y + \mathrm{k}\,z \tag{A.46b}$$

Basically, i, j, and k correspond to $\mathbf{e}_\mathrm{x}$, $\mathbf{e}_\mathrm{y}$, and $\mathbf{e}_\mathrm{z}$, and the quaternion algebra is the reason for the frequent occurrence of cross products in electromagnetism.

In modern notation, the quaternions correspond to the vector formed by the Pauli matrices σ_x, σ_y, and σ_z (2.32, 6.10), and

$$a^2 - \mathbf{b}^2 = (a + \boldsymbol{\sigma} \cdot \mathbf{b})(a - \boldsymbol{\sigma} \cdot \mathbf{b}) \tag{A.47}$$

The Dirac equation is obtained by applying this equation to $\mathsf{H}^2 = m^2c^4 + c^2\mathbf{p}^2$, where $\mathbf{p} = -\mathrm{i}\hbar\nabla$. There are various representations, all related to each other by unitary transformations. Straightforward application of (A.47) to $m^2c^4\psi = (\mathsf{H}^2 - c^2\mathbf{p}^2)\psi$ yields

$$m\,c^2\psi = (\mathsf{H} + c\,\boldsymbol{\sigma} \cdot \mathbf{p})\,\psi' \tag{A.48a}$$

$$m\,c^2\psi' = (\mathsf{H} - c\,\boldsymbol{\sigma} \cdot \mathbf{p})\psi \tag{A.48b}$$

Since $\boldsymbol{\sigma}$ has the character of a 2×2 matrix, the total wave function has the four components ($\psi\uparrow$, $\psi\downarrow$, $\psi'\uparrow$, $\psi'\downarrow$). These wave functions are referred to as spinors. To simplify the interpretation of the wave function, it is convenient to use the Dirac equation in the form

$$\mathsf{H}\,\phi = +mc^2\phi + c\boldsymbol{\sigma} \cdot \mathbf{p}\chi \tag{A.49a}$$

$$\mathsf{H}\,\chi = -mc^2\chi + c\,\boldsymbol{\sigma} \cdot \mathbf{p}\phi \tag{A.49b}$$

Here ϕ and χ describe electrons and positrons of rest energies $\pm mc^2$, respectively.

To describe the interaction with electric and magnetic fields, it is necessary to add the correponding four-vector potential. The electric component is essentially equal to the electrostatic potential, as used in the classical Schrödinger equation (A.29), whereas the magnetic component is the three-dimensional vector potential $\mathbf{A}$ (Section A.4.3). For small relativistic corrections $p/m = v \ll c$, one can start from $\mathsf{H}\phi = +mc^2\phi$ and treat the other terms in (A.49) perturbatively, for example by iteration. The result can be cast into a form similar to (A.49b), known as Pauli expansion (Jones and March 1973). The lowest-order electrostatic contributions are of order $v^2/c^2 \approx \alpha^2$ and correspond to the kinetic and electrostatic energies of electrons in solids. The leading magnetic terms in this expansion are of order v^4/c^4. They include the Zeeman energy proportional to $\mathbf{H} \cdot (\mathsf{L} + 2\mathsf{S})$ notation and the spin-orbit coupling proportional to $\mathsf{L} \cdot \mathsf{S}$. Other higher-order corrections are the Dzyaloshinski-Moriya (DM) interaction and the numerical value

$$g = 2(1 + \alpha/2\pi - 0.301\alpha^2) = 2.0023 \tag{A.50}$$

of the g-factor, which is usually approximated by $g = 2$. Note that (A.50) is unrelated to quenching.

A.4 Electromagnetism

A.4.1 Maxwells equations

Maxwells equations summarize several important findings in electromagnetism. They relate the magnetization $\mathbf{M}$, the magnetic field $\mathbf{H}$, the magnetic flux density $\mathbf{B}$, the electric field $\mathbf{E}$, the electric displacement $\mathbf{D}$, and the electric polarization $\mathbf{P}$ to each other. The also show how these fields are linked to the electric charge density ρ and to the corresponding current density $\mathbf{j}$. In differential form, the equations are

$$\nabla \cdot \mathbf{D} = \rho \tag{A.51a}$$

$$\nabla \cdot \mathbf{B} = 0 \tag{A.51b}$$

$$\nabla \times \mathbf{E} = -\frac{\partial \mathbf{B}}{\partial t} \tag{A.51c}$$

$$\nabla \times \mathbf{H} = \mathbf{j} + \frac{\partial \mathbf{D}}{\partial t} \tag{A.51d}$$

where the current density obeys the continuity equation $\mathrm{div}\,\mathbf{j} + \mathrm{d}\rho/\mathrm{d}t = 0$. Equation (A.51a) is Gauss's law, (A.51b) describes the absence of magnetic flux sources, (A.51c) is Faraday's law, and (A.51d) is Maxwells law, a generalization of Ampère's law (curl $\mathbf{H} = \mathbf{j}$).

In addition, the electromagnetic fields $\mathbf{E}$, $\mathbf{D}$, $\mathbf{H}$, and $\mathbf{B}$ obey the constitutive or *material* equations

$$\mathbf{D} = \varepsilon_o \mathbf{E} + \mathbf{P} \tag{A.52}$$

$$\mathbf{B} = \mu_o (\mathbf{H} + \mathbf{M}) \tag{A.53}$$

where ε_o and μ_o are the permeability and permittivity of vacuum, respectively. The electric and magnetic field energies are $\frac{1}{2}\int \mathbf{D} \cdot \mathbf{E}\,\mathrm{d}V$ and $\frac{1}{2}\int \mathbf{B} \cdot \mathbf{H}\,\mathrm{d}V$, respectively, and electromagnetic interactions of a charge q are described by $\mathbf{F} = q\mathbf{E} + q\mathbf{v} \times \mathbf{B}$. This equation can also be written as a force density, $\mathbf{f} = \rho\mathbf{E} + \mathbf{j} \times \mathbf{B}$. Finally, dissipative transport is described by Ohm's law $\mathbf{j} = \sigma\mathbf{E}$, where σ is the conductivity (inverse resistivity).

In contrast to widespread belief, the material equations are an essential aspect of electromagnetism. For example, replacing (A.53) by $\mathbf{B} = \mu\mathbf{H}$ discards the nonlinear and multivalued character of magnetic hysteresis, and expressing $\mathbf{M}$ in terms of atomic currents neglects the spin. It is often convenient to consider $\mathbf{M}$ as a response to an external field $\mathbf{H}$, without complicating the situation by the use of permeabilities and susceptibilities. Equations such as $\mathbf{B} = \mu\mathbf{H}$ and $\mathbf{M} = \chi\mathbf{H}$ are useful phenomenological approximations for para- and diamagnets, but applied to ferromagnets they tend to fail completely (see exercise on the susceptibility of iron).

It is often convenient to use Maxwell's equations in the form of integrals. For example, (A.51a) means that the electric charge density is the source, $\nabla\cdot$, of the electric displacement. Integrating over ρ and exploiting that Gauss's theorem yields

$\int \nabla \cdot \mathbf{D} \, \mathrm{d}V = \oint \mathbf{D} \cdot \mathrm{d}\mathbf{A}$ we obtain

$$\oint \mathbf{D} \cdot \mathrm{d}\mathbf{A} = Q \tag{A.54}$$

where $Q = \int \rho \, \mathrm{d}V$ is the charge enclosed by an arbitrary surface A. Consider, for example, the electric field created by a vacuum point charge Q. The symmetry of the problem is exploited by choosing a spherical surface of radius R. The surface normal $\mathbf{e}_\mathrm{A} = \mathbf{A}/|A|$ and $\mathbf{D}$ are then parallel and $4\pi R^2 D = Q$. In vacuum, $\mathbf{D} = \varepsilon_\mathrm{o}\mathbf{E}$ and we obtain the familiar result

$$\mathbf{E} = \frac{Q}{4\pi\varepsilon_\mathrm{o} R^3} \mathbf{R} \tag{A.55}$$

Similar simplications and interpretations exist for other Maxwell equations. For example, in the absence of dielectric changes, (A.51d) reduces to $\nabla \times \mathbf{H} = \mathbf{j}$. This means that electric currents create a curl ($\nabla\times$) of the magnetic field. To transform this differential equation into an integral we exploit Stokes's theorem (A2.4), which yields $\int \nabla \times \mathbf{H} \cdot \mathrm{d}\mathbf{A} = \oint \mathbf{H} \cdot \mathrm{d}\mathrm{l}$ and Ampère's law

$$\oint \mathbf{H} \cdot \mathrm{d}\mathbf{l} = \mathrm{I} \tag{A.56}$$

Here I is the current through the surface, that is, the current enclosed by the boundary curve. For a long air-cored solenoid, as shown in Fig. 1.4(a), the integration becomes very simple if performed along the curve C. Outside the solenoid, the field is small, so that $\oint \mathbf{H} \cdot \mathrm{d}\mathbf{l} = HL$ and $H = I/L$. Here H is the field inside the coil, L is the length of the solenoid and I is the total current in the coil, proportional to number of turns.

The field of solenoids with ferromagnetic cores is, essentially, limited by the saturation magnetization of the core material, that is, about 2 T for strong electromagnets. The material with the highest known room-temperature magnetization is the alloy $Fe_{65}Co_{35}$, with $\mu_\mathrm{o} M_\mathrm{s} = 2.43\,\mathrm{T}$. By comparison, the geomagnetic field is of the order of 0.05 mT (0.5 Oe), "fridge magnets" create a field of about 100 mT, and $Nd_2Fe_{14}B$ magnets have a remanence somewhat smaller than the saturation magnetization of 1.61 T. Higher fields are created by superconducting magnets and pulse-field sources.

Exploiting the symmetry of a problem is a standard approach in electromagnetism. It can also be used to investigate the other Maxwell equations, such as Faraday's law. Using Stoke's theorem we obtain the voltage $\oint \mathbf{E} \cdot \mathrm{d}\mathbf{l} = -\partial\Phi/\partial t$ in a current ring, where $\Phi = \int \mathbf{B} \cdot \mathrm{d}\mathbf{A}$ is the magnetic flux. The induced electric current creates a magnetic field that opposes the flux change. This law, named after nineteenth-century German physicist Heinrich Lenz, explains the diamagnetism created by atomic currents and the inductivity of electric circuits. An example is the inductive voltage $U = L_\mathrm{ind}\, \mathrm{d}I/\mathrm{d}t$, which defines the inductance L_ind. In a long air-cored solenoid $L_\mathrm{ind} = \mu_\mathrm{o} N^2 A/L$, where N is the number of turns. Physically, inductivity means that a current needs some time to create a magnetic field of energy density $\frac{1}{2}\mathbf{H} \cdot \mathbf{B}$. Since the energy W of a circuit is equal to $\int UI \, \mathrm{d}t$, we can write $W = \int L_\mathrm{ind} I \, \mathrm{d}I$ and obtain $W = \frac{1}{2} L_\mathrm{ind} I^2$. In solenoids containing a magnetic core, the inductance is field-dependent and enhanced by a factor $\mu_{\text{eff}}/\mu_{\text{o}} = 1 + \frac{1}{2}(M/H + \chi)$. In most cases $\mu_{\text{eff}} \neq \mu$, so that $L_{\text{ind}} = \mu N^2 A/L$ can *not* be used to determine the inductive voltage. An exception is the linear regime, where $\chi = \partial M/\partial H$ is equal to M/H. In high magnetic fields, M saturates, χ is negligible, and $\text{d}W = UI\,\text{d}t$ is smaller than predicted from μ.

A.4.2 Simple magnetostatic solutions

At surfaces, the magnetostatic equations $\nabla \cdot \mathbf{B} = 0$ and $\nabla \times \mathbf{H} = 0$ imply that the respective perpendicular flux-density and parallel field components $B_\perp$ and $H_{||}$ remain unchanged (Fig. 3.4). This is exploited in the calculation of demagnetizing fields for flat and elongated ellipsoids of revolution and toroids.

The linear character of Maxwell's equations means that solutions can be superposed and decomposed. This feature is of limited practical use, because $M(H)$ tends to be nonlinear, but it makes it possible to rewrite complicated magnetization distributions in terms of magnetic poles. Putting $\mathbf{B} = \mu_{\text{o}}(\mathbf{H} + \mathbf{M})$ into $\nabla \cdot \mathbf{B} = 0$ yields

$$\nabla \cdot \mathbf{H} = \rho_{\text{M}} \tag{A.57}$$

where $\rho_{\text{M}} = -\nabla \cdot \mathbf{M}$ is the magnetic pole or magnetic charge density. The positive pole is at the tip of the magnetization vector that creates the field. It is also known as the red or magnetic north pole. The field lines go from + to −, that is, from the red pole to the blue pole.

The integral $\int \rho_{\text{M}}\,\text{d}V$ is the magnetostatic equivalent of the flux Φ created by a coil (Fig. 1.3). Cylindrical magnets and solenoidal coils of cross section $A = \pi R^2$ and length $L \gg R$ obey $\Phi = \mu_{\text{o}} H\text{A}$, and in both cases the dipole moment $m = \Phi L$. However, orbital currents in magnetic materials are not connected to the environment, so we can often restrict ourselves to $\nabla \cdot \mathbf{M}$ and use (A.57). An example of a magnetostatic problem involving external currents is the magnetic field created by a long wire (exercise in Chapter 1).

Magnetostatic and electrostatic phenomena can be treated on a common footing, because $\mathbf{H}$ and $\mathbf{E}$ are curl-free and created by charge densities. Without loss of generality, we will restrict ourselves to the magnetic case, where $\nabla \times \mathbf{H} = 0$ and $\nabla \cdot \mathbf{H} = \rho_{\text{M}}$. The first equation is solved by introducing a magnetic potential ϕ_{M}, so that $\mathbf{H} = -\nabla \phi_{\text{M}}$ and $\nabla^2 \phi_{\text{M}} = -\rho_{\text{M}}$. The solution of the latter equation contains two contributions, namely a potential ϕ_{o} describing the divergenceless external field, $\mathbf{H}_{\text{o}} = -\nabla \phi_{\text{o}}$, and the potential created by the magnetic charges,

$$\phi_{\text{M}}(\mathbf{r}) = \frac{1}{4\pi} \int \frac{\rho_{\text{M}}(\mathbf{r}')}{|\mathbf{r} - \mathbf{r}'|} \text{d}V' \tag{A.58}$$

For uniformly magnetized bodies, the only contribution to $\rho_{\text{M}} = -\nabla \cdot \mathbf{M}$ is from the surface, and (A.58) acquires the character of a surface integral involving $\mathbf{M} \cdot \mathbf{n}$, where $\mathbf{n}$ is the surface normal. However, it is generally convenient to consider $\mathbf{M}$ rather than ρ_{M}. Exploiting the fact that the volume integral over $\nabla \cdot (\mathbf{M}(r)/r) = \nabla \cdot \mathbf{M}/r - \mathbf{M} \cdot \mathbf{r}/r^3$ vanishes for finite magnetic bodies, we obtain

$$\phi_{\text{M}}(\mathbf{r}) = \frac{1}{4\pi} \int \frac{(\mathbf{r} - \mathbf{r}') \cdot \mathbf{M}(\mathbf{r}')}{|\mathbf{r} - \mathbf{r}'|^3} \text{d}V' \tag{A.59}$$

From this equation, the magnetostatic field is obtained as $\mathbf{H} = -\nabla \phi_{\text{M}}$.

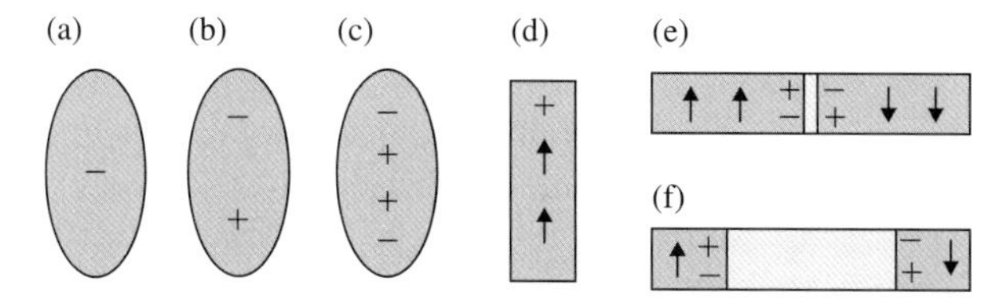

Fig. A.1 Multipole moments: (a) monopole moment, (b) dipole moment, and (c) quadrupole moment. The remaining configurations are magnetostatic approximations: (d) monopole approximation for a long ferromagnetic rod, (e) quadrupole moment associated with a narrow domain wall, and (f) two dipoles separated by a broad domain wall. The configurations (e) and (f) can be interpreted as side views on magnetic thin films with perpendicular magnetization and film thicknesses $t \ll \delta_W$ (e) and $t \gg \delta_W$ (f).

For magnetic charge distributions centered around $\mathbf{r} = 0$, the term $1/|\mathbf{R} - \mathbf{r}|$ can be expanded into powers of $\mathbf{r}$. For $r < R$, the multipole expansion is based on the mathematical identity

$$\frac{1}{\sqrt{R^2 + r^2 + 2Rr\cos\theta}} = \frac{1}{\mathrm{R}} \sum_{\mathrm{m}=0}^{\infty} \frac{r^{\mathrm{m}}}{R^{\mathrm{m}}} P_{\mathrm{m}}(\cos\theta) \tag{A.60}$$

where the P_{m} are the Legendre polynomials, and $\cos\theta = \mathbf{r} \cdot \mathbf{R}/rR$.

Putting (A.60) into (A.59) yields $\phi_{\mathrm{M}}(\mathbf{r})$ as a sum of multipole contributions,

$$\phi_{\mathrm{M}}(\mathbf{R}) = \frac{1}{4\pi\,\mathrm{R}} \sum_{\mathrm{m}=0}^{\infty} \frac{Q_{\mathrm{m}}}{R^{\mathrm{m}}} \tag{A.61}$$

where $Q_{\mathrm{m}} = \int \rho_{\mathrm{M}}(\mathbf{r}')r'^{\mathrm{m}} P_{\mathrm{m}}(\cos\theta)\,\mathrm{d}V'$ is the m-th multipole moment.

Figure A.1 shows some multipole charge distributions. The monopole moment $Q_0 = \int \rho_{\mathrm{M}}(\mathbf{r})\,\mathrm{d}V$, illustrated in (a), is equal to the total charge. The dipole distribution (b) and the quadrupole distribution (c) exhibit $Q_0 = 0$ (charge neutrality), but their higher-order moments are nonzero. In (b), the dipole moment $Q_1 = \int \rho(\mathbf{r})\cos\theta\,\mathrm{d}V$ is nonzero, whereas in (c) the lowest-order nonzero moment is the quadrupole moment Q_2. The actual calculation of multipole moments may be quite cumbersome. Many charge distributions are asymmetric, and even a rotational symmetry axis is not necessarily parallel to $\mathbf{R}$. A very simple approach to the derivation of the dipole moment is to add the monopole potentials $\phi_0(r) = q/4\pi r$ for two charges $+q$ and $-q$ located at $(\mathbf{R} + (\mathbf{a}/2))$ and $(\mathbf{R} - (\mathbf{a}/2))$, respectively. The result is the dipole moment qa or, in vector notation, $q\mathbf{a}$. The corresponding magnetic moment can also be written as $m = \Phi a$, where Φ is the pole strength (see above).

Since there are no free magnetic monopoles, the magnetization can be described as a superposition of dipoles. This is, of course, the origin of the *magnetic moment.* However, monopoles are used in some approximations, such as the description of long magnetic rods as two monopoles of opposite charge located at the ends of the rod, as in Fig. A.1(d). This is possible because north and south poles cancel each other in the middle of the rod. As shown in Chapter 3, electrostatic quadrupole moments

($m=2$) describe the magnetocrystalline anisotropy constant K_1. Second- and third-order anisotropy constants reflect electrostatic hexadecapoles ($m=4$) and hexacontatetrapoles ($m=6$), respectively. Octupole moments ($m=3$) are important in several areas of physics and describe, for example, higher-order vibrations in nuclei.

Multipole potentials are of order $1/r^{\mathrm{m}}$, where r is the distance from the charge distribution. Since the field is a real-space derivative of the potential, multipole fields scale as $1/r^{\mathrm{m}+2}$. Another interesting aspect of multipoles is the *spin* of the associated waves, which is equal to m. Photons are created by dipoles (antennae) and have the spin 1, whereas gravitational waves are created by quadrupoles, such as stars orbiting around each other, and have the spin 2. This feature is closely related to the *symmetry* of the multipoles: m-fold rotational symmetry, that is, invariance under a rotation by angle $2\pi/m$, corresponds to a spin m. For dipoles and quadrupoles, the angles are 360° and 180°, respectively. For electron wave functions (spinors), the unitary rotation matrix (2.35) yields an angle of 720°, corresponding to a spin of 1/2.

Magnetic fields acting on magnetic moments give rise to mechanical torques and forces. This is exploited in many applications, from permanent and soft magnets to experimental techniques such as magnetic force microscopy (MFM, see Fig. 4.13). The underlying energy contribution is the Zeeman energy $E=-\mu_{\mathrm{o}}\mathbf{m}\cdot\mathbf{H}$. To derive the *torque* $T=-\partial E/\partial\theta$, we apply a field in the z-direction and write the z-component of the moment as $m\cos\theta$ and obtain $T=\mu_{\mathrm{o}}mH\sin\theta$. For arbitrary angles, the torque $\mathbf{T}=\mu_{\mathrm{o}}\mathbf{m}\times\mathbf{H}$. An expression for the *force* is obtained by putting a moment $\mathbf{m}=m\mathbf{e}_{\mathrm{z}}$ in a magnetic field $H_{\mathrm{z}}(z)$, so that $F=-\partial E/\partial z$ and $F=-\mu_{\mathrm{o}}m\partial H/\partial z$. This means that the force is proportional to the field *gradient* and therefore zero in a uniform magnetic field.

A.4.3 Simple dynamic solutions

The time dependence of the Maxwell's equation gives rise to electromagnetic waves and to time-dependent interactions with magnetic and dielectric matter. Here we focus on two cases: electromagnetic waves in vacuum and eddy currents in metals.

In free space, Maxwell's equations become $\nabla\cdot\mathbf{E}=0$, $\nabla\cdot\mathbf{H}=0$,

$$\nabla\times\mathbf{E}=-\mu_{\mathrm{o}}\frac{\partial\mathbf{H}}{\partial t} \tag{A.62a}$$

and

$$\nabla\times\mathbf{H}=\frac{\varepsilon_{\mathrm{o}}\partial\mathbf{E}}{\partial t} \tag{A.62b}$$

Substituting (A.62a) into the time derivative of (A.62b) yields $\nabla\times\nabla\times\mathbf{E}=-\mu_{\mathrm{o}}\varepsilon_{\mathrm{o}}\partial\mathbf{E}^2/\partial t^2$ and, with $\nabla\times\nabla\times\mathbf{E}=\nabla(\nabla\cdot\mathbf{E})-\nabla^2\mathbf{E}$ and $\nabla\cdot\mathbf{E}=0$,

$$\nabla^2\mathbf{E}-\frac{1}{c^2}\frac{\partial\mathbf{E}^2}{\partial t^2}=0 \tag{A.63}$$

where $c^2=1/\varepsilon_{\mathrm{o}}\mu_{\mathrm{o}}$. Equation (A.63) means that the electric field propagates as a wave whose speed is $c=(\varepsilon_{\mathrm{o}}\mu_{\mathrm{o}})^{-1/2}\approx3\times10^8$ m/s. The same is true for $\mathbf{H}$ and for the respective scalar and vector potentials V and $\mathbf{A}$, defined by $\mathbf{E}=-\nabla V-\partial\mathbf{A}/\partial\mathrm{t}$

and $H = \nabla \times A/\mu_o$. Note that the electromagnetic wave equation (A.63) is relativistic (four-vector symmetric), as are the electromagnetic potentials (V and $\mathbf{A}$) and the Maxwell equations as a whole.

Abandoning the assumption of vacuum, we obtain a variety of additional magnetooptical effects, whose description goes beyond the scope of this book. One example is the index of refraction, n. It obeys $n^2 = \varepsilon_r \mu_r$ and has two solutions, $n = \pm(\varepsilon_r \mu_r)^{-1/2}$. This is important when ε_r, μ_r, or both are negative, which is often the case near resonances of atomic origin. When only one constant is negative, then n is imaginary, and the waves are strongly damped. However, if both are negative, then the index of refraction is negative (NIR), with a variety of interesting consequences. For example, the angle of refraction changes sign (Shelby, Smith, and Shultz 2001).

Another example is the skin effect in metals, where $\nabla \times \mathbf{H} = \mathbf{j} = \sigma \mathbf{E}$. In the weak-current limit, that is, far away from magnetic saturation, $\mathbf{B} = \mu_o \mu_r \mathbf{H}$, and we obtain

$$\nabla \times (\nabla \times \mathbf{H}) = \mu_o \mu_r \sigma \frac{\partial \mathbf{H}}{\partial t} \tag{A.64}$$

This equation describes the damping of electromagnetic waves in a metallic medium. Dimensional analysis of (A.64) yields a penetration or skin depth $\delta R \approx 1/(\mu_o \mu_r \sigma \omega)^{1/2}$. Physically, electromagnetic waves create currents which are dissipated due to the metallic resistivity $1/\sigma$. After a sudden change of the magnetic field, this gives rise to a magnetic aftereffect of relaxation time $1/\omega$, although this effect is far too small to explain magnetic viscosity.

A.5 Magnetic materials

There are various methods of classifying magnetic materials, based on criteria such as spin structure (ferromagnets vs. ferrimagnets and other structures), electronic structures (for example oxides and transition-metal alloys), real structure (thin films, polycrystalline magnets, nanostructures, sintered alloys), and applications (such as permanent magnets, recording materials, soft magnets, and in various kinds of specialty materials). There are several introductions to general magnetism and magnetic materials, such as Crangle (1991), Craik (1995), Jiles (1998), Skomski and Coey (1999), O'Handley (2000), Spaldin (2003), Liu, Sellmyer, and Shindo (2006), and the purpose of this appendix is to provide a brief introduction.

One criterion is the *magnetic hardness*, as parameterized by the coercivity. The division of ferromagnets into "hard" and "soft" magnets has its origin in mechanical and magnetic hardness of steel, as compared to pure iron. This hardness is due to carbon and other additives, which improve the mechanical hardness and act as pinning centers (Section 4.3.2). *Permanent magnets* (hard magnets) have moderate to high coercivities, usually above $0.25\,\mathrm{T} = 2.5\,\mathrm{kOe} \approx 200\,\mathrm{kA/m}$. They are used for a wide range of applications, such as electromotors, loudspeakers, windscreen wipers, locks, microphones, and toy magnets. They play an important role in computer technology, because hard-disk drives also contain permanent magnets (for rapid mechanical data access). Typical *soft magnets* have coercivities below 1 mT, down to less than 0.0001 mT for supermalloy and some amorphous magnets. They are widely used for flux guidance in permanent-magnet and other systems, in transformer cores, for high-frequency and microwave applications, and in recording heads. Aside from the low coercivity, soft magnets often require high initial permeabilities $\mu_r \sim \partial M/\partial H$ as high as 1000. At high frequencies, eddy-current losses become a serious consideration, and a low electrical conductivity is often more important than a low coercivity.

Magnetic-recording materials are used in audio-visual technology, such as magnetic tapes, on credit cards, and, most importantly, in magnetic hard-disk drives. They are traditionally considered as semihard magnets, where well-defined loop shapes and intermediate coercivities yield stable information storage without requiring powerful and bulky writing facilities. However, the coercivities of some recently developed materials, such as $L1_0$-based recording media, are fairly high. This reflects the need for high anisotropy, to ensure thermal stability (K_1V) in the limit of ultrahigh recording densities (small V). Note that bit sizes are already in the submicron region and will approach the superparamagnetic limit in the foreseeable future.

Other applications are ferrofluids and other fine-particle magnets (Section 6.4.6), magnetostrictive materials, used in ultrasonic and other devices, and spin electronics, aiming at the exploitation of the spin as an additional degree of freedom in various types of electronic devices. The last category includes sensor applications of various kinds (giant magnetoresistive or GMR devices consisting of multilayered and granular materials, nanojunctions), magnetic semiconductors, spin transistors, and materials for distant-future quantum computing.

Table A.1 shows intrinsic magnetic and structural properties for a variety of ferri- and ferromagnetic substances. The table contains both industrially important magnets, such as $Nd_2Fe_{14}B$, and reference materials of primarily scientific interest, such $Y_2Fe_{14}B$. The structures mentioned in the table are: bcc (body-centered cubic), hcp (hexagonal close packed), fcc (face-centered cubic), NiAs (hexagonal), cubic ferrite ($MgAl_2O_4$), *hexagonal ferrite* (M ferrite), garnet (cubic), spinel (disordered spinel, cubic), rutile (tetragonal, TiO_2), $L1_0$ (tetragonal), 1:5 (hexagonal, $CaCu_5$), 2:14:1 (tetragonal, $Nd_2Fe_{14}B$), 1:12 (tetragonal, $ThMn_{12}$), and 2:17 (rhombohedral, Th_2Zn_{17} for the light rare earths and hexagonal Th_2Ni_{17} for the heavy rare earths and Y).

Table A.2 shows micromagnetic and extrinsic magnetic properties for a variety of substances and materials. By definition, substances (elements and compounds) differ from materials by the requirement of chemically purity, but the term material is widely used in the magnetism community to denote both substances and materials. In Table A.1, the basic crystal structures of some of the materials differ from those of the corresponding compound in the first column. These materials are: a-$Fe_{80}B_{20}$ (amorphous), PtCo ($L1_0$), CrO_2 (rutile), Mn-Zn ferrite (cubic ferrite), and sintered 2:17/1:5 Sm-Co magnets (2:17 cells surrounded by 1:5 grain boundaries).

A.5.1 Transition-metal elements and alloys

Iron-based magnets, such as steels, are widely-used magnetic materials. Until the first half of the twentieth century, most permanent magnets were made from steel, and due to the low coercivity of the magnets it was necessary to resort to cumbersome horseshoe shapes. Permanent magnets made from steels are now obsolete, but the high saturation magnetization of $Fe_{65}Co_{35}$ and its pronounced temperature stability continue to be

Table A.1 Intrinsic properties of some ferri- and ferromagnetic elements and compounds. M_s and K_1 are room-temperature values.

Substance	$\mu_o M_s$ (RT) T	T_c K	K_1 (RT) MJ/m^3	Structure*
Fe	2.15	1,043	0.048	bcc
Co	1.76	1,388	0.53	hcp
Ni	0.62	631	−0.0048	fcc
PtCo	1.00	840	4.9	$L1_0$
PtFe	1.43	750	6.6	$L1_0$
PdFe	1.37	760	1.8	$L1_0$
Fe_3O_4	0.60	858	−0.011	cubic ferrite
$MnFe_2O_4$	0.52	573	−0.0028	cubic ferrite
$CoFe_2O_4$	0.50	793	0.270	cubic ferrite
$NiFe_2O_4$	0.34	858	−0.0069	cubic ferrite
$BaFe_{12}O_{19}$	0.48	723	0.330	hex. ferrite
$SrFe_{12}O_{19}$	0.46	733	0.35	hex. ferrite
CrO_2	0.56	390	0.025	rutile
γ-Fe_2O_3	0.47	863	−0.0046	spinel
$Y_3Fe_5O_{12}$	0.16	560	−0.00067	garnet
$Sm_3Fe_5O_{12}$	0.17	578	−0.0025	garnet
$Dy_3Fe_5O_{12}$	0.05	563	−0.0005	garnet
$NdCo_5$	1.23	910	0.7	1:5
$SmCo_5$	1.07	1,003	17.0	1:5
YCo_5	1.06	987	5.2	1:5
$Pr_2Fe_{14}B$	1.41	565	5.6	2:14:1
$Nd_2Fe_{14}B$	1.61	585	5.0	2:14:1
$Sm_2Fe_{14}B$	1.49	618	−12.0	2:14:1
$Dy_2Fe_{14}B$	0.67	593	4.5	2:14:1
$Er_2Fe_{14}B$	0.95	557	−0.03	2:14:1
$Y_2Fe_{14}B$	1.36	571	1.06	2:14:1
$Sm(Fe_{11}Ti)$	1.14	584	4.9	1:12
Sm_2Co_{17}	1.20	1,190	3.3	2:17
Sm_2Fe_{17}	1.17	389	−0.8	2:17
$Sm_2Fe_{17}N_3$	1.54	749	8.9	2:17

*For explanations, see main text.

exploited in alnico permanent magnets. The moderate coercivity of alnico magnets reflects the shape anisotropy of elongated $Fe_{65}Co_{35}$ particles embedded in a Ni-Al matrix. Along with oxides (see below), Fe-, Co-, and Ni-based particles, thin films, and nanostructures are also used in applications such as sensors and ferrofluids. For example, Fe-Rh-N alloys are used in read heads.

Soft magnets are often made from Fe-Co and other iron-containing metallic magnets, such as Fe-Si and permalloy. Permalloys are Ni-Fe alloys and exist in various compositions and modifications, such as 78 permalloy ($Ni_{78}Fe_{22}$) and $Ni_{79}Fe_{16}Mo_5$

Table A.2 Micromagnetic and extrinsic properties at room temperature. Since extrinsic properties are strongly real-structure dependent, the last three columns are merely a rough guide. The distinction between substances and materials, as well as the listed materials, are explained in the main text.

Compound	A $\frac{\text{pJ}}{\text{m}}$	K_1 $\frac{\text{MJ}}{\text{m}^3}$	δ_B nm	γ $\frac{\text{mJ}}{\text{m}^2}$	l_o nm	R_{sd} nm	H_a T	Material	H_c mT	M_r T	M(H) loop (see below)
Fe	8.3	0.05	40	2.6	0.12	6	0.06	Fe (unalloyed)	0.07	0.9	6,000*
								carbon steel	1.8	1.0	0.9**
								cobalt steel	20	0.9	6**
								$Fe_{96}Si_4$	0.05	1.0	7,000*
								alnico (anis.)	70	1.30	50**
								a-$Fe_{80}B_{20}$	0.007	1.0	100,000*
Co	10.3	0.53	14	9.3	0.46	34	0.76	Co (unalloyed)	5	1.6	1,000*
								PtCo	300	0.64	76**
Ni	3.4	−0.005	82	0.5	0.13	16	0.03	Ni (unalloyed)	0.3	0.35	2,400*
								permalloy	0.005	0.6	100,000*
								supermalloy	0.0003	0.5	1,000,000*
$BaFe_{12}O_{19}$	6.1	0.33	14	5.7	1.37	290	1.8	sintered	250	0.39	28**
								CrO_2	50	0.25	10*
								Mn-Zn ferrite	0.03	0.15	2,500***
$SmCo_5$	22.0	17	3.6	77	4.35	764	40	metal-bonded	1,880	0.92	175**
								polymer-bonded	1,000	0.58	60**
								sintered 2:17/1:5	800	1.08	225**
$Nd_2Fe_{14}B$	7.7	4.9	3.9	25	1.54	107	7.6	sintered	1,600	1.33	400**
								polymer-bonded	750	0.55	48**

The last column displays typical hysteresis-loop properties:
*Dimensionless relative permeability $\mu_{max} = 1 + \chi_{max}$.
**Energy product $(BH)_{max}$, measured in kJ/m^3.
***Dimensionless high-frequency permeability μ_{max}, measured at 2.5 MHz.

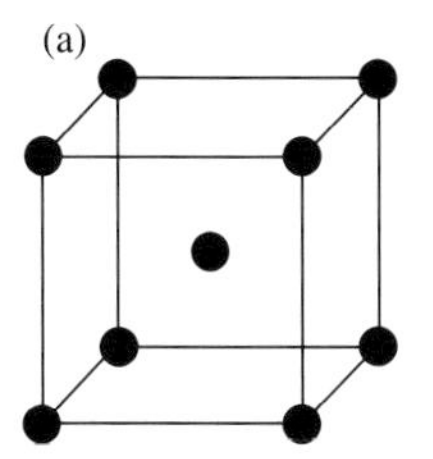

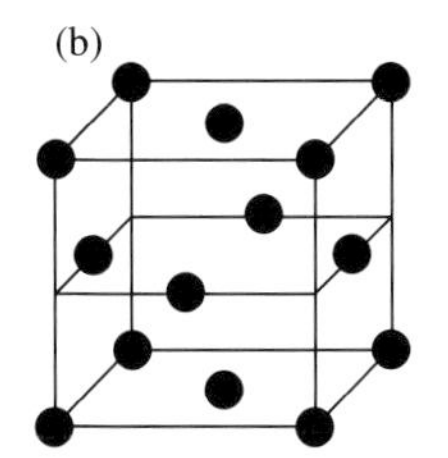

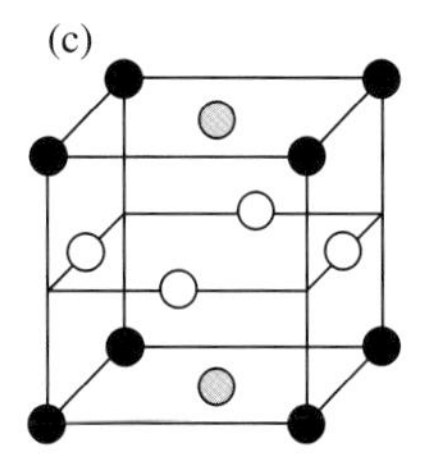

Fig. A.2 Crystal structures: (a) bcc, (b) fcc, and (c) $L1_0$.

(supermalloy, Table A.2). $Ni_{80}Fe_{20}$ has an anisotropy of about 0.15 kJ/m^3, an anisotropy field of about 0.4 mT, and a typical coercivity of order 0.04 mT [0.4 Oe]. By tuning the Ni-Fe ratio, it is possible to reduce the magnetocrystalline anisotropy to zero, but one must then consider the magnetostrictive anisotropy (Section 3.5.1). Since there is always some residual mechanical strain, the minimization of the hysteresis requires additional fine tuning (chemistry and processing). Another iron-based soft-magnetic material is sendust (Fe-Si-Al), which has a coercivity of about 0.025 mT. More recently, amorphous and nanostructured metals have attracted much attention as soft-magnetic materials. Examples are nanostructured Fe-Si-B-Cu-Nb (Yoshizawa, Oguma, and Yamauchi: 1988) and soft-magnetic glasses, such as $Fe_{40}Ni_{38}B_{18}Mo_4$, where $H_c \approx 0.0008$ mT and $\mu_{max} \approx 800{,}000$, and amorphous $Fe_{80}B_{20}$.

Figure A.2 shows typical crystal structures of metallic transition-metal magnets. Fe and Ni crystallize in the cubic bcc and fcc phases, respectively, whereas the room-temperature equilibrium phase of Co is hexagonal (hcp). The hcp phase is closely related to the fcc phase, differing by the stacking of the atomic planes along the cube diagonal. Equiatomic alloys such as PtCo, PtFe, and PdFe crystallize in the tetragonal $L1_0$ structure, Fig. A.2(c). The most general realization of the $L1_0$ structure has three nonequivalent crystallographic sites, 1a (black), 1c (gray), and 2e (white). In equiatomic $L1_0$ compounds, the two atomic species form layers, that is, the black and gray sites are occupied by the same kind of atom. The c-over-a ratio of magnetic $L1_0$ magnets is close to one, but the strongly uniaxial chemical environment (layered structure) makes the magnets strongly anisotropic. The result is a pronounced uniaxial anisotropy, largely but not completely due to the 5d (Pt) or 4d (Pd) sublattice.

In the middle of the twentieth century, $L1_0$ magnets were used as permanent, but they are now largely replaced by the less expensive and better performing rare-earth magnets. However, PtFe (or FePt) continues to be of interest in magnetic recording, as are other transition-metal magnets. A traditional recording material $Co_{80}Cr_{20}$, with coercivities of order 0.1 T [1 kOe]. Recent high-density recording media, characterized by more than 10 Gb/in^2 (1.55 Gb/cm^2), are based on materials such as Co-Cr-Pt-B, where the Pt improves the anisotropy. However, magnetic recording is a developing field, and new materials are constantly being considered. One question is to realize thermal stability (large K_1V) while ensuring writability. This includes not only the chemical composition of the media but also the nanostructure, including the in-plane or perpendicular character of the anisotropy (longitudinal vs. perpendicular magnetic recording).

A.5.2 Magnetic oxides

The magnetizations of oxides tend to be significantly smaller than those of metallic Fe and Co based magnets (Table A.1). One reason the frequent occurrence of antiferromagnetic bonds, which makes most oxides ferri- or antiferromagnetic. Another reason is the large size of the oxygen anions, which dilutes the magnetization. This limits the applicability of oxides as permanent and soft magnets. For example, magnetite or lodestone (Fe_3O_4) is no longer used as a permanent magnets, and a similar trend exists for the use of oxides in magnetic recording (CrO_2, α-Fe_2O_3). However, hexagonal ferrites, especially $BaFe_{12}O_{19}$ and $SrFe_{12}O_{19}$, are widely used as cheap but moderately performing permanent magnets.

Oxides continue to be of great importance in high-frequency applications, where eddy-current losses preclude the use of metals. Garnets, such as $Y_3Fe_5O_{12}$ and $Gd_3Fe_5O_{12}$, and other cubic oxides, such as $Ni_{1-x}Zn_xFe_2O_4$, are used in microwave and other high-frequency devices. Antiferromagnets cannot be used as traditional magnets but have found special applications, such as exchange-biasing in magnetic thin films. Magnetic perovskites (manganites) such as $(La_{0.7}Sr_{0.3})MnO_3$ have attracted interest as potential magnetoresistive sensors, as have half-metallic ferromagnets such as CrO_2 and the semi-Heusler alloy NiMnSb, where one spin channel is metallic and the other (nearly) insulating.

A.5.3 Rare-earth magnets

Today's high-performance permanent magnets are made from rare-earth transition-metal intermetallics, especially Nd-Fe-B and Sm-Co. There are different processing routes, nanostructures, and chemical compositions, but the underlying intermetallics are the tetragonal compound $Nd_2Fe_{14}B$ (Fig. 5.15) and the hexagonal compounds Sm_2Co_{17} and $SmCo_5$. The nitride $Sm_2Fe_{17}N_3$ (Section 7.2.6) is used in form of polymer-bonded magnets for niche applications. For suitably chosen rare earths (Section 3.4 and Table A.1), the rare-earth sublattice yields a large magnetocrystalline anisotropy, created by the crystal field acting on the tripositive rare-earth ions. By contrast, the magnetization is largely provided by the transition-metal sublattice (Fe or Co). The moments of heavy rare earths, such as Dy, are large but couple antiferromagnetic to the transition metal-sublattice. For this reason, permanent magnets contain light rare earths, especially Nd and Sm.

Since the magnetization contribution of the light rare-earth atoms is small and the energy product scales as the square of the magnetization, interest in permanent magnetism is limited to transition-metal-rich intermetallics. However, magnetization is a less important consideration for other applications. For example, cubic Laves-phase intermetallics such as $TbFe_2$ and $SmFe_2$ exhibit huge saturation magnetostrictions and are therefore used as magnetostrictive materials. Another example is magneto-optical recording, where one uses thin-film materials such as amorphous $Tb_{22}Fe_{66}Co_{12}$. The anisotropy of these many-sublattice materials is caused by growth-induced strain and relatively small. However, since $H_c \approx 2K_1/\mu_o M_s$, the low magnetization near a compensation point (Section 5.3.6) ensures a sufficiently high coercivity. The small net magnetization has no negative impact on the reading of the stored information, because the antiparallel sublattices yield different signal amplitudes, but the wavelength of light limits the achievable storage density.

A.6 Forgotten and reinvented

Rediscoveries and intermitted periods of ignorance occur in all branches of science, slow down scientific progress, and lead to a waste of resources. There are many reasons for the "cycles", ranging from accidental individual errors and poor literature search to collective rediscovery in adjacent fields or the wish to add "novelty" to incremental scientific progress. For example, it is common that one branch of magnetism discovers things that have been known in other branches for decades. When these findings are then given a new name, the confusion is complete, and it becomes hard to distinguish scientific progress from repetition. Examples of creative and partially amusing new names are vortex formation for curling, generalized Stoner-Wohlfarth theory for micromagnetism, superferromagnetism for exchange-interaction domains, and magnetic solitons for domain walls. A related issue is the use of high-powered numerical or complicated analytical methods to prove or disprove exact results that are conveniently derived by using simple models or approximations. As noted by de Gennes (1979) in the context of the Flory theory (Section 7.2.3), this approach is also dangerous because a better quantitative understanding of one aspect of a problem does not necessarily translate into a better understanding of the problem as a whole. In some cases, an "improved" model differs from the underlying simple model merely by an additional adjustable variable.

The following incomplete list contains findings that seem to be ignored or rediscovered every 15 to 20 years.

Anisotropy and coercivity are very different phenomena

Poorly understood hysteresis loops are sometimes interpreted as a manifestation of a new or unknown type of magnetic anisotropy. In fact, anisotropy and hysteresis are completely different phenomena of respective atomic-scale (intrinsic) and nanoscale (extrinsic) origins. Even for shape anisotropy, which is realized on a somewhat larger length scale, we can use the definition of magnetic anisotropy as the energy change on magnetization *rotation*. With the exception of coherent rotation, hysteresis involves nonuniform magnetization processes, and the determination of the local magnetization rotations goes far beyond the scope of magnetic anisotropy.

Magnetic interactions in thin-films structures are usually small

It is tempting but ill-based to assume that the magnetostatic interaction between two thin films with perpendicular magnetization, separated by a spacer layer, is ferromagnetic. The normal component $B_z = M_z + H_z$ is zero inside each layer, because $H_z = -DH_z$ and $D = 1$. By continuity of the normal component, $B_z = H_z = 0$ between the layers, so that the coupling is *zero* for infinite films. In reality, there is a small coupling due to features such as domain walls, as in Fig. A.1(e–f), and film edges. Similar considerations apply to films with in-plane magnetization. In ultrathin films, the ratio of film thickness to the lateral lengths of the film, such as feature sizes and domain-wall width, is usually small and sometimes translates into an exponential dependence of micromagnetic lengths such as domain sizes.

Surface anisotropy is unrelated to broken exchange bonds

Broken symmetry is a necessary but not sufficient condition for magnetic anisotropy, and the breaking of exchange bonds is unable to turn the isotropic Heisenberg model into an anisotropic model. The Heisenberg model is generically isotropic, whereas magnetocrystalline anisotropy is due to crystal field interactions and spin-orbit coupling. There is also no point in making assumptions such as easy axes perpendicular to the surface of a spherical particles. Surface anisotropy depends on the indexing of the surface planes and does not exhibit any kind of spherical symmetry.

Thermally activated magnetization reversal has been well understood since the 1930s

The name Néel-Brown theory for thermally activated magnetization reversal is somewhat unfortunate, because the theory was fully developed by the mid 1930s, as discussed in some detail by Becker and Döring (1939). Their description also contains the history of the logarithmic magnetic-viscosity law. Brown's paper (1963b) is actually a part of the second stage of the development, which was initiated by Kramers' approximate solution of the Fokker-Planck equation (1940). Néel's main contribution is perhaps the derivation of the $m = 3/2$ power law, which is closely related to the time-dependent equation (6.69) derived by Kneller (1966). This formula was rediscovered around 1990 and is sometimes referred to as the Sharrock equation (1994). Kneller also derived a relatively unimportant additional factor of about $\ln 2 = 0.693$, which is ignored in our simple derivation.

There is a big difference between single-domain and Stoner-Wohlfarth particles

The single-domain character of a particle means the absence of equilibrium domains, whereas Stoner-Wohlfarth behavior, or coherent rotation, refers to magnetization reversal (hysteresis). Magnetic fields may destroy or create magnetic domains, irrespective of the particle's single-domain character. For example, or most submicron particles used in permanent magnets and recording media are single-domain but reverse by incoherent mechanisms.

A very widespread magnetoresistive mechanism is anisotropic magnetoresistance

Anisotropic magnetoresistance (AMR) is a spin-orbit effect (Campbell, Fert, Jaoul 1970) and means that the electrons are scattered by aspherical charge distributions, such as those shown in Figs (3.8) and (5.23). Since the orientation of the electron clouds depends on the spin direction, this translates into a field dependence of the resistivity or resistance. Almost all materials exhibit some magnetocrystalline anisotropy and, therefore, some AMR. Until about 20 years ago, AMR was seen correctly as a main source of magnetoresistance in ferromagnets, including nanostructures, but then focus on GMR and CMR diverted attention from AMR and led to the discovery of a number of "new" and "unexplained" magnetoresistance contributions.

Figures of merit of magnetic materials refer to well-defined conditions

This point seems trivial, but very often one focuses on the zero-temperature performance of magnetic materials that are used primarily at or above room-temperature. In particular, the magnetic anisotropy exhibits a strong temperature dependence and is zero above the Curie temperature. There are many systems with Curie temperatures below room-temperature, including some with very high anisotropies, such as rare-earth and actinide compounds. In iron-series transition-metal systems, one may try to use nearly degenerate levels to realize huge anisotropies in 3d systems, but this only works at low temperature, because thermal excitations are very effective in overcoming small level splittings. Another example is that permanent magnets and recording media require large volume and areal densities, respectively. Approaches such as adding large amounts of other magnetic materials and the consideration of noninteracting particles may improve selected properties but deteriorates the overall performance of the material. An exception is the two-phase nanostructuring of permanent magnets, where the "filler" has a higher magnetization than the hard-magnetic skeleton.

Electron interactions are important in magnetism

The success of advanced density-functional electronic structure calculations has lead to the wide spread belief that interaction effects (correlations) are unimportant in magnetism or well-described by quantum-mechanical mean-field approaches such as LSDA+U. In fact, such calculations may crudely misinterpret the involved physics. One example is the prediction of narrow bands, as compared to localized states, in strongly correlated systems. Another example is spin-change separation, where the low-lying excitations involve spin changes that leave the charge (U) unaffected.

The Curie transition is an atomic phenomenon

The thermodynamic correlation length diverges at the critical point, but this does not lead to a proper description of macroscopic magnetism. Domains, domain walls, and nanoscale real-structure features all interfer with long-range magnetic order, and in a strict sense it is fair to say that bulk ferromagnets do not exist. However, the thermodynamic concept of ferromagnetism is based on the involvement of many atoms, and a length scale of a few nanometers is usually sufficient to establish "long-range" order.

References

Aharoni, A. (1996) *Introduction to the Theory of Ferromagnetism*, Oxford University Press, Oxford.

Akulov, N. (1936) "Zur Quantentheorie der Temperaturabhängigkeit der Magnetisierungskurve," *Z. Phys.* **100**, 197–202.

Alben, R., M. C. Chi, and J. J. Becker (1978) "Random Anisotropy in Amorphous Ferromagnets," *J. Appl. Phys.* **49**, 1653–1658.

Alefeld, G. (1969) "Wasserstoff in Metallen als Beispiel für ein Gittergas mit Phasenumwandlungen," *Phys. stat. sol.* **32**, 67–80.

Al-Omari, A. and D. J. Sellmyer (1995) "Magnetic Properties of Nanostructured CoSm/FeCo Films," *Phys. Rev. B* **52**, 3441–3447.

Anderson, P. W. (1958) "Absence of Diffusion in Certain Random Lattices," *Phys. Rev. B* **109**, 1492–1505.

Anderson, P. W. (1963) "Exchange in Insulators: Superexchange, Direct Exchange, and Double Exchange," in: G. T. Rado and H. Suhl (eds), *Magnetism I*, Academic Press, New York, pp. 25–85.

Anderson, P. W. (1965) "Concepts in Solids: Lectures on the Theory of Solids," W. A. Benjamin, New York.

Ashcroft, N. W. and N. D. Mermin (1976) *Solid State Physics*, Saunders, Philadelphia.

Astalos, R. J. and R. E. Camley (1998) "Magnetic Permeability for Exchange-spring Magnets: Application to Fe/Sm-Co," *Phys. Rev. B* **58**, 8646–8653.

Asti, G. and S. Rinaldi (1974) "Singular Points in the Magnetization Curve of a Polycrystalline Ferromagnet," *J. Appl. Phys.* **45**, 3600–3610.

Baettig, P. and N. A. Spaldin (2005) "Ab initio Prediction of a Multiferroic with Large Polarization and Magnetization," *Appl. Phys. Lett.* **86**, 012505–1-3.

Baibich, M. N., J. M. Broto, A. Fert, F. Nguyen Van Dau, F. Petroff, P. Eitenne, G. Creuzet, A. Friederich, and J. Chazelas (1988) "Giant Magnetoresistance of (001)Fe/(001)Cr Magnetic Superlattices," *Phys. Rev. Lett.* **61**, 2472–2475.

Ballhausen, C. J. (1962) *Ligand Field Theory*, McGraw-Hill, New York.

Bander, M. and D. L. Mills (1988) "Ferromagnetism of Ultrathin Films," *Phys. Rev. B* **38**, 12015–12018.

Barkhausen, H. (1919) "Zwei mit Hilfe der Neuen Verstärker entdeckte Erscheinungen," *Phys. Z.* **20**, 401–403.

Baxter, R. J. (1982) *Exactly Solved Models in Statistical Mechanics*, Academic Press, London.

Bean, C. P. and J. D. Livingston (1959) "Superparamagnetism" *J. Appl. Phys.* **30**, 120S–129S.

Becker, R. and W. Döring (1939) *Ferromagnetismus*, Springer, Berlin.
Belashchenko, K. D. and V. P. Antropov (2002) "Structure of Macrodomain Walls in Polytwinned Magnets," *J. Appl. Phys.* **91**, 8474–8476.
Berg, H. C. (1993) *Random Walks in Biology*, Princeton University Press, Princeton, NJ.
Berlin, T. H. and M. Kac (1952) "The Spherical Model of a Ferromagnet," *Phys. Rev.* **86**, 821–835.
Bertotti, G. (1998) *Hysteresis in Magnetism*, Academic Press, London.
Bethe, H. (1929) "Termaufspaltung in Kristallen," *Ann. Physik* [5] **3,** 133–208.
Binder, K. and P. C. Hohenberg (1972) "Phase Transitions and Static Spin Correlations in Ising Models with Free Surfaces," *Phys. Rev. B* **6,** 3461–3487.
Binek, Ch. (2003) *Ising-Type Antiferromagnets: Model Systems in Statistical Physics and in the Magnetism of Exchange Bias*, Springer, Berlin.
Bitter, F. (1931) "On Inhomogeneities in the Magnetization of Ferromagnetic Materials," *Phys. Rev.* **38**, 1903–1905.
Bland, J. A. C. and B. Heinrich (eds) (1994) *Ultrathin magnetic structures I*, Springer, Berlin.
Bloch, F. (1929) "Bemerkung zur Elektronentheorie des Ferromagnetismus und der elektrischen Leitfähigkeit," *Z. Phys.* **57**, 545.
Bloch, F. (1930) "Zur Theorie des Ferromagnetismus," *Z. Phys.* **61**, 206–219.
Bloch, F. (1932) "Zur Theorie des Austauschproblems und der Remanenzerscheinung der Ferromagnetika," *Z. Phys.* **74**, 295.
Bloch, F. and G. Gentile (1931) "Zur Anisotropie der Magnetisierung ferromagnetischer Einkristalle," *Z. Phys.* **70**, 395.
Blundell, S. (2001) *Magnetism in Condensed Matter*, Oxford University Press, Oxford.
Boltzmann, L. (1896) "Entgegnung auf die wärmetheoretischen Betrachtungen des Hrn. E. Zermelo," Annalen der Physik **57**, 773–84.
Bornholdt, S. and F. Wagner (2002) "Stability of Money: Phase Transitions in an Ising Economy," *Physica A* **316**, 453–468.
Bozorth, R. M. (1951) *Ferromagnetism*, van Nostrand, Princeton.
Braun, H. B. (1993) "Thermally Activated Magnetization Reversal in Elongated Ferromagnetic Particles," *Phys. Rev. Lett.* **71**, 3557–3560.
Braun, H.-B. and N. H. Bertram (1994) "Nonuniform Switching of Single Domain Particles at Finite Temperatures," *J. Appl. Phys.* **75**, 4609.
Brooks, H. (1940) "Ferromagnetic Anisotropy and the Itinerant Electron Model," *Phys. Rev.* **58**, 909–918.
Brooks, M. S. S. and B. Johansson (1993) "Density Functional Theory of the Ground-State Magnetic Properties of Rare Earths and Actinides," in: K. H. J. Buschow (ed.), *Handbook of Magnetic Materials*, Vol. 7, Elsevier, Amsterdam, pp. 139–230.
Brown, W. F. (1963a) *Micromagnetics*, Wiley, New York.
Brown, W. F. (1963b) "Thermal fluctuations of a single-domain particle," *Phys. Rev.* **130,** 1677–1686.
Bruggeman, D. A. G. (1935) "Berechnung verschiedener physikalischer Konstanten von heterogenen Substanzen (I)," *Ann. Phys.* (5) **24**, 637.
Bychkov, A. and E. I. Rashba (1984) "Oscillatory Effects and the Magnetic Susceptibility of Carriers in Inversion Layers," *J. Phys. C* **17**, 6039–6045.

Callen, E. R. and H. B. Callen (1963) "Static Magnetoelastic Coupling in Cubic Crystals," *Phys. Rev.* **129**, 578–593.

Callen, E., Y. J. Liu, and J. R. Cullen (1977) "Initial Magnetization, Remanence, and Coercivity of the Random Anisotropy Amorphous Ferromagnet," *Phys. Rev. B* **16**, 263–270.

Campbell, I. A., A. Fert, and O. Jaoul (1970) "The Spontaneous Resistivity Anisotropy in Ni-based Alloys," *J. Phys. C: Solid State Phys.* **3**, S95–S101.

Charles, S. W. (1992) "Magnetic Fluids" in: J. L. Dormann and D. Fiorani (eds), *Studies of Magnetic Properties of Fine Particles and their Relevance to Materials Science*, Elsevier, Amsterdam, p. 267.

Chikazumi, S. (1964) *Physics of Magnetism*, Wiley, New York.

Chipara, M. I., R. Skomski, and D. J. Sellmyer (2002) "Magnetic Modes in Nanowires," *J. Magn. Magn. Mater.* **249**, 246–250.

Christensen, R. M. (1979) *Mechanics of Composite Materials*, Wiley, New York.

Chuang, D. S., C. A. Ballantine, and R. C. O'Handley (1994) "Surface and Step Magnetic Anisotropy," *Phys. Rev. B* **49**, 15084–15095.

Chudnovsky, E. M. and L. Gunther (1988) "Quantum Tunneling of Magnetization in Small Ferromagnetic Particles," *Phys. Rev. Lett.* **60**, 661–664.

Chudnovsky, E. M., W. M. Saslow, and R. A. Serota (1986) "Ordering in Ferromagnets with Random Anisotropy," *Phys. Rev. B* **33**, 251–261.

Coehoorn, R. (1989) "Calculated Electronic Structure and Magnetic Properties of Y-Fe Compounds," *Phys. Rev. B* **39**, 13072–13085.

Coehoorn, R., D. B. de Mooij, J. P. W. B. Duchateau, and K. H. J. Buschow (1988) "Novel Permanent Magnetic Materials Made by Rapid Quenching," *J. de Physique* **49** C-8, 669–670.

Coey, J. M. D. (ed.) (1996) *Rare-earth Iron Permanent Magnets*, Oxford University Press, Oxford.

Coey, J. M. D., and R. Skomski (1993) "New Magnets from Interstitial Intermetallics," *Physica Scripta* **T49**, 315–321.

Coey, J. M. D., and H. Sun (1990) "Improved Magnetic Properties by Treatment Iron-Based Rare-Earth Intermetallic Compounds in Ammonia," *J. Magn. Magn. Mater.* **87**, L251-L254.

Coey, J. M. D. and M. Venkatesan (2002) "Half-Metallic Ferromagnets; the Example of CrO_2," *J. Appl. Phys.* **91**, 8345–8350.

Coey, J. M. D., M. Viret, and S. von Molnár (1999) "Mixed-Valence Manganites," *Adv. Phys.* **48**, 167–293.

Coey, J. M. D., K. O'Donnell, Q. Qinian, E. Touchais, and K. H. Jack (1994) "The Magnetization of α"$Fe_{16}N_2$," *J. Phys.: Condens. Matter* **6**, L23–L28.

Coey, J. M. D., A. E. Berkowitz, L. Balcells, F. F. Putris, and A. Barry (1998) "Magnetoresistance of Chromium Dioxide Powder Compacts," *Phys. Rev. Lett.* **80**, 2815–2818.

Comstock, R. L. (1999) *Introduction to Magnetism and Magnetic Recording*, Wiley, New York.

Coulson, C. A. and I. Fischer (1949) "Notes on the Molecular Orbital Treatment of the Hydrogen Molecule," *Phil. Mag.* **40**, 386–393.

Cowburn, R. P. and M. E. Welland (2000) "Room Temperature Magnetic Quantum Cellular Automata," *Science* **287**, 1466–1468.

Craik, D. 1995 *Magnetism: Principles and Applications*, Wiley, New York.
Craik, D. J., and E. D. Isaak (1960) "Magnetic Interaction Domains," *Proc. Phys. Soc.* **76**, 160–162.
Craik, D. J. and R. S. Tebble (1961) "Magnetic Domains," *Rep. Prog. Phys.* **24**, 116–166.
Crangle, J. (1991) *Solid-State Magnetism*, Arnold, London.
Cullity, B. (1972) *Introduction to Magnetic Materials*, Addison-Wesley, Reading MA.
Cyrot-Lackmann, F. (1968) "Sur le calcul de la cohésion et de la tension superficielle des métaux de transition par une méthode de liaisons fortes," *J. Phys. Chem. Solids* **29**, 1235–1243.
Daalderop, G. H. O., P. J. Kelly, and M. F. H. Schuurmans (1990) "First-principle Calculation of the Magnetocrystalline Anisotropy Energy of Iron, Cobalt, and Nickel," *Phys. Rev. B* **41**, 11919–11937.
Davies, J. E., O. Hellwig, E. E. Fullerton, J. S. Jiang, S. D. Bader, G. T. Zimányi, and K. Liu (2005) "Anisotropy Dependence of Irreversible Switching in Fe/SmCo and FeNi/FePt Exchange Spring Magnet Films," *Appl. Phys. Lett.* **86**, 262503, 1–3.
de Gennes, P.-G. 1979 *Scaling Concepts in Polymer Physics*, Cornell University Press, Ithaca.
de Groot, R. A. and K. H. J. Buschow (1986) "Recent Developments in Half-Metallic Magnetism," *J. Magn. Magn. Mater.* **54–57**, 1377–1380.
de Jongh, L. J. and A. R. Miedema (1975) "Experiments on Simple Magnetic Model Systems," *Adv. Phys.* **23,** 1–260.
Della Torre, E. (1999) *Magnetic Hysteresis*, IEEE Press, New York.
Demokritov, S. O., V. E. Demidov, O. Dzyapko, G. A. Melkov, A. A. Serga, B. Hillebrands and A. N. Slavin (2006) "Bose–Einstein Condensation of Quasi-Equilibrium Magnons at Room Temperature under Pumping," *Nature* **443**, 430–433.
Desjonquères, M. C. and D. Spanjaard (1993) *Concepts in Surface Physics*, Springer, Berlin.
de Toro, J. A., M. A. López de la Torre, J. M. Riveiro, J. Bland, J. P. Goff, and M. F. Thomas (2001) "Mossbauer Study of Ni Superspin Glass Transfer in Nanogranular $Al_{49}Fe_{30}Cu_{21}$," *Phys. Rev. B* **64**, 224–421.
Dietl, T., H. Ohno, F. Matsukura, J. Cibert, and D. Ferrand (2000) "Zener Model Description of Ferromagnetism in Zinc-Blende Magnetic Semiconductors," *Science* **287**, 1019–1022.
Dormann, J. L., D. Fiorani, and E. Tronc (1999) "On the Models for Interparticle Interactions in Nanoparticle Assemblies: Comparison with Experimental Results," *J. Magn. Magn. Mater.* **202**, 251–267.
Dowben, P. A. and R. Skomski (2003) "Finite-Temperature Spin Polarization in Half-Metallic Ferromagnets," *J. Appl. Phys.* **93**, 7948–7950.
Ducharme, S., V. M. Fridkin, A. Bune, L. M. Blinov, S. P. Palto, and S. G. Yudin (2000) "The Intrinsic Ferroelectric Coercive Field," *Phys. Rev. Lett.* 84, **175**.
Dyson, F. J. (1956) "General Theory of Spin-Wave Interactions," *Phys. Rev.* **102**, 1217–1229.
Economou, E. N. (1990) *Green's Functions in Quantum Physics*, 2nd edition, Springer, Berlin.
Edwards, S. F. and P. W. Anderson (1975) "Theory of Spin Glasses," *J. Phys. F* **5**, 965–974.

Einstein, A. (1906) "Eine neue Bestimmung der Moleküldimensionen," *Ann. Phys.* **19**, 289–306.

Einstein, A. (1911) "Eine neue Bestimmung der Moleküldimensionen," *Erratum: Ann. Phys.* **34**, 591–592.

Eisenbach, M., B. L. Györffy, G. M. Stocks, and B. Újfalussy (2002) "Magnetic Anisotropy of Monoatomic Iron Chains Embedded in Copper," *Phys. Rev.* B **65**, 144424-1-7.

Engel, H.-A. and D. Loss (2005) "Fermionic Bell-State Analyzer for Spin Qubits," *Science* **309**, 586–588.

Eriksson, O., B. Johansson, R. C. Albers, and A. M. Boring (1990) "Orbital Magnetism in Fe, Co, and Ni," *Phys. Rev. B* **42**, 2707–2710.

Erman, B. and J. E. Mark (1997) *Structures and Properties of Rubberlike Networks*, Oxford University Press, New York.

Eschrig, H., M. Sargolzaei, K. Koepernik and M. Richter (2005) "Orbital Polarization in the Kohn-Sham-Dirac Theory," *Europhys. Lett.* **72**, 611–617.

Eshbach, J. R. and R. W. Damon (1960) "Surface Magnetostatic Modes and Surface Spin Waves," *Phys. Rev.* **118**, 1208–1210.

Ewing, J. A. (1889) "Time Lag in the Magnetization of Iron," *Proc. Roy. Soc.* (London) **46**, 269–286.

Falicov, L. M. and R. A. Harris (1969) "Two-Electron Homopolar Molecule: A Test for Spin-Density Waves and Charge-Density Waves," *J. Chem. Phys.* **51**, 3153–3158.

Farle, M. (1998) "Ferromagnetic Resonance of Ultrathin Metallic Layers," *Rep. Prog. Phys.* **61**, 755–826.

Fast, J. D. (1976) *Gases in Metals*, MacMillan, London.

Fermi, E. (1928) "Eine statistische Methode zur Bestimmung einiger Eigenschaften des Atoms und ihre Anwendung auf die Theorie des periodischen Systems der Elemente," *Z. Phys.* **48**, 73–79.

Fischer, K.-H. and A. J. Hertz (1991) *Spin Glasses*, Cambridge University Press, Cambridge.

Fisher, D. S. and D. A. Huse (1986) "Ordered Phase of Short-Range Ising Spin-Glasses," *Phys. Rev. Lett.* **56**, 1601–1604.

Freeman, A. J., and R. E. Watson (1962) "Theoretical Investigation of Some Magnetic and Spectroscopic Properties of Rare-Earth Ions," *Phys. Rev.* **127,** 2058–2075.

Fulde, P. (1991) *Electron Correlations in Molecules and Solids*, Springer, Berlin.

Fullerton, E. E., S. J. Jiang, and S. D. Bader (1999) "Hard/Soft Heterostructures: Model Exchange-Spring Magnets," *J. Magn. Magn. Mater.* **200**, 392–404.

Gambardella, P., S. Rusponi, M. Veronese, S. S. Dhesi, C. Grazioli, A. Dallmeyer, I. Cabria, R. Zeller, P. H. Dederichs, K. Kern, C. Carbone, and H. Brune (2003) "Giant Magnetic Anisotropy of Single Cobalt Atoms and Nanoparticles," *Science* **300**, 1130–1133.

Garcia-Sanchez, F., O. Chubykalo-Fesenko, O. Mryasov, R. W. Chantrell, and K. Yu. Guslienko (2005) "Multiscale Versus Micromagnetic Calculations of the Switching Field Reduction in FePt/FeRh Bi-Layers with Perpendicular Exchange Spring," *J. Appl. Phys.* **97**, 10J101–3.

Gaunt, P. (1983) "Ferromagnetic Domain Wall Pinning by a Random Array of Inhomogeneities," *Phil. Mag. B* **48**, 261–276.

Gaunt P. (1986) "Magnetic Viscosity and Thermal Activation Energy," *J. Appl. Phys.* **59**, 4129–4132.

Gay, J. G. and R. Richter (1986) "Spin Anisotropy of Ferromagnetic Films," *Phys. Rev. Lett.* **56**, 2728–2731.

Givord, D. and M. F. Rossignol (1996) "Coercivity," in: J. M. D. Coey (ed.), *Rare-earth Iron Permanent Magnets*, Oxford University Press, Oxford, pp. 218–285.

Glaeser, E. L., B. Sacerdote, and J. A. Scheinkam (1996) "Crime and Social Interactions," *Quarterly Journal of Economics* **111**, 507–548.

Glauber, R. J. (1963) "Time-Dependent Statistics of the Ising Model," *J. Math. Phys.* **4**, 294–307.

Goodenough, J. B. (1963) *Magnetism and the Chemical Bond*, Wiley, New York.

Goodstein, D. L. (1975) *States of Matter*, Prentice-Hall, Englewood Cliffs.

Goto, E., N. Hayashi, T. Miyashita, and K. Nakagawa (1965) "Magnetization and Switching Characteristics of Composite Thin Magnetic Films," *J. Appl. Phys.* **36**, 2951–2958.

Gradmann, U. (1993) "Magnetism in Ultrathin Transition Metal Films" in: K. H. J. Buschow (ed.), *Handbook of Magnetic materials*, Vol. 7, Elsevier, Amsterdam, p. 1.

Griffiths, R. B. (1969) "Nonanalytic Behavior above the Critical Point in a Random Ising Ferromagnet," *Phys. Rev. Lett.* **23**, 17–19.

Grünberg, P., R. Schreiber, Y. Pang, M. B. Brodsky, and H. Sowers (1986) "Layered Magnetic Structures: Evidence for Antiferromagnetic Coupling of Fe Layers across Cr Interlayers," *Phys. Rev. Lett.* **57**, 2442–2445.

Gubanov, V. A., A. I. Liechtenstein, and A. V. Postnikov (1992) *Magnetism and the Electronic Structure of Crystals*, Springer, Berlin.

Heine, V. (1980) "Electronic Structure from the Point of View of the Local Atomic Environment," *Solid State Phys.* **35**, 1–127.

Heisenberg, W. (1928) "Zur Theorie des Ferromagnetismus," *Z. Phys.* **49**, 619–636.

Haken, H. (1983) *Synergetics*, Springer, Berlin.

Hámos, L. V. and P. A. Thiessen (1931) "Über die Sichtbarmachung von Bezirken verschiedenen ferromagnetischen Zustandes fester Körper ", *Z. Phys.* **71**, 442–444.

Harris, R., M. Plischke and M. J. Zuckermann (1973) "New Model for Amorphous Magnetism," *Phys. Rev. Lett.* **31**, 160–162.

Herbst, J. F. (1991) "$R_2Fe_{14}B$ Materials: Intrinsic Properties and Technological Aspects," *Rev. Mod. Phys.* **63**, 819–898.

Herzer, G. (1992) "Nanocrystalline Soft Magnetic Materials," *J. Magn. Magn. Mater.* **112**, 258–262.

Herzer, G. (1995) "Soft Magnetic Nanocrystalline Materials," *Scripta Metal.* **33**, 1741.

Hillebrands, B. and K. Ounadjela (eds) (2002) *Spin Dynamics in Confined Magnetic Structures I*, Springer, Berlin.

Hilzinger, H. R. and H. Kronmüller (1975) "Investigation of Bloch Wall Pinning by Antiphase Boundaries in RCo_5 compounds," *Phys. Lett. A* **51**, 59–60.

Himpsel, F. J., J. E. Ortega, G. J. Mankey, and R. F. Willis (1998) "Magnetic Nanostructures," *Adv. Phys.* **47**, 511–597.

Hinzke, D. and U. Nowak (1998) "Magnetization Switching in a Heisenberg Model for Small Ferromagnetic Particles," *Phys. Rev. B* **58**, 265–272.

Hirohata, A., C. C. Yao, H. T. Leung, Y. B. Xu, C. M. Guertler, and J. A. C. Bland (2000) "Magnetic Domain Studies of Permalloy Wire-Based Structures with Junctions," *IEEE Trans. Magn.* **36**, 3068–3070.

Huang, K. (1963) *Statistical Mechanics*, Wiley, New York.

Hubert, A. and R. Schäfer (1998) *Magnetic domains: the analysis of magnetic microstructures*, Springer, Berlin.

Hund F. (1925) "Atomtheoretische Deutung des Magnetismus der seltenen Erden," *Z. Phys.* **33**, 855–859.

Hurd, C. M. (1981) *Electrons in Metals*, Malabar, Florida: Krieger.

Hutchings, M. T. (1964) "Point-Charge Calculations of Energy Levels of Magnetic Ions in Crystalline Electric Fields," *Solid State Phys.* **16**, 227–273.

Huxley, A. F. (1974) "Muscular Contraction," *J. Physiol.* **243**, 1–43.

Imry, Y. and S.-K. Ma (1975) " Random-Field Instability of the Ordered State of Continuous Symmetry," *Phys. Rev. Lett.* **35**, 1399–1401.

Ising, E. (1925) "Beitrag zur Theorie des Ferromagnetismus," *Z. Phys.* **31**, 253–258.

Jack, K. H. (1951) "The Occurrence and the Crystal Structure of α"-Iron Nitride; a New Type of Interstitial Alloy Formed during the Tempering of Nitrogen-Martensite," *Proc. R. Soc. A* **208**, 216–224.

Jahn, L., R. Schumann, and V. Christoph (1985) "The Influence of Texture on the Coercitivity of Polycrystalline magnets," *Phys. Stat. Sol. (a)* **88**, 595–599.

Jiles, D. (1998) *Introduction to Magnetism and Magnetic Materials*, Chapman and Hall, New York.

Jiles, D. C. and D. L. Atherton (1986) "Theory of Ferromagnetic Hysteresis," *J. Magn. Magn. Mater.* **61**, 48–60.

Johnson, M. T., P. J. H. Bloemen, F. J. A. den Broeder, and J. J. de Vries (1996) "Magnetic Anisotropy in Metallic Multilayers," *Rep. Prog. Phys.* **59**, 1409–1458.

Jones, W. and N. H. March (1973) *Theoretical Solid State Physics I*, Wiley & Sons, London.

Kadanoff, L. P., W. Götze, D. Hamblen, R. Hecht, E. A. S. Lewis, V. V. Palciauskas, M. Rayl, J. Swift, D. Aspnes, and J. W. Kane (1967) "Static Phenomena Near Critical Points: Theory and Experiment," *Rev. Mod. Phys.* **39**, 395–431.

Kashyap A., R. Skomski, A. K. Solanki, Y. F. Xu, and D. J. Sellmyer (2004) "Magnetism of $L1_0$ Compounds with the Composition MT (M = Rh, Pd, Pt, Ir and T = Mn, Fe, Co, Ni)," *J. Appl. Phys.* **95**, 7480–7482.

Kersten, M. (1943) "Zur Theorie der ferromagnetischen Hysterese und der Anfangspermeabilität," *Z. Phys.* **44**, 63–77.

Khanna, S. N., B. K. Rao, and P. Jena (2002) "Magic Numbers in Metallo-Inorganic Clusters: Chromium Encapsulated in Silicon Cages," *Phys. Rev. Lett.* **89**, 016803-1-4.

Kim, T. K. and M. Takahashi (1972) "New Magnetic Material Having Ultrahigh Magnetic Moment," *Appl Phys. Lett.* **20**, 492–494.

Kirkpatrick, S. (1973) "Percolation and Conduction," *Rev. Mod. Phys.* **45**, 574–588.

Kittel, C. (1949) "Physical Theory of Ferromagnetic Domains," *Rev. Mod. Phys.* **21**, 541–583.

Kittel, C. (1986) *Introduction to Solid-State Physics*, Wiley, New York.

Klemmer, T., D. Hoydick, H. Okumura, B. Zhang, and W. A. Soffa (1995) "Magnetic Hardening and Coercivity in $L1_0$ Ordered FePd Ferromagnets," *Scripta Met. Mater.* **33**, 1793–1805.

Kneller, E. (1966) "Theorie der Magnetisierungskurve kleiner Kristalle," in: H. P. J. Wijn (ed.), *Handbuch der Physik XIII/2: Ferromagnetismus*, Springer, Berlin, p. 438.

Kneller, E. F. (1991) "The Exchange-Spring Magnet: a New Material Principle for Permanent Magnets," *IEEE Trans. Magn.* **27**, 3588–3599.

Kohn, W., and L. J. Sham (1965) "Self-Consistent Equations Including Exchange and Correlation Effects," *Phys. Rev.* **140**, A1133–A1138.

Komelj, M., C. Ederer, J. W. Davenport, and M. Fähnle (2002) "From the bulk to monatomic wires: An ab initio study of magnetism in Co systems with various dimensionality," *Phys. Rev. B* **66**, 140407-1-4.

Kondorski, E. (1937) "On the Nature of Coercive Force and Irreversible Changes in Magnetisation," *Phys. Z. der Sowj.* **11**, 597–620.

Kooy, C. and U. Enz (1960) "Experimental and Theoretical Study of the Domain Configuration in Thin Layers of $BaFe_{12}O_{19}$," *Philips Res. Repts.* **15**, 7–29.

Kramers, H. A. (1940) "Brownian Motion in a Field of Force and the Diffusion Model of Chemical Reactions," *Physica.* **7**, 284–304.

Krams, P., B. Hillebrands, G. Güntherodt, and H. P. Oepen (1994) "Magnetic Anisotropies of Ultrathin Co Films on Cu(1 1 13) Substrates," *Phys. Rev. B* **49**, 3633–3636.

Kronmüller, H. (1987) "Theory of Nucleation Fields in Inhomogeneous Ferromagnets," *phys. stat. sol. (b)* **144**, 385–396.

Kuch W., L. I. Chelaru, K. Fukumoto, F. Porrati, F. Offi, M. Kotsugi, and J. Kirschner (2003) "Layer-Resolved Imaging of Magnetic Interlayer Coupling by Domain Wall Stray Fields" *Phys. Rev. B.* **67**, 214–403.

Kuhrt, C., K. O'Donnell, M. Katter, J. Wecker, K. Schnitzke, and L. Schultz (1992) "Pressure-Assisted Zinc Bonding of Microcrystalline $Sm_2Fe_{17}N_x$," *Appl. Phys. Lett.* **60**, 3316–3318.

Kumar, K. (1988) "$RETM_5$ and RE_2TM_{17} Permanent Magnets Development," *J. Appl. Phys.* **63**, R13–57.

Landau, L. and E. Lifshitz (1935) "On the Theory of the Dispersion of Magnetic Permeability in Ferromagnetic Bodies," *Phys. Z. Sowjetunion* **8**, 153.

Landau, L. D. and E. M. Lifshitz (1984) *Theoretical Physics 8: Electrodynamics of Continuous Media*, Pergamon Press, New York.

Landauer, R. (1970) "Electrical Resistance of Disordered One-Dimensional Lattices," *Phil. Mag.* **21**, 863–867.

Lee, T. D. and C. N. Yang (1952) "Statistical Theory of State and Phase Equations: Lattice Gas and Ising Model," *Phys. Rev.* **87**, 410–419.

Liechtenstein, A. I., M. I. Katsnelson, V. P. Antropov, and V. A. Gubanov (1987) "Local Spin Density Functional Approach to the Theory of Exchange Interactions in Ferromagnetic Metals and Alloys," *J. Magn. Magn. Mater.* **67**, 65–74.

Liou, S.-H. and Y. D. Yao (1998) "Development of High Coercivity Magnetic Force Microscopy Tips," *J. Magn. Magn. Mater.* **190**, 130–134.

Liu, Y., D. J. Sellmyer, and D. Shindo (2006) *Handbook of Advanced Magnetic Materials*, Springer, Berlin.

Liu, J. P., C. P. Luo, Y. Liu, and D. J. Sellmyer (1998) "High Energy Products in Rapidly Annealed Nanoscale Fe/Pt Multilayers," *Appl. Phys. Lett.* **72**, 483–485.

Lodder, J. C. (2006) "Patterned Nanomagnetic Films," in: D. J. Sellmyer and R. Skomski (eds), *Advanced Magnetic Nanostructures*, Springer, Berlin, pp. 261–293.

Lyberatos, A. and R. W. Chantrell (1997) "The Fluctuation Field of Ferromagnetic Materials," *J. Phys. Condens. Matter* **9**, 2623–2643.

Málek, Z., and V. Kamberský (1958) "On the Theory of the Domain Structure of Thin Films of Magnetically Uniaxial Materials," *Czech. J. Phys.* **8**, 416–421.

Manaf, A., P. A. Buckley, and H. A. Davies (1993) "Microstructure Analysis of Nanocrystalline Fe-Nd-B Ribbons With Enhanced Hard Magnetic Properties," *J. Magn. Magn. Mater.* **128**, 307–312.

Marinari, E., G. Parisi, and J. J. Ruiz-Lorenzo (1998) "Phase Structure of the Three-Dimensional Edwards-Anderson Spin Glass," *Phys. Rev. B* **58**, 14852–14863.

Mathon, J. and G. Bergmann (1986) "Surface Enhancement of Palladium," *J. Phys. F: Met. Phys.* **16**, 887–891.

Mattheiss, L. F. (1972) "Electronic Structure of the 3d Transition-Metal Monoxides: I. Energy Band Results," *Phys. Rev. B.* **5**, 290–306; "II. Interpretation," *Phys. Rev. B.* **5**, 306–315.

Mattis, D. C. (1965) *Theory of Magnetism*, Harper and Row, New York.

Mattis, D. C. (1976) "Solvable Spin Systems with Random Interactions," *Phys. Lett.* **56A**, 421–422.

Mattuck, R. D. (1976) A *Guide to Feynman Diagrams in the Many-Body Problem*, McGraw-Hill, New York.

McCurrie, R. A. (1994) *Ferromagnetic Materials—Structure and Properties*, Academic Press, London.

McHenry, M. E., B. Ramalingum, S. Willoughby, J. MacLaren, and S. G. Sankar (2001) "First Principles Calculations of the Electronic Structure of $Fe_{1-x}Co_xPt$," *IEEE Trans. Mag.* **37**, 1277–1279.

Meiklejohn, W. H. and C. P. Bean (1956) "New Magnetic Anisotropy," *Phys. Rev.* **102**, 1413–1414.

Mermin, N. D. and H. Wagner (1966) "Absence of Ferromagnetism or Antiferromagnetism in One or Two-Dimensional Isotropic Heisenberg Models," *Phys. Rev. Lett.* **17**, 1133–1136.

Mézard, M., G. Parisi, N. Sourlas, G. Toulouse, and M. Virasoro (1984) "Nature of the Spin-Glass Phase," *Phys. Rev. Lett.* **52**, 1156–1159.

Miller, A. and P. A Dowben (1993) "Substrate Induced Magnetic Ordering of Rare Earth Overlayers II," *J. Phys.: Condens. Matter* **5**, 5459.

Millev, Y., R. Skomski, and J. Kirschner (1998) "High-order and Next-nearest-neighbor Néel Anisotropies," *Phys. Rev. B* **58**, 6305–6315.

Mills, D. L. (1971) "Surface Effects in Magnetic Crystals near the Ordering Temperature," *Phys. Rev. B* **3**, 3887–3895.

Mohn, P. (2003) *Magnetism in the Solid State*, Springer, Berlin.

Moodera, J. S., and G. Mathon (1999) "Spin Polarized Tunneling in Ferromagnetic Junctions," *J. Magn. Magn. Mater.* **200**, 248–273.

Moorjani, K. and J. M. D. Coey (1984) *Magnetic Glasses*, Elsevier, Amsterdam.

Mori, H. (1965) "Transport, Collective Motion and Brownian Motion," *Prog. Theor. Phys.* **33**, 423.

Moriya, T. (1985) *Spin Fluctuations in Itinerant Electron Magnetism*, Springer, Berlin.

Mott, N. F. (1974) *Metal-insulator Transitions*, Taylor and Francis, London.

Mott, N. F. and H. Jones (1936) *The Theory of the Properties of Metals and Alloys*, Oxford University Press, Oxford.

Mryasov, O. N., U. Nowak, K. Guslienko, and R. Chantrell (2005) "Temperature-Dependent Magnetic Properties of FePt: Effective Spin Hamiltonian Model," *Europhys. Lett.* **69**, 805.

Müller, K.-H., D. Eckert, P. A. P. Wendhausen, A. Handstein, and M. Wolf (1994) "Description of Texture for Permanent Magnets," *IEEE Trans. Magn.* **30**, 586–588.

Murata, K. K. and S. Doniach (1972) "Theory of Magnetic Fluctuations in Itinerant Ferromagnets," *Phys. Rev. Lett.* **29**, 285–288.

Nalwa, H. S. (ed.) (2002) *Magnetic Nanostructures*, American Scientific, Stephenson Ranch.

Néel, L. (1950) "Le traînage magnétique des substances massives dans le domaine de Rayleigh," *J. de Phys. Rad.* **11**, 49–61.

Néel, L. (1951) "Le traînage magnétique," *J. de Phys. Rad.* **12**, 339–351.

Néel, L. (1954) "Anisotropie Magnétique Superficielle et Surstructures d'Orientation," *J. Phys. Radium* **15**, 225–239.

Newman, D. J. and B. Ng (1989) "The Superposition Model of Crystal Fields," *Rep. Prog. Phys.* **52**, 699–763.

Nieber, S. and H. Kronmüller (1989) "Nucleation Fields in Periodic Multilayers," *Phys. Stat. Sol. (b)* **153**, 367–375.

Nielsen, M. A. and I. L. Chuang (2000) *Quantum Computation and Quantum Information*, Cambridge University Press, Cambridge.

Nowak, U., O. N. Mryasov, R. Wieser, K. Guslienko, and R. W. Chantrell (2005) "Spin Dynamics of Magnetic Nanoparticles: Beyond Brown's Theory," *Phys. Rev. B* **72**, 172410-1-4.

Oepen, H. P. and J. Kirschner (1989) "Magnetization Distribution of 180° Domain Walls at Fe (110) Single-Crystal Surfaces," *Phys. Rev. Lett.* **62**, 819–822.

O'Handley, R. (2000) *Modern Magnetic Materials*, Wiley, New York.

Onsager, L. (1944) "Crystal Statistics: a Two-Dimensional Model with an Order-Disorder Transition," *Phys. Rev.* **65**, 117–149.

Onsager, L. and S. Machlup (1953) "Fluctuations and Irreversible Processes," *Phys. Rev.* **91**, 1505–1512, 1512–1515.

Osborn, J. A. (1945) "Demagnetizing Factors of the General Ellipsoid," *Phys. Rev.* **67**, 351–357.

Padilla, J., W. Zhong, and D. Vanderbilt (1996) "First-principles Investigation of 180° Domain Walls in $BaTiO_3$," *Phys. Rev. B* **53**, R5969–R5973.

Pankhurst, Q. A., J. Connolly, S. K. Jones, and J. Dobson (2003) "Applications of Magnetic Nanoparticles in Biomedicine," *J. Phys. D: Appl. Phys.* **36**, R167–R181.

Panyam, J., D. Williams, A. Dash, D. Leslie-Pelecky, and V. Labhasetwar (2004) "Solid-State Solubility of Drug in PLGA/PLA Polymers Influences Encapsulation and Release of Hydrophobic Drugs from Nanoparticles," *J. Pharm. Sci.* **93**, 1804–1814.

Parkin, S. S. P., K. P. Roche, M. G. Samant, P. M. Rice, and R. B. Beyers, R. E. Scheuerlein, E. J. O'Sullivan, S. L. Brown, J. Bucchigano, D. W. Abraham, Yu Lu, M. Rooks, P. L. Trouilloud, R. A. Wanner, and W. J. Gallagher (1999) "Exchange-Biased Magnetic Tunnel Junctions and Application to Nonvolatile Magnetic Random Access Memory (invited)," *J. Appl. Phys.* **85**, 5828–5833.

Paschen, S., T. Lühmann, S. Wirth, P. Gegenwart, O. Trovarelli, C. Geibel, F. Steglich, Coleman, P., and Q. Si (2004) "Hall-Effect Evolution across a Heavy-Fermion Quantum Critical Point," *Nature* **432**, 881–885.

Pauling, L. and E. B. Wilson (1935) *Introduction to Quantum Mechanics*, New York: McGraw-Hill.

Pecharsky, A. O., K. A. Gschneidner, and V. K. Pecharsky (2003) "The Giant Magnetocaloric Effect of Optimally Prepared $Gd_5Si_2Ge_2$," *J. Appl. Phys.* **93**, 4722–4728.

Pinto, M. A. (1987) "Catastrophe Model for Micromagnetics," *Phys. Rev. Lett.* **59**, 2798–2801.

Preisach, F. (1935) "Über die magnetische Nachwirkung.," *Z. Phys.* **94**, 277–305.

Qi, Q.-N., R. Skomski, and J. M. D. Coey (1994) "Strong Ferromagnets: Curie Temperature and Density of States," *J. Phys.: Condens. Matter* **6**, 3245–3252.

Reddy, B. V., S. N. Khanna, and P. Jena (1999) "Structure and Magnetic Ordering in Cr_8 and Cr_{13} Clusters," *Phys. Rev. B* **60**, 15597–15600.

Richter, J. and R. Skomski (1989) "Antiferromagnets with Random Anisotropy," *Phys. Stat. Sol (b)* **153**, 711–719.

Risken, H. (1989) *The Fokker-Planck Equation*, Springer, Berlin.

Ross, C. A., S. Haratani, F. J. Castaño, Y. Hao, M. Hwang, M. Shima, J. Y. Cheng, B. Vögeli, M. Farhoud, M. Walsh, and Henry I. Smith (2002) "Magnetic Behavior of Lithographically Patterned Particle Arrays," *J. Appl. Phys.* **91**, 6848–6853.

Sabiryanov, R. F. and S. S. Jaswal (1998) "Electronic Structure and Magnetic Properties of Hard/Soft Multilayers," *J. Magn. Magn. Mater.* **177–181**, 989–990.

Sander, D., R. Skomski, C. Schmidthals, A. Enders, and J. Kirschner (1996) "Film Stress and Domain Wall Pinning in Sesquilayer Iron Films on W(110)," *Phys. Rev. Lett.* **77**, 2566–2569.

Sandratskii, L. M. (1998) "Noncollinear Magnetism in Itinerant-Electron Systems: Theory and Applications," *Adv. Phys.* **47**, 91–160.

Sawicki, M., G. J. Bowden, P. A. J. de Groot, B. D. Rainford, J.-M. L. Beaujour, R. C. C. Ward and M. R. Wells (2000) "Exchange Springs in Antiferromagnetically Coupled $DyFe_2$-YFe_2 Superlattices," *Phys. Rev. B* **62**, 5817–5820.

Schofield, A. J. (1999) "Non-Fermi liquids," Contemporary Phys. **40**, 95–115.

Schrefl, T., J. Fidler, and H. Kronmüller (1994) "Remanence and coercivity in isotropic nanocrystalline permanent magnets," *Phys. Rev. B* **49**, 6100–6110.

Sellmyer, D. J. (2002) "Applied Physics: Strong Magnets by Self-Assembly," *Nature* **420**, 374–375.

Sellmyer, D. J. and S. Nafis (1986) "Phase Transition Behavior in a Random-Anisotropy System," *Phys. Rev. Lett.* **57**, 1173–1176.

Sellmyer, D. J. and M. J. O'Shea (1992) "Random Anisotropy and Phase Transitions in Magnetic Glasses," in: D. H. Ryan (ed.), *Recent Progress in Random Magnets*, World Scientific, Singapore, pp. 71–121.

Sellmyer, D. J., M. Zheng, and R. Skomski (2001) "Magnetism of Fe, Co and Ni Nanowires in Self-Assembled Arrays," *J. Phys.: Condens. Matter* **13**, R433–R460.

Sellmyer, D. J., M. Yu, R. A. Thomas, Y. Liu, and R. D. Kirby (1998) "Nanoscale Design of Films for Extremely High Density Magnetic Recording," *Phys. Low-Dim. Struct.* **1–2**, 155–165.

Sellmyer, D. J., C. P. Luo, M. L. Yan, and Y. Liu (2001) "High-Anisotropy Nanocomposite Films for Magnetic Recording," *IEEE Trans. Magn.* **37**, 1286–1291.

Sellmyer, D. J., C. P. Luo, Y. Qiang, and J. P. Liu (2002) "Magnetism of Nanophase Composite Films," in: H. S. Nalwa (ed.), *Handbook of Thin Film Materials, Vol. 5: Nanomaterials and Magnetic Thin Films,* Academic Press, San Diego, pp. 337–374.

Senatore, G. and N. H. March (1994) "Recent Progress in the Field of Electron Correlation," *Rev. Mod. Phys.* **66**, 445–479.

Sharrock, M. P. (1994) "Time Dependence of Switching Fields in Magnetic Recording Media," *J. Appl. Phys.* **76**, 6413–6418.

Shelby, R. A., D. R. Smith, S. Shultz (2001) "Experimental Verification of a Negative Index of Refraction," *Science* **292**, 77–79.

Shen, J., R. Skomski, M. Klaua, H. Jenniches, S. S. Manoharan, and J. Kirschner (1997) "Magnetism in One Dimension: Fe on Cu(111)," *Phys. Rev. B* **56**, 2340–2343.

Sherrington, D. and S. Kirkpatrick (1975) "Solvable Model of a Spin-Glass," *Phys. Rev. Lett.* **35**, 1792–1796.

Skomski, R. (1994) "The Screened-Charge Model of Crystal-Field Interaction," *Phil. Mag. B* **70**, 175–189.

Skomski, R. (1996a) "The Itinerant Limit of Metallic Anisotropy," *IEEE Trans. Magn.* **32**, 4794–4796.

Skomski, R. (1996b) "Interstitial Modification," in: J. M. D. Coey (ed.), *Rare-Earth—Iron Permanent Magnets*, Oxford University Press, Oxford, pp. 178–217.

Skomski, R. (1999) "RKKY Interactions between Nanomagnets of Arbitrary Shape," *Europhys. Lett.* **48**, 455–460.

Skomski, R. (2001) "Micromagnetic Spin Structure," in: M. Ziese and M. J. Thornton (eds), *Spin Electronics*, Springer, Berlin, pp. 204–231.

Skomski, R. (2003) "Nanomagnetics," *J. Phys.: Condens. Matter* **15**, R841.

Skomski, R. (2004) "Nanomagnetic Scaling," *J. Magn. Magn. Mater.* **272–276**, 1476–1481.

Skomski, R. and V. Christoph (1989) "Power Law Behaviour of Magnetic Viscosity," *Phys. Stat. Sol. (b)* **156**, K149–K152.

Skomski, R. and J. M. D. Coey (1993) "Giant Energy Product in Nanostructured Two-Phase Magnets," *Phys. Rev. B* **48**, 15812–15816.

Skomski, R. and J. M. D. Coey (1999) *Permanent Magnetism*, Institute of Physics, Bristol.

Skomski, R. and P. A. Dowben (2002) "The Finite-Temperature Densities of States for Half-Metallic Ferromagnets," *Europhys. Lett.* **58**, 544–548.
Skomski, R., A. Kashyap, and D. J. Sellmyer (2003) "Finite-Temperature Anisotropy of PtCo Magnets," *IEEE Trans. Magn.* **39**, 2917–2919.
Skomski, R., R. D. Kirby, and D. J. Sellmyer (1999) "Activation Entropy, Activation Energy, and Magnetic Viscosity," *J. Appl. Phys.* **85**, 5069–5071.
Skomski, R., J.-P. Liu, and D. J. Sellmyer (1999) "Quasicoherent Nucleation Mode in Two-Phase Nanomagnets," *Phys. Rev. B* **60**, 7359–7365.
Skomski, R., H.-P. Oepen, and J. Kirschner (1998) "Micromagnetics of Ultrathin Films with Perpendicular Magnetic Anisotropy," *Phys. Rev. B* **58**, 3223–3227.
Skomski, R., C. Waldfried, and P. A. Dowben (1998) "The Influence of the Surface on the Spontaneous Magnetization of Gd Thin Films," *J. Phys. CM: Condens. Matter* **10**, 5833–5838.
Skomski, R, J. Zhou, and D. J. Sellmyer (2005) "Relaxation in Magnetic Nanostructures," *J. Appl. Phys.* **97**, 10A702-1-3.
Skomski, R., H. Zeng, M. Zheng, and D. J. Sellmyer (2000) "Magnetic Localization in Transition-Metal Nanowires," *Phys. Rev. B* **62**, 3900–3904.
Skomski, R., J. Zhou, A. Kashyap, and D. J. Sellmyer (2004a) "Domain-Wall Pinning at Inhomogenities of Arbitrary Cross-Sectional Geometry," *IEEE Trans. Magn.* **40** 2946–2948.
Skomski, R., A. Kashyap, K. D. Sorge, and D. J. Sellmyer (2004b) "Multidomain and Incoherent Effects in Magnetic Nanodots," *J. Appl. Phys.* **95**, 7022–7024.
Skomski, R., A. Y. Istomin, A. F. Starace and D. J. Sellmyer (2004c) "Quantum Entanglement of Anisotropic Magnetic Nanodots," *Phys. Rev. A* **70**, Art. No. 062307-1-4.
Skomski, R., A. Kashyap, J. Zhou, and D. J. Sellmyer (2005) "Anisotropic Exchange," *J. Appl. Phys.* **97**, 10B302–1-3.
Skomski, R., O. N. Mryasov, J. Zhou, and D. J. Sellmyer (2006a) "Finite-Temperature Anisotropy of Magnetic Alloys," *J. Appl. Phys.* **99**, 08E916-1-3.
Skomski, R., J. Zhou, J. Zhang, and D. J. Sellmyer (2006a) "Indirect exchange in dilute magnetic semiconductors," *J. Appl. Phys.* **99**, 08D504-1-3.
Slater, J. C. and G. F. Koster (1954) "Simplified LCAO method for the periodic potential problem," *Phys. Rev.* **94**, 1498–1524.
Slonczewski, J. C. (1991) "Fluctuation Mechanism for Biquadratic Exchange Coupling in Magnetic Multilayers," *Phys. Rev. Lett.* **67**, 3172–3175.
Slonczewski, J. C. (1996) "Current-Driven Excitation of Magnetic Multilayers," *J. Magn. Magn. Mater.* **159**, L1–L7.
Smart, J. S. (1966) *Effective Field Theories of Magnetism*, Saunders, Philadephia.
Solzi, M., M. Ghidini, and G. Asti (2002) "Macroscopic Magnetic Properties of Nanostructured and Nanocomposite Systems," in: H. S. Nalwa (ed.), *Magnetic Nanostructures*, American Scientific, Stephenson Ranch, pp. 123–201.
Sommerfeld, A. and H. A. Bethe (1933) "Ferromagnetism," in: S. Flügge (ed.), *Handbuch der Physik, Vol. 24/II*, Springer, Berlin, pp. 334–620.
Sorge, K. D., A. Kashyap, R. Skomski, L. Yue, L. Gao, R. D. Kirby, S.-H. Liou, and D. J. Sellmyer (2004) "Interactions and switching Behavior of Anisotropic Magnetic Dots," *J. Appl. Phys.* **95**, 7414–7415.

Sorge, K. D., R. Skomski, M. Daniil, S. Michalski, L. Gao, J. Zhou, M. Yan, Y. Sui, R. D. Kirby, S. H. Liou, and D.J. Sellmyer (2005) "Geometry and Magnetism of $L1_0$ Nanostructures," *Scripta Mater.* **53**, 459–463.

Spaldin, N. (2003) *Magnetic Materials*, Cambridge University Press, Cambridge.

Stauffer, D. and A. Aharony (1992) *Introduction to Percolation Theory*, Taylor & Francis, London.

Stoner, E. C. (1938) "Collective Electron Ferromagnetism," *Proc. Roy. Soc. A* **165**, 372–414.

Stoner, E. C. and E. P. Wohlfarth (1948) "A Mechanism of Magnetic Hysteresis in Heterogeneous Alloys," *Philos. Trans. R. Soc. London* **A240**, 599–608.

Street, R. and J. C. Wooley (1949) "Study of Magnetic Viscosity," *Proc. Phys. Soc.* **62A**, 562–572.

Sui, Y. C., R. Skomski, K. D. Sorge, and D. J. Sellmyer (2004) "Nanotube Magnetism," *Appl. Phys. Lett.* **84**, 1525–1527.

Sutton, A. P. (1993) *Electronic Structure of Materials*, Oxford University Press, Oxford.

Suzuki, K. and J. M. Cadogan (1999) "The Effect of the Spontaneous Magnetization in the Grain Boundary Region on the Magnetic Softness of Nanocrystalline Materials," *J. Appl. Phys.* **85**, 4400–4402.

Taylor, K. N. R. and M. I. Darby (1972) *Physics of Rare-Earth Solids*, Chapman & Hall, London.

Terris, B. D., L. Folks, D. Weller, J. E. E. Baglin, A. J. Kellock, H. Rothuizen and P. Vettiger (1999) "Ion-Beam Patterning of Magnetic Films Using Stencil Masks," *Appl. Phys. Lett.* **75**, 403–405.

Treloar, L. R. G. (1973) "The Elasticity and Related Properties of Rubbers," *Rep. Prog. Phys.* **36**, 755–826.

Trygg, J., B. Johansson, O. Eriksson, and J. M. Wills (1995) "Total Energy Calculations of the Magnetocrystalline Anisotropy Energy in the Ferromagnetic 3d Metals," *Phys. Rev. Lett.* **75**, 2871–2874.

Tsymbal, E. Y., O. N. Mryasov, and P. R. LeClair (2003) "Spin-Dependent Tunneling in Magnetic Tunnel Junctions," *J. Phys.: Condens Matter* **15**, R109-R142.

Victora, R. H. (1990) " Predicted Time Dependence of the Switching Field for Magnetic Materials," *Phys. Rev. Lett.* **63**, 457–460 (1989); **65**, 1171.

Victora, R. H. and J. M. McLaren (1993) "Theory of Magnetic Interface Anisotropy," *Phys. Rev. B* **47**, 11583–11586.

Walker, L. R. (1957) "Magnetostatic Modes in Ferromagnetic Resonance," *Phys. Rev.* **105**, 390–399.

Wang, D.-Sh., R.-Q. Wu, and A. J. Freeman (1993) "First-Principles Theory of Surface Magnetocrystalline Anisotropy and the Diatomic-Pair Model," *Phys. Rev. B* **47**, 14932–14947.

Wannier, G. H. (1966) *Statistical Physics*, Wiley, New York.

Ward, I. M. and D. W. Hadley (1993) *Mechanical Properties of Solid Polymers*, Wiley, New York.

Wegrowe, J.-E., D. Kelly, Y. Jaccard, Ph. Guittienne, Y. Jaccard, and J.-Ph. Ansermet (1999) "Current-Induced Magnetization Reversal in Magnetic Nanowires," *Europhys. Lett.* **45**, 626–632.

Weiss, M. P. (1907) "L'Hypothese du champ moleculaire et la propriete ferromagnetique," *J. de Phys.* **6**, 661–690.

Weissmüller, J., A. Michels, J. G. Barker, A. Wiedenmann, U. Erb, and R. D. Shull (2001) "Analysis of the Small-Angle Neutron Scattering of Nanocrystalline Ferromagnets Using a Micromagnetics Model," *Phys. Rev. B* **63**, 214414-1-18.

Weller, D. and T. McDaniel (2006) "Media for Extremely High Density Recording," in D. J. Sellmyer and R. Skomski (eds), *Advanced Magnetic Nanostructures*, Springer, Berlin, pp. 295–324.

Weller, D. and A. Moser (1999) "Thermal Effect Limits in Ultrahigh Density Magnetic Recording," *IEEE Trans. Magn.* **35**, 4423–4439.

Weller, D., A. Moser, L. Folks, M. E. Best, W. Lee, M. F. Toney, M. Schwickert, J.-U. Thiele, and M. F. Doerner (2000) "High Ku Materials Approach to 100 Gbits/in^2," *IEEE Trans. Magn.* **36**, 10–15.

Wernsdorfer, W. (2006) "Molecular Nanomagnets," in: D. J. Sellmyer and R. Skomski (eds), *Advanced Magnetic Nanostructures*, Springer, Berlin, pp. 147–181.

Wernsdorfer, W., E. B. Orozco, K. Hasselbach, A. Benoit, D. Mailly, O. Kubo, H. Nakano, and B. Barbara (1997) "Macroscopic Quantum Tunneling of Magnetization of Single Ferrimagnetic Nanoparticles of Barium Ferrite," *Phys. Rev. Lett.* **79**, 4014–4017.

White, R. M. (1970) *Quantum Theory of Magnetism*, New York, McGraw-Hill.

Wijn, H. P. J. (ed.) (1991) *Magnetic Properties of Metals: d-Elements, Alloys, and Compounds*, Springer, Berlin.

Williams, D. S., P. M. Shand, T. M. Pekarek, R. Skomski, V. Petkov, and D. L. Leslie-Pelecky (2003) "Magnetic Transitions in Disordered $GdAl_2$," *Phys. Rev. B* **68**, Art. No. 214404, 1–8.

Wohlfarth, P. (1958) "Relation between Different Methods of Acquisition of the Remanent Magnetization of Ferromagnetic Particles," *J. Appl. Phys.* **29**, 595–596.

Wöhlisch, E. (1940) "Muskelphysiologie vom Standpunkt der kinetischen Theorie der Hochelastizität und der Entspannungshypothese des Kontraktionsmechanismus," *Naturwissenschaften* **28**, 305–312, 326–335.

Yeomans, J. M. (1992) *Statistical Mechanics of Phase Transitions*, University Press, Oxford.

Yoshizawa, Y., S. Oguma, and K. Yamauchi ((1988)) "New Fe-Based Soft Magnetic Alloys Composed of Ultrafine Grain Structure," *J. Appl. Phys.* **64**, 6044–6046.

Yu, P. Y. and M. Cardona (1999) *Fundamentals of Semiconductors*, Springer, Berlin.

Zener, C. (1951) "Interaction between the d-Shells in the Transition Metals. II. Ferromagnetic Compounds of Manganese with Perovskite Structure ", *Phys. Rev.* **82**, 403–405.

Zhang, X. X., L. Balcells, J. M. Ruiz, J. L. Tholence, B. Barbara and J. Tejada (1992) "Quantum Tunnelling Effects in Fe/Sm Multilayers," *J. Phys.: Condens. Matter* **4**, L163–L168.

Zheng, M., R. Skomski, Y. Liu, and D. J. Sellmyer (2000) "Magnetic Hysteresis of Ni Nanowires," *J. Phys.: Condens. Matter* **12**, L497–L503.

Zhou J., R. Skomski, C. Chen, G. C. Hadjipanayis, and D. J. Sellmyer (2000) "Sm-Co-Ti High-Temperature Permanent Magnets," *Appl. Phys. Lett.* **77**, 1514–1516.

Zhou J., R. Skomski, A. Kashyap, K. D. Sorge, Y. Sui, M. Daniil, L. Gao, M. L. Yan, S.-H. Liou, R. D. Kirby, and D. J. Sellmyer, "Highly Coercive Thin-Film Nanostructures," *J. Magn. Magn. Mater.* **290–291**, 227–230 (2005b).

Zhou J., R. Skomski, K. D. Sorge, and D. J. Sellmyer (2005) "Magnetization Reversal in Particulate $L1_0$ Nanostructures," *Scripta Mater.* **53**, 455–458.

Ziese, M. and M. J. Thornton (eds) (2001) *Spin Electronics*, Springer, Berlin.

Ziman, M. (1979) *Models of Disorder*, Cambridge University Press, Cambridge.

Zwanzig, R. (1961) "Memory Effects in Irreversible Thermodynamics," *Phys. Rev.* **124**, 983.

Index

Entries in boldface refer to pages where an extended discussion of the topic is given.